Advances in Spatial Science

For further volumes:
http://www.springer.com/series/3302

Juan R. Cuadrado-Roura
Editor

Service Industries and Regions

Growth, Location and Regional Effects

Springer

Editor
Juan R. Cuadrado-Roura
Applied Economics
University of Alcala - Madrid
Alcalá de Henares
Spain

ISSN 1430-9602
ISBN 978-3-642-35800-5 ISBN 978-3-642-35801-2 (eBook)
DOI 10.1007/978-3-642-35801-2
Springer Heidelberg New York Dordrecht London

Library of Congress Control Number: 2012956222

Printed on acid-free paper

Springer is part of Springer Science+Business Media (www.springer.com)

Contents

Authors' Biographies

William B. Beyers is professor at the Department of Geography, University of Washington, Seattle. He has written a high number of books and articles particularly on the evolution of service industries, new services and sectorial prospects.

Rafael Boix is lecturer in applied economics at the Department of Economic Structure, University of Valencia, Spain. His research is focused on regional and urban economics and local development. He is author of many scientific papers and reports for several national and international institutions.

Roberta Capello is professor in regional and urban economics at the Polytechnic of Milan, Italy. He is past-president of the RSAI and editor in chief of the *Italian Journal of Regional Science* and co-editor of *Letters in Spatial and Resource Science* (Springer-Verlag). He is also author of many scientific papers and a textbook in regional economics, published in Italian and English.

Philip Cooke is research professor at Cardiff University and founding director of its Centre for Advanced Studies. He is author of more than 200 scholarly papers and author or editor of over 20 books. His latest is *Complex Adaptive Innovation Systems* (Routledge).

Juan R. Cuadrado-Roura is professor of applied economics and founder-director of IAES, University of Alcala, Madrid. He is former president of European Regional Science Association (ERSA) and a member of many editorial boards. He has published a high number of books and papers in indexed international journals, mainly on European economic policy, productivity and competitiveness, regional analysis and service industries. He is director of the journal *Investigaciones Regionales.*

Peter W. Daniels is emeritus professor of geography at the University of Birmingham, UK. He has published a number of books and papers that explore the growing role of advanced service industries in urban and regional development and in international trade and FDI with particular reference to the economies of Europe, North America and the Asia-Pacific.

Blanca De Miguel-Molina is associate professor in strategy and economics at the Polytechnic University of Valencia, Spain. Her research interest includes innovation in cultural and creative industries, regional development and sustainable tourism. She is author of several articles in leading journals and book chapters.

Blanca L. Delgado-Márquez She is assistant lecturer and researcher, Department of International and Spanish Economy, University of Granada, Spain.

Ugo Fratesi is associate professor of regional and urban economics at the Polytechnic of Milan, Italy. He is member of the editorial board of the *Italian Journal of Regional Science*. He is treasurer and member of the Council of the AISRe, Italian Section of the RSAI. He is author of three books and several papers published in refereed journals in the field of regional economics.

Faïz Gallouj is professor of economics at Lille University, France, and editor of the Journal "Economies et Sociétés" (EGS Series). He has published more than 100 papers in scientific journals and books and is also author or co-author of 12 books on services.

Camal Gallouj is professor of management sciences at the University of Paris 13 (CEPN-CNRS), France. He is author, co-author or editor of 15 books on services and retailing.

Marcos M. García-Velasco PhD in economics, University of Malaga, is senior lecturer (Titular Professor) at the Department of International and Spanish Economics, Faculty of Economics and Business, University of Granada, Spain.

José Enrique Garcilazo, PhD, University of Texas, Austin, is the head of unit for the Rural and Regional Development Program within the Regional Development Policy Division at the OECD. His work has mainly focused on the areas of regional competitiveness, measuring regional performance and understating the key factors for growth at the regional level and their impact on aggregate performance.

Johannes Glückler is Chair of Economic and Social Geography, Institute of Geography, University of Heidelberg, Germany. His research interests are in the fields of economic geography, social networks and the service economy. His latest books discuss the geographical foundations of the relational economy and of organized inter-firm networks.

Ingmar Hammer is research associate at the Chair of Economic and Social Geography, Institute of Geography, University of Heidelberg, Germany. His research interests are in the areas of economic geography, innovation studies and social networks.

José Luis Hervás-Oliver is associate professor of management and economics of innovation at the Polytechnic University of Valencia, Spain. His research interests include innovation management, economics of innovation, clusters and creative industries and regions. He is author of many articles in leading journals and book chapters.

Henk L.M. Kox is a senior economist at Netherlands' Bureau for Economic Policy Analysis (CPB), the main economic think tank of the Dutch government. In the past, he held positions as lecturer at Tilburg University and the Free University (Amsterdam). He wrote widely about the economics of services markets.

Andres Maroto-Sanchez, PhD in Economics, University of Alcala is associate professor at Autonomous University of Madrid, Spain, and senior researcher at IAES (University of Alcala, Spain). His main research fields are services economy, productivity and efficiency, competitiveness and international trade and innovation theory.

Florence Mouradian is graduated in International and development economics and she is actually a doctoral candidate to the University of Dauphine ["Paris Sciences et Lettres" (PSL). Her research interests are international and development macroeconomics and her thesis is titled "Exchange rate, tertiarization and growth".

Joaquim Oliveira-Martins is the head for the Division Regional Development Policy Division at the OECD. Former Head of the Structural Economic Statistics Division and Head of Desk for emerging markets. PhD in Economics (University of Paris-I) and lecturer at University of ParisDauphine.

Mario Polèse is professor at INRS, Centre Urbanisation Culture Société, in Montreal and holds the Senior Canada Research Chair in urban and regional studies. He is the author of *The Wealth and Poverty of Regions: Why Cities Matter* and he has published a high number of articles on urban and regional problems in refereed journals.

Laura Resmini is associate professor in economic policy at the University of Valle d'Aosta and senior researcher at ISLA, Bocconi University, Italy. He is treasurer of the ERSA and past executive director of the Italian Section of the RSAI. He is author of several papers published in refereed journals in the field of economics of multinational firms.

Fernando Rubiera is senior lecturer in applied economics, University of Oviedo, Spain. His academic publications and research papers are focused on the role of cities in regional development and location patterns of economic activity. He has published a high number of articles in refereed international and national journals.

Simón Sánchez-Moral is "Ramón y Cajal" research fellow at Department of Regional Geographical Analysis of the Complutense University of Madrid, Spain. His research interest focusses mainly on world cities and knowledge-based cities and models in economic geography and regional development.

Patrik Ström is the Staffan Helmfrid Pro Futura Scientia Fellow at the Swedish Collegium for Advanced Study, Uppsala, Sweden, and associate professor at the Department of Human and Economic Geography, University of Gothenburg. His research is focusing on the service economy and regional economic integration, particularly in Asia.

Peter Taylor is director of the Globalization and World Cities (GaWC) Research Network and professor of Human Geography at Northumbria University. His research interest focusses on world city networks, contemporary and historical, and how they relate to the territorial imperatives of states and empires. He is a Fellow of the British Academy.

Chapter 1
Introduction: Objectives, Approach and Main Lessons Learned

Juan R. Cuadrado-Roura

Although the most developed countries are still known as 'industrialised economies' they have all become *service economies*. This is also true in the case of a great number of less developed countries. The data gathered from many European countries, United States, Canada, Japan and Australia are absolutely clear: service activities reach about 70 % of total employment and GDP. According to United Nations statistics, the service sector represents 67.5 % of world GDP in 2008; developed economies are almost 4–5 % above this figure whereas transition and developing economies are 11 % and 16–17 % below it, respectively. Within the most developed economies, United States emerge as the world leader in services (76–77 % of GDP). However, the picture is quite heterogeneous across other groups of world countries. Tertiary activities of the former Eastern European countries (Poland, Hungary, Czech Republic, Slovenia and Slovak Republic), Korea and Finland show the lower share, ranging from 52 % to 62 %. In the case of developing America, the service economy represents 59 % of GDP; Caribbean and some Central American countries show higher tertiary shares than South America, mainly due to the role played by tourism and distributive activities (Di Meglio, 2010). In developing Asia, services account nearly 50 % of GDP, but India leads the ranking reaching almost 60 %. On the contrary, in developing Africa services represent less than 45 % of GDP despite the greater disparities that exist among the countries.

World statistics (UN, OECD) show that service sector has grown worldwide since the 70s. to now, although at different paces. Empirical evidence shows also that, while services growth in developed economies seems to follow a constant pattern, much more irregular trends characterize the developing and transition economies (Fig. 1.1). The latter are experiencing a clear process of tertiarization due not only to the structural changes accompanying all growth processes, but to the implementation of market oriented reforms and the effects of their own internationalization.

J.R. Cuadrado-Roura (✉)
Department of Applied Economics, University of Alcala, Madrid, Spain
e-mail: jr.cuadrado@uah.es

J.R. Cuadrado-Roura (ed.), *Service Industries and Regions*,
Advances in Spatial Science, DOI 10.1007/978-3-642-35801-2_1,

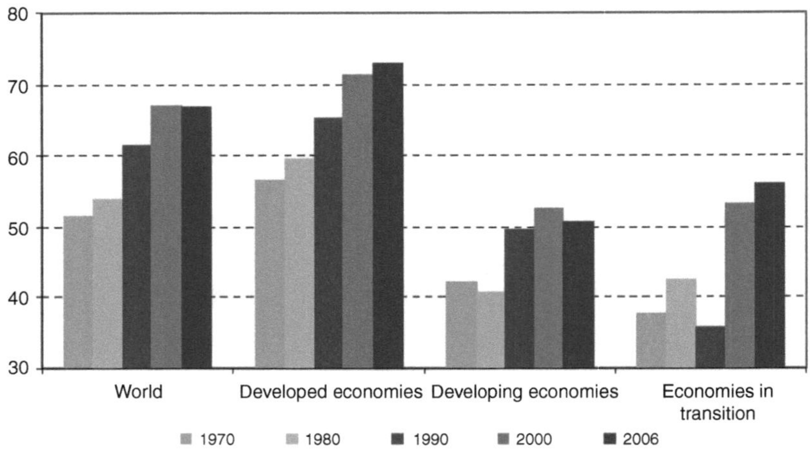

Fig. 1.1 Evolution of the share of service sector (% of GDP) in different groups of economies between 1970 and 2006. *Source*: UNCTAD Handbook of Statistics. Elab. G. di Meglio

An important and well-known fact to be underlined is that the service sector in all countries is composed of a wide variety of market activities, ranging from air transport and retailing, to hotels, restaurants and personal services, plus the non-market services, like public education and health. The shares in total value added and employment of different service industries reflects clearly this fact. Figures on the OECD countries show that Finance, insurance, real estate and business is the largest subsector within services. These activities are followed by community, social and personal services, wholesale and retail trade, and hotels and restaurants. Cross-country analysis shows that important differences exist between economies. In the European Union (EU25), five service industries account for more than 40 % of the total value added generated and near 50 % of employment. Business services were in 2006–2007 the most important industry in terms of value added and appear in the second place in terms of employment. Distributive activities (both retailing and wholesale trade) registered the higher increase according to their weight in GDP and employment. Transport, Financial services and Real estate are located immediately after the first two, also in terms of GDP and the number of people employed.

In recent years some service industries have lost employment due to restructuring processes, like wholesale and retail trade, and simultaneously hotels and restaurants has decreased or stagnated in almost all countries, except Japan, Greece, Canada, México and Switzerland, between others. The share of employment in Transport and Communications in total employment has also stagnated in practically all OECD countries, but their contribution to VA has increased.

All these changes require specific and in-depth analysis to explain why and where they are taking place and, on the other hand, there are territorial en regional impacts and effects which demand more research efforts.

1 A Not Widely-Covered and Underestimated Sector

Literature on service sector has increased substantially along the last two decades, particularly on some specific activities, like trading, business and financial services, transport and communications and tourism. But despite recent and important advances, services are still lacking of considerable attention by researchers and frequently underestimated by politicians. The historical background considering services as unproductive activities still persists. Our societies are characterized by knowledge, information, intangibles and the growth of service industries, but many policy makers continue considering services as second order activities for economic growth.

Nevertheless, it must be recognised that these attitudes are changing rather quickly, particularly in the most developed countries. Statistically the number of books and articles on service trends, problems and policies has greatly increased from the last decade of the past century to nowadays and there are many reasons supporting the need of paying much more attention to the role of service activities in our societies. Firstly, as it has been stated, because they represent the major share of our economies and they are increasingly integrated in the overall production system. All manufacturing activities as well as service industries require services to produce, design and distribute their products. Secondly, service industries are really playing a very active role in market integration and the globalisation process. Thirdly, new jobs creation, added value and income are increasingly related to the good performance of services. In the fourth place, the development of some service activities is directly related to the increasing welfare of our societies (education, health, personal services, leisure...). And last, but surely not least, the opening up to competition is one of the characteristics of recent service sector developments in advanced and emergent economies, diminishing regulations established and breaking traditional public monopolies in activities like communications, transports, trade and the supply of some health and social services.

Quite a few factors can explain the services expansion in modern economies. As it is analysed in Chap. 2 of this book, some demand and supply factors can be particularly underlined. From the demand side, the Engel's law contribute to explain why households tend to increase their demand of services when income grows. But, the demand for services (to produce and distribute their products) by manufacturing, agriculture and the service industries is also important, even having into account that this fact includes processes of externalization of some services previously produced inside the industrial firms. Two more factors that favour the services growth are: the increasing international trade of services and the demand of services by Public Administrations (at state, regional and local levels).

From the supply side, possibly the most accepted factor is the low productivity traditionally allocated to the services sector, which determines the increasing number of people needed to raise the production of a good number of services. This is an idea progressively unaccepted by recent literature (see Chap. 2) but still valid to understand the employment increase in many service activities, e.g. personal services, retailing, hotels and restaurants, etc. Also, from the supply side, some other factors explain services growth, as those non-market services produced by public administrations and linked to the welfare state developments. And, of course, there must be pointed out the more recently decisions on the liberalisation of service markets and the stunning ICT developments which are contributing to create new services and to improve some of the more traditional ones.

The analysis of the aspects and problems previously cited has undoubtedly promoted the increase observed in the literature on services produced over the last two decades. No one can discuss today the strong interactions between services and the other economic big sectors and industries (including the services sector itself). It is also clear that without well-established and dynamic financial, communications, distribution and transport mechanisms, the operation of the economic system cannot be efficient. The importance of business services, from consultancy, to engineering, design or information is undoubtable and the literature on these activities is really plentiful. A wide number of older and more recent references of interest are available for the reader in each chapter of this book, which would be unsuitable to quote in this introduction.

Many services activities have a very long tradition, but recent literature refers frequently to the *new services economy*. This idea is supported by the following reasons. Firstly, there are services which are actually *new*, as the e-services group, and it is also true that many old services—from distribution to personal services—are now provided on a *new basis*. Secondly, because the new services economy can be characterised by a concept of service not conferred any more to the boundaries of an economic sector or to a big box of heterogeneous activities. The new services economy goes beyond the traditional division between services and manufacturing activities and this is logic, at present, because the relationships between industry and services cannot be understood nor analysed separately. Finally, new trends in terms of demand of services by households and individuals are also changing, particularly in the most developed countries. This change is partly owed to the fact that new technologies modify several traditional services and leave space to *new* ones, as well they modify the *way to supply and to use them*. It is also due to demographic trends such as ageing and the evolution of cultural and social habits and practices, mainly because of the internationalisation and globalisation processes.

2 Services, Regions and Cities

From the territorial perspective the expansion of services has a variety of important effects and consequences. These aspects are receiving an increasingly attention by researchers, but it is clearly a field which needs more updated contributions. The reason for publishing a book like this one has to do with the lack of a volume offering not only an overview of the relations between services industries and their territorial effects and disparities, but also to provide an updated analysis of the new services from a regional and urban point of view.

Some of the changes observed in this field in modern economies are taking place too fast for conclusive theoretical approaches and empirical analyses to be set out. The contributions by Christaller (1933) and other pioneering authors on the location and concurrence theories and tendencies to agglomeration are still fruitfully used to explain how and where productive activities, in general, and services industries, in particular, tend to locate. The gravity models developed on the basis of the work carried out by Reilly (1931), the probabilistic models (Huff, 1963), and the spatial interaction models are still useful, although there are recent and interesting contributions regarding the location of business services and KIBS (Knowledge-Intensive Business Services) as well as retailing and wholesale. And, of course, the new approaches generated on the basis of the NEG (New Economic Geography) by Krugman (1991a, 1991b, 1998) and Fujita, Krugman, and Venables (1999)[1] must be taken into consideration. Some contributions to urban economy and the role of big cities and their growth in recent decades have joined the new approaches, as it is widely considered in Chap. 11.

Some years ago, Illeris (1989a, 1989b) summed up some of the most prominent features of services location in the 1980s. According to his analysis: (1) service employment were more concentrated in the big cities than the population as a whole; (2) financial services, business services, and wholesale trade were over-represented in the big cities; (3) retail trade and public services were more evenly distributed following the distribution of total population; and (4) some activities were showing special location patterns; e.g. hotels/restaurants were over-represented in tourist areas.

These trends showed that many services tended generally to be spatially concentrated, particularly in large cities (Taylor, 2004) However, the analysis carried out within the FAST II program (European Commission)[2] revealed that in most countries (in the EU and also in the United States), some services sub-sectors were showing higher growth rates in rural areas and small towns in the periphery than in big cities. Furthermore, the relationships between industry and services and the capacity of large industrial centres to attract services were highlighted. In fact, many papers published in the 1980s (Maroto-Sanchez, 2009a, 2009b) emphasize the role of services in enhancing the prospects of attracting and retaining manufactures.[3]

The contributions to the services-manufacturing-territory relations have grown substantially during the past two decades thanks to the empirical analysis regarding

[1] In Cuadrado-Roura (2011), an assessment of the objectives and approaches as well as the degree of novelty and limitations of this methodological school is carried out.

[2] As part of the FAST II (Forecasting and Assessment in Science and Technology) 1982–1986 programme, numerous documents were prepared (*Fast Occasional Papers,* generally photocopied, although some gave rise to printed but scattered publications). The subject of expansion, employment and location of services, as well as the influence of new technologies were analysed within the framework of this programme.

[3] Pedersen (1986), Bailly and Maillat (1988), Stabler and Howe (1988), Illeris (1989a, 1989b), Hansen (1990).

the location of business services and producer service companies (Aslesen & Isaksen, 2007; Cook, Pandit, Beaverstock, Taylor, & Pain, 2007; Doloreux, Amara, & Landry, 2008a, 2008b), and other studies on the availability of human capital and on the location of services, or the proximity of public agencies (Alvesson, 2004; Koch & Stahlecker, 2006; Sokol, Van Egeraat, & Williams, 2008; Wood, 2002).

The role of services in agglomeration economies seems to lie within the foundations of many of the aforementioned contributions. Yet, the focus on this issue and its better understanding had much in common with the 'New Economic Geography' (NEG) approaches and also with the contribution by Glaeser, Kallal, Scheinkman, and Shleifer (1992) regarding the growth of cities and other later contributions (Duranton & Puga, 2002, 2005; Ottaviano & Puga, 1998; Puga, 1998; Venables, 2001) linking agglomeration economies with ICTs and cities. However, these economies were not only linked to cities as protagonists, but also to the relations between the expansion of services and regional development, as well as to the regional disparities caused by the different growth of services and their composition (Maroto-Sanchez & Cuadrado-Roura, 2007; Messina, 2004).

Is it positive for a region to specialise in services activities from the perspective of its regional growth? In order to answer this question, it is necessary to clarify, or at least to pay attention, to the fact that some regions specialise in services as a consequence of their natural conditions (e.g. conditions enabling tourism) or as a result of the advantages which stem from the geographic situation of a region or city, which promotes the opportunities to expand certain services (e.g., the large European ports and the railway hubs and the high development of services linked to the activities having advantage from these infrastructures). Some recent works have provided empirical results that offer different answers to this question. Paci and Usai (2006) consider that the results are positive, but they underline the appropriateness of diversification. A clearer answer refers to the role of big cities, in general, and the fact that their conditions attract a significant amount of different services instead of becoming excessively specialised (Desmet & Fafchamps, 2005; Doloreux et al., 2008a, 2008b).

Regional asymmetries exist in the geographical distribution of services in many countries. As it is shown in Table 1.1, services-oriented regions are almost systematically leaded by a capital city or some metropolitan regions in Europe (Di Meglio, 2010; see also Chap. 11). The share of service activities in total employment is the highest in regions located in the United Kingdom, Belgium, Sweden, France, Germany, Luxemburg and Czech Republic. On the contrary, the weight of this sector is much less important in regions of Romania, Poland, Greece and Portugal, except their capital state. Of course, regional disparities are important: the share of services of Inner London in total employment of the area is above 86 %, as it is also in Brussels and Stockholm, and the lowest share corresponds to Nord-East region, Romania, with 29.4 %. Statistics show that practically all European regions have experienced an increase in the share of employment in services between 2000 and 2008, but, simultaneously, the process of concentration continues in terms of persons occupied in service industries as well as activities qualified as 'advanced services'. In the last years this trend seems to be slightly stopped due to the economic and financial crisis.

Table 1.1 Regional specialization in services in the EU-27

Country	Top ten regions in 2007	Share of services in total employment (%)	Country	Top ten regions movers	Annual growth rate, share of services in employment, 2000–2007 (%)
UK	Inner London	87.7	RO	Sud-Vest Oltenia	5.6
BE	Règion de Bruxelles-Capitale	86.7	RO	Sud-Muntenia	4.2
SE	Stockholm	86.2	RO	Centru	3.9
FR	Corse	84.9	RO	Nord-Est	3.7
DE	Berlin	83.2	GR	Dytiki Ellada	3.5
FR	Île de France	82.2	GR	Anatoliki Makedonia. Thraki	3.4
CZ	Praha	81.5	RO	Nord-Vest	3.1
LU	Luxembourg (Grand-Duchè)	81.2	PL	Slaskie	2.7
UK	Outer London	81.1	PT	Região Autónoma da Madeira (PT)	2.7
UK	Surrey, East and West Sussex	80.9	RO	Sud-Est	2.6

Source: Eurostat. Services includes NACE codes G to Q

The services location is explored in depth through several chapters of this book, including the overview offered in Chap. 11. This is undoubtedly an open subject about which no conclusive statements can be made. The fast development of ICTs and their incorporation to services, together with the creation of new services and the organisational changes in companies (manufacturing and service firms) are altering some of the conclusions that seemed to be valid in the 1980s, although some changes that have occurred recently were already foreseen. Any case, it is evident that the continuous expansion of services in advanced economies has important territorial effects and such effects are unquestionably characterised by heterogeneity.

Cities, and large metropolitan areas in particular, are the major protagonists regarding the concentration of services, both because they generate the highest percentage of demand and also because they attract the location of new services activities. At the same time, some medium-sized cities and rural areas have also managed to attract services providers, whose market does not depend on the surrounding population. This is the case of professionals dedicated to activities such as design, engineering, leisure, architecture, etc., although this phenomenon cannot be generalised. A more specific case, generally covering more extensive areas, is the concentration of tourist supply and demand, which has provided, in many cases, regions previously registering a medium or even low development very important growth advantages.

3 Structure and Objectives of the Book

The preparation of this book has been long. Over the last years the need for a book dedicated to the expansion of services, its location and its importance for Regional Economics has been increasingly justified, due to the lack of a compiling study on these subjects. Therefore, our main objective was to offer a book gathering a wide group of works on the evolution of services activities, its stimulating factors and its dynamics, with a special emphasis on the urban and territorial trends. The adequacy of including authors from different countries, giving some preference to team works, was considered as a second objective.

The starting point to have some interesting contributions for the book were the 'special sessions' held during the ERSA (European Regional Science Association) Congresses of 2010 (Jönköping, Sweden) and 2011 (Barcelona, Spain). An important group of researchers specialized in the services sector participated in these two sessions, which allowed to select some 'papers', according to the subject and the value added of the texts. To complete the wide scope of subjects to deal with, some well-known experts on services were also invited to present additional papers. Afterwards, a timetable was set, so that the authors could prepare or rewrite their contributions. All manuscripts received were submitted to external evaluation and the suggestions or changes to be taken into account were sent to the authors to prepare the final version of their papers.

The papers finally selected have been organized in three sections. Within the first one, there are those contributions offering a wide and general panorama of the evolution of the services activities, their stimulation factors, the incidence of globalization, a general panorama of regional differences according to services provision in OECD countries and to the importance of the processes of innovation in course or to be carried out in the following years. Chapter 2, whose author is the coordinator of this book, offers a general panorama of the growth of the service sector, of the main explanatory factors for their growth and of the future prospects of the sector, taking as a reference the most developed economies. The analysis, from the regional point of view, of the services globalization (Chap. 3) was carried out by Profs. R. Capello and U. Fratesi, who were in their turn responsible of an important research on this subject for the European Commission. A team of the Regional Policy Division of the OECD, directed by J.E. Garcilazo, has dedicated its paper to the deep analysis of patterns and trends in services related activities in OECD regions (Chap. 4). It is a work which uses very recent and not previously published data, reinforcing the value of its conclusions. On his turn, Professor P. Cooke, one of the most outstanding international experts on innovation, provides with some reflections and data on global innovation networks and their relation with services innovation and territory (Chap. 5). Finally, Prof. P. Daniels, a true pioneer of the study of services, gives some answers to the question if there is a rebuilding of the global Service Economy (Chap. 6), which leads to study the specific case of the so-called BRICs countries, their expansion and the role of services, particularly the more advanced ones.

The second part consists of the works dedicated to some relevant aspects of the behaviour of the services sector which have important effects on a territorial scale, like the ones of foreign direct investments in services in Europe, studied by Laura Resmini, who points out their impact on European regions (Chap. 7), and the firms' decisions to export services (Chap. 8) by H. Kox, one of the most renowned experts in the study and monitoring of the changes in the services industries.

Another subject of importance was the study of the productivity behaviour of services and if there are important differences at a regional level (Chap. 9); the work carried out by professors A. Maroto and J.R. Cuadrado, belongs to a research scope in which the authors have already published several articles and books, with the added value of the study of what has happened in the European regions. Finally, W. Beyers, from the University of Washington (US), makes a contribution on the trends of employment and unemployment in services in relation with the effects of the current Great Depression (Chap. 10), taking the OECD countries as a reference.

Although in the previous two parts many regional aspects of services development are analyzed, the third one gathers all the works dedicated specifically to the analysis of the services location and to some case-studies. Chapter 11, by this book's editor, is its framework, since there appear not only the available explaining theories for services location, but also the way this location is carried out in the practice, taking as reference the retailing activities, hotels and restaurants and business services, adding a final section on the role of the capital-cities attracting the headquarters of the most important services firms. A group of researchers directed by Professor P.J. Taylor provides a comparative analysis based on the location strategies of advanced producer services (Chap. 12), which reinforces the conclusions of a very well-known research topic on the role of the big cities on a global scale and their relation with the services attraction, especially those with an advanced business character. The contribution of B. Delgado and M.G. Velasco (Chap. 13) completes and analyses in depth the former ones. These authors study the geographical distribution and regional specialization of KIBS, one of the most expanding and dynamic group within the services sector, which includes all the business services with a knowledge basis.

J. Glückler and I. Hammer present a interesting work by designing a new services typology for the diversity and dynamics of the German services economy (Chap. 14). It is a very well- known fact that there are notable difficulties when trying to compare the evolution and localization of services among countries.. M. Polèse y F. Rubiera (Chap. 15) have faced this problem taking Canada and Spain as a reference. The results are the expected (there are difficulties) but also some facts and trends seem to be common to both countries (which should not be surprising). Boix, De Miguel and Hervas, following some recent works referred especially to the manufacturing sector and to the professional services, present an excellent text (Chap. 16) on creative services firms.. Their analysis is not only interesting because they clarify what and where the creative services firms are, but also because the authors make emphasis on the relation of the subject with the wealth of regions and explain the influence of those services on the regional improvement. P. Ström, who is specialized in the study of services in Asia, contributes to the book with a text (Chap. 17), where the Japanese

Services Industry is analyzed as a very significant example of the behaviour of the Japanese firms in terms of localization, competitiveness and internationalization. Finally, C. Gallouj and F. Gallouj study the role of services in the regional development policies. France is set as example and their work points out the various existing options for using services in order to boost regional and local development. To do so, services markets must be analyzed on a regional scale. There must be also analyzed the problems inherent to the implementation of a services policy in a certain territory, where the balance between supply and demand is a key matter since promoting supply will have no effects if the chosen region lacks in demand.

Undoubtedly, the book does not consider all the subjects which could be of interest and it doesn't close or conclude the options of analysis, because every advance opens new guidelines of study. Nevertheless, the book offers a wide array of new texts, analyses and reflections which will be very useful for many professionals, academics and policy makers. We expect that this book will promote further research in such an attractive and emergent field: the relations between the expansive and continuous advance of services and the heterogeneity of trends and effects observed on a territorial scale.

4 Some Lessons Learned and Final Remarks

The papers included in this book offer the possibility of underlying some key ideas and remarks that not only can be considered by the researchers to pull along their future works, but also by policy-makers interested in regional development and the role of services. The points suggested below are not the conclusions of the book contents. The aim is rather more modest: it is to point out some of the key ideas launched by the authors of this book.

1. *Increasing weight of services activities in almost all economies.* The services sector has gained importance in a great number of economies, both as a proportion of workers employed in this sector to the total of workforce and as a proportion of total value added produced by this sector. In the OECD countries, service industries employ close to 70 % of total workforce occupied and add around 70 % of total output. Figures of emergent and less developed economies rang below these last, of course, but their growth rates have been always positive. Essentially, the expansion of services has to do with the economic structural changes experienced or in process in almost all countries, with a reduction of the comparative size of primary activities (mainly agriculture) and the manufacturing sector and increases in services sector. The share of this last in the economies is also related to the evolution of relative prices between the tradable and non-tradable sectors, the latter being closely linked to the real exchange rate.
2. *Factors underlying services growth.* Two issues have become clear regarding the factors explaining the expansion of services, particularly in the most

advanced economies. The first is that there is no single factor, of course, not even a main factor, but several of them. The second is that, besides the economic ones, other types of factors are in place. Although the increase of expenditure on services by individuals and households is primarily related to the increase of their incomes, other factors clearly affect the changes in the families' expenditure structure; f. ex.: the urbanization processes, the entry of women into the labor market, the ageing of the population and the incorporation of young people as consumers. Additionally to the demand of services by the households, other factors pushing services from the demand side are the inter-industrial demand of services by manufacturing (linked to the outsourcing processes), the international trade and the demand of market services by the Administrations. From the supply side some factors are also helping to increase the process of tertiarization of the economies. Undoubtedly two are particularly important: on the one hand, the low productivity of many service activities which must use always more workforces to increase their production and, on the other, the non-market services supplied by the state and other public authorities, linked to the welfare state developments. Other factors are also influencing the production of both traditional and new services, of course, as the ICT developments and the markets de-regulation.

3. *Services expansion will continue in the future*. The future and the expansion of services industries will essentially continue to be marked by the influence and behaviour of the factors previously pointed out. In terms of demand, the four main driving forces of the sector's growth have been and continue being the consumption by families, the increasing use of services to produce goods and services, the demand by public administrations and the international trade of services. The analysis developed leads to the conclusion that once the worst phase of the current recession should be over, the growth of services will resume, although possibly in a more moderate path. Yet, we are bound to witness restructuring processes within the services sector that should lead to a reduction in the number of companies operating in some industries, accompanied by an increase of their dimension, while in some individual cases, those firms whose profitability and viability in terms of efficiency and market are highly doubtful, such as some segments of the hotel and catering sector, retailing and conventional transport may disappear in the coming years.
4. *Globalization and regions*. The globalization of tasks rather than sectors, the off-shoring and outsourcing of service functions, the de-industrialization in favour of services and the decentralization of intertwined functions are reshaping the division of labour in the subnational economies of advanced countries. On the other hand, regional economies are increasingly competing to seize the opportunities which these new trends offer. All these changes affect the services sector. It is from the services sector that most outsourcing of tasks takes place; it is the services sector that is the most engaged in off-shoring of functions; it is in the services sector where jobs and productivity are most affected by the new globalization trends. The empirical analysis carried out

distinguishing (Chap. 3) between 'highly global regions'—those regions with above-average (economic and physical) connectivity with the rest of the world—and 'local regions' has shown that higher productivity growth rates are associated with advanced services industries in global regions. Global regions in Western countries are the only ones where a significantly higher specialization in advanced services activities is associated with the virtuous regions. And the overall conclusion is that the services sector is increasingly important for regions to be able to compete in the global world.

5. *International trade and Foreign Direct Investments in services.* The growth in services has been accompanied by a rising of services in international transactions: trade in services has grown faster than trade in goods and now it represents about 12 % of world GDP. Moreover, there has been a marked shift of foreign direct investment, FDI, from the manufacturing sector towards the services sector worldwide. The share of services in total FDI stocks has increased to about 63 % in 2010, as compared to 49 % in 1990. The European Union has played and still plays a dominant role in international transactions of services. It is the largest exporter in the world for services and the largest market for FDI in services. Single Market Programme provided impetus for the expansion of FDI in services sectors and for an EU-wide restructuring of several services industries, accelerating intra-EU services FDI. Moreover, the enlargements of the EU have offered new opportunities for FDI in services given the opening up of new markets and the liberalization processes (telecommunications, banking and transport in particular). From the regional point of view, empirical analysis carried out (Chap. 7) indicates that the distribution of FDI flows in Europe has changed. From geographical and sectorial point of view FDI in services are mainly driven by market reasons. Local demand seems to be more important than market potential, indicating that fragmented service markets across Europe still exist. Only FDI in transportation and communication services seems to be affected by the settlement structure of EU and regions hosting larger cities seem to be less attractive than less urbanized regions, because of stronger competition effects.

6. *Services productivity and regional growth.* A wave of economic literature on the productivity in services sector has been supporting the conventional thesis that the continuous increase of services in the economic structures and the low productivity of these activities, compared with manufacturing industries, entail a clear threat for future growth, while its rates should be pushed down. Nevertheless, the relationship between growth of services and labour productivity, comparing different samples of advanced countries and time periods has been revised in the recent literature. Some service branches show an increase in productivity which is comparable to, or even higher, than those of manufacturing. At a regional level, the results obtained lead to conclude (Chap. 9) that structural change still plays an important role in the improvement of productivity of each region as a whole and that some services industries (financial and insurance; some transport branches and telecommunications) are at least so productive as

the most dynamic manufactures. The analysis carried out demonstrated that the growth of services and productivity is positive and significant. Moreover, it has been verified that there is a process of convergence between those regions registering higher productivity levels at the beginning and the most backward regions. It is also confirmed that those regions highly specialized in services register more positive dynamics regarding productivity growth. Finally, as expected, services branches subject to market conditions have greater impact on the variation of productivity and this is contrary to the case of non-market services.

7. *Innovation as a key of the new services economy.* Unlike a widespread perception that the services industries are somewhat overshadowed by manufacturing in relation to innovation, an increasing amount of contemporary innovation actually occurs in services. This is related to technological change, but also to the application of recombined or 'modular' knowledge bundles (Chap. 5), which occurs but is more rarely written about from the perspective of manufacturing industry. Service innovation, especially in more advanced economic platforms like ICT services takes one or a combination of three forms—architectural, meaning a major reconfiguration of the key elements of the innovation network; modular, meaning recombination of separate but related elements to contribute to the implementation of innovations; and exploration innovation where the result of knowledge exploration, or research, can be the catalyst for innovation on a large, including global scale as with the other modes. Service innovations are opening new opportunities to regions to modernize their services and regional policies must pay much more attention to this field, strongly related to increasing regional productivity and to improving their supply of traditional and new services.
8. *Spatial concentration of services.* The analysis of geographical concentration of services industries in the OECD regions reveal that the services sector is more concentrated in space than economic activities in all sectors. Financial and business services are particularly concentrated amongst service subsectors. Additionally, specialization in financial and business services appears to be higher in capital regions or regions with large cities and such specialization has increased among OECD regions more than other services. Nevertheless, it is clear that services sector activities are not homogeneous in space. Some regions can be quite more dependent on services, in some cases representing about 90 % of their total GVA (e.g. New York, Hawai, Brussels, Oslo, Ile de France, Attiki, Madeira and Lazio). In contrast, services in some regions can represent less than one third of total GVA and employ less than 45 % of the workforce (e.g. Campeche and Tabasco in Mexico, Atacama and Bio-Bio in Chile, Gyeonbuck in Korea and Central Transdanubia in Hungary).
9. *Understanding the location of service industries and firms.* The main conclusion drawn from the analysis developed in Chap. 11 about the available theoretical contributions to services location is that their number is not very

high. Moreover, it is worth mentioning that changes that have taken and are taking place in recent years are not easy to be understood on the basis of some traditional approaches. In any case, it is clear that Geography and Urban Economy do currently feature conceptual and theoretical elements which contribute to understand the services location processes. The analysis of three of the most interesting cases: retailing and distribution, hotel and restaurant activities and business services and KIBS, offer conclusions of interest. The main characteristic of the latter is spatial concentration, not only at a wider territorial level—countries, regions—but also within metropolitan areas. However this does not exclude some movements towards dispersion observed in business services and in the new retailing structures. In the case of the hotels, a wide dispersion is combined with concentration in the geographical areas offering attractive resources for tourism, leisure and specific sports. At the same time, some new forms of travelling and transport facilities, the undertaking of business activities and the increase of urban tourism, meetings and congresses are bringing about a high concentration of hotels in large cities.

10. *The attraction of State capitals and major cities.* Some contributions included in the book suggest interesting remarks on this point. At an international level, but also within each country, globalization process gives advantages to the large cities, and particularly capital cities, boosting the concentration of the headquarters of large services companies. Agglomeration economies favour increasing concentrations of services and headquarters, as the human capital supply, good connections and easy accessibility at an international scale contribute to do. In Europe, this is clearly the case of London and Paris, but also of Brussels, Stockholm, Copenhagen, Madrid, Warsaw, Dublin, Lisbon, Wien or Prague. Nevertheless, those cities with a high economic importance, sometimes even higher than the one of the capital cities, are also very likely to attract this type of firms and the concentration of advanced services. Some European examples include Milan, Amsterdam, Frankfurt, Zurich and Barcelona. Connection to the world cities network (Chap. 12) is a key factor. Europe as a whole has numerous cities comparatively well connected within such a network due to its multiple states. The analysis of UK, Germany and Spain, despite major differences in national urban structures, shows that all the leading cities have reasonable overall connections and the major cities have strong connections to the competing centres of economic globalization.

11. *The case of Knowledge Intensive Business Services (KIBS).* KIBS constitute a relevant source of employment, production, investment and knowledge dissemination, particularly in industrialized areas. Thus, they may have a relevant role in the spatial allocation of economic activities and regional development. Whilst previous studies have shown that KIBS were not equally distributed across European regions, is this description still valid nowadays? The data analyzed reveal that in 2010 KIBS were already the most geographically concentrated activity. The greatest concentration has been found in Hi-tech and Financial KIBS. Nevertheless there are significant differences among the

groups of regions. These activities were mainly concentrated in capital city regions, as well as in Central and Northern regions of EU-27. Employment concentration in those areas was highly due to the agglomeration of economic activity and to a greater accessibility to markets and knowledge resources in these regions. On the contrary, less developed European regions of EU-15 and the new member countries (except capital cities regions) have obtained the lowest indexes. Nevertheless, results on the evolution of spatial concentration during the period 1995–2007 show a spatial diffusion process in KIBS location. Traditional regions where KIBS were agglomerated have lost relative participation in favour of peripheral regions, as well as regions where these services were not very important in the past (e.g., in France, Germany, Sweden and UK). Therefore, it seems that there has been a process of catching up and geographical diffusion during the economic expansion period, reflecting possibly some of the effects of cohesion policy. Some policy implications of the analysis are underlined (Chap. 13) from the research results.

12. *Creative services. Their contribution to the wealth of regions.* The analysis of the 'creative' industries has received increasing attention during the last decade, recognizing the particular role they play in some cities and regions. The hypothesis that creative services can be a 'growth driver' that promotes wealth in the regions where they are located is the departing point of one of the contributions of this book (Chap. 16). This is possibly due to the fact that firms in creative industries introduce new ideas that subsequently transferred to other firms of the economy, increasing the output of the whole economy. The analysis carried out offers interesting results. One of them is that an increase of 1 % in the percentage of employment in creative industries in the region translated into an increase of 0.39 % in GDP per capita, that is to say 1.479 Euros per capita. This was a higher effect than the one that was found for the presence of 'rest of knowledge-intensive services', manufacturing or agglomeration economies. All subsectors in creative services proved to be positively and significantly correlated with GDP p.c., and several patterns of co-location between these sub-sectors have also been detected, which generated diverse profiles across the regions. Other impacts have also been explored with positive results. Thus, if creative services impact basically on wealth and have highly local effects, they could be a significant objective for regional-driven policy. However, if the geographical effects are supra-regional, then national policy or coordination between regions could play an important role. Other interesting suggestions for regional policy can be extracted from this new approach to creative services.

13. *The analyses on services in different countries still offer interesting results.* The previous points are not the only contributions that can be extracted from the book. In fact, one of the most interesting aspects is that the researches included in this book use comparative data and information on a big number of countries, especially the ones integrated in the OECD. The book includes analysis referred to specific countries and it also compares a reduced number

of countries by studying some particular services activities. Japan, Germany, UK, the Netherlands, US, Canada, Poland, Spain, France and the BRICs are object of special reference in various chapters.

The same thing happens with the effects in terms of employment and unemployment generated by the Great Crisis, where US is taken as a reference, but always comparing the results with other countries; or the analysis of the spatial distribution of services industries in Canada and Spain in search of coincidences and disparities; the study of differences among countries in the localization of specific activities such as retailing and business services; or, finally, the regional policies for services in France.

Undoubtedly, this volume enriches all these aspects and it wouldn't be wise to underline just a couple of them. Nevertheless, it is undeniable that it is still very interesting and necessary, of course, to study in depth the services with reference to a specific country, their regions and the metropolitan areas, although other countries are taken as comparative reference. The main reason is that, although it is or should be possible to put forward a good number of conclusions they have in common, especially among the most developed countries, there still are very obvious differences that grant value to the analyses of cases referring to specific countries.

References

Alvesson, M. (2004). *Knowledge work and knowledge-intensive firms*. Oxford: Oxford University Press.

Aslesen, H. W., & Isaksen, A. (2007). Knowledge-intensive business services and urban industrial development. *Service Industrial Journal, 27*(3), 321–338.

Bailly, A. S., & Maillat, D. (1988). *Le secteur tertiare en question*. Paris: Economica.

Christaller, W. (1933). *Die Zentralen Orte in Süddeutschland*. Iena, G. Fischer. English version: *Central Places in Southern Germany*. Englewood Cliffs: Prentice Hall, 1966.

Cook, G. A., Pandit, N. R., Beaverstock, J. V., Taylor, P. J., & Pain, K. (2007). The role of location in knowledge creation and diffusion: Evidence of centripetal and centrifugal forces in the city of London financial services agglomeration. *Environment and Planning, 39*, 1325–1345.

Cuadrado-Roura, J. R. (2011). ¿Es tan nueva la Nueva Geografía Económica? *WP Institute of Economic and Social Analysis* n.1/12, Madrid.

Desmet, K., & Fafchamps, M. (2005). Changes in the spatial concentration of employment across US counties: A sectoral analysis 1972-2000. *Journal of Economic Geography, 5*(3), 261–284.

Di Meglio, G. (2010). *The role of public and private services in European economies*. Doctoral thesis, University of Alcala, Madrid (polic.).

Doloreux, D., Amara, N., & Landry, R. (2008). Mapping regional and sectorial characteristics of knowledge-intensive business services: Evidence from the province of Quebec (Canada). *Growth and Change, 39*(3), 464–496.

Duranton, G., & Puga, D. (2002). Diversity and specialisation in cities. Why, where and when does it matter? *European Urban and Regional Studies, 9*(4), 331–341.

Duranton, G., & Puga, D. (2005). From sectorial to functional urban specialisation. *Journal of Urban Economics, 57*(2), 343–370.

Fujita, M., Krugman, P., & Venables, A. (1999). *The spatial economy. Cities, regions and international trade*. Cambridge: MIT Press.

Glaeser, E. I., Kallal, H. D., Scheinkman, J. A., & Shleifer, A. (1992). Growth in cities. *Journal of Political Economy, 100*, 1126–1152.
Hansen, N. (1990). Innovative regional milieu, small firms and regional development: Evidence from Mediterranean France. *The Annals of Regional Science, 24*, 107–123.
Huff, D. (1963). A probabilistic analysis of shopping centre trade areas. *Land Economics, 53*, 81–90.
Illeris, S. (1989a). Producer services: The key sector for future economic development. *Entrepreneurship and Regional Development, 1*, 267–274.
Illeris, S. (1989b). *Services and regions in Europe*. Aldershot: Avebury.
Koch, A., & Stahlecker, T. (2006). Regional innovation systems and the foundation of knowledge-intensive business services. A comparative study in Bremen, Munich and Stuttgart (Germany). *European Planning Studies, 14*(2), 123–146.
Krugman, P. (1991a). *Geography and trade*. Cambridge: MIT Press.
Krugman, P. (1991b). Increasing returns and economic geography. *Journal of Political Economy, 99*(3), 483–499.
Krugman, P. (1998). What's new about economic geography? *Oxford Review of Economic Policy, 14*(2), 7–17.
Maroto-Sanchez, A. (2009a). Relationships between services, growth, productivity and welfare. The state of the art. *WP, ServPINN programme.*
Maroto-Sanchez, A. (2009b). *La productividad en el sector servicios*. Madrid: Marcial Pons.
Maroto-Sanchez, A., & Cuadrado-Roura, J. R. (2007). Productivity and tertiarization in industrialized countries: A comparative analysis. *Efficiency Working Series*. Oviedo: University of Oviedo.
Messina, J. (2004). Institutions and service employment: A panel study for OECD countries. *European Central Bank Working Paper*, 320. Frankfurt: European Central Bank.
Ottaviano, G., & Puga, D. (1998). Agglomeration in the global economy. A survey of the new economic geography. *The World Economy, 21*, 707–731.
Paci, R., & Usai, S. (2006). Agglomeration economies and growth. The case of Italian Local Labour Systems 1991–2001. *Contributi di Recerca Crenos*, 12.
Pedersen, P. O. (1986). The role of business services in regional development: A new growth centre strategy. *Scandinavian House and Planning Research, 3*, 167–182.
Puga, D. (1998). Urbanisation patterns: European vs. less developed countries. *Journal of Regional Science, 38*(2), 231–252.
Reilly, W. J. (1931). *The law of retail gravitation*. New York: Knickerbocker Press.
Sokol, M., Van Egeraat, C., & Williams, B. (2008). Revisiting the informational city: Space of flows, policentrity and the geography of knowledge-intensive business services in the emerging global city-region of Dublin. *Regional Studies, 42*(8), 1133–1146.
Stabler, J., & Howe, E. (1988). Service exports and regional growth in the post-industrial era. *Journal of Regional Science, 28*, 303–316.
Taylor, P. J. (2004). *World city network. A global urban analysis*. London: Routledge.
Venables, A. J. (2001). Geography and international inequalities: The impact of new technologies. *Journal of Industry, Competitions and Trade, 1–2*, 135–139.
Wood, P. (2002). Knowledge-intensive services and urban innovativeness. *Urban Studies, 39*(5–6), 993–1002.

Part I
Service Industries: Growth Factors, Globalization and Innovation

Chapter 2
Towards Increasingly "Tertiarised" Economies: Facts, Factors and Prospects

Juan R. Cuadrado-Roura

1 Introduction

Although countries with a medium-high level of development are still being often considered as "industrialised economies", the large majority of them, if not all, are "tertiarised" economies.

Nowadays, nobody questions that all developed countries have become *service economies,* particularly when we measure this in terms of employment in service activities and even more so when employment in service occupations is considered (Schettkat & Yocarini, 2006). The data gathered from some countries of the European Union, the United States, Canada, Japan, and Australia are absolutely clear in this regard. Nevertheless, could this be applicable to regions? Are there any differences regarding the development of services within advanced countries? One of the objectives of this book is to contribute to answer both questions appropriately by providing the necessary elements for it, or by presenting at least analyses, data and points for consideration that will allow gaining deeper insights into the problem and the prevalent trends.

The main aim of this chapter is to describe the trends identified in the growth of services in the most developed countries, as well as to clarify which are the factors boosting the almost ever-increasing importance of services. This would lead to a better understanding not only of the current prevailing trends, but also of their effects and/or impacts at territorial level and the differences observed in this regard. To this end, the text firstly provides in Sect. 2 data showing the almost unstoppable trend of the service sector to increase its weight in the so-considered most advanced economies. In Sect. 3, the question of why services have grown and are currently growing, both in terms of employment and GVA, is raised. The most outstanding factors from the demand side have always been the following: (1) the increase in the

J.R. Cuadrado-Roura (✉)
Department of Applied Economics, University of Alcala, Madrid, Spain
e-mail: jr.cuadrado@uah.es

J.R. Cuadrado-Roura (ed.), *Service Industries and Regions*,
Advances in Spatial Science, DOI 10.1007/978-3-642-35801-2_2,

demand of services by the households when their incomes rise; (2) the growing use of services in inter-industrial relations; and (3) the high demands for employment required to produce many services, which is directly related to its low productivity. However, these are not the only explanatory factors, as some on-going changes also contribute to understanding the upward trend of services. Section 4 focus particularly driving factors from the supply side, paying special attention to the productivity of services (an aspect which is analysed in-depth in chapter 9), but considering also some other factors actually pushing services growth. Finally, the chapter includes some reflections upon the continuing expected growth of services in the following years (Sect. 5). There are some reasons for a positive outlook, but this will of course entail some changes in the structure and dynamics of the sector.

2 The Shift of Service GVA and Employment

Some decades ago, one of the pioneering authors in analyzing the services sector, R. Fuchs, in an outstanding study for the NBER described the quiet transition towards a service society which was taking place in the main western economies. "The transition from an agricultural to an industrial economy, which began in England and has been repeated in most of the Western world, has been characterized as a 'revolution'. The shift from industrial to service employment, which has advanced furthest in the United States but is evident in all developed economies, has proceeded more quietly, but it too has implications for society, and for economic analysis, of 'revolutionary' proportions" (Fuchs, 1968, p. 2).

The evolution towards total employment in the activities of the service sector has increased almost constantly in all countries still being paradoxically called as "industrialized". The United States have nearly always been ahead in this process, and those employed in service activities in this country now account for more than 75 % of the total number of employees. However, many other countries have followed the American trail. Figure 2.1 shows the trends followed by employment in the tertiary sector from 1970 to date, taking various countries as a reference (although many others could also be included). In fact, employment in this sector is very close to or exceeds 70 % of total employment in quite a large number of the most advanced economies.

A similar pattern to that of employment is observed when taking into account GVA in current monetary values (Fig. 2.2). In all selected countries, but also in many others considered to be developed, the same trend prevails. However, it should be noted that, when GVA is measured in constant values (Fig. 2.3), the increase registered is much more moderate. As it is well-known, this is due to the fact that, in the majority of the countries, the prices of services have tended to rise above the average increase in the rest of prices. In fact, the data show clearly that in virtually all OECD countries, the increases in the prices of services have exceeded those registered by manufacturing and agricultural sectors. As a result,

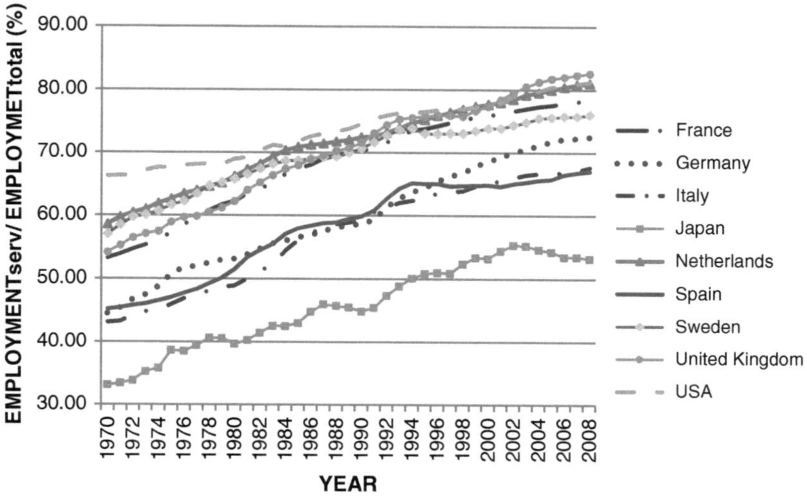

Fig. 2.1 Trends of service sector employment (% of total employment). *Source*: Prepared by the author. EU KLEMS, database

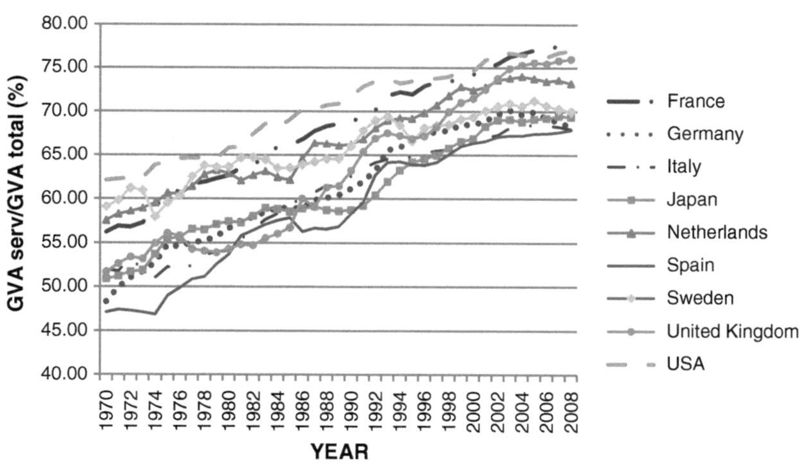

Fig. 2.2 GVA of the service sector as a function of total GVA (current prices). *Source*: Prepared by the author. EU KLEMS database

the increase of service GVA is partly due to this price differential (OECD, 2005a, 2005b).[1]

The advance of services reflects to a large extent the structural changes brought about almost inexorably in the economies by economic growth. Kaldor (1961),

[1] This behaviour of the prices of services has driven up the general price index of many economies, with the corresponding inflationary impact and a negative influence on their competitiveness.

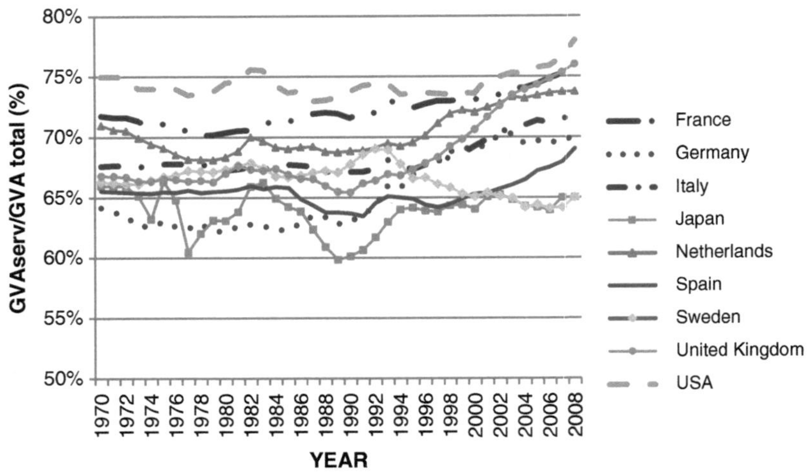

Fig. 2.3 Evolution of GVA of the service sector (constant prices). *Source*: Prepared by the author. EU KLEMS database

Kuznetz (1971), and Maddison (1980), among others, already demonstrated this fact more than three decades ago this fact, i.e., the existence of empirical regularities in the structural transformation of advanced economies. More specifically, Kuznetz and Maddison underlined the canonical shifts of output and labour from agriculture to industry, and later on from manufacturing to services. Both authors also highlighted a fact which has generated a fair amount of literature and a broad discussion. They stated, among other things, that productivity increases in the service sector were much lower than in other sectors of the economies (manufacturing, but also primary productions), and service prices tended to increase more rapidly. This is an important point underlined also by Baumol (1967) in order to establish the theory known as the "cost disease". According to this approach, the expansive dynamics of services should result, as a consequence of the low productivity rate of most of the activities in the sector, in a global growth tending to decrease. That is to say, the expansion of the tertiary sector in developed countries should lead to lower growth rates than in other previous stages of its development.

This theory, accepted by many for years, has been questioned of late after delving into the behaviour of the different and heterogeneous industries comprising the service sector (Maroto & Cuadrado, 2009; Triplett & Bosworth, 2000, 2001, 2002; Wölfl, 2003; among others). This has led to the statement that some service activities register quite high productivity growth rates comparable to or even higher than those of some manufacturing activities.

Nevertheless, without entering into this debate—which would of course apply to the case of regions—it would be worth considering those factors that apparently best explain the almost constant growth of services in the economies. We will focus our attention on the most advanced economies because, as discussed below, the possible "causes" or factors are more clearly observed in these economies. However, we should remember that services also have a significant weight in less developed economies, although the internal composition of the sector is quite different from the advanced economies. Against the development of activities such as finance, transport, business services, personal services, and even those offered by the public sector observed in the most developed countries, services feature a much more traditional composition in the developing and more backward countries, with high underemployment and black economy rates, as well as the predominance of not very technically advanced and much smaller companies.

3 Why Do Services Grow? Factors Behind the Expansion of the Sector

Once the near-constant growth of services in developed economies has been accepted as a fact, it is necessary to consider which are the main factors underlying this process. In fact, the explanations provided are quite varied, although some of these are agreed to be much more influential than others, such as the evolution of per capita income or the productivity of services. However, it is clear that there are various independent factors explaining the expansive process of the service sector.

It was initially proposed that a determining reason was simply the changes undergone in the consumption patterns of citizens as their incomes increase. Almost at the same time, and together with this proposal, it was highlighted that the labour demands required by many services brought about a lower productivity compared to the case of the production of goods. Fourastié (1949) already referred to this fact as the basic factor in the expansion of tertiary activities. This hypothesis then gave rise, as previously mentioned, to the theory of the reallocation of resources from the most to the least productive activities, as in the case of services (Baumol, 1967; Baumol & Bowen, 1965), with a corresponding impact on the rates of change in total product, costs, and prices.

The relationship between the growth of services and the increases in terms of per capita income was stated early on as one of the possible reasons for growth of the tertiary sector as a whole (Clark, 1940; Fourastie, 1949). Nevertheless, there have been other arguments aiming to explain the expansion of services by linking it to the decline of industry, which has become apparent in almost all of the most developed countries, particularly as a result of the consequences of the 1970s crisis, which originated extensive literature regarding the *deindustrialisation* of the most advanced economies and their role in favour of the continuous increasing importance of services in the productive structures of such economies (Blackaby, 1978; Cairncross, 1978; Gemmell, 1982; Gershuny, 1978; OECD, 1975; among others). This hypothesis was well supported by two arguments that are still valid: firstly, the organisational changes taking place in industrial companies, which determined the *outsourcing* of many services previously produced "within" the manufacturing

Table 2.1 Factors explaining service growth

EXPLAINING FACTORS

- **DEMAND-SIDE FACTORS**
  - **Household Expenses (→ Engel's Law; evolution of individual and social behaviour)**
  - **Service demand as production inputs (→ inter-industrial relationships)**
  - **Demand by Gov't and Public Authorities**
  - **International Trade**
- **SUPPLY-SIDE FACTORS**
  - **TRADITIONAL**
    - **Low productivity of services**
    - **Service supply by the State (Welfare State)**
  - **NEWS**
    - **ICT developments**
    - **Markets liberalization**
    - **Human Capital**

Source: Prepared by the author

companies; and secondly, the off-shoring process of industries occurring in recent decades, which has resulted in the relocation of manufacturing companies from the most advanced to poorer countries, with lower salaries and costs.

Consequently, two issues have become clear in recent years regarding the factors explaining the expansion of services. The first is that there is no single factor, not even a main factor, but several of them. The second is that, besides the "economic" factors, other types of factors are in place. Therefore, although the increase of expenditure on services by individuals and families is primarily related to their higher incomes, other factors clearly affect the changes in the families' expenditure structure, for example the urbanisation processes, the entry of women into the labour market, the ageing of the population, and the incorporation of young people as consumers.[2] Moreover, the outsourcing processes have not only affected the growing demand for services by companies, but also other productive and organisational changes, as well as the growing complexity of the business environment, both from the fiscal, legal, or technical perspectives and also due to the increasing competition and the need to internationalise.

The factors that have had an impact and still affect the expansion of services may be arranged according to different approaches. However, we consider that the clearest way to organise them is according to their comparative influence from the demand and supply viewpoints, as it is synthetically shown in Table 2.1.

[2] See Mañas, Gabaldon, and Cuadrado-Roura (2002).

This division, as useful as it may be, should not be taken too rigidly, as the growth of some service activities is very often due to demand and supply factors at the same time. Moreover, the fact that the factors shown in this table are always included within a specific legal-institutional framework affecting the conditions related to the establishment, production, and distribution of the services must be also taken into consideration. In this sense, it should be noted that services are a sector that has been and still is subject to several regulations by central, regional, and local governments.

Factors described in Table 2.1 deserve some explanatory comments, which are presented following the same order as in the table.

4 Driving Factors of Services Growth from the Demand Side

4.1 The Increase in the Household Consumption of Services: Demographic and Social Changes

The rise in the levels of per capita income has been undoubtedly a significant factor that boosts the demand for services. We must first turn necessarily to the well-known "Engel's law" when searching for the reasons accounting for this fact. The law establishes that the evolution of household incomes gives rise to changes in the expenditure structure due to the different income elasticity values.[3] The result is that the expenditure on basic necessities tends to decrease proportionally to the increase of income level, whereas other expenses—including many services—not included previously or rather included in the structure of household consumption since they were considered unnecessary or even a luxury, absorb higher levels of expenditure as income rises (Falvey & Gemmel, 1996; Kravis, Heston, & Summers, 1981; Maddison, 1980). This happens, for example, in the case of education, health, transport, or leisure and culture expenses, whose relative costs have also tended at times to decrease and become more affordable.

Numerous authors have analysed the household consumption performance and consumption patterns of goods and services. See, for example, the works carried out by Fuchs (1968), Gershuny (1978), Gershuny and Miles (1983), Gregory, Salverda, and Schettkat (2007), among others. Moreover, there exists a fair amount of researches referring to a specific country or comparing different countries (Mañas et al., 2002; Schettkat, 2004).

However, it is worth highlighting that the only cause for this higher level of expenditure on services is not the behaviour of income and price elasticity, but also a series of demographic, social and cultural changes. Some examples of these changes include the ageing process of the population of some advanced countries

[3] Fisher (1935) and Clark (1940) applied this law to the changes in the demand for manufactured products and stated that the elasticity of demand for many goods was less than one, while that of services and luxury items was more than one.

(mainly in Europe), the past and present processes of growing urbanisation (involving changes in lifestyles and in the type of expenses to be made), the entry of women into the labour market (Mañas et al., 2002; Yang and Magrabi, 1989), the greater attention paid to personal care, and the evolution of some more general patterns of social behaviour due to specific sociological and cultural changes.

The higher demand for services by individuals and families is not, obviously, the only factor that has boosted the expansion of services. However, it has been and still is one of the main protagonists in the process of tertiatisation of economies. The *surveys on household income and expenditure* conducted by the majority of advanced countries reveal the continuous increase of the relative weight of services in the household expenditure structure. At the beginning of the 1990s, this expenditure meant on average less than 30 % of income, a percentage that nowadays is around 45 % of the average monetary expenditure per household. By contrast, the average expenditure of households on food has fallen from 33–34 % in 1970 to approximately 20 % in 2010, though there exist obviously wide differences due to income levels and the places of residence.

Prices have also affected the weight of services in household expenditure as they have tended to rise to a larger extent rather more than manufacturing products and raw materials. On the other hand, we should remember that the increasing importance of consumer services is not only quantitative, but also qualitative. In fact, the household consumption of services may be understood both as a consequence of development and social modernisation and also as an explanatory cause of these processes. Certain services constitute a paradigmatic tool of the improvement in standards of living, which is translated into the increasingly general possibility of enjoying services related to leisure (entertainment, travel, sport...), culture, or aesthetics, for example. Yet, it is also worth mentioning that some services seem to have become essential elements for families, such as those related to private training and education, childcare and elderly care, or assistance with household chores. Finally, as previously pointed out, the significant entry of women into the labour market has too resulted in the increase in consumption of specific services (meals outside the home, laundry, transport, etc.). This was already highlighted many years ago by Shaninger and Allen (1981), Jacobs, Shipp, and Brown (1989), Magrabi et al. (1991) and the idea has become consolidated.

Furthermore, the empirical analysis of expenditure on services by individuals and families enables us to demonstrate the existence of very clear differences in the dynamics of the various services. Thus, there is evidently a group of services on which expenditure has increased proportionally more than on others from the 1990s to date. This is the case of insurances (featuring an average annual growth of 6.5 %), telecommunications (above 5 %), hotels and travelling (3.9–5 %), personal care services (3.5–4 %), and services related to leisure and culture (around 5 % in the European countries, although with significant differences by countries, but always showing an increase).

At the other end of the spectrum we find home and repair services (with a negative variation above −7 %)[4] and private transport services (−1.2 %), which, according to Eurostat, registered a fall in expenditure during the period under analysis (1990–2008). Similarly, although medical and transport services show slight increases in final expenditure, a loss of their relative weight is also observed regarding the total household expenditure. In addition, this fact is not disconnected from the supply of these and other services by the public sector. Finally, there is a group of services (regular meals outside the home and nurseries) which have slightly increased the average family expenditure, although their weight remains constant compared to total expenditure.

4.2 Increase of Intermediate Consumption of Services by All Productive Activities

In all advanced economies, services have clearly increased their importance as inputs in the production functions of the productive sectors, such as manufactures, energy, primary production, or the majority of services themselves, which also use other services as inputs to produce the services they offer. In short, what has been happening for several years is that intersectoral demand for services has increased substantially in the most developed economies (Elfring, 1988a, 1988b, 1989). This has occurred partly as a result of the organisational and productive changes that have taken place in companies, so that some services that were previously "produced" in-house have been gradually outsourced to other companies, more specialised in services, which has led to reduce costs (Camacho & Rodríguez, 2007; Cuadrado-Roura and Rubalcaba, 2000; Kox, 2002; OECD, 2005a, 2005b) and also to improve the quality of business and producer services. In addition, the growing complexity of the environment in which companies operate (legal and fiscal problems, requirements for exporting, design and advertising difficulties, etc.) has also boosted the demand for external services (Rubalcaba, 2007).

There is no question that modern companies are large consumers of services, but this issue requires further clarification. The outsourcing processes mentioned above have also brought about a "statistical effect" that has caused the service sector to increase its dimension in all economies due to the creation of specialised companies that did not previously exist or whose employees and production were *within* the manufacturing or service companies themselves. Nevertheless, these changes in business organisation are not just a transfer of employment from one sector to another (from the industrial sector to the service sector), but are generally associated with a process defined by a greater specialisation and extension of the services requested, and also by a better quality of supply and the demand requirements. As a

[4] Gershuny and Miles (1983) already underlined the trend to self-produce services in the homes as a consequence of the cost increase involved by their demand to external producers (repairs, leisure, domestic service, etc.).

result of this, for example, a high growth of the business service subsector or, more broadly, of producer services,[5] has occurred in recent years.

Today, services account for 38 % of total intermediate consumption in numerous countries. However, there appear differences when comparing various countries and, of course, when the analysis is carried out by industries, both from the viewpoints of services and manufactures and other demanding sectors.

Some empirical analyses show that there are a very large number of tertiary activities whose products are mainly assigned to intermediate demand. Among these, overland transport and other services connected to transport, commercial distribution, financial intermediation services, and machinery and equipment renting services, as well as other business services stand out. The production of retail commercial distribution and hotel and catering activities is also included in the intermediate consumption of the remaining sectors, although their production is mainly assigned to final demand.

From the perspective of service-demanding sectors, percentages vary widely from construction (about 7.5 %) to manufacturing industries (between 26 % and 35 %) and services themselves (58–65 %, depending on the country). In any case, these weights are quite similar in all developed European countries. On average, services account for 14.1 % of the productive structure of manufactures in the EU, whose intermediate consumption mainly comes from the industrial activity itself. This average percentage is slightly lower than that of countries[6] such as Germany (15.9 %), France (18 %), or Ireland (17.2 %), and similar to that registered by Italy (13.5 %), The Netherlands (14.4 %), Finland (13.7 %), and the United Kingdom (12.3 %), among others.

One of the consequences of the aforementioned facts is that services and the rest of the productive system are progressively interwoven. This also causes differentiating between goods and services, as was done in the past, to become increasingly difficult (Greenfield, 2002; Pilat & Wölfl, 2005). Furthermore, another important fact is that it is not easy to determine the contribution of services to the improvement of productivity in the rest of sectors, particularly in the industry sector and in several service activities which also employ services as production inputs. This issue is still the subject of open research and discussions regarding the measurement of productivity (Bosworth & Triplett, 2000; Kuroda, Motohashi, & Kazushige, 1996; Maroto, 2009, 2010; Maroto & Cuadrado, 2009; Nordhaus, 2000).

4.3 International Trade of Services

The expansion of international trade of services (transport in its various forms, financial services, tourism, consultancy and technical assistance, etc.) has also

[5] Including transport and storage, renting of machinery and distribution services.

[6] Figures estimated from the National Input–output Tables, 2005. Eurostat databases.

played a growing role in driving the expansion of services on the demand side.[7] However, the differences between countries are remarkable. The Netherlands is an extraordinary example showing the importance the export of services has held in its economy. A rather similar case is the United Kingdom, particularly in the areas of business services, financial services, and transport. In other countries—Italy, Spain, and even France—tourism has been a significant driver for the expansion of the hotel and catering sector, transport services and other activities linked to the reception of visitors. In the latter countries, however, the export of other services represents a significantly lower percentage.

When it comes to assess the role of international trade in favour of the expansion of services, it is worth noting that the expansion of the supply of many services is not carried out via commerce, but through direct investments in other countries, by creating subsidiaries and acquiring or owning a share in local companies (Cuadrado-Roura, Rubalcaba, & Bryson, 2002; Illeris, 1989; OECD, 2005a). This means that their production does not appear any longer in the national accounts of the country. This has been the case for sectors such as banking, telecommunications, surveillance, water distribution, and other services. Only the revenues derived from transferred benefits and royalties will appear in the balance of payments.

Some recent contributions reduce the importance of international trade as a driving factor for the expansion of services (Savona & Lorentz, 2009). The reasons adduced are technical—the need for many services to be provided in situ—and also institutional and regulatory due to the barriers and controls hindering the international trade of many services. The WTO has estimated based on data from 2007 that, while services account for around 66 % of the world GDP, they just represent about 17 % of the world trade. Figure 2.4, which compares the coverage of the trade of goods and services by countries, shows, as already mentioned, that differences among countries are significant (Di Meglio, 2010). Scandinavian countries, together with The Netherlands and Belgium are net exporters of commercial goods and services (upper right quadrant). However, other countries (Germany, Japan, Canada) are net exporters of goods, but net importers of services. Finally, there is a broad group of countries (from the United States and the United Kingdom to Spain and Portugal) characterised as being oriented towards the export of services and the import of goods, although with significant differences among them. Due to their importance and level of specialisation, the exports of services vary from tourism (Italy, Spain, Greece, Portugal) to financial services (Luxembourg, the United Kingdom), or also to a growing specialisation (ICTs) achieved through off-shoring (India).

[7] Obviously, countries also acquire services from other countries, but we refer here to the role of the demand for services received by a country, which boosts this sector of its economy.

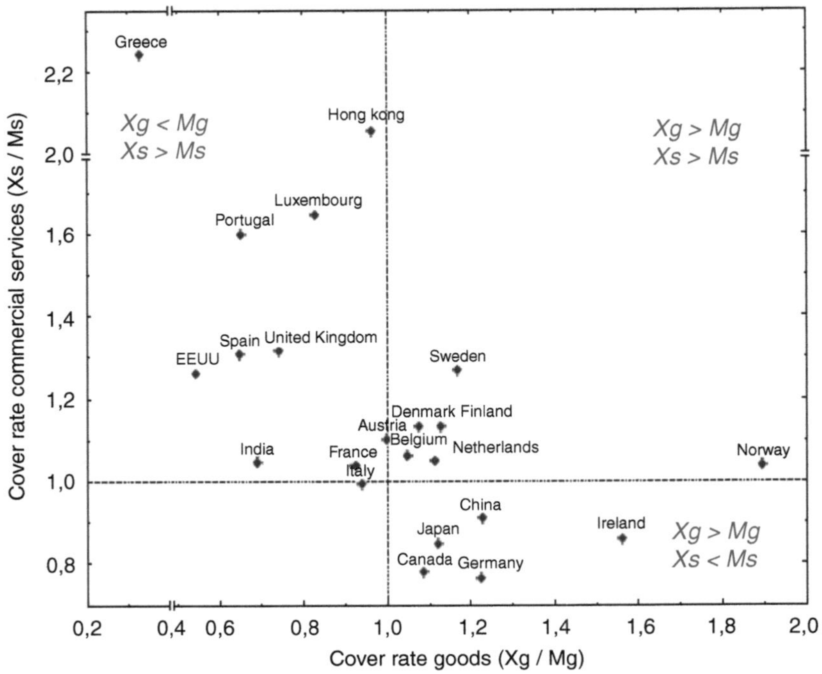

Fig. 2.4 International perspective on goods and commercial service trade. *Source*: Di Meglio, 2010, Figures from WTO (World Trade Organization). Note: X = exports, M = imports, *g* = goods, *s* = commercial services. Commercial services are services sector excluding government services

4.4 Demand for Services by Governments and Public Administrations

Administrations, either national, regional/state or local governments, have also contributed and continue to contribute to promote the demand for services as they need a wide variety of services in order to operate, for example, advice, reports, project assessments, policy impact assessments, transport, etc. However, rather than a quantitative increase in the demand for these services, two clear trends are observed: on the one hand, a transfer of the economic activity from public administrations to private companies (outsourcing), which is justified by the increasing complexity of many decisions and actions of the public sector; and on the other hand, cuts in the increase of public expenses and bureaucracy, which have recourse to demand that the private sector collaborate with public authorities and contribute to solve certain problems, supplies and necessities, often through a public-private collaboration.

5 Driving Factors from the Supply Side

Table 2.1 also illustrates the main factors promoting the expansion of services from the perspective of supply. Undoubtedly, the most important factor is the one that relates the production of services with productivity and the high employment requirements by the majority of service industries.

5.1 Low Productivity of Services as a Driving Factor

Data from the most advanced economies show that, generally speaking, there is a clear relationship between the weight of services in terms of employment over the total amount of people employed and the growth of productivity. Therefore, the greater the significance of services, the lower the rates of increase of their aggregate productivity by worker (Fig. 2.5).

This trend would confirm the theory developed by Baumol et al., which we have already commented upon. The explanation is apparently simple: the production of many services requires a high use of the labour factor which, unlike what happens in manufacturing, is not easily replaceable by either capital equipment or by technology. There are many services—from the so-called "personal" services to trade, hotel and catering sector, and the majority of business services, for example—in which the production of the service in question requires the use of a great amount of staff and any increase in the supply results in a parallel increase in labour. This clearly happens when a service involves simultaneity of production and consumption,[8] but also to some extents when it is not exactly like this (e.g., hotel and catering sector). This happens even when possibilities of technification and greater capitalisation exist, as in distribution and education sectors. When more capital and technology are applied, the quality of service tends to improve, but, on the whole, the rate of change in productivity does not.[9]

However, the rates of change in average productivity of the main service industries show remarkable differences, and while some register generally low productivity indicators (hotel and catering sector, trade, other services, and

---

[8] Such is the case of many personal services.

[9] This statement immediately suggests that, in many cases, the measurement of service productivity is made according to the same pattern as in the case of manufacturing, i.e., by linking the GVA to the number of employees or to the number of hours employed. Undoubtedly, in sectors such as health, education and others, this form of measurement is not exact. The employment of more and better equipment and technological developments enables an offering of higher quality services. However, it is arguable that the output of such activities may be identified just with GVA and not with other very different product indicators related to health or levels of education. See the excellent research carried out by Maroto (2010) regarding the productivity measurement issues in the services sector and the different positions in this regard.

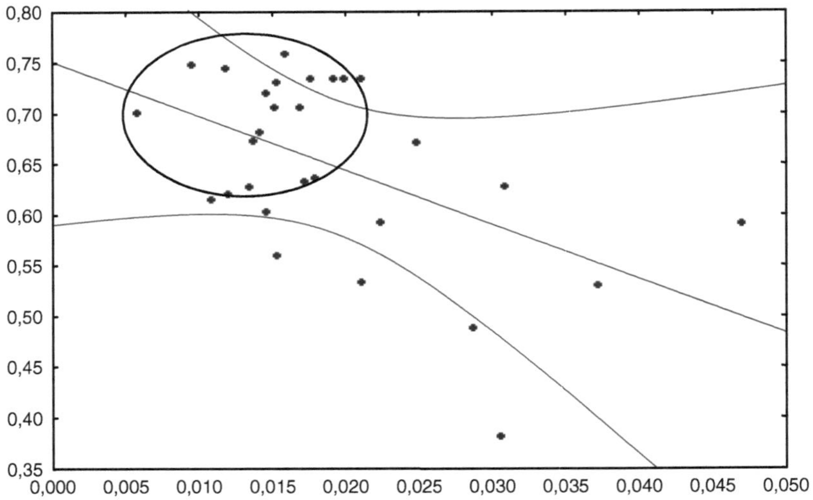

Fig. 2.5 Relationship between service weight and productivity increase average (1980–2007). *Source:* Cuadrado-Roura and Maroto-Sanchez (2010). Figures: EU-KLEMS, 2009. Notes: X: productivity average annual growth (1980–2007); Y: average weight of the service sector in the economies

health and education, for example), other industries record comparatively high rates (transport, communications, and financial services). Several recent works ((O'Mahony & van Ark, 2003; Oulton, 2001; Maroto & Cuadrado, 2009; Wölfl, 2003, 2005 among others) have highlighted that, within the service sector, very different behaviours in labour productivities can be observed by industries. Some of these endorse the theory that their productivity is low and that their rates of change are always highly reduced. But, in other cases, the indicators evidence that some industries have registered much higher rates of change in productivity and are comparable to or higher than the average in manufacturing. This is the case of activities in which the incorporation of technological developments and the substitution of employment by capital are not only feasible, but essential for the efficiency of those companies operating in such activities to achieve high levels of efficiency. The analyses referring to several advanced countries indicate that the financial sector, communications and some transport activities record very high productivity changes related to capital per worker and technological developments, which also determine cuts in staff. Additionally, in ICT-related services and some business services there are positive effects on the productivity of other sectors and activities (Baumol, 2000; Maroto & Cuadrado, 2009; Rubalcaba & Kox, 2007; Triplett & Bosworth, 2002; van Ark & Piatowski, 2004; Wölfl, 2003, 2005).

Of course, the final conclusion does not consist in denying any validity to the theory that the average rates of change in aggregate productivity of services are rather low and that this is the case in quite a number of their industries. This fact is linked to the high correlation existing between the delivery of many services and their growth with the labour requirements demanded by their production. However, this theory must be questioned when data disaggregated by service industries are used. Furthermore, we must also remember that the measurement of productivity of services raises very serious issues, both regarding the assessment of their output and the measurement of the factors involved in the respective production functions (Maroto, 2010).

5.2 Services Delivered by the State and Other Administrative Levels as a Driver of the Weight of the Sector

Public Administrations have been and still are playing a leading role in the growth of services. In fact, the public sector is a supplier and producer of services at its various levels. However, being a supplier does not mean being a producer at the same time (using its own staff and facilities) because the services from private companies may be hired (accounted for in market services). But, in any case, the State, the regions, and the municipalities are clear producers of services.

The development of the Welfare State (WS) boosted, particularly from the Second World War, the expansion of public services and, therefore, of the positions that these services require if directly produced by public entities. Such services include public education and health, social and assistance services, as well as others which are more conventional, such as the army, legal services, etc. Due to their nature, the majority of these are often qualified as "non-market" services. However, it is evident that the extent and extension of these services have also driven up the weight of those employed in the services sector, particularly in the European countries.

From the 1980s to date, many countries have tended to reduce their role as service providers, either by ceasing to render these and transferring their possible demand to the private sector, or through their outsourcing to private entities and companies. The privatisation processes have also resulted in the fact that some previously public services are now produced by private companies, although these may be subject in some cases to regulations in the delivery of the service in question, as well as to subsidies to cover deficit services considered as socially necessary. Nevertheless, at least in relative terms, the weight of the number of employees from Public Administrations has not fallen.

According to EU-KLEMS figures, there are numerous countries where the employment in non-market services accounts for more than 25 % of total employment (Table 2.2) and the trend even indicates that, in the last 30 years, not only the number of people employed in non-market services has increased, but also the

Table 2.2 Employment in non-market services (% of total employment)

Country/area	1980	1990	2000	2007
Belgium	24.14	26.55	27.83	29.32
Germany	20.79	22.39	24.55	25.40
Spain	15.46	19.63	20.94	20.51
France	26.24	29.97	31.18	31.17
Italy	22.90	24.98	24.48	22.30
Netherlands	28.27	28.09	25.96	28.58
United Kingdom	22.02	24.97	25.56	28.14
Sweden	34.38	36.29	34.80	35.20
USA	27.43	28.12	28.18	30.42
Japan	11.20	12.22	14.40	16.66
Eurozone	**22.24**	**24.73**	**25.79**	**25.89**
Euro-15	**19.63**	**22.15**	**23.51**	**24.12**

Source: Own elaboration, EU KLEMS data base

percentage they represent over total employment. Consequently, the upwards contribution to service employment by services included within the "non-market" group is clear.

5.3 Other Factors Having an Impact on the Supply and the Expansion of Services

Besides the two factors having the greatest impact on the growth of services—productivity and public services—other influential factors exist, although these are not easily quantified. Among these we find new technologies (ICTs) and the higher average qualifications of people employed in many services, a fact that has also promoted the expansion of their supply. Moreover, the liberalisation of service markets is another of these factors.

New technologies are having an important impact on the offer and diversification of services, not only from the supply, but also from the demand side. In terms of supply, their main influence is reflected in the innovations that ICTs enable to develop. This does not only give rise to new services but also to new forms of producing those already existing. Their impact is particularly significant in industries such as finance, distribution services, health-related services and, obviously, the transformation of telecommunications, but their influence extends to virtually all service activities and their production: press, leisure, transport logistics, etc.

The development of ICTs not only expands the production of services, but also their productivity (Pilat, 2004), and leads to improvements in the quality of services and the emergence of new ones, the reduction in the time required for their production and the narrowing of the distance between producers and those requiring the service. In short, technological changes boost a "new" tertiary sector and, at the same time, improve the supply of many of the most traditional services while they promote changes in the organisation of companies and extend the market areas that they may supply, either within each country, across borders, or even at a global level.[10]

The influence of ICTs is also present from the perspective of the demand for services, as it is diversifying the range of services available and the easier access to them through telecommunications in many cases.

The overall improvements in **education and human capital** are too playing a significant role in the supply of services. There are many service industries in which human capital is vital (Messina, 2004; OECD, 2005a, 2005b). In fact, despite the existence of routine services which do not require highly qualified staff—industrial cleaning, surveillance and personal services, for instance—many other activities require more and more qualified personnel. As pointed out by the OECD (2000) in the 1990s, the proportion of university and non-university staff in service activities was then three times that of manufacturing and all the signs are that this proportion will tend to rise. Furthermore, the improvement of human capital also has effects on the demand for services, as preferences for culture, travelling, sports, personal care, gastronomy, etc. tend to increase when the education level is higher.

6 Final Comments and Remarks: The Expansion Prospects of Services and Their Territorial Implications

The near future of service industries will essentially continue to be marked by the influence and behaviour of the factors indicated as drivers of the expansion of services in recent years. Some of these are currently hinge on the effects of the financial crisis, which is affecting virtually all of the most "industrialised" economies.

In terms of demand, the four main driving forces of the sector's growth have been and continue to be the consumption by families, the increasing use of services to produce goods and services, the demand by Public Administrations, and the international trade of services. Based on this, everything suggests that all these factors will continue to boost the production of services once the current recession is overcome. Under normal conditions, it is be expected that per capita income reaches the level recorded before the actual economic and financial crisis. Therefore, the fulfilment of Engel's law, along with the social and cultural changes that contemporary societies are undergoing, will cause the demand for services by

[10] Amazon is one example of a company offering services at an international scale. Others devoted to sport activities and material have also changed their offer as they incorporate customer loyalty programmes including connection with and information on important events, etc. The example shown by Barrabés Cº demonstrates how a company from a village in the Pyrenees can provide services and offer high-mountain and climbing equipment in the United States, Kuwait, or South-East Asia.

families to continue to grow in order to cover personal, leisure, and cultural services, as well as transport, private health and education, among other services.

In the case of inter-industrial relations, it is also clear that the demand for services as inputs to produce new goods and services will continue, as derived from the analyses based on input–output tables.[11] Furthermore, Public Administrations are not envisaged to eliminate or further reduce the services delivered to them by the private sector. By contrast, it is expected that public-private collaboration will increase in response to the reduction of the direct role of Administrations as service providers. And, finally, foreign trade in services, which has been growing significantly in recent years, is not expected to become stagnant (again once the current global situation should be overcome). The international agreements signed and the liberalisation processes of the international trade of services will contribute to this end. In this regard, the EU is developing new initiatives under the directive on the liberalisation of services and other directives on transport and communications oriented towards the promotion of the internal market for services and a greater flexibility and competence in their delivery.

The final conclusion is that, once the worst phase of the current recession would be over, the growth of services will resume, although perhaps in a more moderate way. In this sense, it seems safe to state that all advanced countries will exceed the threshold of 70 % of GVA and of the number of people employed in tertiary activities.

Yet, at any rate we are bound to witness restructuring processes "within" the service sector that should lead to a reduction in the number of companies operating in some industries, accompanied by an increase of their dimension, while in some individual cases, those firms whose profitability and viability in terms of efficiency and market are highly doubtful, such as some segments of hotel and catering trade, retail trade, and conventional transport, may disappear in the coming years.

There are **several other questions** that should be taken into consideration regarding the future of services. Will the internal market for services in Europe become a reality? Will services be more productive? Will the external investment in services be progressively more significant? Should an increase in the off-shoring of service companies be expected?

All these questions cannot be answered concisely. However, some ideas can be anticipated and we can refer to other recent studies and works providing quite solid data and elements in this regard.

Regarding the progress towards an **internal service market in the EU**, the following observations can be made. Firstly, it is evident that the EU should move towards an internal service market, as this is an essential requirement to continue progressing towards an increasingly complete and effective European integration. However, although some sectorial directives (on transport, banking and others, besides a more general directive) that point in the right direction have been approved,

[11] To this regard, see the research carried out by Pilat and Wölfl (2005) published by the OECD: http://www.oecd.org/dataoecd/43/33/34946920.pdf

the difficulties arising during the approval of the last directive on services[12] highlighted once again that the "national" interests and pressures are still very strong and that eliminating some of the existing barriers will continue to be difficult.

The question of **whether services will be more productive or not** in the future may have a more positive answer than the previous question. When highly aggregated values and indicators are used, it is difficult to deny the theory that services are a less productive sector than the rest, or at least below manufacturing and energy. Increasing the productivity per worker across services is and will remain difficult as a result of the high workforce requirements for the expansion of many of their industries, in which the labour factor is hard to be replaced with capital or technology.

This does not mean that "quality" improvements are occurring in several services and that "innovations" are not being introduced to become more efficient. In fact, as previously indicated, there are doubts about whether the "productivity" variable in services is being measured correctly. However, a large number of recent contributions to the analysis of services, already cited, has demonstrated that there are service industries where productivity records rates of change which are comparable to, or even higher than, many manufacturing activities. These are always activities where capital requirements and technological progress enable the production not only of enhanced quality services, but also services with a higher added value, such as communications, many transport activities, financial services, and even some business services. There is no reason to believe that these trends will not continue in the future.

On the other hand, we have to pay more attention to the differences between service "**innovations**" manufacturing innovations. Many services innovations rarely appear in official statistics, as the patent system, for instance, is not well adapted to the production of numerous services. Something similar occurs with the proliferation of networks, franchises, and other formulas. Yet, these innovations are a fact and the consequence of several factors: the improvement of human capital in the sector, the growing competition (even in the case of services where the producer-client contact is essential and should happen simultaneously and at a specific place), the returns of scale and external economies generated by the networks of service companies, and the increase in the size of companies. And all this appears as part of a process of "internationalisation" of services—through trade and investments—which we have already referred to.

Finally, there is another issue that should be taken into consideration when analysing the future of the service sector, particularly because it is directly linked to the services-territory relationship: the **processes of geographic "offshoring"**[13]

---

[12] Directive 2006/123/EC of the European Parliament and of the Council of 12 December 2006. Its implementation in all countries has been slower than anticipated.

[13] That is, the displacement of the production of some services to countries where production costs and regulations in force are more advantageous. This has been the case of computer treatments (where India has absorbed a large production volume), the production of books, and material for media and leisure, for example.

that have been taking place in many services. It may also be questioned whether this trend will continue in the future or not, although it seems that the answer can only be affirmative. Some recent documents analyse the offshoring processes of services at an international scale and their implications.[14] In addition to this, however, the flows of investments in service activities occurring at an international level indicate that the trends initiated in recent years will continue. These investments are often justified by the nature of a large number of services involving a direct contact between the producer and the consumer for their production and delivery. On the other hand, the ICTs developments also favour some specific off-shoring as these enable some tasks to be carried out in locations/countries lying very far from the service user, although the contact is fully kept and the production costs are reduced, as it already happens in computing activities, data treatment, publishing, and other activities.[15]

To conclude, and even considering the difficulties involved by the current economic situation, services offer quite clear prospects for growth in the coming years. Together with new possibilities to create jobs in the sector, productivity improvements in many of its industries, improvements in the quality of the delivery of almost all services,[16] and an increasing internationalisation of service companies, a trend existing for more than a decade (Cuadrado-Roura et al., 2002), can be expected.

References

Baumol, W. (1967). Macroeconomic of unbalanced growth: The anatomy of urban crisis. *American Economic Review, 57*(3), 415–426.

Baumol, W. (2000). *Services as leaders and the leader of the services*. Inaugural Lecture, International Conference on the Economics and Socio-economics of services, Lille, June 22–23.

Baumol, W. J., & Bowen, W. G. (1965). On the performing arts: The anatomy of their economic problem. *The American Economic Review, 55*(1/2), 495–502.

Blackaby, F. (1978). *Deindustrialisation*. London: Heinemann.

Bosworth, B., & Triplett, J. (2000). Numbers matter. The US statistical system and a rapidly changing economy. *Policy Brief, 63*, 5–6.

Cairncross, A. (1978). What on deindustrialisation? In F. Blackaby (Ed.), *Deindustralisation*. London: Heinemann.

Camacho, J. A., & Rodríguez, M. (2007). Integration and diffusion of KIS for industry performance. In L. Rubalcaba & H. Kox (Eds.), *Business services in European economic growth* (pp. 128–143). London: MacMillan/Palgrave.

Clark, C. (1940). *The conditions of economic progress*. London: MacMillan.

[14] See http://www.mckinsey.com/mgi/publications/emerginggloballabormarket/index.asp and http://www.mckinsey.com/mgi/publications/emerginggloballabormarket/index.asp.

[15] The examples of computing services, call centres, book publishing processes, etc. are well known and numerous.

[16] We could maybe exclude the so-called 'routine' services, such as industrial cleaning and domestic service, for example, where innovation is highly limited.

Cuadrado-Roura, J. R., & Maroto-Sanchez, A. (2010). *Servicios y regions en España*. Madrid: Funcas.
Cuadrado-Roura, J. R., & Rubalcaba, L. (2000). *Los servicios a empresas en la industria española*. Madrid: Inst. Estudios Económicos.
Cuadrado-Roura, J. R., Rubalcaba, L., & Bryson, J. (Eds.). (2002). *Trading services in the global economy*. Cheltenham: Edward Elgar.
Di Meglio, G. (2010). *The role of public and private services in European economies*. Doctoral thesis. University of Alcala, Madrid (polic.).
Elfring, T. (1988a). *Service sector employment in advanced economies*. Aldershot: Gower.
Elfring, T. (1988b). *Service employment in advanced economies: A comparative analysis of its implications for economic growth*. Groningen: Rijksuniversiteit Groningen.
Elfring, T. (1989). New evidence on the expansion of service employment in advanced economies. *Review of Income & Wealth, 35*(4), 409–440.
Falvey, R., & Gemmel, N. (1996). Are services income-elastic? Some new evidence. *Review of Income and Wealth, 42*(3), 257–269.
Fisher, A. G. B. (1935). *The clash of progress and security* (2nd ed.). Paris: Le Seuil.
Fourastie, J. (1949). *Le Grand Espoir du XXe Siecle*. Paris: Presses Universitaires de France.
Fuchs, V. R. (1968). *The service economy*. New York: National Bureau of Economic Research.
Gemmell, N. (1982). Economic development and structural change: The role of the service sector. *Journal of Development Studies, 19*(1), 37–66.
Gershuny, J. (1978). *After industrial society? The emerging self-service economy*. London: MacMillan Co.
Gershuny, J., & Miles, I. (1983). *The new service economy. The transformation of employment in industrial societies*. London: Frances Pinter.
Greenfield, H. (2002). A note on the goods/services dichotomy. *Services Industries Journal, 22*(4), 19–21.
Gregory, M., Salverda, W., & Schettkat, R. (Eds.). (2007). *The US-European gap in service demand and employment*. New York: Princeton University Press.
Illeris, S. (1989). *Services and regions in Europe*. Aldershot: Avebury.
Jacobs, J., Shipp, S., & Brown, G. (1989). Families of working wives spending more on services no durable. *Monthly Labor Review, 112*(2), 15–25.
Kaldor, N. (1961). Capital accumulation and economic growth. In F. A. Lutz & D. C. Hague (Eds.), *The theory of capital* (pp. 177–222). New York: St Martins Press.
Kox, H. (2002). *Growth challenges for the Dutch business services industry: International comparison and policy issues*. The Hague: CPB Netherlands Bureau for Economic Policy Analysis.
Kravis, I., Heston, A., & Summers, R. (1981). *World product and income*. Baltimore: Johns Hopkins University Press.
Kuroda, M., Motohashi, K., & Kazushige, S. (1996). Issues on the international comparison of productivity: Theory and measurement. In OECD. *Industry productivity. International comparisons and measurement issues* (pp. 49–95). Paris: OECD.
Kuznetz, S. (1971). *Economic growth of nations: Total output and production structure*. Cambridge, MA: Harvard University Press.
Maddison, A. (1980). Economic growth and structural change in the advanced countries. In I. Leveson (Ed.), *Western economies in transition: Structural change and adjustment policies in industrial countries* (pp. 41–60). Boulder: Westview Press.
Magrabi, F. M., et al. (1991). *The economics of household consumption*. London: Praeger.
Mañas, E., Gabaldon, P., & Cuadrado-Roura, J. R. (2002). *El consumo de Servicios en los hogares españoles*. Madrid: IEE.
Maroto, A. (2009). *Relationships between services, growth, productivity and welfare. The state of the art*. WP, ServPINN programme.
Maroto, A. (2010). Productivity in the service sector: Conventional and current explanations. *The Service Industries Journal, 30*, 1–29.
Maroto, A., & Cuadrado, J. R. (2009). *Structural Change and Economic Dynamics, 20*(4), 254–265.

Messina, J. (2004). *Institutions and service employment: A panel study for OECD countries*. Working paper. Frankfurt: European Central Bank.
Nordhaus, W. (2000). *Alternative methods for measuring productivity*, NBER Working Paper, 8095. Cambridge, MA: National Bureau of Economic Research.
O'Mahony, M., & van Ark, B. (2003). *EU productivity and competitiveness. An industry perspective. Can Europe resume the catching-up process?* Brussels: Enterprise Publications, European Commission.
OECD. (1975). *Adjustment for trade*. Paris: OECD.
OECD. (2000). Service employment. In OECD. *Employment Outlook 2000*. Paris: OECD.
OECD. (2005a). *Enhancing the performance of the service sector*. Paris: OECD.
OECD. (2005b). *Growth in services. Fostering employment, productivity and innovation*. Paris: OECD.
Oulton, N. (2001). Must the growth rate decline? Baumol's unbalanced growth revisited. *Oxford Economic Papers, 53*, 605–627.
Pilat, D. (2004). *The ICT productivity paradox. Insights from microeconomic data (OECD Economic Studies, 38)*. Paris: OECD.
Pilat, D., & Wölfl, A. (2005). *Measuring the interaction between manufacturing and services*. OECD Science, Technology & Industry Working Papers 2005/5. Paris: OECD.
Rubalcaba, L., & Kox, H. (2007). *Business services in European economic growth*. London: MacMillan/Palgrave.
Savona, M., & Lorentz, A. (2009). *Demand and technology contribution to structural change and tertiarization: An I-O structure decomposition analysis*, SerPINN. http://www.servpinn.com/.
Schettkat, R. (2004). *The two blades of the scissors: The interaction between demand and supply in market economies*. Mimeo, University of Wuppertal.
Schettkat, R., & Yocarini, L. (2006). The shift to services employment. A review of the literature. *Structural Change and Economic Dynamics, 17*, 127–147.
Shaninger, C. M., & Allen, C. T. (1981). Wife's occupational structures as a consumer behaviour constract. *Journal of Consumer Research, 8*, 189–196.
Triplett, J., & Bosworth, B. (2000). *Productivity in the service sector*. Brookings Institution, Mimeo. Available at www.brookings.edu/views/papers/triplett/20000112.pdf.
Triplett, J., & Bosworth, B. (2001). Productivity in the services sector. In D. Stern (Ed.), *Services in the international economy*. Ann Arbor: University of Michigan Press.
Triplett, J., & Bosworth, B. (2002). *Baumol's disease has been cured. IT & multifactor productivity in US services industries*. Paper presented in the Brookings Workshop on Services Industry Productivity, Brookings Institution, September, Washington, DC.
van Ark, B., & Piatowski, M. (2004). *Productivity innovation and ICT in old and new Europe*. Groningen: Research Memorandum, GD-69, GGDC.
Wölfl, A. (2003). *Productivity growth in service industries. An assessment of recent patterns and the role of measurement*. STI Working Paper, 2003–7. Paris: OECD.
Wölfl, A. (2005). The service economy in OECD countries [Chapter 2]. In *Enhancing the performance of the service sector*. Paris: OECD.
Yang, S.-J., & Magrabi, F. (1989). Expenditures for services, wife's employment and other household characteristics. *Home Economics Research Journal, 18*(2), 133–147.

Chapter 3
The Service Sector in the New Globalization Phase: Evidence from European Regions

Roberta Capello and Ugo Fratesi

1 Introduction

Qualitative rather than quantitative changes have characterized the reshaping of the global economy over the past decade. The globalization of production no longer only involves the off-shoring and outsourcing of production phases in developing countries; it no longer simply affects the division of labour between emerging and advanced countries and developed and developing economies; and it is no longer confined to manufacturing activities (Baldwin, 2006).

The globalization of tasks rather than sectors, the off-shoring and outsourcing of service functions, de-industrialization in favour of services, and the decentralization of intertwined functions (manufacturing and related services) are reshaping the division of labour in the sub-national economies of advanced countries, and regional economies are increasingly competing to seize the opportunities which these new trends offer (Capello, Fratesi, & Resmini, 2011; Fontagné & Lorenzi, 2005).

All these changes affect the service sector. It is from the service sector that most outsourcing of tasks, rather than of whole functions, takes place; it is the service sector that is the most engaged in the off-shoring of functions; it is in the service sector where jobs, and productivity, are most affected by the new globalization trends. Most of the challenges and growth opportunities related to globalization trends are expected to be related to the presence of the service sector in the economy.

The aim of the chapter is to analyse the relationship between the trend of the service sector in European regions and the existence of globalization conditions. This aim is achieved in a purely descriptive way by analysing the economic trend of macro sectors and the regional degree of openness to the external world.

This chapter first presents the recent qualitative changes that have taken place in the service sectors of advanced economies (Sect. 2). Secondly, the analysis concentrates on productivity, employment and GDP dynamics in European regional

R. Capello (✉) • U. Fratesi
ABC Department, Politecnico di Milano, Milan, Italy

J.R. Cuadrado-Roura (ed.), *Service Industries and Regions*,
Advances in Spatial Science, DOI 10.1007/978-3-642-35801-2_3,

economies, and descriptively links these trends to different regional globalization conditions (Sect. 3) by grouping regions according to their degree of globalization with a methodology already suggested by the authors in a previous study (Capello et al., 2011). Moreover, the different regional performances are associated with service specialization so as to describe whether virtuous growth rates are linked to specific service rather than manufacturing specialization (Sect. 4). Some concluding remarks on the importance of the service sector in a period of new spatial globalization patterns are made in Sect. 5.

2 New Globalization Trends in the Service Sector

2.1 *Deindustrialization and the Rising of Service Economy*

Globalization is generally associated with deindustrialization. In advanced countries, and in regions specialized in manufacturing, the new forms of production organization adopted by firms entail a shift of functions and tasks outside the area, with the expected consequence of job losses in industrial employment and, at the same time, productivity increases in the manufacturing sector.

Deindustrialization is too often a process defined only in terms of industrial employment losses. Yet purely industrial employment decrease is not enough to identify a deindustrialization process, which takes place when industrial employment losses are associated with industrial productivity losses, and with a real industrial GVA decline. While strategies of outsourcing and off-shoring easily impact on employment by eliminating some blue-collar activities in the traditional industrial regions of advanced countries, their effects on industrial productivity are contradictory. It may be the case that industrial productivity increases due to the dropping of inefficient functions and tasks, or due to the region's specialization in higher value-added functions. However, this latter process must be efficient enough to guarantee an increase in industrial GVA at the local level (Affuso, Capello, & Fratesi, 2011; Camagni, 1991).

In coping with these processes, regional economies must strike an important balance through the spatial reorganization of production; the losses in industrial employment must be counterbalanced locally by a more than proportional increase in industrial productivity so as to guarantee at the same time an increase in real industrial GVA. Achievement of this goal depends on the capacity of regional economies to re-orient their specialization to new-growth industries and activities in related sectors. Examples of such transitions include the switch from telephone handset production to mobile internet system design, or from vehicle production to GPS, road sensing and safety equipment (OECD, 2007). Industrial regions endowed with command and control and creative functions are probably those best able to exploit this globalization trend.

More importantly, in a period of the rapid service fragmentation of production, a shift to the service sector has been highlighted as a possible counterbalance to industrial employment losses. An OECD Report (2007) shows that, between 1998 and 2004, most regions experienced large job losses in manufacturing and that these job losses were usually, although not always, offset by growth in service employment (OECD, 2007). This substitution process between industry and service employment represents a challenge and a possible threat for regional economies, because it imposes the search for a balance between industry and service employment. In fact, major effects on the real local economy are registered when the new service jobs are high-value added jobs, generally in "producer services" (working for industries from outside). On the other hand, when regional specialization shifts towards low-value added services, mainly in "consumer-oriented" activities or "low-profile functions" (e.g. call centres), the net advantage for the regional economy may be limited or even negative. At least part of the present slowdown of aggregate productivity growth in advanced countries is linked to a trend of this kind.

Moreover, service activities (both low and high-value added functions) exhibit a slower pace in innovation trends than manufacturing does. This element represents another challenge associated with the move towards service activities, which imposes a slower innovation pace on local economies specialized in services compared with those specialized in manufacturing, with a consequent reduction in productivity increases. Regional economies are obliged to strike a balance between industrial and service sectors so as to maintain a certain rate of innovation and productivity. The mere quantitative substitution between numbers of jobs lost and re-created is a dangerous strategy: high-quality skilled jobs must be protected in order to achieve productivity gains.

2.2 *Off-Shoring and Outsourcing of Service Functions*

New globalisation trends are reflected in the new spatial trends of FDI. Most of these investments are directed to developed countries (80 % in 1986–1990, around 60 % in 1993–1997, and 65 % more recently in 2006), and they seem particularly attracted by accelerations in economic integration processes: in fact, EU15 countries, at the end of the process of creation of the Single Market in 1991–1992, received up to 50 % of world FDI, and similar accelerations were evident in the case of Eastern European countries after their accession.

Moreover, since 1990 services have accounted for the majority of total FDI; in 2005 they accounted for almost two thirds of the total, while manufacturing represented 30 % and primary sectors less than 10 %. Services still maintain a large share of greenfield FDI (42 % in 2006, with manufacturing accounting for 54 %), and greenfield FDI representing one-third of total FDI.

The world's inward stock of services quadrupled between 1990 and 2002, from an estimated 950 billion US dollars to over 4 trillion US dollars (UNCTAD, 2004). This explosion was certainly linked to the liberalization of FDI policies, which began in the mid-1980s and gathered momentum during the 1990s. This process has had important consequences if one considers that services constitute the largest productive sector in most economies, and that their competitive (and efficient) production is crucial for the welfare of a society as a whole.

The growth of service FDI has gone hand in hand with the industry mix of such FDI (Golub, 2009). Until the 1990s, services FDI were concentrated in trade and finance, accounting for 25 % and 40 %, respectively, of total inward FDI stock in services (UNCTAD, 2004). Since the 1990s, other services have undergone more dynamic FDI growth, among them telecommunications and electricity, water supply and business services. This increasing tendency to off-shore services is likely to be a major trend in the next few years if one considers that service off-shoring is, compared to manufacturing off-shoring, simpler in terms of resources, space and equipment requirements, and may therefore be more 'footloose' given the lower sunk costs involved. It affects firms in all sectors, and may therefore have greater implications for the host economy than the fragmentation of manufacturing. It mainly affects white-collar workers, while manufacturing off-shoring primarily involves blue-collar workers and generally creates jobs of this latter kind in the host area without destroying them in the home area.

This change of service mix also reflects the different reasons for off-shoring (Davies & Guillin, 2011; Riedl, 2008). Finance and retail trading used to be the traditional host-country market-oriented services; today, more complex strategies are put in place in order to obtain efficiency gains based on an inter-affiliate division of labour whereby foreign affiliates produce components not necessary for their parent firms but for other affiliates specialized in other components. Therefore apparent in services as well is the breaking up of service activities into components produced wherever it is more convenient to do so, with the result that certain foreign affiliates perform back-office functions of various kinds for their parent company, or for other foreign affiliates.

Whilst in Europe 45 % of the largest firms with off-shoring experience have off-shored activities to their foreign affiliates, 48 % of the companies have outsourced activities to third-party service providers (Lejour, 2007; UNCTAD, 2004), which evidence that the phenomenon of service outsourcing is also common. The choice between off-shoring and outsourcing service activities in favour of the former depends primarily on the need to maintain strict control on those activities. For example, the financial service industry appears to rely almost exclusively on internalized models of off-shoring. Moreover, off-shoring is preferred when the level of internal interaction with other functions matters. Service, manufacturing and R&D activities require strong interaction if the firm is to be efficient; by contrast, back-office functions and customer interaction services can be easily outsourced. Out-sourcing, in any case, is strongly conditional on the existence of capable local firms; there are several examples of cases in which off-shoring has been chosen because of the lack of efficient and reliable local companies in the host country.

The global shift in services offers large potential benefits for regions at both ends of the process: receiving countries gain jobs, skills, access to foreign skills; while the sending ones improve their competitiveness by moving to higher-level activities.

Since most off-shoring and outsourcing has taken place among developed countries, this underscores that this process does not primarily represent a "North/South" divide, and that it mainly affects regional economies in developed countries.

It is clear from what has been said that the service sector plays an important role in local economies specialized in service activities. For these economies, the service sector is a source of structural changes brought about by the new globalization trends; but for those regions able to adjust their economies to the structural changes, it offers great opportunities of growth.

A descriptive analysis of the major economic growth measurement, namely productivity, employment and GVA trends, is presented in the chapter, so as to highlight whether it is true that in regions with higher involvement in the globalization process the service sector has demonstrated a different economic performance. Before entering the descriptive analysis, the logic with which "open regions" are classified is now presented in detail.

3 Economic Performance in the Service Sector: Global vs. Local Regions

3.1 Measuring Globalization at Regional Level

The aim of this part is to build a typology of European regions according to their degrees of exposure to globalization; in particular, three different groups of European regions identified in Capello et al. (2011) and Fratesi (2012) according to:

- An economic dimension, measured in terms of regional specialization in open (through international trade and FDI) growing sectors; and
- A functional/territorial dimension, measured through higher-than-average scores in a globalization index based on structural—urban—material and non-material connectivity indicators.

Measuring the involvement in globalization is, in fact, a difficult task, because globalization involves a large number of processes which take place simultaneously and are related to each other. It is especially difficult to capture it at regional level owing to the low availability of data: for most other indicators, especially trade, while regional data are missing or are available for only a small sub-set of European countries. Moreover, the regional dimension of globalization cannot be captured by flow variables alone (FDI, trade, and migration flows all belong to this category) since the structure of the regional economy is fundamental for explaining the role that a region can play in the global economy and what flows it is able to attract. Since the only reliable data available at EU-27-wide NUTS 2 regional level are those on FDI flows (see Resmini, 2013, in this volume), the lack of direct statistical sources entails that an indirect method must be used to measure globalization.

Table 3.1 Taxonomy of regions according to their degree of integration into global markets

	Functional/territorial dimension	
Economic dimension	Openness above average	Openness below average
Specialization in open growing sectors	1 Global players	2 Regional players
De-specialization in open growing sectors	4 Pure gateways	3 Local players

Source: Capello et al. (2011)

In particular, the approach is based on two main dimensions that reinforce and complement each other in capturing the different aspects of integration. They derive from two main streams of literature: the first oriented to the territorial/functional structure of the local economy in order to capture integration processes, the second to economic integration processes. The former strand of analysis identifies the competitive advantages of regions undergoing global processes in the presence of a large city in which the international headquarters of multinationals, high-value service functions (like international-level finance and insurance), and high-qualified human capital attracted from outside find an efficient location thanks to agglomeration externalities and physical accessibility. The feature shared by all these concepts is the idea that one way to be integrated into the global economy, and to gain advantages from it, is to comprise international high-value functions, qualified human capital, increasing returns to production activities, and physical accessibility. The second strand of analysis with which to measure a local economy's degree of integration into the world market is a pure economic dimension captured by the degree of that local economy's specialization in activities that are particularly open to international markets. This dimension explains the capacity of a region to grow by virtue of the presence in it of dynamic open sectors. It captures a MIX effect of a traditional shift-share analysis (Perloff, 1957; Perloff et al., 1960).

Only those regions well endowed with physical connections and possessing the appropriate specialization in competitive and dynamic sectors have the potential to be *global players*, these being defined as *regions where globalization's impact is felt first and most strongly*.

Table 3.1 contains the conceptual taxonomy obtained if the two dimensions of integration into global markets—the territorial/functional and the economic dimensions—are cross-referenced: on the vertical axis is the degree of openness to globalization, i.e. a *globalisation index*; on the horizontal axis the *regional specialization in open growing sectors* (belonging either to services or manufacturing).

The territorial/functional dimension (vertical axis) requires a synthetic indicator for the openness of regions, which affects their participation in global networks. The economic dimension (horizontal axis), in the absence of trade data at regional level, requires identification of the degree of specialization in open growing sectors of each region. This horizontal dimension is therefore the result of a two-step procedure which first identifies those sectors which are more open at European level, and then identifies the regions which are specialized in them.

A synthetic indicator was constructed to capture the various components that define structural openness to globalization; in fact, there exist a number of indicators, not statistically independent from each other but normally positively correlated one another. The synthetic indicator was built using a principal component analysis (PCA) on five available relevant indicators. Each of these indicators captured a different element in the functional/territorial integration of European regions with the extra-European world and economy. The five indicators were:

- Extra-European born population, as a proxy for the attraction of foreign labour;
- Extra-European airflow connections, bound to represent the integration of a region with global networks;
- Number of offices of advanced services firms, expected to capture the presence of value-added functions;
- Headquarters of transnational corporations, as a proxy for the attraction of international high-value functions;
- Extra-European FDI in the region, representing the attraction of extra-European capital.

The second dimension of the external openness of regions is their industrial specialization. In fact, being specialized in sectors which are relatively more open to trade and perform better than average in periods of sustained globalization is an important channel through which regions can take advantage of globalization trends. By contrast, specialization in closed and/or declining sectors makes a region less able to play a role in globalization processes, and hence to take advantage of external opportunities.

Regional sectoral trade data would be extremely useful here. However, given the lack of sectoral trade data at regional level for all EU countries, sectorally open regions had to be identified by means of a two-step procedure. As in a traditional shift-share analysis (Perloff et al., 1960), in fact, regional specialization in more dynamic sectors is a factor which, ceteris paribus, enables regions to benefit from the global processes of which the same sectors are the principal beneficiaries. The first step is therefore to determine which sectors are the open and growing ones; and the second step is to determine which regions are specialized (i.e. have a location quotient higher than 1) in those sectors.

Since pure gateways, which are theoretically puzzling, do not exist empirically (Capello et al., 2011; Fratesi, 2012) three groups of European regions (NUTS 2 level) were identified in this way, and are presented in order of involvement in global flows, namely:

1. Global players. These are regions at the core of globalization processes: they are structurally open and have all the necessary physical and functional linkages with the rest of the world; moreover, they are specialized in sectors which are open and growing, so that their role in world trade flows and FDI attractiveness is maximum. These regions are therefore expected to be able to lead Europe and drive patterns of response to globalization also for the other regions of the EU.
2. Regional players. These regions are specialized in open growing sectors but have below-average physical and functional connectedness with other areas in the world. These regions are therefore expected to take advantage of their specialization,

but they are also expected to be somewhat penalised with respect to global regions because their good sectoral mix does not take advantage of a strong and efficient territorial settlement structure, and does not exploit the agglomeration advantages guaranteed by a city-region. The economic dynamics of these areas are expected to be due to a MIX effect deriving from the presence in the region of sectors that are more dynamic and more open than average at regional level because of increasing demand in those sectors. The label "regional" is attached to these players because their sectoral specialization would allow them to play a worldwide role, but, given their lack of an urbanised settlement structure, they normally have to resort to global regions as gateways to world markets. The term "regional" is hence to be understood in its trade literature meaning, which interprets Europe as a region of the world. At the same time, the term recalls the limited physical accessibility to and from the world.

3. Local players. This category consists of regions which have neither the functional/territorial elements to connect with the world nor the appropriate specialization in open growing sectors. These regions are rather peripheral to globalization processes and will hence be used as a control category by all the analyses conducted in the following chapters. Trends that pertain to globalization forces are expected to be limited in this category. We label them "local" players because their markets are expected to be local, i.e. normally limited to their own region and, possibly, country.

The result of the empirical taxonomy is presented in Map 3.1. The three groups of European regions will be at the basis of the following empirical analyses.

3.2 *Employment, Productivity and GVA Dynamics: Manufacturing vs. Service Sectors*

As mentioned in Sect. 2, globalization is often associated with a shift to services, sometimes even involving an outright deindustrialization process, which implies not simply that industrial employment decreases but that industrial employment losses are associated with insufficient productivity gains, and with a decline in industrial GVA in real terms. For this reason, analyses of employment, GVA and productivity must be performed simultaneously.

Table 3.2 shows the patterns followed in the period 1997–2007 by the three main indicators of regional growth, namely employment, productivity and value added, for the three types of region illustrated in Sect. 3.1: the global players, the regional players, and the local players. Only two of these indicators are really independent, and data on the third of them have been obtained by combining the other two.

In the first part of Table 3.2, each indicator is represented separately for agriculture, manufacturing and services, plus construction, in order to capture the different patterns followed by the macro-sectoral activities of the economy.

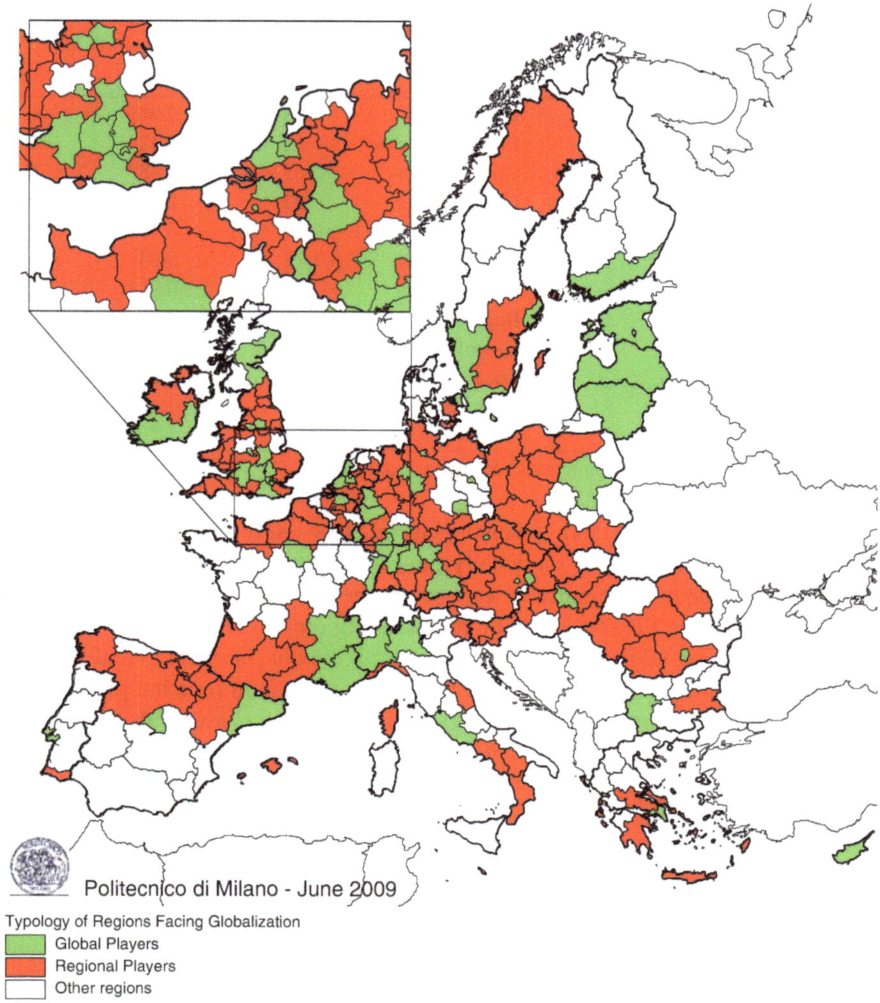

Map 3.1 Typology of regions in regard to globalization. Source: Capello et al. (2011)

All patterns are represented as average annual (real) growth rates in the 10 years of fast globalization before the big economic crisis, i.e. for the period 1997–2007.

Employment growth shows a clear shift to service activities: employment has rapidly grown in services, whereas employment in manufacturing and agriculture has decreased. Service employment growth is sizeable in all regional typologies and especially, though not sizeably, in global players.

Manufacturing employment, by contrast, has decreased in the three typologies, but especially in global and, second, regional players. Being specialized in open growing sectors, therefore, is not enough to maintain manufacturing employment levels. Manufacturing employment has remained almost stable in local players, especially

Table 3.2 Annual average growth rates of employment, productivity and GVA by macro-sector in the period 1997–2007. EU27 Nuts2 regions, by typology

Sector (Ateco code in parenthesis)	Type of region	Employment	Productivity	GVA
Agriculture (A+B)	Local players	−2.76	2.88	0.04
	Regional players	−4.12	4.17	−0.12
	Global players	−2.63	3.48	0.76
Energy and manufacturing (C+D+E)	Local players	−0.22	2.39	2.16
	Regional players	−0.79	3.16	2.35
	Global players	−1.17	3.04	1.83
Construction (F)	Local players	2.67	−0.61	2.05
	Regional players	1.39	0.25	1.65
	Global players	2.16	−0.39	1.76
Services (from G to P)	Local players	2.00	0.82	2.84
	Regional players	1.73	1.46	3.21
	Global players	2.04	1.52	3.60

Source of data: Cambridge Econometrics Regional Database

if compared with global and regional players. It is likely that this employment stability can be explained by the locally protected nature of local players' markets with respect to the other regions, at least before the economic crisis.

The gap between the service and manufacturing employment growth rates may represent a first signal of the shift to service activities of the most global regional economies, as this is much stronger in global players than in regional players, and even more so with respect to local players. We can therefore assume that a long-run shift in the globalization process towards the services sector in the European economy is indeed taking place, and that it is affecting global regions much more than the others.

The second column of Table 3.2 represents productivity growth. First to be observed is that, in European regions of all types, manufacturing productivity has been growing more than service productivity.[1] The slow growth of productivity in services—the sector accounting for almost all employment growth—signals that there is an ongoing process of manufacturing restructuring, and that the service sector also acts as a channel for job creation and absorption of shed workers.

Productivity increases have been large in global players both in services and manufacturing, but manufacturing productivity has been growing even more in regional players which are specialized in open growing sectors, whereas productivity increases have been consistently lower in the local players, signaling that globalization forces have been drivers of productivity increases, probably because of technological and productive competition.

The third column of Table 3.2 presents the patterns followed by gross value added. Even if GVA is simply the product of employment and productivity, its

[1] We acknowledge that measuring service GVA and productivity is not straightforward (Mark, 1982; OECD, 2001), especially for public services, but we believe that aggregate data are in any case able to capture the general trends though they might not be extremely accurate.

pattern is perhaps the most interesting because one can understand whether some effects are purely statistical or whether they hide important economic trends. In fact, productivity increases could be obtained by cutting the less productive jobs, and in this case total value added would also decrease; but it is also possible that, thanks to technological or organizational innovations, the restructuring process can yield higher total value added with lower employment levels.

It is the second possibility that applies in manufacturing in all three types of regions, where total value added has increased owing to productivity increases and despite employment decreases. Even if the GVA pattern appears to be similar, global forces may have played a role in this regard: local players obtain high manufacturing GVA growth through much lower productivity increases, and almost a maintenance of employment levels. Local players, de-specialized in open growing sectors, have therefore probably been the collectors of lower manufacturing production phases, whereas the most open regions have had to shift to phases with higher value added, and to cut and delocalize the lower phases.

Globalization, however, is a process closely linked to services: global players have been growing faster in both employment and productivity, so that also their GVA has grown more than in any other group of regions. The second performance in terms of GVA growth is that of regional players, which have had slow service employment growth but a high service productivity growth. Finally, local players record the lowest GVA growth owing to much lower productivity growth, so that also in services these non-global regions are mainly specializing in low-value added functions.

Therefore, the descriptive analysis shows that the strong position and socio-economic structure of global players is associated with a more decisive development of the service activities, and with larger shares of the economy in this sector and the higher value added functions. The other regions, by contrast, have experienced lower service employment growth, limited to the lower value-added functions, those that can be more easily decentralized.

3.3 Employment, Productivity and GVA Dynamics by Types of Services

The descriptive analysis has shown that the most global regions are also those with the best performance in service sectors, and that therefore the shift towards the service activities is a process strictly linked to globalization. An analysis of sectorally disaggregated data is helpful to determine whether the patterns of different types of services are indeed differentiated. The expectation is that the different functions performed by different types of regions in the global economy will also be reflected by different economic specializations.

As Table 3.3 shows, service employment growth has been larger in non-market services in all regional typologies, signaling the role of the service sector as the

Table 3.3 Annual average growth rates of employment, productivity and real GVA by service sector in the period 1997–2007. EU2 Nuts2 regions, by typology

Sector (Ateco code in parenthesis)	Regional type	Employment	Productivity	Real GVA
Market Services (G+H+I+J+K)	Local players	1.49	0.90	2.40
	Regional players	1.12	1.89	3.03
	Global players	1.42	2.07	3.52
Non-Market Services (L+M+N+O+P)	Local players	3.29	−0.01	3.28
	Regional players	3.13	0.24	3.38
	Global players	2.96	0.67	3.65
Distribution (G)	Local players	1.47	0.51	1.99
	Regional players	1.06	1.88	2.96
	Global players	1.04	2.03	3.09
Hotels and restaurants (H)	Local players	2.28	−0.83	1.43
	Regional players	2.07	−0.44	1.61
	Global players	2.43	−0.17	2.26
Transport, storage and communications (I)	Local players	0.94	2.67	3.63
	Regional players	0.60	3.16	3.78
	Global players	1.64	2.96	4.65
Financial intermediation (J)	Local players	0.20	2.69	2.90
	Regional players	0.22	3.49	3.72
	Global players	0.58	3.69	4.29
Real estate, renting and business activities (K)	Local players	5.00	−1.25	3.68
	Regional players	4.40	−0.96	3.40
	Global players	4.16	−0.39	3.75

Source of data: Cambridge Econometrics Regional Database.

recipient of employment losses created by manufacturing restructuring. Global players are those regions which have been less reliant on these protected services. Market service employment, at the same time, has been growing less, and not only in global players but also in local players, while regional players have created less jobs.

More than employment, however, service productivity is more closely linked to global flows. It is in fact clear that the productivity of services in global regions and, to a lesser extent, in regional players, has been growing much more than in local players. Especially higher is the productivity growth of global players in market services, which are obviously those more concerned with the globalization processes described in Sect. 2.

As can be observed from the patterns of GVA, growth in global players has been driven by market services, which have a growth rate similar to that of non-market services, while in less open regions the growth of public-sector-related services is significantly higher.

It is also possible to disentangle the patterns of individual sectors within market services. In particular (Table 3.3), it is possible to observe that in distribution (G), a sector where FDI are mainly market-seeking, the GVA performance of global and regional players is the highest, and this is accompanied by efficiency gains in terms of productivity, whereas in local players the sector has been growing especially in terms of employment, absorbing jobs from restructuring sectors.

Hotels and restaurants, which are endogenously linked to people flows, have also been growing more than anywhere else in global players (see GVA). Moreover, this pattern is accompanied by larger employment creation and, especially, by a productivity pattern which sees global players as the only type of regions able to maintain almost constant productivity in this sector.

Transport, storage and communications is a mixed sector in which activities from logistics to ICTs are classified. Also in this sector, global players have been outperforming the other regions of Europe in terms of employment and GVA; but in this case the largest productivity increases are in the most manufacturing regions, the regional players—those where manufacturing has had the strongest performance.

Given the urban nature of most global player regions, it is unsurprising that financial intermediation further concentrates in these regions, so that they have the largest growth rate in terms of employment, productivity and value added. This service sector is consequently strictly linked to regions able to play a role in global flows, as also testified by the fact that regional players come second in all three indicators.

The last sector, real estate, renting and business activities is another mixed sector. This has been acting as a very large creator and collector of employment, as shown by the very high employment growth rates. Interestingly, employment growth has been stronger in less global regions, the same ones where productivity decreases have been more marked. As a consequence, the GVA performance is very similar across globalization typologies. But again this is obtained in a different way, i.e. by simply creating new jobs in globalized regions and by increasing the employment base in the less global ones.

4 GDP Growth and Service Specialization in European Regions

The previous section underlined the relative performance of service sectors with respect to manufacturing and agricultural sectors, in terms of GVA, employment and productivity dynamics. It is of interest to conduct further analysis on whether the best performing regions in Europe are associated with some specific service specialization, or if instead manufacturing specialization still characterizes relatively well performing regions.

The first step of such analysis requires the identification of regions that record a GDP growth higher than the European average. Table 3.4 shows the number of regions that have higher-than-average GDP growth in Europe for each specific category, namely global, regional and local regions.

An interesting result emerges from a simple exercise like this one: regions that record higher-than-average GDP growth rates are evenly distributed among categories. Global players more often fall short of the average GDP growth, even

Table 3.4 Number of higher-than-average performance regions for each category[a]

		Global players Old 15	Regional players Old 15	Local players Old 15	Old 15	Global players New 12	Regional players New 12	Local players New 12	New 12
Regions with higher than EU average GDP growth (virtuous regions)	– Total number	16	44	33	93	8	18	5	31
	– Percentage	34	45.4	54.1	45.4	80	52.9	41.7	55.4
Total number of regions		47	97	61	205	10	34	12	56

Source: authors' calculations

[a]Annual average real GDP growth rates are calculated for Western and Eastern Europe separately

Table 3.5 Annual average real GDP growth rates of the three types of regions, 1997–2002 and 2002–2007

	Global players	Regional players	Local players	F
All European regions				
Growth rate 1997–2002	3.29	2.28	2.06	9.40***
Growth rate 2002–2007	2.84	3.04	2.53	2.14
Differential growth with respect to the nation 1997–2002	0.53	−0.50	−0.75	12.74***
Differential growth with respect to the nation 2002–2007	0.12	−0.09	−0.41	5.33***
Old 15 country regions				
Growth rate 1997–2002	2.94	2.45	1.96	6.03***
Growth rate 2002–2007	2.26	2.44	2.21	1.00
Differential growth with respect to the nation 1997–2002	0.19	−0.31	−0.79	7.11***
Differential growth with respect to the nation 2002–2007	−0.08	0.02	−0.19	1.59
New 12 country regions				
Growth rate 1997–2002	6.06	1.79	2.60	11.07***
Growth rate 2002–2007	7.33	4.80	4.20	8.83***
Differential growth with respect to the nation 1997–2002	3.14	−1.06	−0.57	17.42***
Differential growth with respect to the nation 2002–2007	1.72	−0.41	−1.51	14.85***

Source: calculated by the authors on Cambridge Econometrics Regional Database
***$p < 1\%$; **$p < 5\%$; *$p < 10\%$

if the comparison is made with the EU15 average rather than with the total European average, which also contains fast-growing Eastern regions.

If the regional growth rates are calculated, a slightly different picture emerges (Table 3.5). In the first period of time, i.e. 1999–2002, global players significantly outperformed the other types of regions in terms of GDP performance. This was the case of both regions in the Old 15 member countries and in the New 12 ones. Interestingly, in Western regions regional players were the second performers, close to global players, whereas in Eastern regions global players by far outperformed local players (the second best performers) as well as regional players. In the second period of time (2002–2007), global players were again significantly the best performers among Eastern European regions, but not significantly different overall and in Western countries.

National effects were controlled for once regional growth had been analysed with respect to its national average. The results show that global players have generally been leading their respective countries in terms of growth rates. Being a global player appears significantly to increase the possibility of being a region growing more than the average and to lead the country in terms of growth. In Eastern countries, the differential of global players with respect to their countries is high and significantly different from that of the other regions in both periods. In Western countries the differential growth rate is larger and significant in the first period, while in the second period, regional players perform better, but not significantly.

Understanding causally the differential growth rates recorded by global and regional players requires in-depth analysis of the structural features characterising virtuous regions with respect to non-virtuous ones in each group of regions, through the use of multivariate econometric regressions.[2] This is not the aim of this chapter, which instead focuses on a descriptive analysis of the structural features of a local economy which are all local assets and are conceptually linked to regional growth patterns from the quality and quantity of human capital to entrepreneurship, innovation, productive functions, transport infrastructure endowment.[3] In particular, since this if the focus of the book, our interest is to associate the role of service specialization with virtuous patterns of growth. Structural features, in fact, also encompass the mix of sectors in the region, the regional sectoral specialisation and its spatial concentration, as well as policy measures like structural funds.

An analysis of variance makes it possible to compare the values that structural features assume between virtuous and non-virtuous regions, and to calculate the statistical differences among these values for those European regions, namely the global and the regional players, which have a role in the global economy.[4] The comparison is made between higher-than-average and lower-than-average growing regions, keeping global and regional players as well as Western and Eastern regions separate from each other, given the "two growth models" hidden behind the economies of the two blocks of countries.

Unfortunately, the number of global players in the East is too small. Moreover, almost all these regions are virtuous, so that this precludes the use of this type of analysis for Eastern global players, and the results will be presented for only Western global players, Western regional players and Eastern regional players.

The results for Western global players are presented in Table 3.6, where only the statistically significant differences in the structural characteristics between virtuous and non-virtuous global players are given.

The sectoral specialization is quite different between the two groups of regions. The more virtuous regions exhibit higher specialization in advanced private services, Transport, storage and communication and Financial intermediation, and, interestingly enough, are characterised by an above-average presence of dynamic sectors, in both manufacturing and services. Specialization in particular manufacturing sectors, on the contrary, seems not to be associated with a virtuous regional growth pattern; the strong specialization of the region in manufacturing (captured by an Herfindal index) seems to play a role.

[2] For an exercise like this see Capello et al. (2011) and Capello and Fratesi (2011).

[3] For a review on regional growth theory, see Capello and Nijkamp (2009).

[4] In that this is simple statistical analysis, the results presented have no cause-effect chain, and their purpose is only to describe the statistical differences in structural feature endowment among the three types of regions. Care is taken to ensure that the structural features analyzed have nothing to do with the way in which the typology has been created. This allows circular reasoning to be avoided.

Table 3.6 Selected characteristics of global players in Western countries

Variables	Virtuous regions[a] 2002–2005	Non-virtuous regions[b] 2002–2005	F	Sig.
Sectoral specialization				
Location quotient in sector J Financial intermediation (2002)	1.11	1.61	4.28	*
Location quotient in sector I Transport, storage and communications (2002)	1.24	1.04	4.26	**
Location quotient of growing service sectors (2002)	1.40	1.07	7.49	***
Location quotient of growing manufacturing sectors (2002)	1.17	0.87	3.14	*
Herfindal index in manufacturing sectors (2002)	0.16	0.14	3.43	*
Functional specialization				
Share of legislators and senior government officials (average value over 3-year period 1999–2001)	0.008	0.005	4.47	**
Share of managers in SMEs (average value over 3-year period 1999–2001)	0.039	0.027	8.07	***
FDI penetration index (average value over 3-year period 1999–2001)	4.94	3.767	0.33	

Source: authors' calculations on Eurostat data
[a]Regions with higher than EU average GDP growth rate
[b]Regions with lower-than-EU-average GDP growth rates
***$p < 1\%$; **$p < 5\%$; *$p < 10\%$

The presence of command and control functions in SMEs (measured as the share of managers in SMEs) makes a difference with respect to non-virtuous regions (Table 3.6). An unexpected result is obtained for FDI penetration. Despite being greater in global regions than the rest of EU global regions, FDI penetration does not appear to differ between virtuous and non-virtuous global regions.

The characteristics which enable regional players to be competitive are expected to be different from those that enable global players to grow, given the lack of world accessibility that characterises regional players.

The results of the analysis for regional players in Old 15 member countries are reported in Table 3.7. The dynamics of virtuous regional players in the West are not driven by manufacturing specialization, neither in high nor in low-tech activities, but rather by specialisation in traditional service sectors, such as those linked with tourism (H Hotels and restaurants) and the public sector (L Public administration and defence; compulsory social security). The success of these regions can probably be explained by their ability to innovate in mature sectors, offering new and attractive services in traditional activities (e.g. agri-tourism, balanced coastal tourism). The Herfindal index is higher in regional virtuous than in regional non-virtuous players; this result testifies that regional virtuous regions draw advantage from localization economies stemming from spatial concentration in manufacturing sectors. Moreover, regional virtuous regions are more assisted than their non-virtuous counterparts by public policies, and structural funds in particular.

Table 3.7 Selected characteristics of regional players in Western countries

Variable	Virtuous regions[a] 2002–2005	Non-virtuous regions[b] 2002–2005	F	Sig.
Sectoral specialization				
Location quotient in sector D Manufacturing (2002)	0.880	1.196	15.4	***
Location quotient in sector H Hotels and Restaurants (2002)	2.106	0.834	16	***
Location quotient in sector L Public Administration and Defence (2002)	1.116	1.006	2.84	*
Location quotient in High-tech manufacturing sectors (2002)	0.607	1.076	13.67	***
Location quotient in Medium High-tech manufacturing sectors (2002)	0.849	1.282	12.08	***
Location quotient in Medium-Low manufacturing sectors (2002)	0.929	1.315	8.72	***
Herfindal index in manufacturing sectors (2002)	0.089	0.080	6.72	**
Functional specialization				
Share of legislators and senior government officials (average value over three-period 1999–2001)	0.106	0.082	8.85	***
Share of managers in SMEs (average value over 3-year period 1999–2001)	0.053	0.028	31.19	***
Share of physical, mathematical and eng. science professionals (average value over 3-year period 1999–2001)	0.023	0.029	6.59	**
Share of people with second-level educations (share of people in EGP-2 professions)	21.30	19.20	8.6	***
Share of people with postgraduate educations (Isced 5 and 6) (average value over 3-year period 1999–2001)	0.830	0.976	4.61	**
FDI penetration index (average value over 3-year period 1999–2001)	0.466	0.837	0.83	***

Source: authors' calculations on Eurostat data
[a]Regions with higher-than-EU-average GDP growth rate
[b]Regions with lower-than-EU-average GDP growth rate
***p < 1%; **p < 5%; *p < 10%

The economies of virtuous regional players are characterized to a more than average extent by control functions (legislators, senior officials and managers), and in particular those of SMEs (share of managers in SMEs), while the scarce presence of physical, mathematical and engineering science professionals may be due to their service specialisation. This datum is corroborated by the scant presence of people with post-graduate degrees (ISCED 5 and 6) and the higher presence of people with second-level qualifications (high share of people in EGP-2 professions). Overall, it appears that, among Western regional players, the virtuous ones are those characterised by intermediate-level service functions and by high functions in the public service sector.

The last analysis performed is for regional players in Eastern regions (Table 3.8). Here a large number of characteristics emerge which differentiate virtuous from

Table 3.8 Selected characteristics of regional players in Eastern countries

Variable	Virtuous regions[a] 2002–2005	Non-virtuous regions[b] 2002–2005	F	Sig.
Sectoral characteristics				
Agricultural productivity (2002)	6.90	2.84	36.92	***
Industry productivity (2002)	7.77	9.68	3.79	*
Service productivity (2002)	7.51	11.3	8.46	***
Growth of service employment (2000–2002)	0.32	1.48	4.02	*
Loc. Quot. in sectors A Agriculture, hunting and forestry, B Fishing (2002)	3.45	2.00	4.66	**
Location quotient in sector D Manufacturing (2002)	1.56	1.25	10.41	***
Location quotient in sector F Construction (2002)	1.08	0.90	17.29	***
Location quotient in sector I Transport, storage and communications (2002)	1.23	0.95	8.72	***
Location quotient in sector J Financial intermediation (2002)	0.31	0.61	52.21	***
Location quotient in sector K real estate, renting and business activities (2002)	0.50	0.63	18.31	***
Location quotient in Medium Low-tech manufacturing sectors (2002)	1.78	1.29	4.99	**
Location quotient in Low-tech manufacturing sectors (2002)	1.65	1.38	3.17	*
Herfindal index in manufacturing sectors (2002)	0.13	0.11	6.07	**
Lawrence index in all sectors (1995–2002)	0.15	0.21	10.75	***
Functional specialization				
Share of legislators and senior government officials (average value over 3-year period 1999–2001)	0.04	0.06	7.13	**
Share of physical, mathematical and engineering science professionals (average value over 3-year period 1999–2001)	0.012	0.017	3.69	*
Share of clerks (average value over 3-year period 1999–2001)	0.054	0.087	11.04	***
Share of craft and related trade workers (average value over 3-year period 1999–2001)	0.214	0.183	6.72	**
Location quotient of growing manufacturing sectors (2002)	1.475	1.044	5.53	**
Location quotient of growing service sectors (2002)	0.965	1.080	2.93	*
Share of people with postgraduate educations (Isced 5 and 6) (average value over 3-year period 1999–2001)	0.733	0.950	5.95	**
FDI penetration index (average value over 3-year period 1999–2001)	0.950	0.158	3.8	*

Source: authors' calculations on Eurostat data
[a]Regions with higher- than-EU-average GDP growth rate
[b]Regions with lower-than-EU-average GDP growth rate
***$p < 1\%$; **$p < 5\%$; *$p < 10\%$

non-virtuous regions. Firstly, a convergence process seems to take place. Virtuous regional players in Eastern countries are poorer than the rest of Eastern regional players, and they record lower productivity levels in both services and manufacturing only partly off-set by greater agricultural productivity with respect to non-virtuous regions. The virtuous regional players in the East are specialized in Agriculture, hunting and forestry (A) Fishing (B), Manufacturing (D) and Construction (F), and they are able to maintain their specialization over time, as evidenced by the high manufacturing Herfindhal index and the low Lawrence index. The latter measures the changes in a region's specialisation: the lower the index, the lower the changes in the sectoral specialisation of a region.

Among the service sectors, virtuous regions in Eastern countries are only specialized in traditional sectors like Transport, storage and communication (I), and they are particularly de-specialized in some advanced services, namely Financial intermediation (J) and Real estate, renting and business activities (K), with respect to non-virtuous regions.

Interestingly, the virtuous regional players in the East are specialized in the low- and medium-tech manufacturing sectors, with few physical, mathematical and engineering science professionals and a low share of people with post-graduate degrees (Isced 5 and 6). Low shares of basically service workers, like clerks, are off-set by a high percentage of craft and related trade workers. This sectoral/functional specialization again shows that, in Eastern countries, virtuous regions are the less developed ones that start up a convergence process.

In general, what emerges from this analysis is that the regional specialization in service sectors is associated with virtuous patterns of growth. However, the specialization in services changes according to the degree of openness of regions. In global regions in Western countries, what emerges is the specialization in advanced and dynamic service sectors. Interestingly, this is true only for these regions: in fact, virtuous global regions in Eastern countries do not register any specialization in advanced services, and are more related to manufacturing specialization. Virtuous regional players in Eastern countries are even associated with de-specialization of service activities.

5 Conclusions

The role of the service sector in the global economy has greatly increased in importance over the past decades. All major qualitative changes in globalization trends have affected the service sector. Deindustrialization processes call for an increase in service employment. Higher competition, in its turn, calls for specialization in advanced and private services, which are the most productive activities.

This chapter has presented the qualitative change in the globalization process and the effects that raise new challenges for regional economies. An empirical analysis has investigated at regional level the trend of GDP, employment and productivity in the different macro-industry, agriculture, manufacturing and service; the distinction between highly global regions—those regions with above-average (economic and physical) connectivity with the rest of the world—and local regions has shown that higher productivity growth rates are associated with advanced service sectors in global regions.

Moreover, the chapter has descriptively analysed the relationship between specialization in services and higher relative regional performance. Global regions in Western countries are the only ones where a significantly higher specialization in advanced service sectors is associated with the virtuous regions. In all other cases, the association of virtuous patterns of growth is associated with manufacturing specialization and/or low service specialization.

The overall conclusion is that the service sector is increasingly important for regions to be able to compete in the global world, because it characterizes those European regions more open to the external world. As a consequence, its transformations should not be underestimated in the years to come.

References

Affuso, A., Capello, R., & Fratesi, U. (2011). Globalization and competitive strategies in European vulnerable regions. *Regional Studies, 45*(5), 657–675.

Baldwin, R. (2006). *Globalisation: the Great Unboundling*. Prime Minister's Office, Economic Council of Finland.

Camagni, R. (1991). Regional deindustrialization and revitalization processes in Italy. In H. Sazanami & L. Rodwin (Eds.), *Industrial change and regional economic transformation* (pp. 137–167). London: Harper Collins.

Capello, R., & Fratesi, U. (2011). Crescita Regionale in un Contesto di Globalizzazione: l'Eterogeneità Spaziale nei Fattori di Successo. In L. Resmini & A. Torre (Eds.), *Competitività Territoriale: Determinanti e Politiche* (pp. 129–158). Milano: FrancoAngeli.

Capello, R., Fratesi, U., & Resmini, L. (2011). *Globalisation and regional growth in Europe: Past trends and scenarios*. Berlin: Springer.

Capello, R., & Nijkamp, P. (Eds.). (2009). *Handbook of regional growth and development theories*. Cheltenham: Edward Elgar.

Davies, R. B., & Guillin, A. (2011). *How far away is an intangible? FDI, space and services*, working paper, University College Dublin.

Fontagné, L., & Lorenzi, J.-H. (2005). *Déindustrialisation Délocalisation*. Paris: La Documentation Française.

Fratesi, U. (2012). A globalization-based taxonomy of European regions. *Regional Science Policy and Practice, 4*(1), 1–23.

Golub, S. S. (2009). Openness to foreign direct investment in services: An international comparative analysis. *The World Economy*. doi:10.1111/j.1467-9701.2009.01201.x.

Lejour, A. (2007). FDI in services: Recent developments and prospects in Europe. In K. Liebscher, J. Christl, P. Mooslechner, & D. Ritzberger-Grunwald (Eds.), *Foreign direct investment in Europe: A changing landscape* (pp. 175–189). Cheltenham: Edward Elgar.

Mark, J. A. (1982). Measuring productivity in service industries. *Monthly Labour Review*, pp. 3–8.

OECD (2001), *Measuring productivity*. Paris

OECD. (2007). *Globalisation and Regional Economies*. Paris, chapters 1 and 2.

Perloff, H. (1957). Interrelations of state income and industrial structure. *The Review of Economics and Statistics, 39*(2), 162–171.

Perloff, H., Dunn, E., Lampard, E., & Muth, R. (1960). *Region, resources and economic growth*. Baltimora: John Hopkins.

Resmini, L. (2013). FDI in services in European regions: An overview. In J. R. Cuadrado-Roura (Ed.), *Service industries and regions*. Heidelberg: Springer (in this volume).
Riedl, A. (2008). *Contrasting the dynamic patterns of manufacturing and service FDI: Evidence from transition economies*, working paper n. 117, Vienna University of Economics and Business Administration.
UNCTAD. (2004). *World investment report: The shift towards services.* Washington: UN.

Chapter 4
Patterns and Trends in Services Related Activities in OECD Regions

Enrique Garcilazo, Florence Mouradian, and Joaquim Oliveira-Martins

1 Introduction

Over the past decades, OECD countries have experienced profound economic structural changes with a reduction in the relative size of the agricultural and manufacturing sector and increases in the service sector. The forces driving the structural change, among others are a process of modernisation in the agricultural sector and a gradual dispersion of the value chain within manufacturing toward lower wage economies triggered by a decline of trade costs and improved communications. The share of services in the economy is also related to the evolution of relative prices between the tradable and non-tradable sectors, the latter being closely linked to the real exchange rate. The appreciation of the real exchange rate is a feature of economic development, as it has been abundantly documented in the economic literature since the seminal work by Balassa (1964) and Samuelson (1964). An appreciation trend in the real exchange makes the production of non-tradables more profitable that in the tradable sector, and provides as such an incentive for resources to move from the latter to the former sector. By and large, the service sector is still mostly a non-traded sector and most service activities have a strong local component. This is why the connection between regional development and the service economy—the topic of this paper—seem particularly relevant.

Against this backdrop, service related activities in OECD countries have gained importance in recent years and currently represent around 70 % of total output and employ around the same proportion of workers. Therefore understanding what drives productivity in services is critical for understanding overall

E. Garcilazo (✉) • J. Oliveira-Martins
Regional Development Policy Division, OECD, Paris, France
e-mail: joseenrique.garcilazo@oecd.org

F. Mouradian
University Paris-Dauphine, Paris, France

J.R. Cuadrado-Roura (ed.), *Service Industries and Regions*,
Advances in Spatial Science, DOI 10.1007/978-3-642-35801-2_4,

growth. This structural change however, has not come homogenous in space particularly at the sub-national level due to differences in local labour markets within countries and the strong non-tradable component of services making them quite dependent on local labour market conditions. Therefore understanding the geographical patterns in services-related activities and key factors associated with higher productivity in services at the sub-national level is critical for understating the overall trends in services.

This aim of this chapter is to provide an overview of the main patterns in the service economy in OECD countries, its evolution over time and to identify the spatial patterns associated with these broad trends among OECD regions. The regional analysis focuses on the period 1995–2008. Our sample of regions include 335 OECD Territorial Level 2 regions and the analysis focuses on three sectors within services (1) wholesale and retail trade, repairs, hotels and restaurants, transport (2) financial intermediation, real estate, renting and business activities and (3) public administration and social services. The chapter is structured in three sections.

The first section provides a brief overview on the increasing importance of the service sector among OECD countries in recent years. The analysis carried at the national level, focuses on trends in the value added produced by services to total economic output and the proportion of the workforce employed in services highlighting the growing importance of services in recent years. This analysis then examines trends and patterns within the service sector, focusing on three subsectors. It then measures unit labour costs at the national level and finally compares the productivity of services vis-à-vis other sectors in OECD countries.

Section two examines the spatial patterns associated with these broad patterns at the sub-national level. We first analyse the spatial dimension of GVA and employment shares in services vis-à-vis other sectors among 335 OECD TL2 regions. The section then proceeds to examine whether service sector activities have become more or less concentrated in space in recent years among all OECD regions and among regions within countries. Specialisation patterns are then analysed identifying the regions with the highest degree and the highest gains in service specialisation.

Section three measures the trends of service sector productivity at the regional level. It analyses the links between service sector productivity and overall regional productivity aiming at determining whether regions becoming more productive in services are also becoming more productive overall or not. After measuring productivity trends, the section investigates the contribution of services to overall service growth. Lastly the section analyses key regional characteristics associated with high productive in services, focusing on human capital, density and innovation.

After these three sections, a discussion summarises the main findings and its associated policy implications.

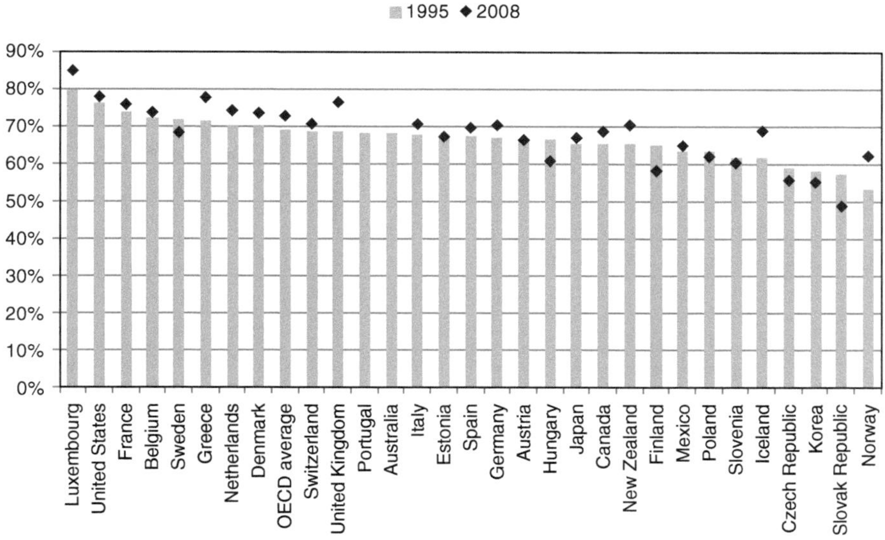

Fig. 4.1 Share of value added by the service sector to total output, 1995 and 2008. *Source*: OECD STAN Database for Structural Analysis

1.1 The Service Sector Has Been Gaining Importance in OECD Countries

The service sector has gained importance in OECD economies over the past decades. Both as a proportion of workers employed in this sector to the total workforce and as a proportion of total output produced by this sector. Currently the service sector employs close to 70 % of the total workforce in OECD economies and adds 70 % of total output (Fig. 4.1).

These proportions however vary among countries, with Luxemburg and the United States recording the largest share of value added, close to 80 %, and the former Eastern European countries (Poland, Hungary, Czech Republic, Slovenia and Slovak Republic), Korea and Finland the lowest, ranging from 50 % to 62 %. Over the period 1995–2008, ten OECD countries (Slovak republic, Finland, Hungary, Sweden, the Czech Republic, Korea, Slovenia, Poland Estonia and Austria Mexico) decreased the relative share produced by services to total output in value added while the rest increased the share (Fig. 4.1).

In terms of employment, the proportions are in similar range with the United States, Netherlands, the United Kingdom and Belgium employing close to the 80 % of their workforce in the service sector and the former Eastern European Countries (Poland, Slovenia, the Czech Republic, Estonia, Hungary and the Slovak Republic) in addition to Portugal and Mexico displaying the lowest share with values ranging around 55–65 % (Fig. 4.2). Over the period 1995–2008, the proportion of workers employed in services increased in all OECD countries.

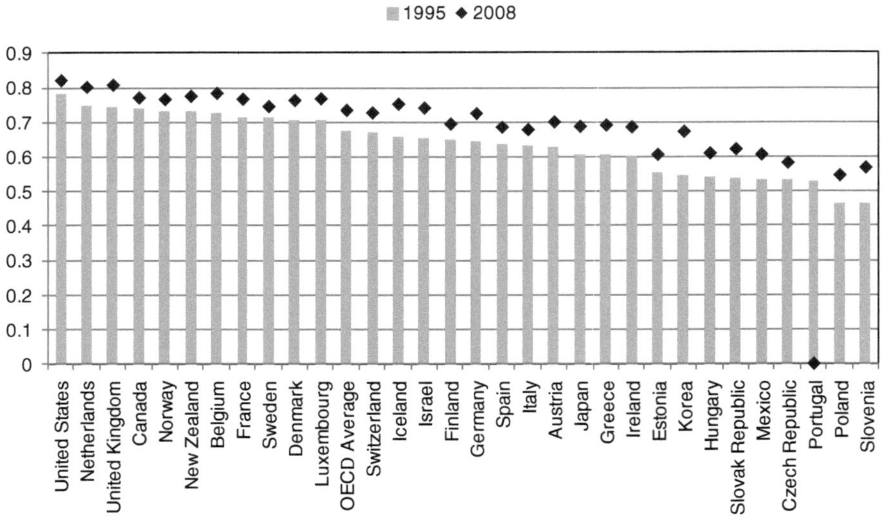

Fig. 4.2 Share of employment in the service sector to total employment, 1995 and 2008. *Source*: OECD STAN Database for Structural Analysis

1.2 Financial Intermediation and Business Services Have Gained Ground Years in Terms of GVA Growth

Within services, its largest sub-sector is finance, insurance, real estate and business (from now on labelled as financial and business services) representing 37.4 % of the value added in services among OECD countries (Fig. 4.3). It is followed by community, social and personal services (30.4 %) and wholesale and retail trade, restaurants and hotels (21.4 %).

These proportions however vary. In Luxembourg, finance and business services represents more than 56 % of its value added in services, while this proportion is just higher than 40 % in Israel, France, Ireland, United States, Germany and United Kingdom. It falls close to 30 % in the former Eastern European countries and in Mexico.

The increasing importance of services has been largely driven by the expansion of finance and business services. The share of value added in community, social and personal services has risen much more slowly, and even stagnated, in nine OECD countries (Germany, Denmark, Korea, Sweden, Slovenia, Switzerland, Finland, Luxembourg, Hungary, Austria, Norway and Mexico). The relative weight of transport, storage and communication and wholesale and retail trade—hotels and restaurants has decreased in around half of the OECD countries (Fig. 4.4).

In terms of employment; community, social and personal services is the largest service sub-sector in all OECD countries, with the exception of Luxembourg, Greece, Japan and Korea, employing around 41 % of its workforce in services. It is followed by wholesale and retail trade, restaurants and hotels (34.2 %) and finance, insurance, real estate and business services in third place (15.4 %) (Fig. 4.5).

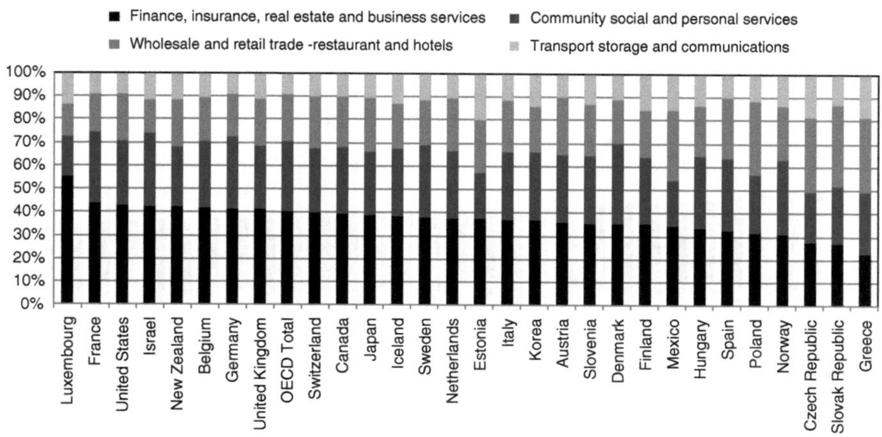

Fig. 4.3 Share of value added in service sub-sectors to total value added in services, 2008. *Source*: OECD STAN Database for Structural Analysis

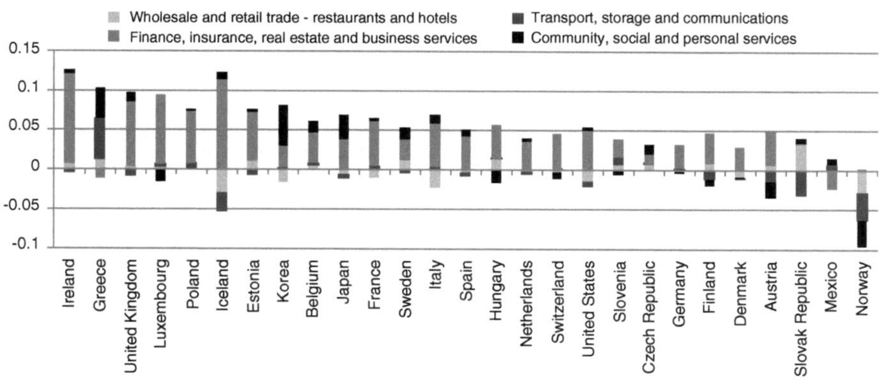

Fig. 4.4 Change of share in value added of service sub-sectors in total value added, 1995–2008. *Source*: OECD STAN Database for Structural Analysis

These proportions also vary by countries. Employment in community, social and personal services represents close to 50 % of total employment in services in Sweden, Iceland and Norway, and less than 37 % in the former Eastern European countries, in New Zealand and in Canada. Finally, wholesale and retail trade, restaurants and hotels is the first service sub-sector in terms of employment in Greece, Japan and Korea and finance, real estate, and business services is the first one only in Luxembourg.

In recent years the largest employment gains within services has been driven by, finance and business services increasing in almost all OECD countries with the exception of Denmark, Sweden, Finland, Norway, Spain, United States and Korea. In contrast employment of the wholesale, trade and retail trade—restaurants and hotels sector has decreased or stagnated in all others OECD countries except in Japan, Greece, Canada, Switzerland and Mexico. Finally, the share of employment

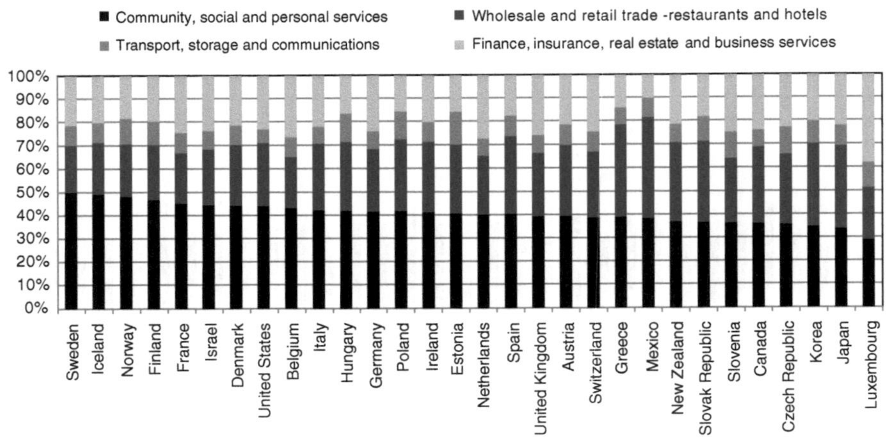

Fig. 4.5 Share of employment in service sub-sectors to total employment in services, 2008. *Source*: OECD STAN Database for Structural Analysis

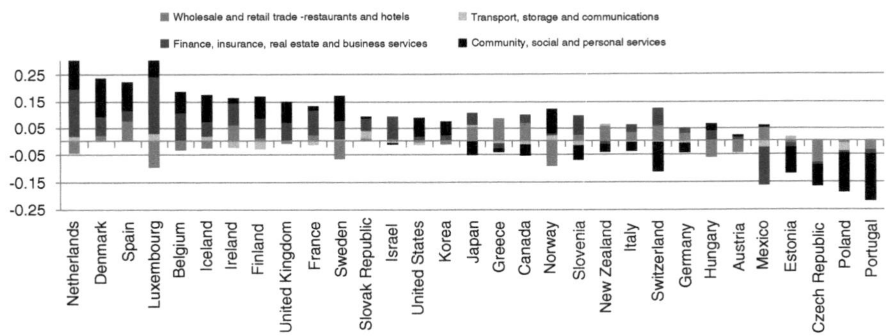

Fig. 4.6 Change in share of employment in service sub-sectors in total employment, 1995–2008. *Source*: OECD STAN Database for Structural Analysis

in transport, storage and communications in total employment has stagnated in almost all OECD countries (Fig. 4.6).

1.3 Relative Prices and the Economic Structure Between Tradable and Non-tradable Sectors

One of the main drivers of the allocation of resources between the service sector and the rest of the economy is the relative price between non-tradable (P_n) and tradable (P_t) sectors. This relative price is close to the real exchange rate ($E.P/P^*$). An increase in this ratio (a real appreciation) indicates that production in non-tradables is likely to be more profitable that in tradables, and provides as such an incentive for resources to move from the latter to the former sector (Edwards, 1989). Note that, by and large, the service sector is still mostly a non-traded sector and, above all, many service activities have a strong local component.

Over the long-run, the real exchange rate and the relative price (P_n/P_t) tend to follow an equilibrium upward trend due to the so-called Balassa–Samuelson effect.[1] This trend is generated by a higher productivity in tradable than in the non-tradable sectors. Given that wages tend to equalize throughout the economy, the productivity differential generates an upward pressure in the unit labour costs (wages divided by the productivity) in the non-tradable sector. This pushes up the relative prices of non-tradable to tradables and appreciates the real exchange rate. This real exchange rate appreciation can be viewed as compatible with the underlying productivity trends and is a stylized fact of economic development.

However, a "misalignment" of relative prices is possible caused by different factors, such as government spending, capital inflows or terms of trade shocks. The use of fixed exchange rate regimes to stabilize inflation typically also induced excessive shifts in relative prices, as it was observed in the Latin America in the 1990s (Baldi & Mulder, 2004). More recently, the integration of less-developed economies of Southern Europe in the euro area also led to an appreciation of their real exchange rate and a shift towards the service non-tradable sectors that could have gone beyond what was allowed by productivity fundamentals (Darvas & Pisani-Ferry, 2011).

The basic macroeconomic transmission mechanisms work as follows. When a small open economy experience an exchange rate appreciation combined with large capital inflows (and current account deficit), this often translates into an expansion of domestic credit, increasing domestic demand for tradables and non-tradables. To increase the supply of non-tradables, a rise in the price of non-tradables is needed to shift labour from tradables to non-tradables. In the new equilibrium the size of the non-tradable sectors has increased.

While the tertiarization of economies is desirable over the long-run, an excessive or too rapid shift towards the non-tradables sectors may have important implications for aggregate productivity. Indeed, services often have lower performance in this respect than industrial sectors (see below). This is due to the non-tradability that reduces competitive pressures (both in terms of prices and innovation) compared with the ones experienced in the tradable sector. Several service sectors are also affected by significant restrictions and regulatory issues compared with the more open sectors of the economy (Conway et al., 2006).

To illustrate the productivity performance of service sectors compared with the industry, Fig. 4.7 displays the evolution of relative labour productivities for three European countries (Germany, Poland and Portugal) compared with the US.

[1] Note that this result also depends on the wage equalisation across sectors and the fact that productivity increases in the tradable sector are typically higher in the less developed countries.

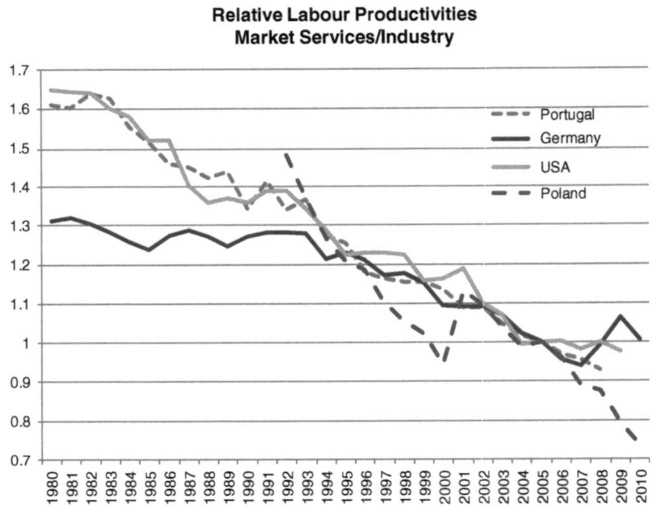

Fig. 4.7 Evolution of relative labour productivities, 1995–2008. *Source*: OECD Unit Labour costs database

It can be seen that relative productivity in the service sectors has steadily declined since the early 1980s, in particular in recent years the new members of the EU and the euro, such as Poland and Portugal, compared with Germany and the US. This relative decline in productivities translates into an increase in the relative unit labour costs in service sector compared with the industry (Fig. 4.8). Again this increase has been particularly rapid in countries such as Poland and Portugal compared with Germany.

What are the implications of these trends for regional economics and policy? As service sectors are often defined regionally or locally, the question on how to boost labour productivity growth in the non-tradable sector could have important regional dimensions. The relative price shocks are also likely to affect regional economies in rather asymmetric way, as the shares of service sectors varies significantly across different types of regions (urban, intermediate, rural). The next sections examine the spatial dimension of these broad macroeconomic trends.

1.4 Productivity in the Service Sector has Increased but to a Lesser Degree than in Manufacturing

Productivity with the service sector is on average similar to the productivity in total economy. It is 61 % higher than in agriculture and 22 % higher than in construction, but 29 % lower than in industry. Nevertheless, productivity within the service sector varies significantly among sub-sectors. Productivity in financial and business activities is 88 % higher than overall country-wide productivity, and 20 % higher in transport, storage and communication. Conversely, productivity in public

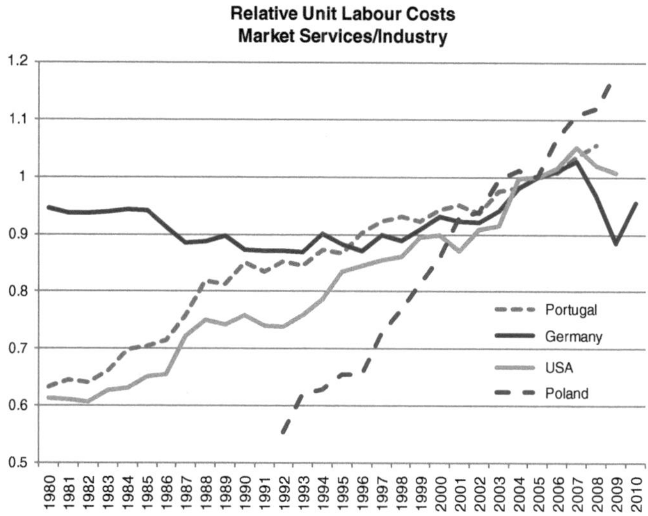

Fig. 4.8 Evolution of relative labour costs, 1995–2008. *Source*: OECD Unit Labour costs database

administration and social services and in wholesale and retail trade are respectively 26 % and 33 % below the average productivity.

Over the 1995–2008 period, productivity in services decreased relative to other sectors, from being 3 % above the average productivity to 0 %. Productivity in financial and business services has particularly decreased, but still fares far above the average productivity (Table 4.1).

Examining trends in productivity and multi-factor productivity over 1995–2008 reveal that service related activities display a lower productivity growth and lower multi-factor productivity growth than in industry.

Within services in almost all OECD countries, labour productivity and multifactor-productivity growth has been lower in wholesale and retail trade, restaurants and hotels, as compared to productivity growth in industry (including energy). Multi-factor productivity growth rate has also been lower in finance, insurance and business services than in industry illustrated in Fig. 4.9.

The main findings reveal that indeed the service sector economy has gained importance over the past decades contributing to around 70 % of the total output and employing the same proportion of workers. In all but two OECD countries the service sector has gained in terms of value added relative to other sectors and in all but five OECD countries employment in services has also increased. Despite these gains, productivity growth in services still trails below productivity growth in manufacturing raising concerns, especially given their growing contribution to total economic output. Many service sector actives are non-tradable and therefore highly depend on the regional and local markets to which they service, making them highly dependent on space. The following sections explore the spatial

Table 4.1 Productivity per person employed, average OECD

	In PPP US$ constant 2005		Percent to total economy		Growth
Sectors	1995	2008	1995	2008	1995–2008
Agriculture	15,984	28,510	35 %	39 %	78 %
Construction	36,655	56,363	80 %	78 %	54 %
Industry (including energy)	54,627	92,907	119 %	129 %	70 %
Wholesale and retail trade	33,704	48,515	73 %	67 %	44 %
Transport, storage, communication	56,522	86,686	123 %	120 %	53 %
Finance, insurance, real estate, business services	94,440	135,752	206 %	188 %	44 %
Community, social and personal services	35,582	53,848	77 %	74 %	51 %
Total services	**47,523**	**72,177**	**103 %**	**100 %**	**52 %**
Total economy	**45,944**	**72,292**	**100 %**	**100 %**	**57 %**

Source: OECD STAN Database for Structural Analysis

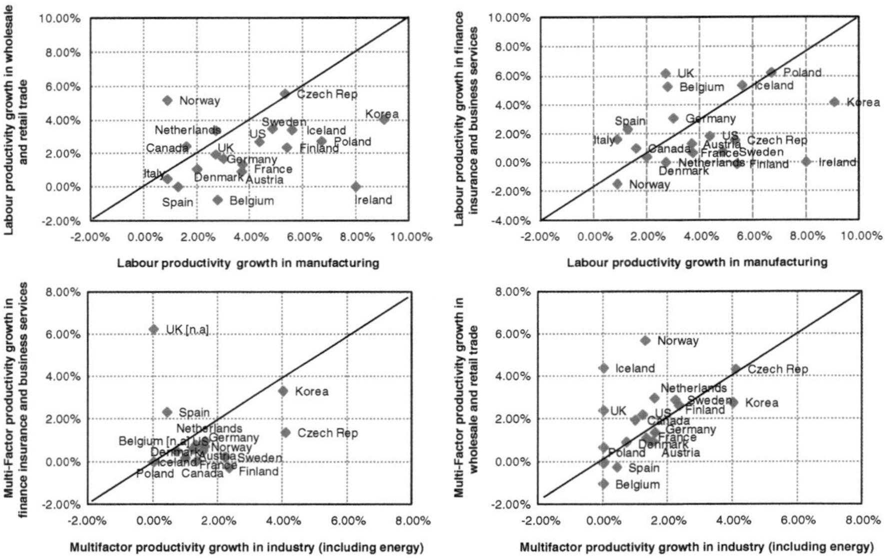

Fig. 4.9 Annual average growth of labour and multi-factor productivity in industry and services, 1995–2008

pattern in service-related activities, examine productivity in services at the regional level, links between labour productivity in services and regional productivity and the key factors associated with high service sector productivity at regional level in order to gain some insight on the critical factors driving overall productivity growth.

2 The Geography of Service Sector Activities

As documented in the previous section, service sector activities have gained significance in recent years in OECD countries. These broad structural changes as we will see are by no means homogeneous in space and moreover have a very strong spatial dimension with some regions displaying a much higher concentration of service sector activities as opposed to other regions.

This section first displays the broad spatial patterns in services at the sub-national scale by measuring trends of concentration among all OECD regions and second by measuring concentration in services among regions within countries. The section then identifies the regions with the highest share of service sector activities and explores the patterns of specialisation in service-related activities among OECD regions benchmarking specialisation relative to all OECD regions and to regions within each country.

2.1 Regional Data by Sector Resembles Data Provided at the National Level

Our unit of analysis is based on TL2 regions, defined as the largest territorial administrative tier at the sub-national level. The OECD regional database defines 355 regions of this kind contained in the 34 OECD countries. Although very detailed databases at the region-industry level do exists for a selected number of OECD countries (e.g. Sweden), a database covering all OECD TL2 regions is confined to a smaller coverage of regions and the following six sectors: (1) public administration social services; (2) construction; (3) agriculture; (4) financial intermediation; (5) mining quarrying and manufacturing and (6) wholesale and retail trade, among which three belong to the category of services: financial intermediation, public administration social services and wholesale and retail trade. Data for gross value added and employment at the regional level for the six sectors are broadly available for 255 TL2 regions with a slightly broader coverage for GVA than for employment. Our data coverage covers the period 1995–2008 with more data availability in the latter years and our data source is the OECD Regional Database.

Despite the data source at the regional level is different than that provided at the national level the indicators computed at the regional level resemble those provided at the national level, with close to 70 % (68.5 %) of total value added activity produced by services (Table 4.2) and a slightly higher proportion of the total workforce (69.3 %) employed in services. Among the three service subsectors considered, all three represent a similar share (around 22 %) in terms of value added. In terms of the employment share, financial services employ half of the workforce (11.6 %) in wholesale and retail trade (29.1 %) and in public administration and social services (28.5 %).

Comparing changes over time, the service sector has been the fastest growing sector over the period 1995–2008, increasing its GVA share by 1.56 percentage

Table 4.2 Sectoral distribution of value added among OECD TL2 regions, 1995–2008

GVA shares 2008	Min	Av	Max
Agriculture	0.0 %	3.4 %	23.9 %
Mining quarrying manuf.	1.5 %	21.2 %	88.2 %
Construction	1.0 %	6.9 %	16.3 %
Services	2.1 %	68.5 %	54.3 %
Wholesale and retail trade	2.9 %	21.8 %	44.7 %
Financial Intermediation	2.1 %	22.5 %	53.9 %
Public administration social services	2.3 %	24.3 %	54.3 %
Change in GVA shares 1995–2008	**Min**	**Av**	**Max**
Agriculture	−11.93	−1.16	12.33
Mining quarrying manuf.	−48.48	−1.52	28.37
Construction	−42.92	0.85	28.37
Services	−35.96	1.56	30.32
Wholesale and retail trade	−48.48	−3.78	17.93
Financial Intermediation	−48.48	3.76	30.32
Public administration social services	−16.55	1.41	16.52

Source: OECD Regional Database (2012)

Table 4.3 Sectoral distribution of employment among OECD TL2 regions, 2008

Employment shares 2008	Min	Av	Max
Agriculture	0.0 %	6.5 %	70.1 %
Mining quarrying manuf.	1.8 %	17.3 %	51.1 %
Construction	1.0 %	7.0 %	17.0 %
Services	24.9 %	69.3 %	97.2 %
Wholesale and retail trade	13.7 %	29.1 %	64.3 %
Financial Intermediation	1.5 %	11.6 %	29.4 %
Public administration social services	5.4 %	28.5 %	56.4 %
Change in employment shares 1995–2008	**Min**	**Av**	**Max**
Agriculture	−22.57	−2.56	18.37
Mining quarrying manuf.	−9.87	−2.39	6.21
Construction	−8.35	1.07	7.98
Services	−25.21	3.89	20.32
Wholesale and retail trade	−15.61	0.81	29.51
Financial Intermediation	−5.11	2.08	7.92
Public administration social services	−39.13	0.83	18.38

Source: OECD Regional Database (2012)

points on average among all OECD regions and its employment share by more than twice that amount (3.89 pp). Within the service sector, financial and business services recorded the largest increase in GVA share (3.76 pp) contrary to wholesale and retail trade experiencing the largest declines (−3.89 pp).

In terms of employment, financial intermediation, despite employing a lower proportion of the workforce (11.6 %) has been increasing its employment share the most amongst the service subsectors considered, followed by construction, public administration social services and wholesale and retail trade (Table 4.3).

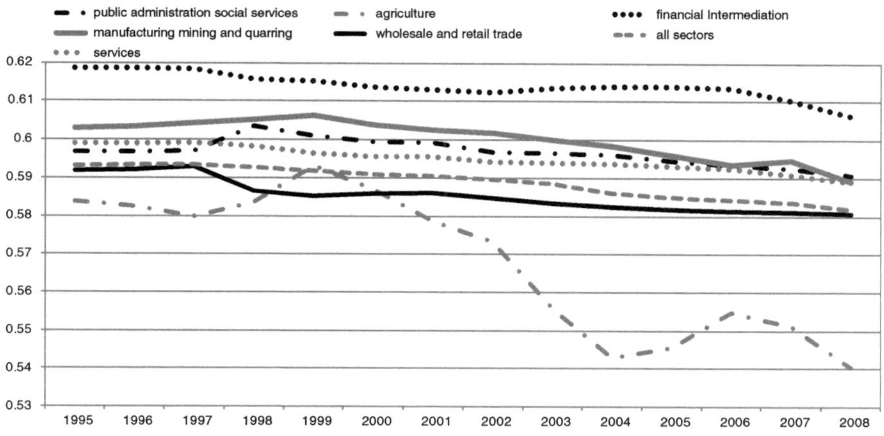

Fig. 4.10 Index of geographic concentration among OECD TL2 regions, 1995–2008. *Source*: OECD Regional Database (2012)

These figures broadly resemble the national figures and confirm the growing importance of the service sector activities in OECD countries in recent years against the backdrop of the relative decline in agriculture and manufacturing.

2.2 *Concentration in Services has Declined Among all OECD Regions but Increased Within Countries*

We apply the Index of Geographic Concentration for measuring trends in concentration at the sub-national level (see Appendix 1). First we consider all OECD regions as one geographic area and compute concentration among each of the six sectors allowing us to compare the degree of geographic concentration amongst them and its evolution over time. The calculations reveal that service sector economies are more concentrated in space than economic activities in all sectors with financial and business services displaying the highest degree of concentration.

Figure 4.10 reveals an interesting pattern over time displaying a gradual decline in the degree of concentration among all sectors considered with a particular drop in agriculture. By treating the OECD universe as one geographic area, this trend may be driven by country-wide effects in particular the catching up-effect of former eastern European countries stimulating economic activities in this geographic area and as a result reducing overall concentration.

Concentration within countries among TL2 regions however reveals a different but consistent pattern. It appears the degree of concentration by service sector activities also tends to be higher than in all economic activities. Services are more concentrated in all OECD countries except in Italy, and Korea (left column

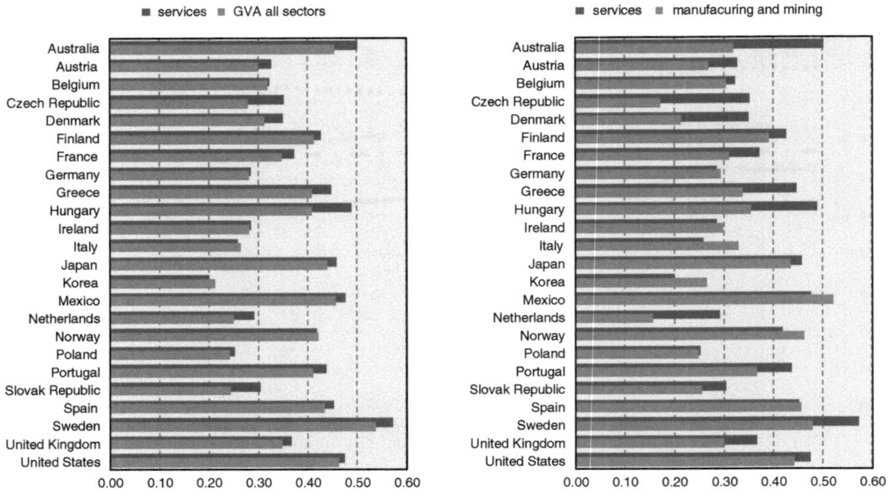

Fig. 4.11 Index of geographic concentration of GVA in services and in all sectors (TL2), 2008. *Source*: OECD Regional Database (2012)

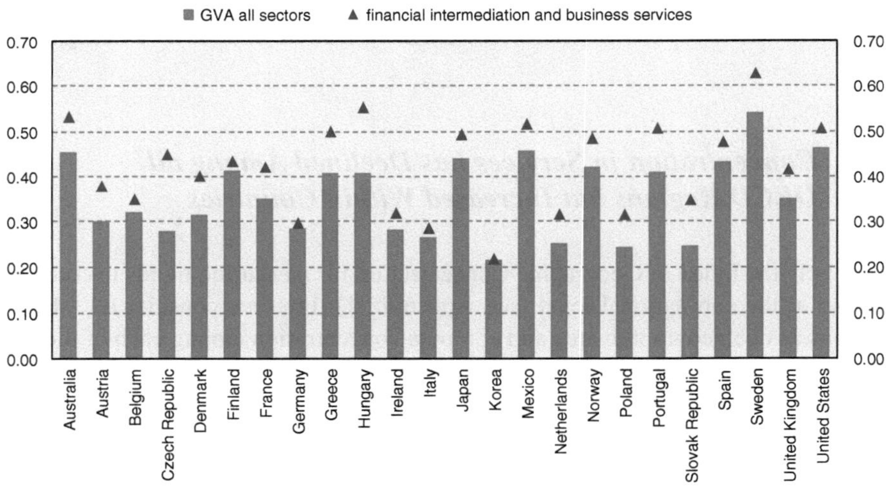

Fig. 4.12 Index of geographic concentration of GVA in financial intermediation and business services (TL2), 2008. *Source*: OECD Regional Database (2012)

in Fig. 4.11). With respect to manufacturing activities, services also tend to be more concentrated in space in all OECD countries except in Mexico, Norway and Spain (right column in Fig. 4.11).

The degree of concentration within each of the three service sectors considered varies however. Economic activities in financial and business related services reveal the highest degree of geographic concentration within countries. In all OECD countries, concentration in financial intermediation and business services exceeds concentration in all economic activities except in Korea (Fig. 4.12).

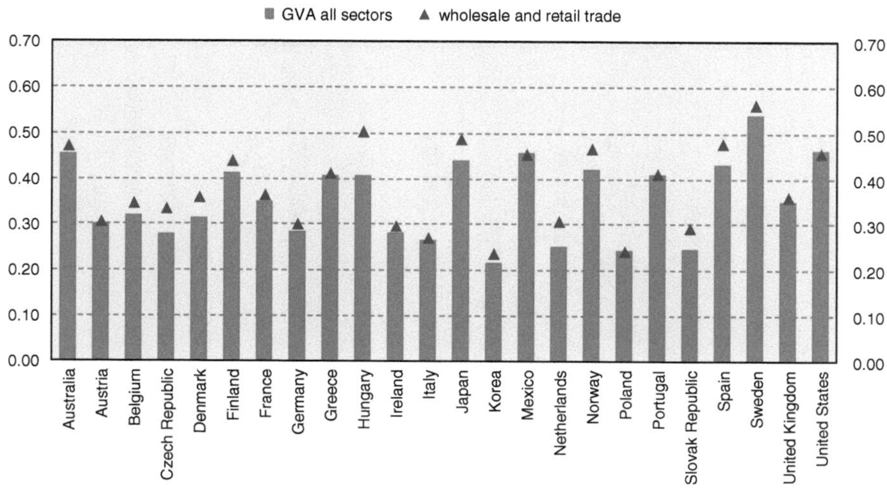

Fig. 4.13 Index of geographic concentration of GVA in wholesale and retail trade (TL2), 2008. *Source*: OECD Regional Database (2012)

Activities in wholesale and retail trade are more concentrated in space than the average of all economic activities within OECD countries with the exception of Mexico (Fig. 4.13), nevertheless wholesale and retail trade is a less geographically concentrated than is financial service with only four OECD countries (Spain, Germany, Finland and Korea) displaying a higher degree of concentration.

Public administration and social services tends to be less concentrated in space amongst the three service subsectors. In fact it is even less concentrated within countries than all economic activities with only five OECD countries (Mexico, Austria, Australia, Slovak Republic and Greece) displaying a higher degree of concentration in public administration and social services than in overall economic activity (Fig. 4.14).

The evolution of the concentration index within countries displays an increase in concentration within countries among regions in all six sectors considered and particularly in service related activities. While concentration in public administration and social services has increased in around half of OECD countries over 1995–2008 (left column in Fig. 4.15), wholesale and retail trade activities has become more concentrated in all OECD countries with the exception of Denmark, Australia, Austria, Portugal, the United States and the Netherlands (right column in Fig. 4.15), with a pronounced rise in the former eastern European countries of Hungary, Czech Republic, Slovak Republic and Slovenia.

The increase of concentration in financial and business services was even higher, increasing in all OECD countries with the exception of Ireland, Germany, United States the Netherlands and Australia (left column of Fig. 4.16). The concentration in service related activities among the three subsectors was much more pronounced than observed in manufacturing activities with an increase in concentration in only eight OECD countries (right column in Fig. 4.16).

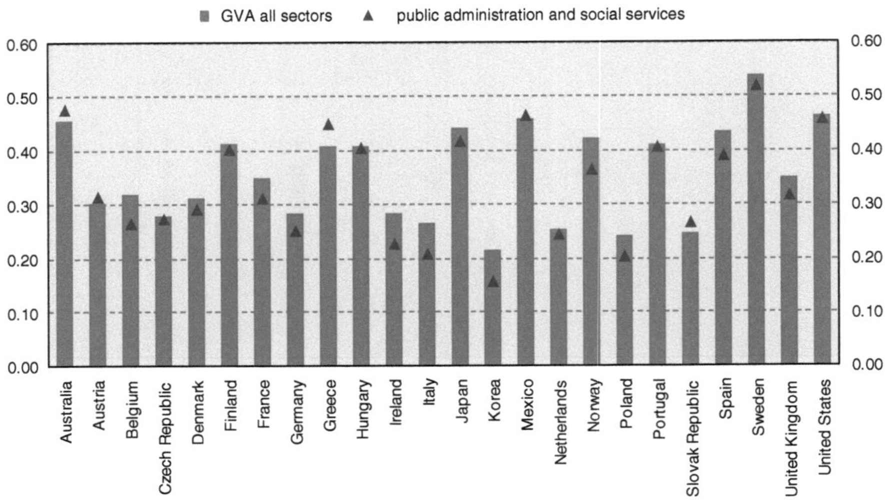

Fig. 4.14 Index of geographic concentration of GVA in public administration and social services (TL2), 2008. *Source*: OECD Regional Database (2012)

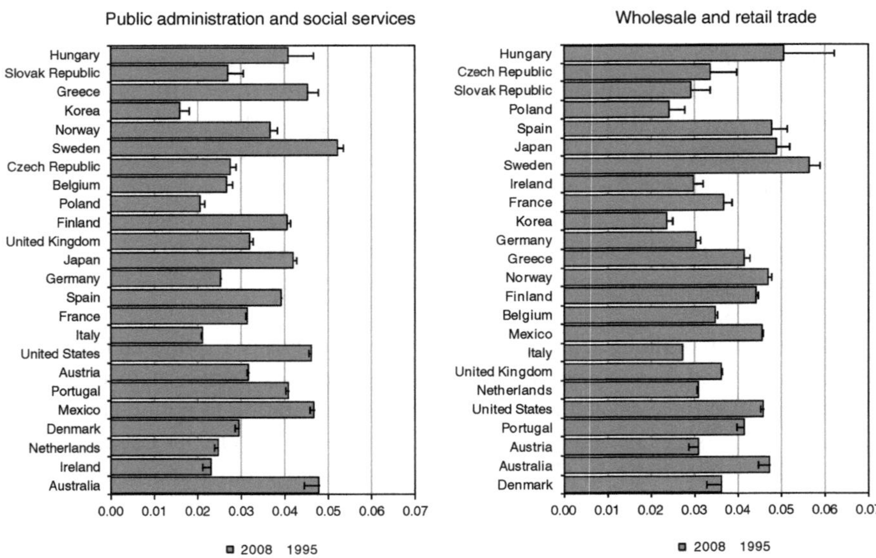

Fig. 4.15 Index of geographic concentration within countries among TL2 regions, 1995–2008. *Source*: OECD Regional Database (2012)

In sum the spatial pattern of concentration reveal a decline in overall concentration across all OECD regions and an increase in concentration within countries. This trend is fairly consistent with the pattern we observe in inequality across space with a decline in inequality among OECD countries and an increase of inequality within the majority of countries (OECD, 2009). The higher level of concentration in

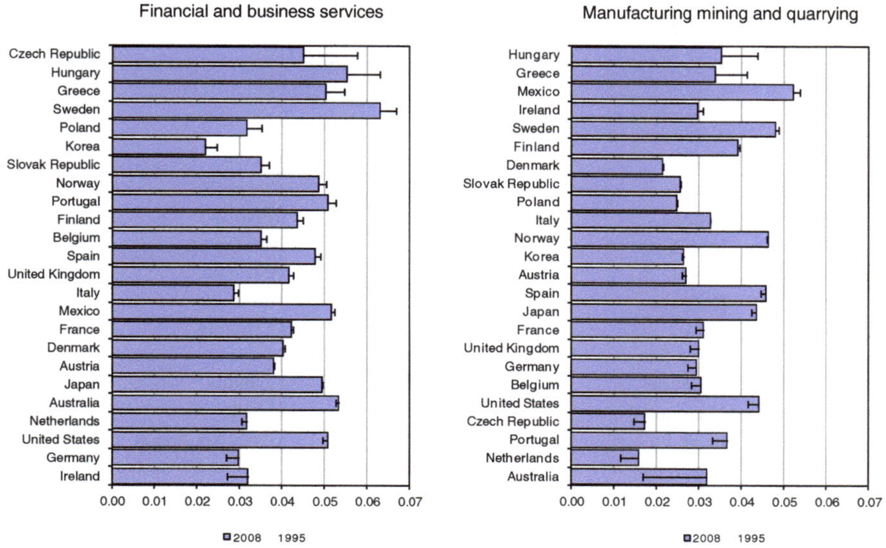

Fig. 4.16 Index of geographic concentration within countries among TL2 regions, 1995–2008. *Source*: OECD Regional Database (2012)

service related activities than overall economic activity suggests the importance of space and geography in services providing a heterogeneous picture.

2.3 Service Activities are Heterogeneous Across OECD Regions

Service sector activities, as previously discussed are not homogenous in space. At the sub-national level some regions can be quite more dependent on services and in some cases represent 90 % of their total GVA and an employ an even larger share of workers. In other regions however services can represent less than one third of total GVA and employ less than 40 % of the workforce (Table 4.4).

Table 4.4 displays the top (and bottom) 25 regions with highest (and lowest) share of service GVA to total GVA in each OECD region. The top group of regions which range from almost (97 %) entirely producing service related actives to 83 %, contain a noticeable presence of regions from within the United States (44 % of them or 11 out of 25), capital regions and or regions with international cities (Brussels, Wien, London, New York, Ill de France). This suggests that in addition to country-wide factors—driven by the high number of regions from the United States—regions with large international cities appears to be a common characteristic.

Among the group of regions with the lowest GVA share in services (right column in Table 4.4), there are many regions from Mexico, Chile and former eastern European countries including Hungary, Czech Republic and Slovak Republic, with no more than half of total GVA produced by services.

Table 4.4 Top 25 OECD TL2 regions with the highest share of GVA in services to total GVA, 2008

Top 25 regions			Bottom 25 regions		
Country	Region	Share of services to total GVA	Country	Region	Share of services to total GVA
United States	District of Columbia	97 %	Mexico	Campeche	7 %
United Kingdom	London	90 %	Mexico	Tabasco	25 %
United States	New York	89 %	Chile	Antofagasta	25 %
United States	Hawaii	89 %	Chile	Atacama	37 %
Belgium	Brussels	89 %	Chile	Bio–Bio	44 %
Norway	Oslo	88 %	Chile	O'Higgins	44 %
Australia	Australian Capital	88 %	Chile	Maule	46 %
United States	Delaware	87 %	Mexico	Coahuila	46 %
United States	Maryland	87 %	Mexico	Hidalgo	46 %
France	Ile de France	86 %	Korea	Gyeonbuk region	46 %
United States	New Jersey	86 %	Slovak Republic	Zapadne Slovensko	46 %
Greece	Attiki	86 %	Korea	Chungcheong region	47 %
United States	Florida	86 %	Korea	Gyeongnam region	47 %
United States	Virginia	85 %	Hungary	Central Transdanubia	47 %
Luxembourg	Luxembourg	85 %	Chile	Tarapacá	48 %
Spain	C.A. de Melilla	85 %	Czech Republic	Moravskoslezko	49 %
United States	Massachusetts	85 %	Australia	Westem Australia	50 %
United States	Rhode Island	85 %	Czech Republic	Stredni Morava	51 %
Spain	C.A. de Ceuta	84 %	United States	Wyoming	51 %
Portugal	R.A. de Madeira	84 %	Hungary	Westem Transdanubia	51 %
Denmark	Hovedstaden	84 %	Korea	Jeolla region	51 %
Finland	Aland	84 %	Chile	Valparaiso	51 %
ITAly	Lazio	84 %	Czech Republic	Severozapad	51 %
United States	Connecticut	83 %	Mexico	Tamaulipas	52 %
Austria	Wien	83 %	Czech Republic	Stredni Cechy	52 %

Note: Data are missing from Turkey
Source: OECD Regional Database (2012)

In terms of employment, the top 25 regions with the largest employment share in services as a proportion of total employment (left column of Table 4.5) also contain a strong presence of regions from the United States (80 % of them or 20 out of 25). There are however fewer regions containing international cities (New York) and more regions without any international cities such as Montana, New Mexico, Colorado, South Dakota and C. A. de Melilla amongst the group of regions with the highest employment share.

Amongst the regions with the lowest employment share (right column of Table 4.5) the dominant group are from Turkey, followed by Poland, Czech Republic, Mexico and Portugal.

Table 4.5 Top 25 OECD TL2 regions with the highest employment share of services to total employment, 2008

Top 25 regions			Bottom 25 regions		
Country	Region	Share of services to total emp.	Country	Region	Share of services to total emp.
United States	Delaware	94 %	Turkey	Zonguldak	33 %
United States	Rhode Island	92 %	Turkey	Samsun	35 %
United States	Nevada	92 %	Turkey	Kastamonu	35 %
United States	Maryland	91 %	Turkey	Gaziantep	35 %
United States	Hawaii	91 %	Turkey	Trabzon	35 %
United States	New York	90 %	Turkey	Manisa	38 %
Australia	Australian Capital Territory	90 %	Turkey	Erzurum	40 %
United States	Florida	90 %	Turkey	Balikesir	41 %
United States	Maine	90 %	Poland	Swietokrzyskie	42 %
Spain	C.A. de Ceuta	90 %	Turkey	Hatay	44 %
United States	Montana	90 %	Turkey	Bursa	44 %
United States	New Mexico	89 %	Turkey	Konya	44 %
United States	Vermont	89 %	Turkey	Tekirdağ	45 %
United States	New Jersey	89 %	Mexico	Chihuahua	45 %
United States	Virginia	89 %	Turkey	Kayseri	47 %
Mexico	Quintana Roo	89 %	Turkey	Kocaeli	48 %
United States	Alaska	89 %	Portugal	Centro	48 %
Spain	C.A. de Melilla	89 %	Poland	Wielkopolskie	48 %
United States	Colorado	89 %	Poland	Lubelskie	48 %
United States	North Dakota	89 %	Turkey	Malatya	49 %
United States	Arizona	89 %	Czech Republic	Severovychod	49 %
United States	Massachusetts	88 %	Poland	Podkarpackie	49 %
United States	South Dakota	88 %	Poland	Kujawsko-Pomorskie	49 %
Norway	Oslo	87 %	Mexico	Baja California Norte	50 %

Source: OECD Regional Database (2012)

2.4 A Large Number of Regions are More Specialised in Public Administration and Social Services

The analysis examining and comparing the relative shares of value-added and employment in services does not taken into account the overall industrial composition in the regions. We account for these factors through a regional specialisation index measured by the Balassa–Hoover index (see Appendix 1) among the six sectors in each of our 297 TL2 OECD regions where data are available. We first compute the index of specialisation among all TL2 regions allowing us to benchmark the degree of industry specialisation across all regions. This indicator permits us to capture structural changes common to all regions in the OECD. We then compute the specialisation of regions within each OECD country with the aim of capturing trends of specialisation controlling for national factors.

Figures 4.17 and 4.18 display the degree of specialisation among all OECD TL2 regions from Europe and from North-America in financial intermediation and business services, and in wholesale and retail trade respectively. The figures reveal the strong heterogeneity in specialisation among regions despite not accounting for national factors such as the industry composition of countries.

Figure 4.17 also reveals an interesting pattern where specialisation in financial intermediation and business services appears to be higher in capital regions or regions with large cities such as Stockholm, Madrid, Rome and Berlin and Prague.

Figure 4.18 also displays a strong heterogeneity present in regional specialisation in wholesale and retail trade among North American regions with a higher degree of specialisation in regions from the western coast and lower specialisation in regions from the eastern coast.

Table 4.6 compares levels and trends in regional specialisation over 1995–2008 using both benchmarks—the first with respect to all regions and the second with respect to regions within countries. In the first benchmark the average (un-weighted) specialisation among all OECD regions in 1995 was highest in agriculture followed by public administration and social services and construction. Over time the largest gains in specialisation have occurred in service activities, both in financial and business services and in wholesale and retail trade at the backdrop of the largest declines occurring in agriculture, public administration, social services and manufacturing. In 2008 the three most specialised sectors belong to services.

Within countries the relative changes have been smaller. Notwithstanding this fact, financial and business services experienced the largest relative shifts over 1995–2008 followed by manufacturing and public administration and social services on average. In contrast the largest declines have occurred in wholesale and retail trade, agriculture and construction.

Despite the larger relative increase by financial and business services, there are more regions becoming more specialised in the remaining service sub-sectors considered. Appendix 2 displays the number of regions becoming more specialised over 1995–2008 in the three service sub-sectors, and Fig. 4.19 depicts the percent of regions experiencing an increase in specialisation in public administration and social services and in financial and business services. The trend reveals that in most OECD countries the majority of regions have become *more* specialised in public administration and social services contrary to the trend in financial and business services where in all but seven OECD countries the majority of regions have become *less* specialised in financial and business services. In wholesale and retail trade (Appendix 2) the proportion is quite balanced.

These trends reinforce the previous patterns highlighting the presence of two simultaneous factors: globalisation and localisation effects influencing the spatial distribution of service sector activities. While the relative increase in specialisation within countries on average has been larger in financial services amongst the six sectors, the rise has been driven by fewer regions when comparing to the other subsectors. In fact the region with the highest specialisation in financial and business services tends to be more specialised than the most specialised region in wholesale and retail trade and in public administration (see maximum value in Table 4.10 from Appendix 2).

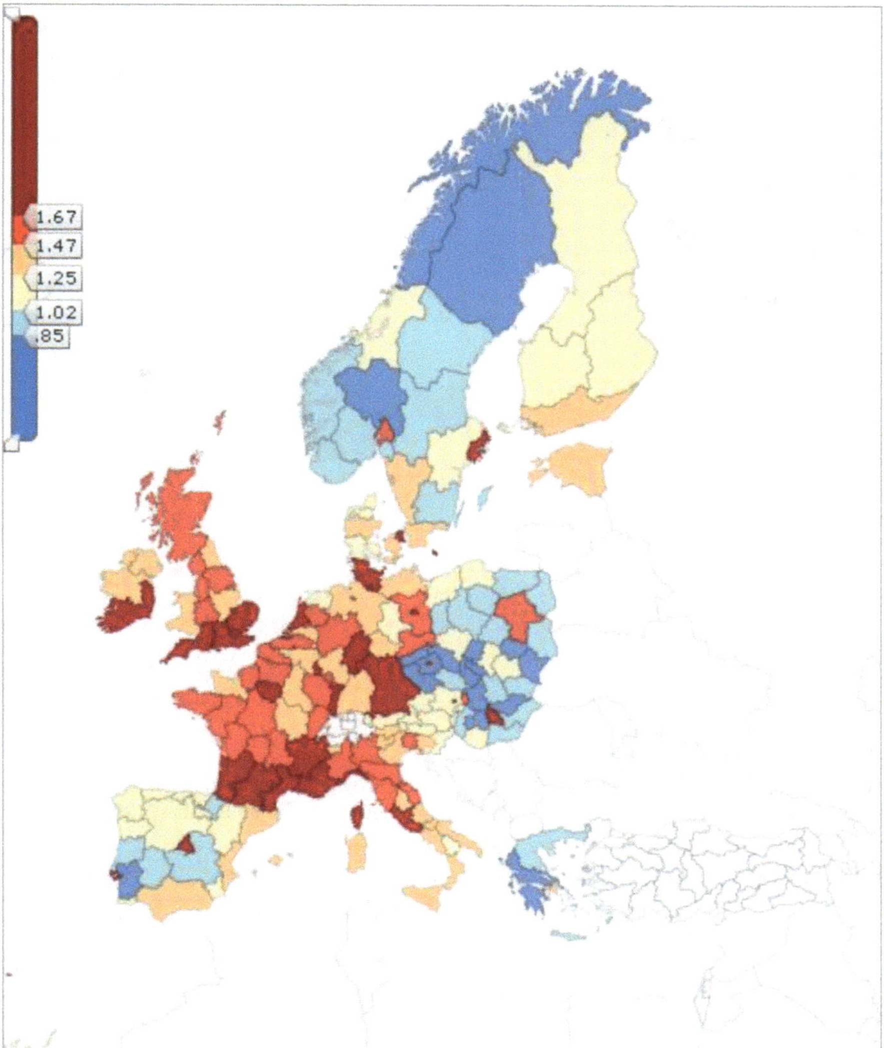

Fig. 4.17 Regional specialisation in Europe in financial intermediation and business services, TL2 2008. *Note*: GVA expressed in PPP 2005 base year. *Source*: OECD Regional Database (2012)

3 Service Sector Productivity and Key Regional Factors Associated with Productivity

The gradual structural shift towards services sectors activities in OECD represents a shift toward non-tradable activities with implications for productivity growth as depicted in section with service sector productivity growth trailing productivity growth in manufacturing.

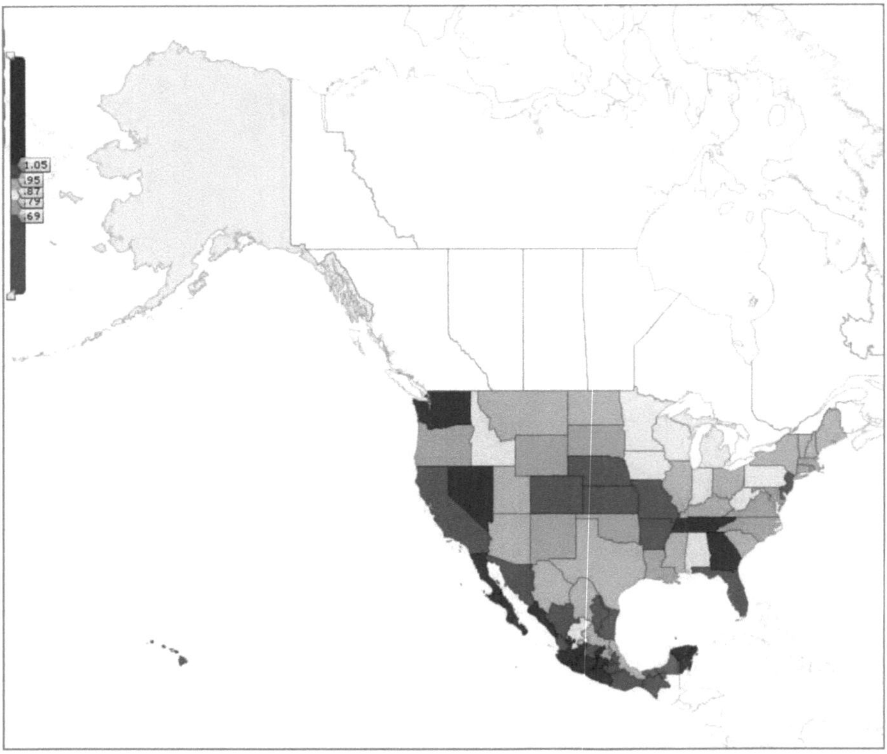

Fig. 4.18 Regional specialisation in North America in wholesale and retail trade, TL2 2008. *Note*: GVA expressed in PPP 2005 base year. *Source*: OECD Regional Database (2012)

Table 4.6 Trends in specialisation index among TL2 regions, 1995–2008

	Specialisation among all OECD regions			Specialisation among regions within countries		
Sector	1995	2008	1995–2008	1995	2008	1995–2008
Public administration social services	2.428	1.602	−0.825	1.072	1.076	0.001
Construction	1.852	0.950	−0.907	1.078	1.073	−0.007
Agriculture	5.070	0.967	−4.137	1.419	1.421	−0.019
Financial interm. & business services	0.906	1.273	0.376	0.848	0.889	0.043
Manufacturing mining and quarrying	1.235	0.666	−0.574	0.983	1.017	0.036
Wholesale and retail trade	0.541	0.895	0.358	1.004	0.974	−0.027

Note: Average values represent the un-weighted value by regions
Source: OECD Regional Database (2012)

At the regional level our proxy for labour productivity in each of the six sectors is derived by ratio of value added (expressed in constant PPP terms) to total employment for each sector. This allows us to compute labour productivity (levels and growth) for each TL2 region among the six sectors.

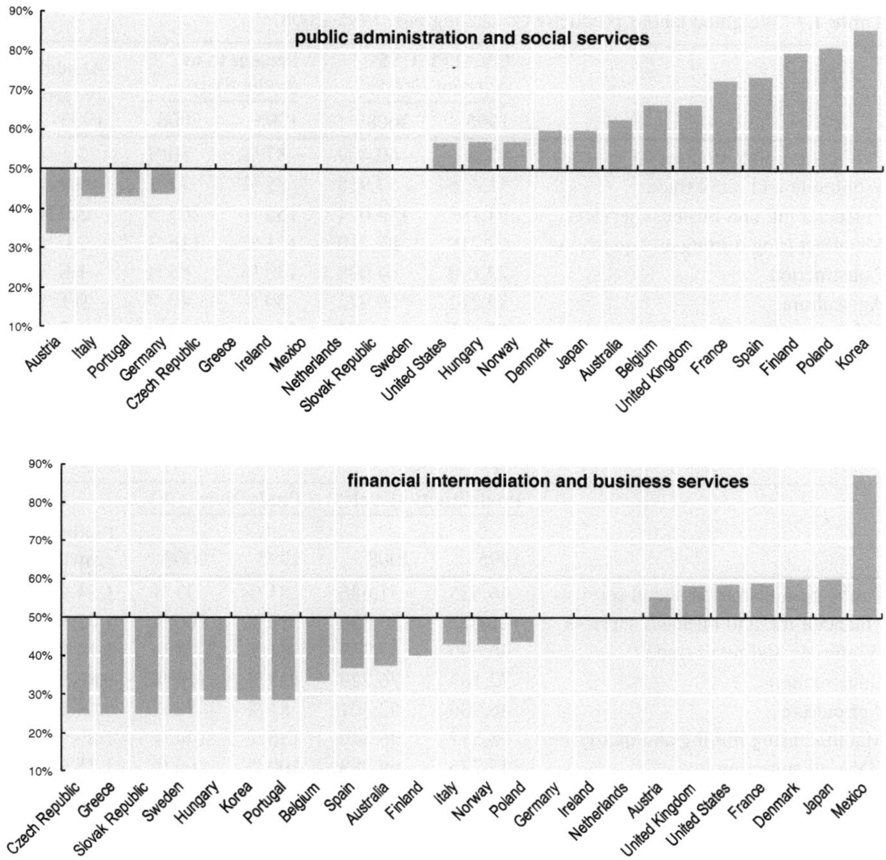

Fig. 4.19 Percent of regions increasing in specialisation index, TL2 1995–2008. *Source*: OECD Regional Database (2012)

We first compute the weighted average labour productivity among OECD regions in 2008 in Table 4.7. The weighed figures which account for the size of each region display a higher productivity in all three service sub sectors than in agriculture. However, only the financial intermediation and business service sector productivity appears to be higher—around twice—than in industry and construction. Wholesale and retail trade section has the lowest productivity within service sub-sectors, around 13 % below the average productivity.

Over the period 1995–2008 the productivity gap has widened within services driven by a strong increase in productivity in financial and business services against the lower increase in wholesale and retail trade. Consequently productivity in financial and business services has increased relative to all sectors, from being 50 % above the average in 1995 to around 101 % above in 2008, or differently said

Table 4.7 Weighted labour productivity TL2 regions, 1995–2008

	GVAPPP US$ constant (2005)		Percent to av. productivity		Annual av. growth
	1995	2008	1995	2008	1995–2008
Public administration social services	52,566	60,409	87 %	80 %	1.1 %
Wholesale and retail trade	51,426	57,925	85 %	77 %	0.9 %
Financial int. and business services	94,189	154,024	155 %	205 %	3.9 %
Manufacturing mining and quarrying	67,219	87,210	111 %	116 %	2.0 %
Construction	79,190	64,098	130 %	85 %	−1.6 %
Agriculture	30,212	29,972	50 %	40 %	−0.1 %
Services	61,485	76,784	101 %	102 %	1.7 %
All sectors	60,741	75,108	100 %	100 %	1.6 %

Source: OECD Regional Database (2012)

Table 4.8 Unweighted labour productivity TL2 regions, 1995–2008

	Productivity (levels)		Productivity (%)		
	1995	2008	1995	2008	Productivity growth
Public administration social services	46,625	51,686	81 %	77 %	1.24 %
Financial int. and business services	121,072	133,820	210 %	200 %	1.05 %
Wholesale and retail trade	46,690	52,238	81 %	78 %	1.14 %
Construction	72,183	76,229	125 %	114 %	0.88 %
Agriculture	46,906	52,301	81 %	78 %	0.64 %
Manufacturing mining and quarrying	78,577	96,340	136 %	144 %	1.89 %
GVA all summing	57,776	66,783	100 %	100 %	1.55 %

Source: OECD Regional Database (2012)

twice as high in 2008. In contrast productivity of wholesale and retail trade has decreased, from 11 % below the average to 26 % below in 2008.

The unweighted average labour productivity provided in Table 4.8 measures productivity on average in all OECD regions treating each region as a single observation. These calculations confirm the higher productivity in financial and business services, around twice as high as average productivity. In contrast productivity in the public administration social services and in wholesale and retail trade appears to be the lowest according to these calculations, with level around 28 % below the average productivity.

In terms of growth, manufacturing on average displays the highest growth followed by the three service sub-sectors, amongst which financial and business services records the lowest growth on average. The lower growth recorded by financial and business services in the un-weighed calculations (as opposed to the weighted) suggests the presence of few regions with an important weight and high productivity growth and a substantial number of regions with low productivity growth.

This is confirmed in Table 4.9 which displays the distribution of growth rates among regions in the six sectors. Financial intermediation indeed contains around 20 % of regions with annual growth rates exceeding 3 %, but also around 30 % of regions with a negative productivity growth over the period 1995–2008. These proportions are significantly higher in both ends of the tail distribution when comparing to the other two service sub-sectors. The presence of 30 % of regions with negative productivity growth pulls down the unweighted average.

A similar distribution is also present in agriculture and in manufacturing with a significant proportion of regions at the lower and at the upper end of the growth distribution. In contrast the distribution is much more balanced in public administration and social services and in wholesale and retail trade.

While some service related activities do depend on international markets (e.g. international consultancies or logistic clusters among others), the majority of service-related activities depend strongly on local conditions and the local market to which they serve. This suggests that performance in service productivity will likely be heterogeneous.

Figure 4.20 confirms an important degree of heterogeneity in service sector productivity among OECD regions over the period 1995–2008, however it appears that regions have entered into a process of convergence in which regions with lower initial (1995) levels of productivity display higher annual average productivity growth rates all things equal.

Figures 4.33, 4.34, 4.35 in Appendix 3 depict the performance of service productivity within each of the three service sub-sectors and display a similar pattern of convergence for all three of them. This pattern highlights the existence of growth potential for service productivity in OECD lagging regions especially for public administration and social services and for wholesale and retail trade, experiencing stronger forces of convergence given that they depend more on regional and local labour markets as opposed to financial and business services. This growth potential is also present in the latter sector.

3.1 Regional Productivity is Highly Dependent on Service Sector Productivity

With the structural changes described in section one during the past decade, most OECD countries have focused their production structures in service sector activities, which nowadays contribute to around 70 % of total output in OECD countries and employ around the same proportion of workers. Given that the distribution of service activities is quite heterogeneous in space, understanding productivity at the regional level is critical and how it relates to total productivity in regions and to overall productivity.

Table 4.9 Growth distribution among OECD TL2 regions, 1995–2008

	Gr. below 0	0–0.5 %	0.5–1 %	1–1.5 %	1.5–2 %	2–2.5 %	2.5–3 %	Gr. above 3 %	Sum
Public administration social services	14 %	15 %	22 %	15 %	9 %	9 %	6 %	10 %	100 %
Financial int. and business services	28 %	9 %	12 %	9 %	11 %	8 %	5 %	19 %	100 %
Wholesale and retail trade	19 %	14 %	19 %	13 %	14 %	6 %	5 %	10 %	100 %
Construction	34 %	9 %	14 %	10 %	7 %	5 %	6 %	15 %	100 %
Agriculture	48 %	5 %	3 %	5 %	4 %	7 %	5 %	24 %	100 %
Manufacturing mining and quarrying	18 %	8 %	8 %	13 %	10 %	9 %	7 %	26 %	100 %
GVA all summing	9 %	8 %	20 %	20 %	14 %	9 %	6 %	13 %	100 %

Source: OECD Regional Database (2012)

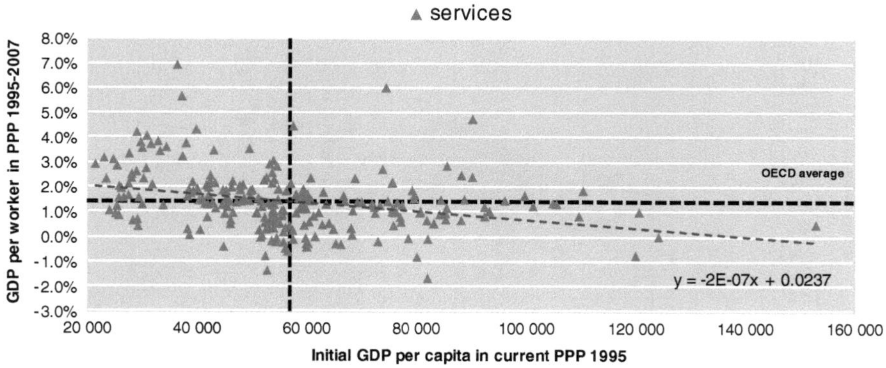

Fig. 4.20 Initial level and annual average growth rates of service sector productivity OECD TL2 regions, 1995–2008

Productivity growth in services sub-sectors is highly correlated with the overall performance in OECD regions. For each of the three service sub-sectors we explore the links between productivity growth in each of the three subsectors with overall regional productivity growth and with regional GVA growth in Fig. 4.21, 4.22 and 4.23. The trends reveal a linear relationship between high productivity growth in service sub-sectors and regional performance, which is not surprising given the significant share of the service-related activities in the overall economy. The links with respect to regional productivity growth appear to be higher in financial and business services and in wholesale in retail trade (top row of Figs. 4.21 and 4.23).

The link between productivity growth in service related activities and overall regional output is also present and positive, albeit the correlation—estimated by the slope of the fitted line—is a bit weaker as with regional labour productivity.

In sum this suggests that higher productivity growth in service related activities will contributed to higher overall output at the regional level highlighting the importance of services in the OECD regions and countries.

3.2 Aggregate Service Sector Activities Depend on Few Regions as Well as on the Performance of Many

Analysis on regional contributions to growth can improve our understanding on how the regional dimension maps into the aggregate picture. For the case of services this is particularly important given that the bulk (70 %) of economic activity in OECD countries nowadays depend on service related activities.

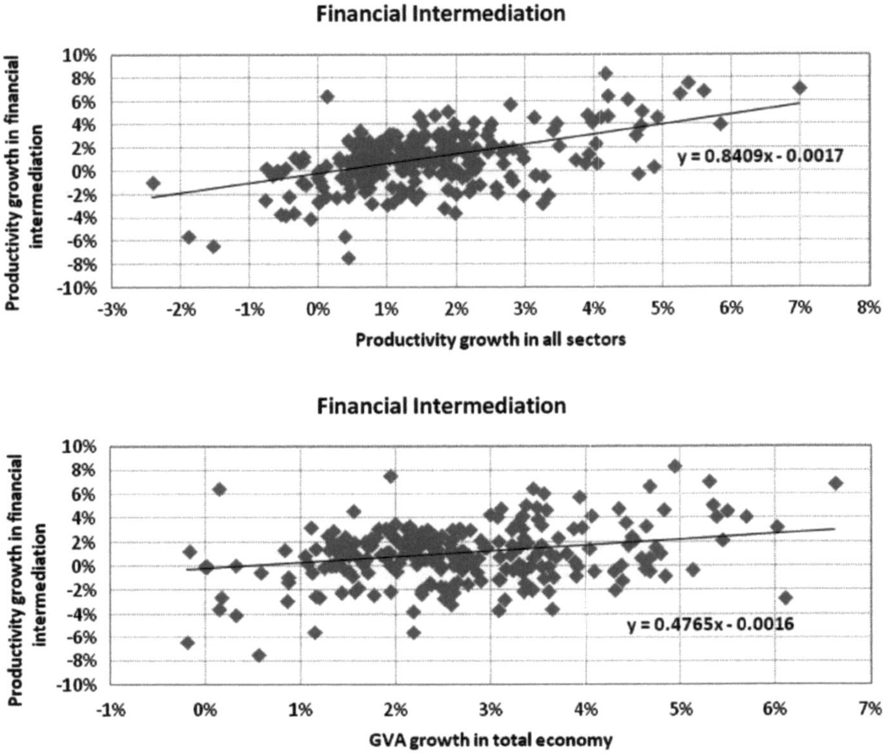

Fig. 4.21 Annual average productivity growth and GVA growth in financial and business services, 1995–2008. *Source*: OECD Regional Database (2012)

Therefore understanding where the bulk of this activity is produced is critical in the policy design whether aggregate services are mainly driven by few large regions with international cities, by few intermediate regions with second tier cities or by many regions. In order to address these questions there is a need to understand where aggregate growth occurs. The distribution of contributions to growth can help in this task.

Contributions to growth depend on two factors: the growth effect capturing the performance of regions, and the size effect capturing the size of regions. Multiplying both factors yields the contributions to aggregate growth. Therefore large regions with high growth rates will have the largest impact to aggregate growth, while small regions with low rates will have the lowest impact. In the middle, the impact of a large region displaying very low growth rates could be equivalent to the one of a small region displaying very fast growth rates. The net effect will depend on balance between the two elements.

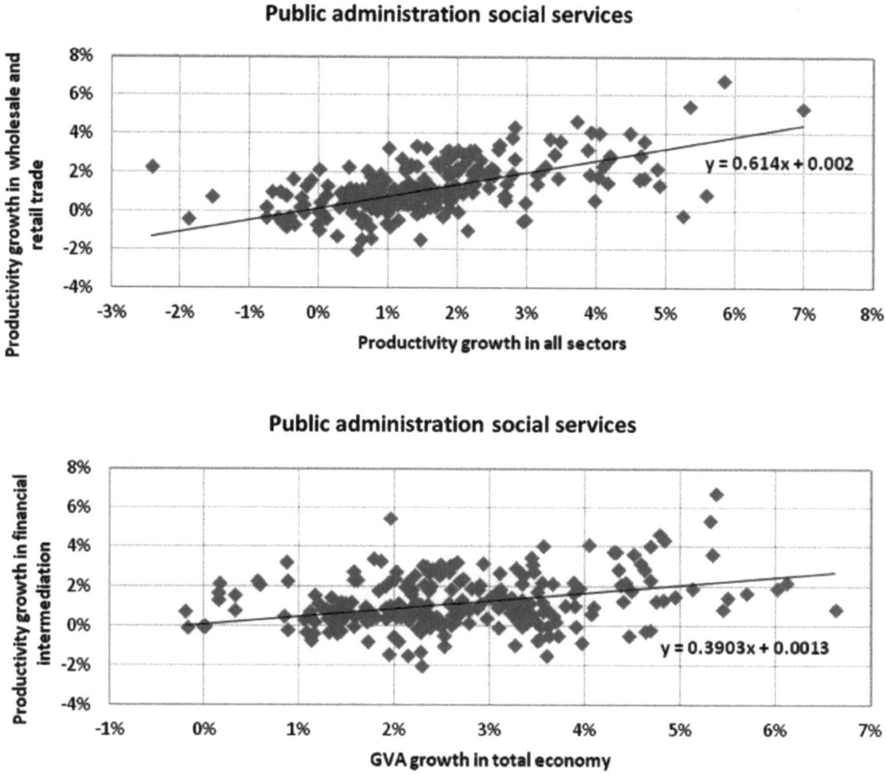

Fig. 4.22 Annual average productivity growth and GVA growth public administration and social services, 1995–2008. *Source*: OECD Regional Database (2012)

Mapping the regional contributions to growth in services reveal a power law type of relationship displaying that on the one hand few regions with a disproportional contribution to overall service sector growth—the big hub regions—and on the other hand the majority of regions do not contribute to the aggregate growth significantly individually but their compounded effect is quite large, even larger than that of the big hub regions. According to our calculations around four percent of regions (listed in Fig. 4.24) contribute a bit less than one third (30 %) of the aggregate growth in services while the rest of regions (96 % of them) contribute to 70 % of aggregate growth.

The shape of this distribution is quite intuitive confirming the importance of few regions with international cities such as California, Tokyo, New York, Paris, London have on overall service sector growth. Despite this significant effect, the shape also reveals the importance of the many regions each servicing their respective local economies to aggregate service sector growth.

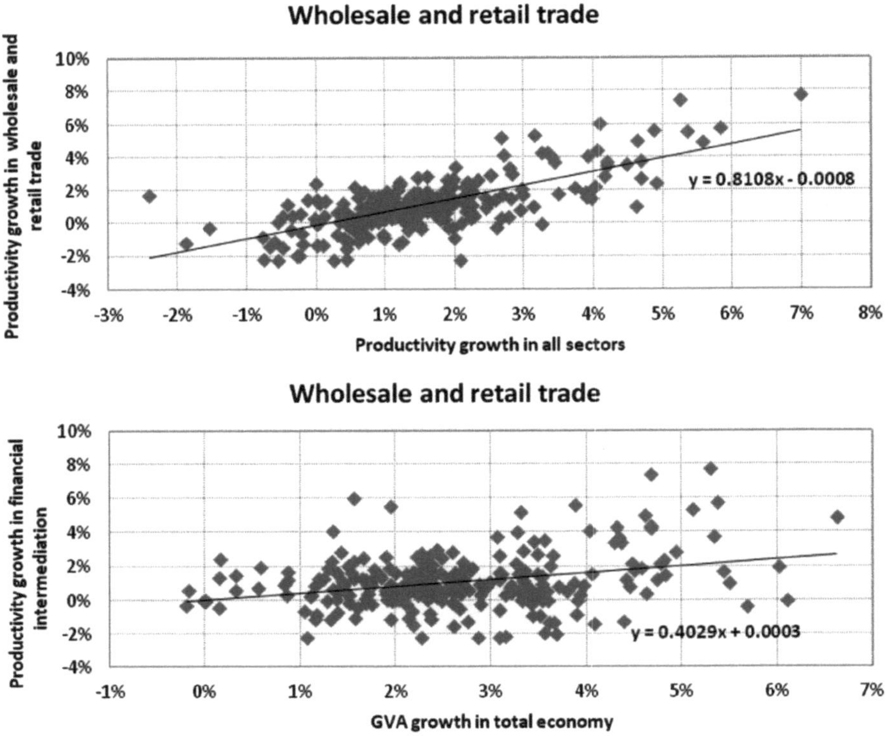

Fig. 4.23 Annual average productivity growth and GVA growth in wholesale and retail trade, 1995–2008. *Source*: OECD Regional Database (2012)

This graph carries important policy implications linking the performance of aggregate growth in services to the many local labour markets across regions. Therefore from a policy design point of view it is important to design policies that reap the benefits of the growth potential of the many regions, particularly lagging regions experiencing a process of convergence, rather than just a few. For this task it is important to understand the main drivers of growth at the regional level and in particular the key factors driving service sector productivity.

3.3 Human Capital, Density and Innovation are Critical Factors for Service Productivity Among Regions

Service-related activates have a strong non-tradable component and therefore by definition depend both on global and on local forces to which they service, or in

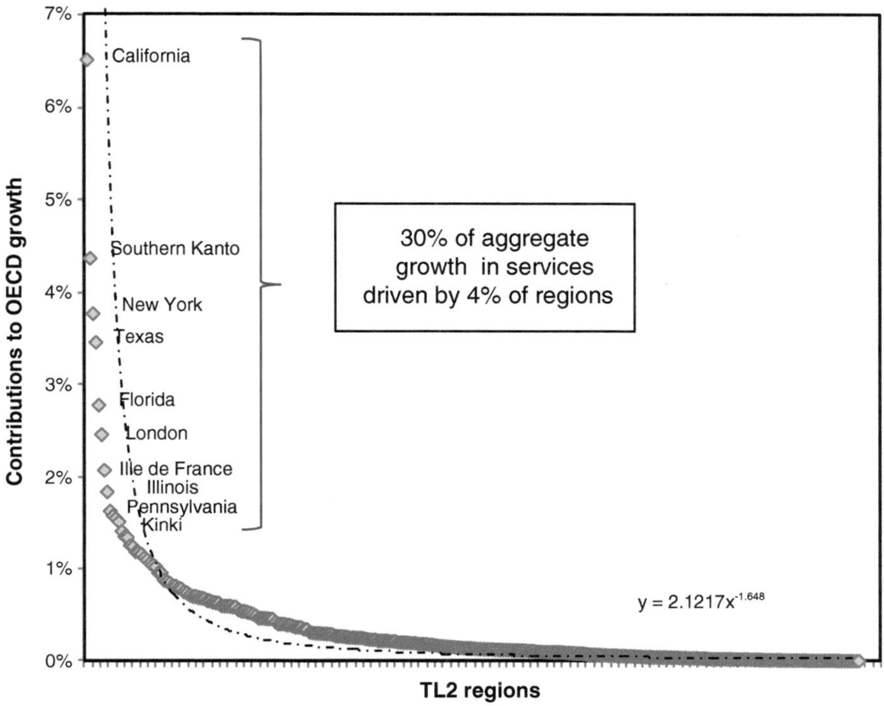

Fig. 4.24 Contributions by TL2 regions to service growth, 1995–2008. *Source*: OECD Regional Database (2012)

other words on general economic conditions (e.g. in expansionary periods there will be more demand for services and in recessionary less demand) and on region-specific and local factors driving the heterogonous picture across space depicted previously. We focus next on critical regional factors driving the heterogeneous picture of service-related activities.

One of the central findings of previous OECD work on regional growth is that there is potential growth in all types of regions. A second key finding is that most of the important determinants of growth are things that policy can address in particular human capital, infrastructure and innovation stand out (Box 1), although these should not be seen as isolated factors, since the analysis in OECD (2009) points to important interactions among them. Therefore it is to no surprise that the functioning of local labour markets, policies and institutional factors are also critical elements (OECD, 2012) by facilitating (or impeding) these interactions which can lead to important synergies. We focus attention to three elements for service-related activities at the regional level: human capital, density and innovation intensity.

Box 1. Drivers of Regional Growth

OECD analyses of regional performance identify a number of critical factors as key drivers of regional growth, including infrastructure, human capital, innovation, economies of agglomeration and accessibility to markets. More importantly, these factors complement each other, highlighting the benefits of an integrated approach.

The first critical factor is *human capital*. The presence of highly skilled workers in a region's workforce and the absence of low-skilled workers can both have a positive influence on regional growth. The effects of human capital also appear to persist for around a 5-year time span. Strikingly, although policy-makers tend to focus attention on the high end of the education/skills spectrum, recent OECD analyses point to the particular importance of the second human-capital variable—the absence of large populations of workers with very low skills.

The second factor is infrastructure. Improving infrastructure will not automatically lead to higher regional growth rates; investment in infrastructure needs to be combined with improvements in education and innovation. One possible explanation for this is that investments in public infrastructure do not stimulate growth in the absence of workers with the requisite levels of education and innovation activity. This suggests that it could be productive to co-ordinate policies for building human capital, enhancing innovation and providing infrastructure. The effects of infrastructure appear to last around 3–5 years.

The third critical element driving growth is innovation, as measured by focusing on the science and technology component of innovation, for which data are available. Innovation appears to produce positive effects over a longer time span, of approximately 10 years.

Economies of agglomeration also have a positive impact on growth, although they will not by themselves generate growth, and do not constitute a sufficient condition for sustaining high growth rates.

Finally, accessibility to markets has a positive effect on regional growth, although this effect is not very robust among the different model specifications.

Source: OECD (2009), *How Regions Grow: Trends and Analysis*, Paris, OECD.

3.4 Human Capital and Service Sector Productivity

Human capital appears to be the most robust factors driving regional growth (OECD, 2009), both the presence of high-skilled labour and the absence of low-skilled labour. Recent regression analysis however underscores the importance of

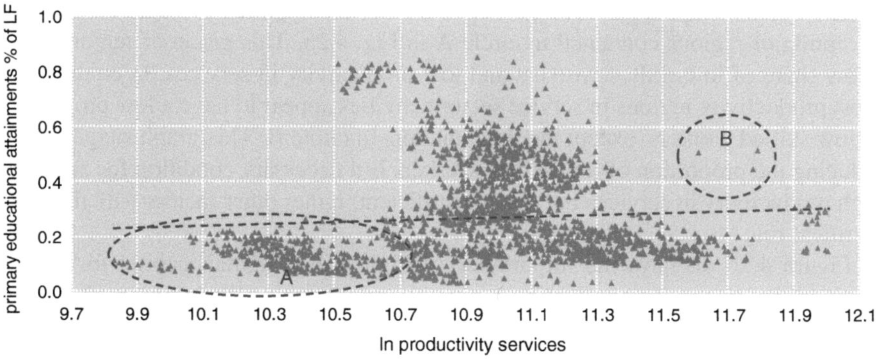

Fig. 4.25 Productivity in services and primary attainments in TL2 regions, 1995–2008. *Source*: OECD Regional Database (2012)

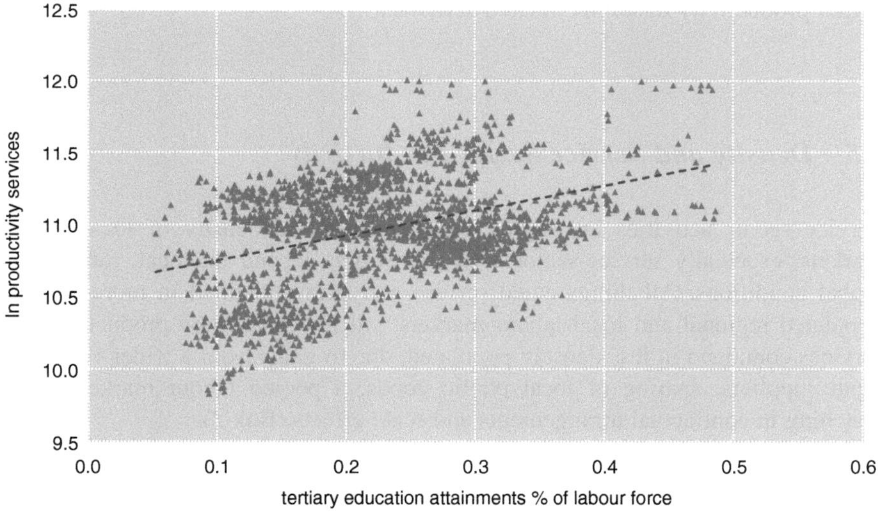

Fig. 4.26 Productivity in services and tertiary attainments in TL2 regions, 1995–2008. *Source*: OECD Regional Database (2012)

the "drag" effect of low human capital acting as an important bottleneck for development in all types of regions (OECD, 2012).

Figures 4.25 and 4.26 examine the relationship between service sector labour productivity and human capital among OECD regions considering both the presence of high skilled workers in the workforce and the presence of low-skilled workers.

Low-skilled human capital is an important factor influencing service sector productivity as depicted in Fig. 4.25, in particular the absence of low-skilled workers appears to be a necessary but not a sufficient condition for high service sector productivity. Low skilled workers are indexed by the proportion of the workforce with only primary attainment rates with higher values representing a larger proportion of low skilled workers in the labour force.

The relationship appears to be negative for the majority of regions with the exception of regions contained in circle A in Fig. 4.25. This group of regions have a low share of low skilled workers and low productivity in services. In contrast the most productivity regions in service sector activities appear to have a low proportion of low skilled workers, except the three outliers in circle B. This graph suggests that reducing the proportion of low skilled workers is a necessary condition for attaining high productivity in services but it is not sufficient; rather other factors will also play a critical role.

Figure 4.26 displays the importance of high-skilled human capital for labour productivity in services indicating a positive relationship. We proxy high skilled workers by the indicator measuring the proportion of working age population with tertiary education attainments. Higher values capture a higher proportion of high-skilled workers and lower values the contrary. Although these graphs just capture correlations, the fact that human capital is a key input for productivity regions improving their share of high-skilled workers in the workforce will likely attain higher productivity in service-related activities.

3.5 Density and Service Sector Productivity

Service sector activities are highly dependent on the density of local labour markets, especially service-sector activities which depend on local, rather than global conditions. All things equal service activities contained in more densely populated regional and local labour markets will likely be more productive than services contained in less densely populated, due to gains from a wider variety of input suppliers, sharing of local public goods, a pooled labour market, better matching in contractual arrangements and scale effects (Box 2).

Box 2. Service Productivity Gains Due to Density of Labour Markets

Economic activity is not naturally dispersed; rather it tends to concentrate in some geographic spaces as opposed to others, mainly due to the benefits associated with economies of agglomeration. People want to live where firms—and therefore job opportunities—are concentrated, and firms want to locate where demand—and therefore population—is large. Service sector activities are particularly sensitive to the location of firms and of consumers and therefore will also benefit from the main mechanisms producing agglomeration economies and in particular from

1. Mechanisms that deal with sharing of:

- Indivisible facilities such as local public goods or facilities that serve several individuals or firms. Some examples, other than public goods, are facilities

(continued)

Box 2. (continued)

such as laboratories, universities and other large goods that do not belong to a particular agent but where some exclusion is implicit in providing them.

- The gains from the wider variety of input suppliers that can be sustained by a larger final goods industry. In other words, the presence of increasing returns to scale along with forward and backward linkages allow firms and service providers to purchase intermediate inputs at lower costs. In addition service providers can purchase a wider variety of inputs and diversify their activities.
- Risks. This refers to Marshall's idea that an industry gains from having a constant market for skills; in Krugman's words, a pooled labour market. If there are market shocks, firms and service providers can adjust to changes in demand if they have access to a deep and broad labour market that allows them to expand or contract their demand for labour.

2. Matching mechanisms by which:

- Delays are alleviated. There is a possibility that contractual problems arising from renegotiation among buyers and suppliers result in one of the parties losing out by being held up by the other party in a renegotiation. This discourages investment. However, if the agglomeration is extensive enough, agents can find an alternative partner.

3. And particularly from scale effects:

Service providers that can enjoy a large and diversified pool of clients on a consistent basis can take advantage of scale effects. These will naturally be present where firms and consumers are located or in economies of agglomeration

Indeed agglomeration effects appear to be positively correlated to the productivity in services as depicted in Fig. 4.27. We proxy the agglomeration effects by the indicator of GDP density expressed in logarithmic terms for each OECD TL2 region. The positive relationship appears to be non-linear with a stronger positive relationship after a threshold value in GDP density.

The relationship between density and productivity among the three service subsectors (Fig. 4.28) appears to be stronger for public administration and social services and for wholesale and retail trade than for financial and business services. This result can be driven by the higher dependence of financial and business services to markets outside the local labour markets.

We next proceed with a dynamic analysis, examining whether the highest growth in service GVA and the most productive growth is occurring in regions that are more densely populated over time. For this task we employ a technique proposed by Cuberes (2010) which follows a three step computational approach:

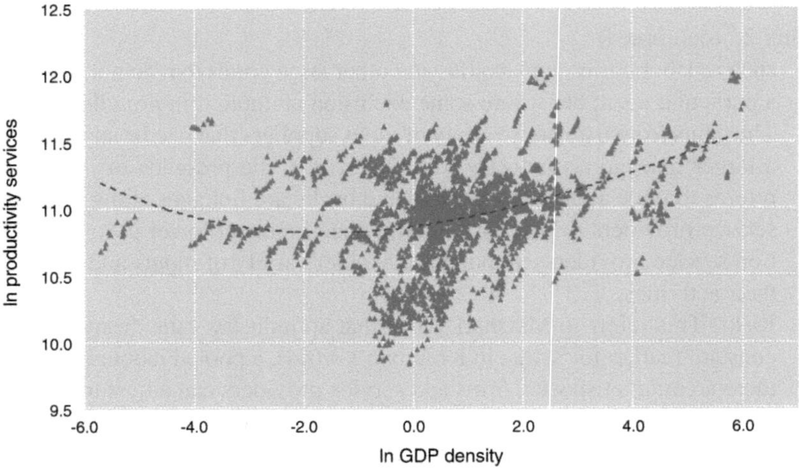

Fig. 4.27 Productivity in services and GDP density in TL2 regions, 1995–2008. *Source*: OECD Regional Database (2012)

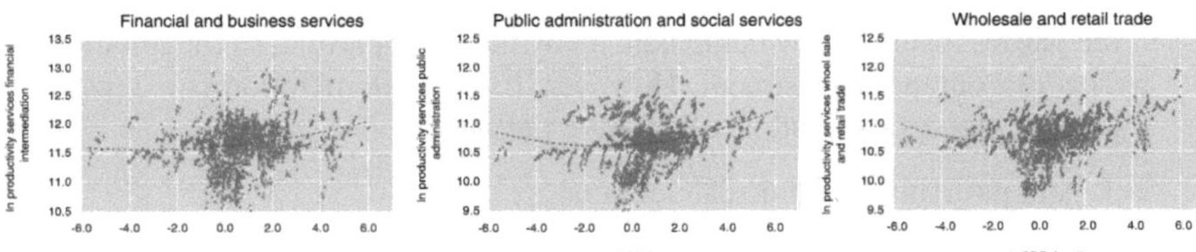

Fig. 4.28 Productivity in service subsectors and GDP density in TL2 regions, 1995–2008. *Source*: OECD Regional Database (2012)

- We first rank regions by population density from high to low. Rank 1 corresponds to the region with the highest population density and rank n corresponds to the region with the smallest.
- Second, for each year (over the 12 year period) we compute the 75th percentile of regions' growth rates (in service GVA growth and in service sector productivity growth) and consider regions whose growth rate is larger or equal to the this threshold capturing thus the fastest growing quartile of regions.
- In the final step we compute the average rank of the fastest growing regions (from step 2) for each the 11 years.

Therefore if the average rank value *decreases* over time it would suggest that densely populated regions are growing faster—than the rest—over time. Vice versa if the average rank *increases* it suggest faster growth is gradually occurring in less densely populated regions.

We first carry this analysis for GVA growth over the 12 year period, permitting us to measure whether the most dynamic regions in service GVA growth are becoming more densely populated or not. We then replicate the analyses with productivity

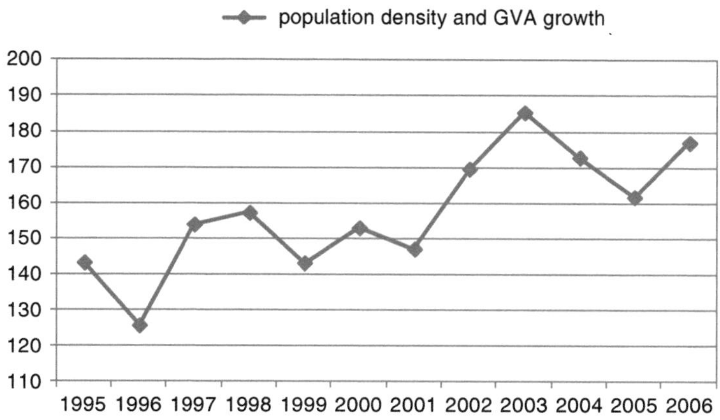

Fig. 4.29 Population density in the 25 % of fastest growing TL2 regions in services GVA 1995–2007. *Source*: Calculations based on OECD Regional Database (2010)

growth over a shorter period due to data availability allowing us to determine whether the most productive regions are indeed becoming more densely populated or not.

We find that over roughly three business cycles there is a clear upward trend in the Y-axis (measuring the average rank of the fastest 25 % growing regions) among the first combinations we consider—GVA growth in services and population density. An upward trends suggests that the fastest growing regions in service GVA are gradually become *less* densely populated. The effect is connected to the fat tail in the power law distribution and the convergence forces in lagging regions suggesting that the localization forces in services are spreading out to the tail of the distribution in regions with a lower population density (Fig. 4.29).

Examining the evolution of productivity growth and its relation to population density over a shorter period, contrary to the previous upward trend we find that over the 5 year period 2001–2005 there is a downward trend in the Y-axis in the combination between labour productivity growth in services and population density. A downward trends suggests the regions with the highest productivity growth in services are indeed become *more* densely populated over time. The 1 year increase from 2000–2001 may very well driven by the shock of the 2000 recession unfolding quicker in more densely populated regions (Fig. 4.30).

3.6 Innovation and Service Productivity

Finally the relationship between innovation intensity in regions and service sector productivity appears to be positively correlated (Fig. 4.31). Our indicator capturing innovation intensity is patent applications per thousand inhabitants therefore capturing the science and technological innovation component.

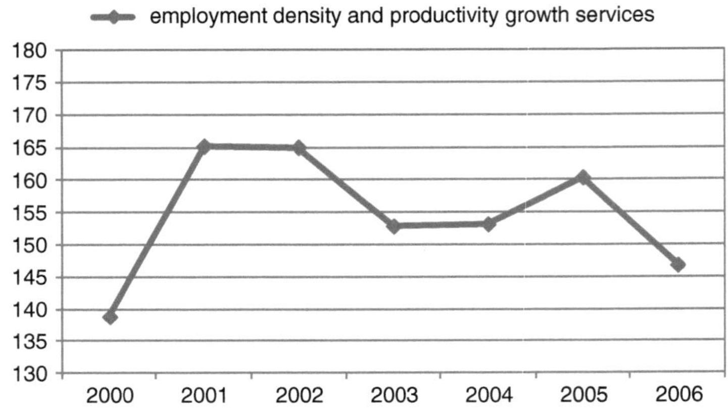

Fig. 4.30 Employment density in the 25 % of fastest growing TL2 regions in service sector productivity 1995–2007. *Note*: The sequential growth analysis necessitates to have the same number of observations in each year for meaningful comparisons. Due to missing employment data for regions from the United States, Finland and Mexico, we restrict the analysis to the period 2000–2008. *Source*: Calculations based on OECD Regional Database (2010)

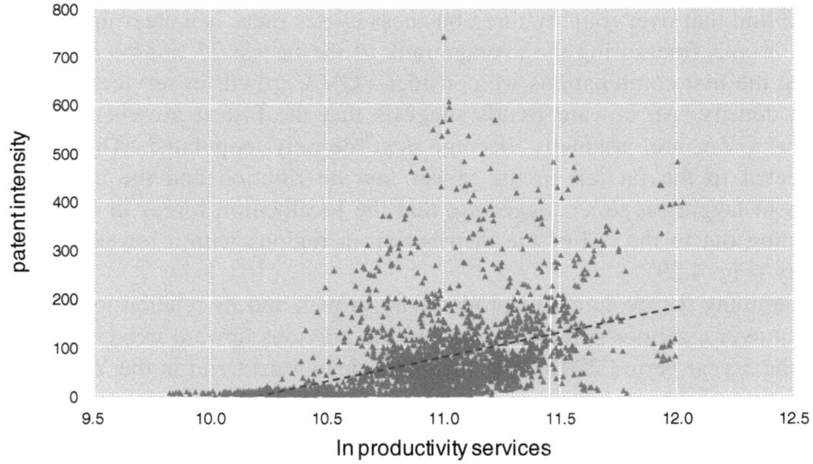

Fig. 4.31 Productivity and patent intensity in TL2 regions, 2008. *Source*: Calculations based on OECD Regional Database (2010)

Despite the positive trend observed by the fitted line, the positive relationship appears weaker than between high-skilled human capital and service sector productivity. The graph displays a high variation of innovation intensity in regions around average levels of productivity and a positive tendency at higher levels of the service sector productivity. The graph also displays that there are no regions with very low levels of productivity and high levels of innovation.

This section measures trend in service sector productivity among OECD regions revealing a strong relationship between overall regional productivity and service sector productivity particularly in financial and business services. The productivity of services is in turn highly related to region-specific factors, in particular human capital, density and innovation intensity. Amongst these the strongest relationship appears to exist between high-skilled labour and density.

4 Conclusions and Policy Recommendations

This chapter examines the spatial patterns in service sector activities and links them to the overall trends in the service sector in OECD countries. Services now are an integral part of OECD economies contributing to the bulk of overall output and employing the majority of workers (e.g. 70 %) in the labour force. Services however display lower productivity growth than manufacturing and have a strong spatial dimension linked to non-tradable nature of services sector activities making them dependent on local conditions.

Understanding what drives productivity growth in services at the regional level and how the regional dimension relates to the aggregate service GVA can help our understanding on the critical factors driving overall productivity and by extension help in the policy design.

The spatial pattern reveals that concentration in services has decreased among all OECD regions but increase among regions within countries making service sector activities more heterogeneous in space over time. Financial and business services are particularly concentrated amongst service subsectors. At the same time specialisation in financial and business services appears to be higher in capital regions or regions with large cities.

Specialisation in financial and business services has increased among OECD regions more than in other services. Within countries however regions have specialised more in public administration and in social services. This latter trend is important given that productivity in public administration and social services and in wholesale and retail trade has a growth potential due to forces of convergence.

Overall regional productivity appears to be highly dependent on service sector productivity. Rather than just focusing on improving the service sector productivity of few regions it is critical to design policies that target all regions given that overall services are more dependent on the many local labour markets rather than few large service markets.

Place based policies can play a critical role in this task by improving human capital, density and innovation intensity. These three factors at the regional level are closely linked with higher productivity of services. Low human capital is a necessary but not a sufficient condition for productivity growth in services. The links are stronger in high skilled human capital. Density appears to be positively related and to a lesser degree innovation intensity.

Appendix 1

Geographic Index of Concentration

The index of Geographic Concentration measuring concentration of GVA is defined as:

$$\left(\sum_{i=1}^{N} \left|GVA_{j,i} - a_i\right| \Big/ 2.\right) * 100$$

where $GVA_{j,i}$ is the GVA share of industry j and region i, a_i is the area of region i as a percentage of the country or OECD area, N stands for the number of regions and || indicates the absolute value.

The index lies between 0 (no concentration) and 1 (maximum concentration) in all countries and is suitable for international comparisons of geographic concentration.

Specialisation Index

Specialisation in an industry measures the ratio between the weight of an industry in a region and the weight of the same industry in the country (Balassa–Hoover index). Formally it is defined by:

$$BH_i = \frac{GVA_{ij}/GVA_j}{GVA_i/GVA}$$

where GVA_{ij} is the total GVA of industry i in region j, GVA_j is total GVA in region j of all industries, Y_i is the national GVA in industry i, and Y is the total national GVA of all industries. A value of the index above 1 shows specialisation in an industry and a value below 1 shows despecialisation.

Appendix 2

Table 4.10 Number of regions increasing and decreasing in specialisation within countries, 1995–2008

	Public administration social services					Financial and business services					Wholesale and retail trade				
	↑	↓	↑(%)	Av	Max	↑	↓	↑(%)	Av	Max	↑	↓	↑(%)	Av	Max
Australia	5	3	63 %	1.25	2.77	3	5	38 %	0.87	1.20	4	4	50 %	0.90	1.12
Austria	3	6	33 %	0.98	1.20	5	4	56 %	0.93	1.37	5	4	56 %	1.00	1.35
Belgium	2	1	67 %	1.07	1.22	1	2	33 %	1.05	1.35	2	1	67 %	0.97	1.06
Czech Republic	4	4	50 %	1.01	1.16	2	6	25 %	0.85	1.82	3	5	38 %	0.95	1.29
Denmark	3	2	60 %	1.03	1.19	3	2	60 %	0.93	1.24	4	1	80 %	0.96	1.13
Finland	4	1	80 %	1.10	1.29	2	3	40 %	0.88	1.12	1	4	20 %	1.08	1.87
France	16	6	73 %	1.07	1.22	13	9	59 %	0.86	1.34	9	13	41 %	0.97	1.17
Germany	7	9	44 %	1.09	1.44	8	8	50 %	0.97	1.27	9	7	56 %	1.05	1.58
Greece	2	2	50 %	0.97	1.10	1	3	25 %	0.93	1.22	2	2	50 %	1.03	1.23
Hungary	4	3	57 %	1.02	1.17	2	5	29 %	0.80	1.30	2	5	29 %	0.87	1.20
Ireland	1	1	50 %	1.11	1.28	1	1	50 %	0.92	1.05	1	1	50 %	0.97	1.02
Italy	9	12	43 %	1.14	1.63	9	12	43 %	0.94	1.17	13	8	62 %	1.00	1.27
Japan	6	4	60 %	1.02	1.27	6	4	60 %	0.92	1.26	2	8	20 %	0.93	1.15
Korea	6	1	86 %	1.20	1.81	2	5	29 %	0.76	1.37	4	3	57 %	0.87	1.26
Mexico	16	16	50 %	1.09	1.49	28	4	88 %	0.89	1.80	11	21	34 %	1.05	1.83
Netherlands	2	2	50 %	1.00	1.12	2	2	50 %	0.91	1.13	2	2	50 %	0.91	1.11
Norway	4	3	57 %	1.07	1.35	3	4	43 %	0.88	1.42	5	2	71 %	0.91	1.31
Poland	13	3	81 %	1.07	1.28	7	9	44 %	0.88	1.46	5	11	31 %	0.98	1.11
Portugal	3	4	43 %	1.03	1.36	2	5	29 %	0.89	1.34	3	4	43 %	1.08	1.47
Slovak Republic	2	2	50 %	1.01	1.18	1	3	25 %	1.01	1.50	2	2	50 %	1.00	1.28
Spain	14	5	74 %	1.18	2.52	7	12	37 %	0.87	1.32	5	14	26 %	0.94	1.57
Sweden	4	4	50 %	1.04	1.12	2	6	25 %	0.84	1.48	6	2	75 %	0.96	1.11
United Kingdom	8	4	67 %	1.06	1.30	7	5	58 %	0.88	1.48	10	2	83 %	1.00	1.09
United States	29	22	57 %	1.05	1.88	30	21	59 %	0.92	1.78	28	23	55 %	0.97	1.32

Source: OECD Regional Database (2012)

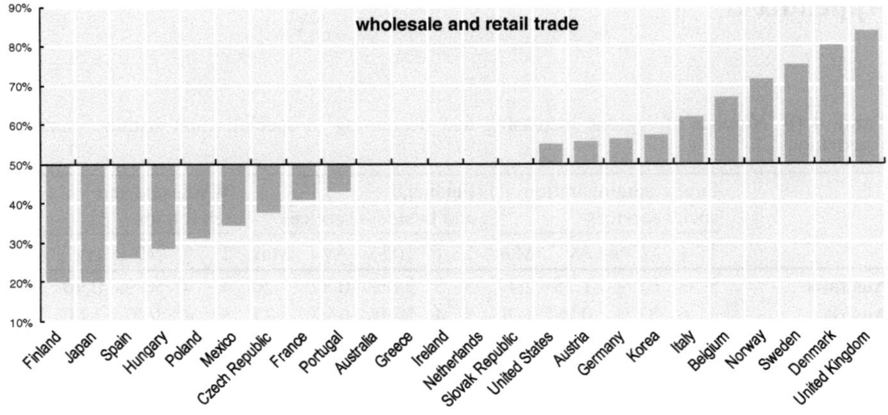

Fig. 4.32 Percent of regions increasing in specialisation index in wholesale and retail trade, TL2 1995–2008. *Source*: OECD Regional Database (2012)

Appendix 3

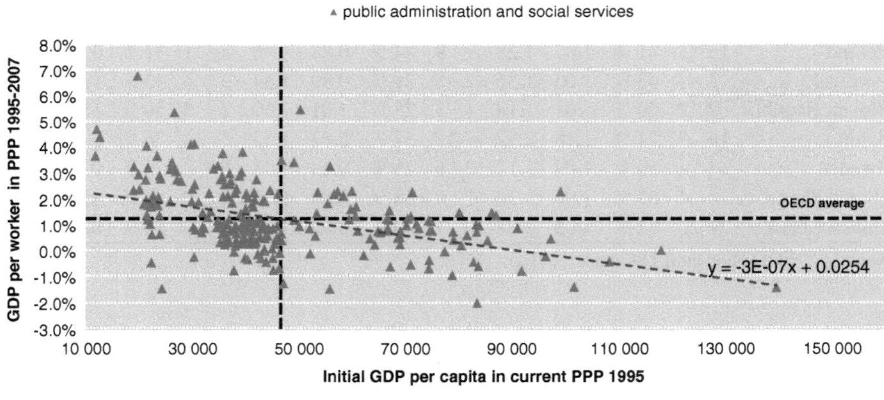

Fig. 4.33 Initial level and annual average growth rates of productivity in public administration and social services OECD TL2 regions, 1995–2008. *Source*: OECD Regional Database (2012)

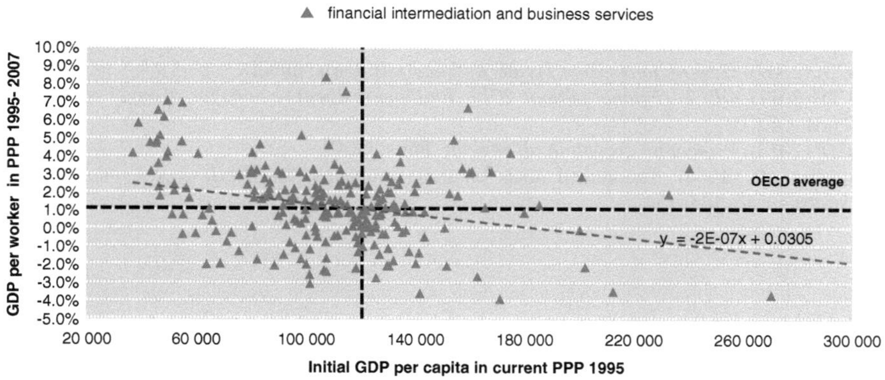

Fig. 4.34 Initial level and annual average growth rates of productivity in financial and business services OECD TL2 regions, 1995–2008. *Source*: OECD Regional Database (2012)

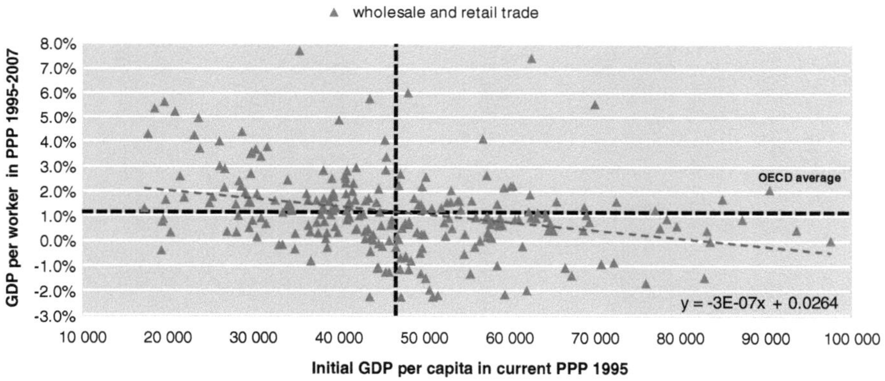

Fig. 4.35 Initial level and annual average growth rates of productivity in wholesale and retail trade OECD TL2 regions, 1995–2008. *Source*: OECD Regional Database (2012)

References

Balassa, B. (1964). The purchasing-power parity doctrine: A reappraisal. *The Journal of Political Economy, 72*, 584–596.

Baldi, A.-L., & Mulder, N. (2004). The impact of exchange rate regimes on real exchange rates in South America. In *Trade and competitiveness in Argentina, Brazil and Chile: Not as easy as ABC*. Paris: OECD.

Paul Conway, Donato de Rosa, Fiuseppe Nicoletti and Faye Steiner (2006). *Regulation, competition and productivity convergence* (OECD economics department working paper no. 509).

Cuberes. (2010). *Sequential city growth: Empirical evidence* (Instituto Valenciano de Investigaciones Económicas working paper 2010-05).

Darvas, Z., & Pisani-Ferry, J. (2011, October 13). *Europe growth emergency* (*Bruegel policy contribution*).

Edwards, S. (1989). *Real exchange rates, devaluation and adjustment: Exchange rate policy in developing countries*. Cambridge: MIT.

OECD. (2009). *How regions grow: Trends and analysis*. Paris: OECD.

OECD. (2011). *OECD regional outlook 2011: Building resilient regions for stronger economies*. Paris: OECD.

OECD. (2012). *Promoting growth in all regions*. Paris: OECD.

Samuelson, P. (1964). Theoretical notes on trade problems. *The Review of Economics and Statistics, 46*, 145–154.

Chapter 5
Global Innovation Networks, Territory and Services Innovation

Philip Cooke

1 Introduction

The study of services innovation is in its infancy. However services, like manufacturing, are becoming increasingly globalised. Accordingly, this chapter will pay special attention to globalised services innovation. The subjects of the chapter require the combination of theoretical concepts. The first of these refers to design theory related to services innovation (Lester & Piore, 2004; Martin, 2009). This involves research on business models, especially around the distinction between exploration and exploitation knowledges. The second perspective involves examination of the extent to which services fit the idea of global innovation networks. This is the latest evolution in thinking beginning with global value chains (GVC) that later became elaborated as global production networks GPN) and has now been re-theorised as global innovation networks (GIN). Finally, for services related to advanced technologies, it has been found that the global connections relate to regional and national—hence 'territorial' innovation systems (TIS) and this is the final element of the proposed conceptual framework.

Each concept refers particularly strongly to a strand of the GIN. Thus design of devices and services used in ICT has both a history of globalisation, but more recently the 'emergence' of a global innovation network for ICT services utilising new 'convergent' communication design and technological services. Contrariwise, financial services, especially the securitisation innovations that brought the Great Recession to us, are the most globalised services of all. Its crisis has spatial origins in US 'sunbelt' housing developments, but how that was 'securitised' then globalised holds important lessons for the future. A key one of these is that a great deal more 'examination' knowledge needs to be introduced to augment the

P. Cooke (✉)
Centre for Advanced Studies, Cardiff University, Cardiff CF10 3BB, Wales, UK
e-mail: cookepn@cf.ac.uk

J.R. Cuadrado-Roura (ed.), *Service Industries and Regions*,
Advances in Spatial Science, DOI 10.1007/978-3-642-35801-2_5,

somewhat instant exploration into exploitation mode of implementing financial services innovation.

So the empirics of the chapter in the last main section will compare and contrast services innovation in two fields. The first is the 'smartphone' ICT globalisation of telecom services. These include 'apps', social networking and the regionalised design of these. Other service elements include chip-sets and systems from ICT design 'hotspots' like Cambridge, with 'apps' in places like Sweden, Canada etc., assembly by Taiwanese firms in China and global sales revenue to US proprietors like Apple and Google. The second and final study focuses on the financial services GIN, the apotheosis of innovation 'at the edge of chaos,' as it is known in complexity theory. The chapter thus proceeds with theoretical and literature review sections and moves to conclusions via accounts of the 'emergence' of GINS in two important global, innovative services platforms; ICT and finance.

2 From Complexity to Simplicity: Emergence, Self-organization and Modularisation in GIN Evolution

Writing about the dynamically changing manner in which economic activity now evolves globally is an exercise in the analysis of complexity. The principal aim of the student of such processes has always to be twofold namely to understand what is really occurring beneath the surface appearance (which can often look chaotic) and to communicate the results of such analysis with simplicity and clarity. Because so much of design and production service innovation occurs, like innovation in general, as non-linear, contextual, networked, emergent, distributed and apparently 'self-organizing' elements of definable global systems, the analytical toolbox must be suited to the task in hand. As Beinhocker (2006) argues at length, this means adoption of an Evolutionary Complexity Theory (ECT) not a Traditional Economics(TE) approach since the latter is linear, equilibrium, reductionist and non-systemic: hence it will not, despite its pretensions to predictability, be able to grasp satisfactorily the fundamentals of the complex processes under inspection.

So what are some of the key concepts that *will* assist analysis of the evolutionary dynamics we have chosen to try to explain? First, let us recognise that ECT, like TE and the variety of perspectives in between are metaphors. They are conceptual models of reality deployed as a means to grasp understanding of core elements and processes of interest. Thus a hybrid approach to understanding the rise of globalised relationships in manufacturing when they started to become widespread was the concept (and metaphor) of the global value chain (GVC). By now, and in the light of the foregoing discussion, the very language of the descriptors reveals the notion to be both linear and reductionist (to value). Ultimately, these characteristics can be traced to the idea of the value chain as promulgated by neoclassical corporate strategists like Michael Porter (1980). The approach was betrayed primarily by its equilibrium assumptions of market stability, a failure to account for change over time in representations of competitiveness, and reification of the large multinational corporation as the strategic 'global controller' of the globalisation process. To try to improve things, adherents of the GVC idea like Gereffi, Humphrey, and Sturgeon (2005, pp. 84–86) developed a theory of value chains based on three factors: the complexity of the knowledge transfer required to sustain a particular transaction; the extent to which this knowledge could be codified; and the suppliers' capabilities in relation to such transactions. On the basis of these three factors, they identified five different GVC patterns:

1. Market-based chains characterized by low complexity of transactions, simple and easily codified product specifications and capable potential suppliers
2. Modular chains characterized by highly codified links simplified by technical standards, where suppliers make products to a customer's specifications and take full responsibility for process technology
3. Relational chains characterized by complex transactions and highly idiosyncratic relationships which are difficult and time-consuming to re-establish with new value chain partners
4. Captive chains characterized by suppliers with low capabilities, dependent on larger, dominant buyers, who exert a high degree of monitoring and control
5. Hierarchy, implying vertical integration when transactions are complex and not easy to codify and the competence of suppliers is low

While it introduced important nuances to the metaphor, notably modularisation, it remained linear, hierarchical (MNC-dominated) and still generic among increasingly diverse industries and services.

Ripe for critique and re-appraisal in light of certain points raised in describing the objects of interest of ECT noted above, GVC began giving way to a new metaphor of Global Production Networks (GPN) after Henderson, Dicken, Hess, Coe, and Yeung (2002). This brings at least three advantages over the GVC metaphor: first GPNs recognise globally co-ordinated interconnected practices of firms and *non-firm* institutions in producing and distributing goods and services. Second, as noted above for ECT it recognizes the centrality of networks. These are conceived as both co-ordinating firms into relations that may cross public-private organizational boundaries and integrating territorial economies in ways that may enhance their developmental potential. Thus in a more systemic way the technological paradigm relates to a territorial regime for developmental purposes. Third, this process is interestingly complex since 'regimes' are territorially specific governance elements while production networks are global. Hence, GPNs interact with territorial 'regimes' in distinctive ways, influenced by incentive, subsidy and regulatory elements of their 'regime' as well as local 'conventions' that contribute to network interactions in *different* relational ways.

This is an improvement in that it allows for some non-linearity in the notion of networks and process-governance interactions, which in turn allow for variety in the modules that make up the articulation of global relations. However, despite its recognition of difference arising from this, its focus remains on *production* in the main and that principally at the behest of MNCs in a still linear, hierarchical and dominating relationship with other incumbents, whether firms or regimes. Most of all, it fails to recognise the dynamic element in such relationships which is accounted for by innovation. Essentially it shares Porter's assumptions of equilibrium, stable market conditions and neglect of change, which through its metric of time measures *entropy*, which marks the erosion of all three in reality. Erosion but, through negative feedback—negentropy—also the aftermath of the creative destruction process in the form of and as driven by innovation. So the GPN like the GVC was locked in a phase of the development of globalisation but both missed the key change element in the broader economic development process, which is capitalism's innovative impulse. Most radically, this impulse is not eternally located in the R&D laboratories of 'global controller' MNCs, rather it can come from anywhere in a self-organizing system, which leads to the need to replace both GVC and GPN metaphors with a more appropriate one which recognizes this, namely the GIN.

The GIN metaphor keeps a few of the GVC ad GPN notions such as *modularization* from the former alongside *networks* and diverse *governance* from the latter (on the contribution of *diversity* to economic growth, see Page, 2006, 2011). However, in order to articulate these in manageable yet clear ways it is useful to introduce two more heavy duty concepts from ECT. The first of these is *emergence* which allows us to situate *modularity* in an innovation process that relates to the second, which links *different* territorial innovation systems (TIS) together across the globe. The nomenclature TIS is needed because some are *regional*, some are *national* and some are strictly-speaking indeterminate, as we shall see.

To summarise the GIN representation, it consists of the following key elements:

1. A definable economic system with global reach in its innovation, production and distribution elements
2. Within that system different territorial innovation systems (TIS) that relate to the technological paradigm in question
3. Networks of innovators and non-innovators interacting in the system
4. Innovation modules that 'self-organize' into successful knowledge combinations
5. Processes of 'emergence' that organize innovation modules into higher order commercial products and services

In what follows, after a brief services innovation literature review, this chapter will show how this module works out for the selected GIN paradigms of ICT design services and financial securitization services. In each case there will be background reference to certain hardware innovations that relate to the service innovations in question. It may be considered unoriginal to admit that services are not freestanding in the economy but retain an intimate relationship with hardware production especially where innovation is concerned. This 'rule' is captured in the insight that IBM's innovative re-invention of itself as a services business was built on the back of its globally extensive installed hardware base and that 'nobody ever got fired for buying IBM' (Raymond, 1999).

3 Services Innovation Analytical Review

3.1 Architectural Innovation: IBM's Transition to Services

It was suggested earlier that services innovation has, with a few notable exceptions, been largely overlooked and this has tended to be even more the case with regional services innovation and its TIS relationships (but for a recent exception, see Cooke, 2011). To begin with, it is instructive to elaborate on Eric Raymond's (1999) insightful point about the relationship of services innovation to manufacturing innovation. He, like others, noted how the apparently radical decision by IBM to concentrate its core competence upon ICT services *exploited* IBM's previous hegemony in the installation of computer *hardware* in many of the largest corporations in the world in the postwar years. Accordingly, although successful, it was not a typical radical innovation decision to escape risk and uncertainty but more in the nature of an architectural 'no brainer'. That is, IBM's hardware markets were being assailed by still stronger competition at the mainframe end of the market than hitherto by the 'BUNCH' (Burroughs, Univac, NCR, Control Data and Honeywell). The new competition came with the rise of minicomputer alternatives such as those marketed by Amdahl, Digital, Data General, Prime and Wang. On top of that was even more acute competition from what IBM considered the 'hobbyist' PC end of the market represented by the likes of Apple, Hewlett Packard, Compaq and Dell and their Asian 'clones' (e.g. Toshiba, Sony, Acer).

Former food and tobacco firm RJRNabisco chief executive Lou Gerstner was hired to take over as chief executive and chairman on April 1, 1993. He overturned predecessor John Akers' plan to divide IBM into separate entities, envisaging a superior future. By 1995 the R&D budget had been reduced and the Lotus software company bought for the company as a supplier of comprehensive business solution services to existing and new clients. Now began the architectural reconfiguration of IBM from a manufacturing to a services innovator. Ill-fitting parts of IBM were divested and new service 'modules' were bought such that new path creation by knowledge recombination became 'emergent'. Thus in 2002 IBM bought the consultancy arm of PriceWaterhouseCoopers (PWC) for $3.5 billion. By 2004 IBM had departed the PC segment, selling its Personal Systems Group to Chinese flagship Lenovo for $1.75 billion. It also sold its PC factories in Research Triangle Park, North Carolina to Sanmina-SCI of Taiwan, its hard disk drive (HDD) facility to Hitachi and it also sold its memory chips, printers and IBM Networks divisions. By 2006 IBM had acquired a further 31 software firms: these gave the company the broad-based portfolio to allow it to build so-called services-oriented architectures (SOAs). 'Architectural innovation' of this kind is described by Henderson and Clark (1990) as *disruptive* because although it may utilise standard elements in its design these are re-configured in ways that make preceding configurations redundant. By then, IBM was competing with Oracle, SAP, Microsoft and Sun Microsystems as a one-stop shop for corporate customers. SOA was an innovative way to build 'back-end' systems to industry standards integrating modular systems and forcing the competition to reconfigure their systems around SOA too. Indicatively, one IBM modular acquisition, Webify, enabled the building of a framework of pre-written software code for specific applications in industries like banking or insurance, reducing development costs accordingly. Other IBM architectural innovations included Radio Frequency Identification (RFID), a printed electronics consignment tracking system, developed to manage its own logistics but sold to Wal-Mart as what is widely perceived to be their key global supply chain management advantage.

3.2 Service Innovation by 'Modularisation'

Another service innovation that originated in the ICT hardware industry as described by Grove (1996) from his time as CEO of Intel is 'modularisation' (see also Sturgeon, 2002). While 'architectural innovation' is emergent through the dispensing of redundant elements and reconfiguration of pre-existing with new elements, 'modular innovation' is emergent through the assembly for a higher purpose of new elements. Even in hardware terms there may be little scope for recombination with pre-existing modules. A good example of modularisation in the software and systems design elements of an innovation is demonstrated by reference to Apple's origination of the iOS system used since its earliest iPod 'smartphone'. This involved co-evolving new combinations of modular elements that would come to characterise smartphones in general. This was achieved through integration of wireless communication, powerful core processors, optical systems, music, video, software 'apps', flat panel display, touchscreens and the various system controls to implement interactions among these. To achieve this, Apple had to make a number of acquisitions of small, specialist companies and decide on key members of its GIN for reliable delivery of robust components. With respect to the latter, a core processor supplier was ARM, the Cambridge (UK) 'fabless' chip design company specialising in cellphone systems, with chipset assembly from the likes of Taiwanese innovator Mediatek. This is a good example of modular emergence requiring new hardware and software elements since Intel had been Apple's preferred supplier for computer chips but it was deemed 'too slow' logistically as well as insufficiently integrative technically for more complex 'smartphone' inter-operability and mobility requirements (Isaacson, 2011; Sturgeon, 2002).

To realise the smartphone's innovative, technically convergent services design, involving telephony (by now the least challenging element), digital camcorder, 'apps', music, TV and video communication capabilities, Apple had to put together new modules. These included the following acquisitions:

- **Emagic**—a music software and hardware company based in Rellingen, Germany with a satellite office in Grass Valley, California. Purchased in 2002, the company was best known for its *Logic* music sequencer, used in Atari and the Apple Macintosh since 1992.

- **Lala**—an online music store acquired by Apple in 2009, whose Palo Alto music engine service allowed members legally to upload their own music for sharing, accessing Apple iTunes and MP3 content. Lala claimed availability of over eight million songs.
- **FingerWorks**—a touchscreen spinout from the University of Delaware known mainly for its *TouchStream* multi-touch keyboard. Initially designed as a service to help sufferers of repetitive strain injury (RSI) the company founded in 1998 and was acquired in 2005.
- **PA Semi**—acquired by Apple in 2009, this Palo Alto fabless semiconductor SME specialised in making powerful and power-efficient Power Architecture processors. The acquisition also added lead designer engineering experience of designing StrongARM processers to Apple's workforce to implement custom chips for the iPod, iPhone and iPad.
- **Intrinsity**—an Austin, Texas based fabless semiconductor design SME acquired by Apple in 2010. Its service was for advanced semiconductor logic design for proprietary ARM, MIPS and Power Architecture cores. Specifically, it enhanced high performance microprocessors by implementing fewer transistors and low power consumption (typically required in smartphones).
- **Nothing Real**—from Venice Beach, California was acquired by Apple in 2002 for its *Shake* advanced digital effects software applications for feature film, broadcast and interactive gaming services.
- **Siri**—acquired by Apple in 2010, San Jose-based Siri was a spinout from Stanford Research International (SRI) Artificial Intelligence Centre at Stanford University funded by DARPA. It specialised in human-computer voice communication. It was launched as a service of the Apple iPhone 4S in November 2011.

Hence we see how modular innovation in this advanced software development-based service industry fits very well with the complexity theory perspective upon innovation. This, it will be recalled, proposes innovation occurs by assembly through recombination of existing, distributed knowledge modules. These combine to implement a higher-level purpose of integrated service provision in the form of a new system. In this respect, inside the system of the firm, *downward causation* is exerted on subsidiary elemental levels in an endogenous developmental manner. This counters the physico-chemical reductionist metaphor of causality, which is that causality is always upwards from the atomic and/or molecular level.

Downward causality of this kind is, in fact, normal in service and other industrial or even public policy innovation, as only the following paragraph-length summary testifies. Thus in pharmaceuticals services, for example, evolutionary transition in business strategy typically produces comparable 'modular innovation' results. The Swiss pharmaceuticals company Roche, third biggest in the world, is in 2012 engaged in a transition from expensive cancer drug research and design towards expansion of the diagnostics side of its business. This led to it acquiring tissue diagnostics firm Ventana Medical Systems of Tucson, Arizona, expert in determining effective treatments for cancer and other infectious diseases, for $3.4 billion in 2008. In 2010, it was announced that Ventana, as a member of the Roche Group, would acquire BioImagene, a digital pathology company based in Sunnyvale, California. Digital pathology is a suite of dynamic, image-based technologies that enable image capture, information management, image analysis and virtual sharing of patients' tissue samples. Other acquisitions at the time included Almira and Allied Medical. In October 2011, the company announced a $17 million cut in its R&D budget because of concerns that government and academic institutions would reduce laboratory drug research funding. In July, the company bought cervical-cancer diagnostic maker mtm laboratories of Heidelberg, Germany for €190 million. In the same year Roche announced the acquisition of PVT (*Probenverteiltechnik*) based in Waiblingen, Germany and of PVT Lab Systems, based in Atlanta, Georgia. PVT was a global market leader in providing customised automation and workflow solutions for in-vitro diagnostic (IVD) testing in large commercial and hospital laboratories. In December 2011 Roche Diagnostics bought Verum Diagnostica, based in Munich, a leading company in platelet function testing, the fastest-growing field in the coagulation diagnostics market. By 2012 Roche was bidding to acquire San Diego gene-sequencing firm Illumina for $5.7 billion. This was to meet accelerating demand for gene-based treatments and speed up the use of DNA research in medical diagnostics.

3.3 *Innovation by Exploration*

This is the third type of service innovation model to be explored in this analytical review. It builds on the distinction utilised widely in ECT made by March (1991) in regard to firm strategies when confronted by circumstances of severe uncertainty bordering on 'creative destruction'. In this circumstance they should engage in *exploration* strategies. In times of stability between the span of the business cycle or crisis punctuation points they should engage in strategies of knowledge *exploitation* of the fruits of the exploration phase. Much organisational learning literature finds firms do this and many are in more or less permanent tension between the professional cadres associated with either side. Martin (2009) shows this for firms like Lucent and Ericsson in telecom services where the accounts-led professionals on the exploitation side found the exploration engineering practices on the exploration side to exist 'on the edge of chaos' and to wish to re-exert system control on the company as soon as possible once the crisis was over (Cooke, 2012). The distinction is also discussed in relation to business innovation by Lester and Piore (2004). While, from an ECT perspective in general the relationship is discussed by Page (2011) and from an evolutionary economics viewpoint by Beinhocker (2006). Here, we seek only to consider the *exploration* dimension and that from the viewpoint of a style of service industry innovation. The exemplar is Microsoft Research, a key division of Microsoft, having six distributed research centres around the world.

The first of these was established at Microsoft's Redmond, Seattle homebase in 1991. Redmond researchers developed Wearable Multitouch Interaction, which turns any surface in the user's environment into a touch interface, while PocketTouch enables users to interact with smartphones inside small surfaces like a pocket or purse. A different team developed much of the computer graphics software that is implemented in the modern visualisation and simulation systems used in films, games and serious games for training and remote monitoring (including military ‘drones’). Established in 1997, Microsoft Cambridge has innovated mathematical proofs for verifying programming logic (the Four Colour Theorem) and Kinect, the controller-free interface that enables users to interact with the XBox 360 by the wave of a hand or the sound of the voice, with a system for programming computers to recognise skeletal movements and body parts in gaming. Microsoft Silicon Valley, opened in 1998, also contributed to Kinect, and developed a landmarks-based shortest drive path algorithm for routing European journey directions. Opened in 2008, Microsoft Research New England works on computational biology and social networking. A further centre, established in 2005 in Bangalore, India makes fundamental contributions to software checking and verification while the centre for Microsoft Research Asia, based in Beijing, China since 1998 has worked on face recognition and visualisation, innovating the Kinect Identity player recognition tool-set for the X-Box 360 gaming device. Across its twelve research groups, the Adaptive Systems & Interaction group is also associated with innovations in 3D virtual reality, while the Programming Languages group has evolved TouchDevelop allowing programming on a Microsoft-enabled smartphone (e.g. Nokia Lumia). All in all, exploration innovation at Microsoft Research has produced some significant successes. For example, many of those listed above became modules of the Kinect device, although Microsoft acquired the 3DV firm behind the Z-Cam camera technology from Israel. This is a good example of how acquired and indigenous, long-term *exploration* innovations like surface touch, facial recognition, voice recognition, and skeletal mapping make their way into the marketplace through a variety of mechanisms, including a dedicated technology-transfer team, product incubations, IP licensing, and the sale of Microsoft Research products through the online Microsoft Store. However, Microsoft suffers from the ‘innovator’s dilemma’ (Christensen, 1997) which occurs when a company focuses on protecting existing markets, rather than trying to create new ones, worried that new markets may eat into existing revenue streams. Accordingly, with more than twice Apple’s patents, Microsoft is less than half as innovative.

3.4 Innovation in Other Services

Deliberately concentrating on obviously innovative services companies like IBM, Apple and Microsoft gives the lie to the claim that services are not innovative. They clearly are. For the first time, in the present account, an original explanation of the nature of distinctive modes of service innovation has been worked out and demonstrated. However, it can be that this gives a misleading impression of just how innovative service industries actually are. Accordingly, with reference to recent reviews a brief attempt is now made to estimate the innovativeness of other services than those evolved from high-tech manufacturing. In this account, after Aoyama and Horner (2011) we shall summarise the following: retail, logistics, contract R&D and public services innovation. In Cooke (2011) a review is offered of the views of the main experts on regional services innovation and it is summarised and updated in the paragraph which follows these specific accounts. It will be seen that the first two involve both architectural and modular innovation due to the re-architecting of the retail and logistics industries into an improved digital fit. Meanwhile, the second two include re-architecting, modular and exploration innovation for R&D services and modular innovation for public services. In each case, modular innovation involves reconfiguration by 'bundling' of technologies and procedures of an emergent nature. For example, mobile telephony is configured with positioning (GPS), scanning and imaging modules in innovative ways in one of the chosen cases of public services innovation while electronic point of sale, consignment tracking, printed electronics and data-mining were 'emergent' in retail and logistics innovation, as will be seen below.

3.4.1 Retail Services

Retail innovation is common, nearly ubiquitous among retail chains and largely defines price and non-price retail competitiveness. Such innovation can also involve branding or re-branding strategies. Cases in point are BP replacing its traditional 'British Petroleum' tagline with 'Beyond Petroleum' in 2002 as it both globalised even further and tried to re-position itself as a 'greener' corporation—more successfully in the former aspiration than the latter. Another would be Wal-Mart that similarly sought to change a bad, exploitative image as an arch-discounter that drove its suppliers to the wall (Fishman, 2007) into an environmentally friendly corporation that painted its stores a solar reflective white in place of its traditional battleship grey(MacDonald, 2008). Technological innovation advanced Benetton's global entry into the fashion retail market in the 1980s, its electronic point-of-sale (EPOS) programming enabling rapid shelf-replenishment from instantly informed supply-chain management. By the 2000s Spanish emulator Zara had successfully augmented Benetton's EPOS system with a fashion forecasting facility that further boosted 'quick fashion' emanating from elsewhere in Italy as the 1990s 'pronto moda' innovation. Japan's convenience stores were something of a role-model here, making larger IT investments than other retailers to abbreviate restocking time. The first EPOS system was developed by 7-Eleven Japan in 1982 to rationalise delivery trips with a view to raising efficiency in overall cost reduction (Aoyama & Horner, 2011). In the UK, Tesco the world's third largest retailer, gained enormous efficiencies from a related innovation that connected loyalty cards to customer profiling. Tesco tripled in size after 1995 when it agreed to work with a start-up company established by Edwina Dunn and Clive Humby. Their contract followed a now legendary presentation to the board in which Lord MacLaurin, Tesco's chairman, replaced Green Shield stamps with the Tesco customer-loyalty ClubCard. The service innovation in question used data-mining to analyse and predict customer purchases The couple's company, Dunnhumby, is now 90 %-owned by Tesco, its most important client. Clearly, retail innovations of the kind described tend to be implemented in large corporations for whom resulting efficiencies can often justify heavy up-front ICT investments. As with Wal-Mart and Tesco, observed efficiencies are achieved through such discount retailers using online networks to drive down supplier costs by adopting the innovation thereby contributing to network efficiencies as returns to innovation diffusion (Miles, 2000). Of course, such modular innovations swiftly migrate and coalesce as drivers of recombinant knowledge among competitors seeking temporary advantage in global markets.

3.4.2 Logistics Innovation

Many of the innovations in retailing are forms of logistics innovation but there are others that belong to the world of logistics itself. Wal-Mart is widely praised for the efficiency of its Radio Frequency ID (RFID) stock consignment logistics system which, as noted earlier, was actually invented by IBM to control its own supply chain. In the case of IBM this also involved out-sourcing logistics below a nominal value to external suppliers. In Germany, for example, IBM's preferred logistics supplier for low-value items was for a time Bertelsmann, the global media corporation, considered more efficient in the 1990s than the majority of pre-logistics 'haulage' firms then prevailing (Cooke & Morgan, 1998). Later, haulage and general transportation companies modernised through innovations utilising ICT and eventually the Internet to consolidate delivery loads. These were absolutely inefficient in the earlier period and, it is widely understood that, even today, some 40 % of average truck volume consists of air rather than goods (McKinnon & Piecyk, 2010). Accordingly, the logistics industry has emerged somewhat as a strategic and knowledge-intensive industry that at its best provides crucial services to many sectors of the economy. This was testified to above in respect of ICT logistics requirements of flagship marketers in ICT who demand 98-2 service: that is 98 % of an order of, for example, Taiwanese chipsets, must reach their Cupertino, California destination in the case of Apple, within 2 days. As Yeung (2011) makes clear the ability of firms like DHL to fulfil these requirements goes a long way to explain the feasibility of long GIN (global innovation networks) and the profitability of firms that can achieve or surpass 98-2.

Aoyama and Horner (2011) argue that the Internet revolutionized logistics through providing new on-line ordering tools and stimulating the rise of new e-logistics providers. The latter can be divided into those that provide logistics services exclusively in virtual space (non-asset-based), and those that provide services in both virtual and geographic spaces (asset-based). The rise of e-logistics suppliers is witness to the organizational decomposition of the physical movement of goods (see also Schmitz and Strambach (2009). Meanwhile, the related transmission and processing of information has been accelerated by the introduction of B-to-B e-commerce. Largely virtual e-commerce providers include internet brokers, online auctions, and online exchanges of the kind routinely utilised by aerospace clients often purchasing from global single sources (Cooke & Ehret, 2009). Such businesses estimate at least one-third of supplies are purchased on-line, including deliveries of special alloy blocks that may only be available from a single source located in, for example, Wichita, Kansas.

3.4.3 R&D Services

It was a standard belief in innovation studies that R&D and, to some extent, innovation associated with it would be the 'stickiest' part of any business to offshore in any evolving global value chain (Pavitt, 1984). This view has been shown to be a sound judgement but the implication that R&D might never move has proved less reliable. Already by the 1990s Chesbrough (2003) showed that outsourcing and indeed, offshoring of R&D had moved ahead apace. In markets like pharmaceuticals, the onset of biotechnology meant that large firms, who initially tried to master the new technology by acquiring smart start-up businesses, soon found they were ill-equipped to conduct non-chemical research and changed their business model accordingly (Cooke, 2007). By the end of the 2000s most 'Big Pharma' was suffering from drying pipelines for new chemical entities from traditional fine chemistry sources and was becoming ever-increasingly reliant upon biotechnology innovations from dedicated biotechnology firms and biotechnology R&D from university biosciences and medical schools to fuel their businesses. Thus in early 2012 AstraZeneca announced 7,350 R&D redundancies over the 2012–2014 period because its 'pharma' innovation model was broken. As noted above, a celebrated global pharmaceutical services company like the Swiss firm Roche had been re-positioning itself as a diagnostics more than a therapeutics medicaments company, while its global neighbour Novartis had been investing enthusiastically in generic drug companies such as Lek (Slovenia), Hexal (Germany) and Eon Labs alongside Bristol Myers Squibb's (US) over-the counter drug divisions. This is because both, like their global rivals, find medical services markets more profitable than drug production markets in the new biopharmaceuticals era. This is also re-architecting innovation, involving modular recombination and exploration anew in unfamiliar fields.

Other, longer-established R&D services innovation markets include those associated with contract companies like Cambridge Consultants and PA Consultants, Cambridge who pioneered the software and systems design capabilities that took advantage of, for example, the kind of 'fabless' chip design industry that now dominates the ICT industry and in which virtual designs are fed by the likes of ARM to countries like Taiwan who invested heavily in silicon foundries to produce and 'chipstack' their products for implementation by the likes of Apple in their iOS smartphone systems. This externalisation model was subsequently exported to emerging market economies like India and China, especially in the former case regarding software checking, de-bugging (as with Y2K work) and back-office design contracts. One belief is that Indian 'body shops' like Infosys and Wipro were the mechanism that started this process because the Y2K scare required in the short-term many programmers familiar with COBOL, the computer language that set the clocks and other controls at the time, and India was the only place that had enough (Yeoh & Willis, 2004). It still remains the case that much software-related work in innovation-related services in the Indian centres in Bangalore, Hyderabad and Mumbai involves back-office software services exemplified by the fact that China's telecom flagship Huawei joined western outsourcers in contracting out its telecom services software checking and de-bugging requirements to such former 'body shop' providers. Singapore's HDD sub-contractors and inward investors like Seagate and Western Digital have similar Indian software development clients. But this is hardly R&D, of course—though it is somewhat indicative that much of the growing outsourcing of such services to China and India involves more checking, de-bugging and following blueprints than truly inventive or even innovative work. Space does not permit further investigation of this fascinating element of 'open innovation' in R&D services but reviewer advice suggests the following should be consulted by interested readers (Castellaci, 2010; Martinez-Noya, Garcia-Canal, & Guillen, 2011)

3.4.4 Public Services

Innovation in public services may seem rare but the reality is that much innovation occurs in public services and, on balance, probably more of it is directly helpful to the health and education of users than much other service innovation that has been discussed thus far. The subject itself is globally under-researched; hence this review is even more reliant on a few indicative examples than in other fields. Let us start with the 'bottom-of-the-pyramid' innovation inspired by Prahalad (2005) who pointed to the vast amount of individually small profits that could be made by global corporations selling services to the poorest strata of societies in developing countries. One such innovation that has proved successful for global information and medical technology firm General Electric is described by Immelt, Govindarajan, and Trimble (2009) as a form of retro-innovation. The case study GE CEO Jeffery Immelt contributed to in the Harvard Business Review concerned body scanners, for which GE is a leading global producer and marketer. The price of an MRI scanner varies, depending on the strength of the scanner. Scanners with more strength produce more detailed images; therefore, these scanners cost more. MRI machines can thus range in cost between $1 and $3 million. Such costs mean installing MRI equipment is unthinkable in most developing country contexts. Alerted to this, GE began exploring the bottom of the pyramid and developed a hand-held device connected to a mobile phone for transmitting scans to a central imaging facility at a cost of some $1,000 per item. This sold very well in developing country markets and saved sufficient lives that a market opened up for such low-cost hand held scanners on the part of police forces and paramedics dealing with highway and other accidents back in the US.

Perhaps the most remarkable platform of innovations to have reduced death rates in western countries has been that part of the medical services dealing with heart disease. Once the number one killer, heart disease has now declined by half 2002–2010 to one of the lesser life-threatening diseases. Research showed that just over half this decline was caused by fewer people having heart attacks and just under half by more people who had heart attacks surviving. In the latter case, angiogram technology that improves diagnosis, angioplasty that facilitates strengthening blocked arteries with medical 'stents', and improved cardiac by-pass survival rates explain much of the improvement. Moreover fewer heart attacks have occurred in recent years and of those that did occur, fewer were fatal. As the authors conclude, the evidence suggests that these remarkable results are caused by healthier lifestyles, better prevention for those at risk and improved medical treatment for heart attack patients (Smolina, Wright, Rayner, & Goldacre, 2012). Thus it can be seen that services innovation can be relatively straightforwardly found in the public services; in the two cases given these were linked to technological change but not all such change relies on increased technological innovation (viz. GE's hand scanner) although angiograms and angioplasty are different. The key is that both service innovations had a clear purpose and attention was devoted in technical and pedagogic means to achievement of the service innovation *through* technological innovation in a demand-driven rather than a traditional technology-push manner. Thus it can be seen that the borderline between services and manufacturing can be quite fragile in connection with high-tech services like control system software design. In this respect, as Metcalfe and Miles (2000) say—technology-based services are not so dissimilar from high-tech manufacturing. However, other services operate differently, with fewer investments in R&D and intellectual property tools like patents. Nevertheless, some services both invest in patents and, more frequently, copyrights and trademarks. Metcalfe and Miles (2000) also argue, correctly, that there is greater focus upon organizational innovation involving training and skills upgrading (Cooke, 2011).

4 The Question of Territory

The chapter turns now to reflections on the territorial dimension of what has been disclosed about GINs for services innovation. Like much other innovation, services innovation has the objective of 'annihilating space with time'. This is especially clear in retail and logistics innovation but is also true of Apple's 98:2 demands described above to achieve exact time-to-market scheduling. IBM's shift to services increased its turnover and its turnover time compared with its mainframe computer business, and so on. Paradoxically, however, achievement of these time for space economies requires an increased spread over and utilisation of spatial locations of economic activity. Furthermore, GINs rely on modularization which is far more complex organizationally than vertical integration. Accordingly, the space of location for service innovators in GINs is extensive but granular. We will see this for financial services later in the chapter, but for the moment consider the pre-crisis advantage to, for example, the Morgan Stanley investment bank, of its Mumbai back-office where some 2,000 employees worked in IT, finance and accounting. Among the 2,000 were 500 knowledge process outsourcing (KPO) workers. A crisis-invoked task for KPO was to value the tranches of millions of sub-prime mortgages Morgan Stanley had invested in. This was achieved, in effect, over a weekend at a billed cost of $95,000, anything between a tenth and a hundredth of what it could have cost in the same timeframe in New York. Accordingly, time annihilated space with low-cost skills, which is a key asset in Mumbai's locational offer for what is described in another US investment bank JPMorgan's website as often innovative KPO work (Cohan, 2010).

What is new in this otherwise unremarkable tale of financial services outsourcing is the emphasis investment banks put on the *innovation* capabilities of a portion of their back-office functions. It remains to be seen precisely what such innovation comprises and, for Morgan Stanley, it involves equity research, complex financial modelling and portfolio analysis while for JPMorgan it is a KPO call centre. These are merely illustrative vignettes of the emergence of at least some elements of novelty in what have hitherto been mainly locations for more humdrum tasks. It is something, as we shall see, displayed also in the changing work content in the ICT services GIN where locations that have self-organizing capabilities to evolve capabilities in knowledge *exploration* in addition to the more familiar and routine knowledge *exploitation* (March, 1991) nowadays win out as the GVC, or its evolved sub-form, the GPN, transitions into a more demanding GIN. Such self-organizing capabilities as those that engage governance accomplishment, on the one hand, with entrepreneurial or innovative attributes, on the other, are scarce and, accordingly, highly valued. In the GIN analysis offered here, they are referred to as territorial innovation systems (TIS). As noted, they engage an exploratory sub-system composed of knowledge and connectivity infrastructures that help form research and technical talent, with a commercialisation or knowledge exploitation sub-system that includes opportunities for creative as well as routine professional or corporate employment in addition to innovative entrepreneurship. While in India, mention must be made of such exemplars as Bangalore and Hyderabad where financial ICT services also proliferate alongside an embryonic biotechnology platform. Other such 'rising' as well as 'setting' TIS locales concomitant with the shift from GPN to GIN in ICT and financial services are discussed in the section which follows.

5 Global Innovation Networks in Services

We now come to the final main section of this chapter, which explores the emergence of global innovation networks (GINs) in a few representative service industries that are not primarily local, yet whose outlines can be readily inferred. Such inferences arise in part from the foregoing discussion of architectural, modular and exploration innovation in advanced technology services like ICT and biotechnology, on the one hand, and, on the other, certain other global services, like retail, logistics and R&D that often sustain them in particular ways. Innovation,

Table 5.1 Main debtor nations, 2011

Country	Foreign debt to GDP (%)	Government debt to GDP (%)
Ireland	1,093	109
Iceland[a]	1,003	100
UK	436	81
Spain	284	67
Greece	252	166
Portugal	251	106
France	235	87
Germany	176	83
US	101	100
Japan	50	233

Source: Bank for International Settlements; IMF; World Bank
[a]NB 2009; Reinhart and Rogoff (2010); IMF

in general, is by now a global business and there is every reason to highlight those global service industry innovations that often drive such emergent processes forward because of service industry demand, as we have seen. We shall give two brief accounts of quite complex GINs in ICT-related services of the kind alluded to earlier, followed by a mirror-image 'scowling curve' GIN derived from recent experience of financial services innovation in securitisation where the upper part of the smile is represented by the kind of debt to GDP ratios revealed in Table 5.1. This is also a markedly more hierarchically (though without a single 'global controller') structured GIN driven mainly by Wall Street, notably that part connected to securitisation of assets that was the proximate cause of the current global economic downturn. In the first of these, we see complex adaptive systems of globalised service innovation emerging in diverse locations from earlier forms of more value chain and production-centred system hierarchies.

5.1 *Innovative ICT Services GINS*

It is important in this section to try to separate global production networks (GPNs) which are still mainly related to production of hardware like hard disk drives in places like Singapore from global innovation networks (GINs) in ICT services. Such centres as Singapore have long animated their own production sub-networks in neighbouring countries like Malaysia, Philippines and Thailand alongside US MNCs like Seagate and Western Digital, something just beginning in ICT services GINs. The ICT services GIN works as follows, with a result sometimes known as the 'smiling curve' of global value appropriation (see also, Mudambi, 2008). The highest value attractors in this model remain the western design, software and systems companies, while the lowest value attractors are the 'world factory' assemblers in China and even lower-wage economies. In truth, the latter do not

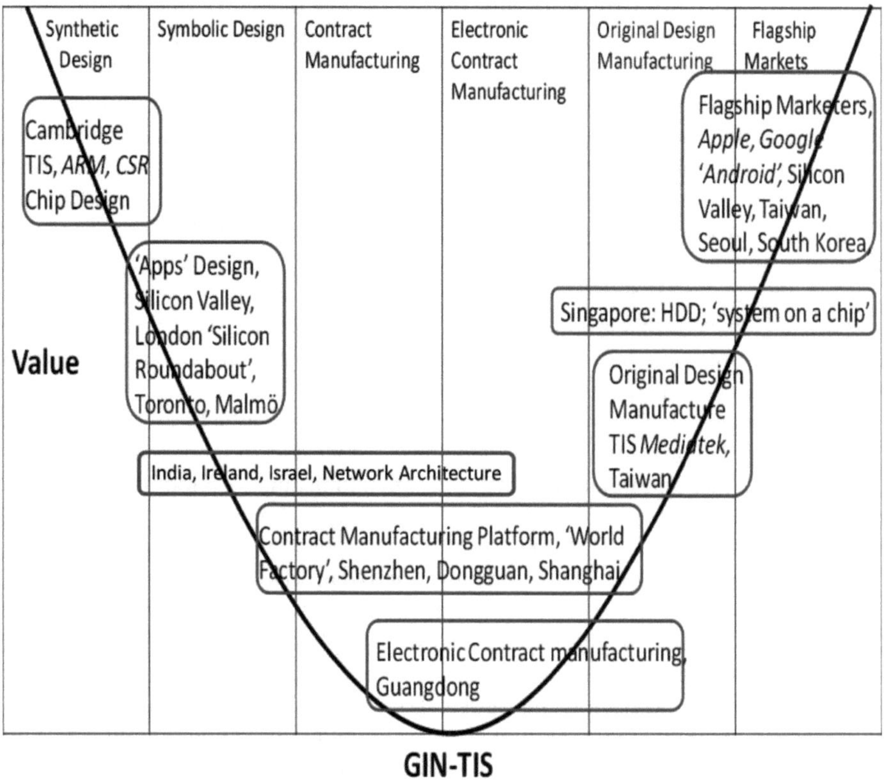

Fig. 5.1 'Smiling curve' of value in ICT Global Innovation Network (GIN)

specialise significantly in services but rather assemble the componentry that allows service innovation to be carried out either in production or the final consumption of such devices as smartphones and tablets that enable ICT services to be accessed. The basic picture is characterised first by the demise of some North American and the main European mobile telephony pioneers like Motorola, Research In Motion (BlackBerry), SonyEricsson and Nokia who are following Siemens and Alcatel into oblivion. Meanwhile the GIN displays the rise in hardware but also services of Asian innovators. Added to this is a question mark about the continued significance of the Singapore part of the old GPN in the new GIN world because its model was very tied to PCs rather than smartphones and market analysts expect the growth rate of tablets to be greater than PCs just as smartphones have outpaced cellular telephony-only devices. As noted earlier, the latter transition caught nearly all declining incumbents unawares.

So how is the ICT services innovation GIN structured? Each sub-unit of the GIN in Fig. 5.1 has some degree of involvement in the service innovation dimension of the GIN for ICT. Even at the bottom of the value curve are B-to-B network relations between the assemblers and the recipient firms in locations such as Guangdong that assemble the simplest parts of the final device. Above them, at a marginally higher point in the ICT services innovation network are assemblers of more complex chipsets or, increasingly, 'chipstacks' that integrate more efficiently relationships between microprocessors (increasingly 'power chips' that are both powerful but economic in their power usage) produced by diverse—often European—companies like ARM, Infineon and ST Microelectronics or US ones like Broadcom, National or Qualcomm. Such system design for chipstacks is a market currently cornered by Taiwanese firms like Mediatek and Wintek who have innovated around Taiwan's early investment in the world's most advanced silicon foundry. This advantage is now waning as South Korea and China, in particular, invest in their own silicon foundries. However, this is a key part of the ICT GIN where Asian producers have implemented software and systems design innovations that Broadcom, Qualcomm and Texas Instruments can no longer compete with.

Approximately level with such original design manufacturers (ODMs) as Mediatek and Wintek from Taiwan are India, Israel and Ireland. India is, as has been indicated, an important back office design and testing location for outsourced software and systems implementation initiated, first, in Bangalore by western firms like Texas Instruments, IBM and Cisco Systems and more recently by Chinese telecom giants like Huawei. This company is active in all spheres of telephony from traditional landline infrastructure through ground stations for cellphones to the Chinese TD-SCMDA standard, lower-end mobile phones and, increasingly, more expensive smartphones. As noted, Huawei has developed offshoring software links to Indian software companies (the former 'body shops') as well as making inroads in European markets (e.g. traditional infrastructure upgrading in the UK and Finland) and hiring redundant telecom engineers from Ericsson in Sweden (Lund, Gothenburg and Stockholm) and possibly in future Nokia in Finland. Israel is expert in software and systems design, especially in security software ('firewalls') and optical systems utilised in smartphones and gaming devices. Ireland hosts software development (e.g. Customer Relations Programming/Management—CRP/CRM; SAP, Symantec), administrative functions for the likes of Google, PayPal and McAfee, and 'cloud' computing services (Hewlett-Packard, Dell).

Above Ireland in value appropriation are the many small and micro-firms that specialise in the writing of software and systems for the 'Apps' that predominate in smartphone and tablet 'culture'. There are concentrations of 'Apps' writers although they can easily be located more or less ubiquitously. Continuing up the symbolic and synthetic (engineering) part of the curve are the power chip designers like ARM (formerly Acorn) and CSR from Cambridge plus others either acquired by 'flagships' like Apple or co-located in Silicon Valley. ARM also provides chip designs to Samsung, LG, Microsoft and Fujitsu while Samsung and LG have launched 'smart' Internet-enabled TVs that use ARM processors (Buncombe, 2012). On the other 'markets' side of the GIN service innovation 'smiling curve' are the flagship providers themselves like Apple and Google (Android) with the latter having twice the market share of the former, and Asian flagships like Samsung and LG from South Korea and HTC from Taiwan, each of which combine competitive ICT service functionality vis à vis the US flagships (Chen & Wen, 2011; Ernst, 2009).

5.2 Financial Securitisation GINs

For this service GIN, the equivalent of the 'smiling curve' presented above would be its inversion to a 'scowling curve' of indebtedness incurred first by various banks in specific locations and then by the countries in which the majority of such debtor institutions were to be found during the post-2008 credit crunch era. The whole of this sorry picture rests upon service innovation in the finance industry, especially the innovation of the credit default swap (CDS) a financial derivative that evolved (mutated) into the subsequent innovation of the collateralised debt obligation (CDO) and other futures and options-derived financial products and services. There are even names and companies associated with the innovation biographies in question. The principal locus of creativity was Wall Street, New York City where a few innovators noticed an opportunity to make huge profits. Traditionally, home loans (mortgages) were in the US provided by community-based 'savings and loans' companies. A loan was made by the bank and stayed with the bank until it was paid off 25 years or some other agreed term later. However, with the growth in the market for owner-occupation fuelled by the postwar 'baby boomer' generation, market observers noticed the 'savings and loans' companies had inadequate capital resources to satisfy demand for new mortgages—especially in the fast-growing Sunbelt states of California and Florida. Meanwhile, in the 'Rustbelt' local banks were faced with the opposite problem of too much capital and too little demand. Economic geography thus lies at the heart of the prevailing global financial crisis that began in 2008 and remains unresolved 4 years later at this writing.

The 'emergent' solution to the double supply and demand problem was perceived by a Salomon Brothers bond trader, Bob Dall, to be their recombination through the mechanism of securitisation. This would be a major re-architecting of the home loans business in which the modules involved in the recombination were the following. Salomon would be the catalyst, or first module, shifting inactive assets, second module, from Rustbelt to Sunbelt, garnering a transaction fee in reward as a third module, then securitising the resulting repackaged transfers into bonds—fourth module—to be sold, fifth, around the world. To achieve the last requirement of selling the newly created bonds, Dall turned to Salomon colleague Lew Ranieri who persuaded banks and legislators at state and federal level to adjust to the new model. Everyone benefited: more loans were available where demand was highest; interest rates were lower due to greater demand for loans to securitise from Wall Street; loan companies had shifted default risk to bondholders; the banks received fees; and investors could acquire relatively low-risk assets. Incremental innovation ensued as Salomon's 'rocket scientists' or 'quants' first evolved collateralised mortgage obligations (CMOs), which sliced a number of already securitised mortgages in a bond according to risk, with the greatest risk yielding the greatest return (Patterson, 2010). Then, securitisation began to be applied to varieties of loans—for students, car purchase, credit cards etc.—before by the late 1990s credit default swaps (CDSs) emerged. These were insurance certificates against mortgage defaults which could be bought and sold on the derivatives market. The next innovation was the securitisation of these derivatives in which banks bundled the securitised loans into collateralised debt obligations (CDOs) which were slices of all kinds of debt reincarnated as bonds. Eventually 'slicing and dicing' of pieces of other CDOs emerged, known as CDO-squared. Finally, Wall Street bank J.P. Morgan produced a 'synthetic' CDO composed of CDS 'swaps'—an insurance on insurances. This allowed the bank to move its own under-performing loan inventory off its balance sheet and onto the market. The CDOs were priced using an algorithm that calculated probabilistic interaction effects of defaulters—in the form of a multidimensional bell curve (the Gaussian copula). The law of large numbers meant that this curve was expected to remain relatively stable, which it was until banks began filling CDOs with the sub-prime mortgages that were by the mid-2000s the main or only source of new mortgage demand. Under these circumstances, volatility began rising as mortgagees began defaulting when repayment terms ended or mortgage costs otherwise began to increase (Tett, 2009).

Such innovation was clearly exploratory, architectural and modular in its recombination of derivative securitisation knowledge. Moreover, there was no single 'global controller' of the system but rather competitive emulation, imitation and incremental innovation in evidence, albeit focused on the Wall Street TIS. The key question is where did the emerging volatility, leading to the bankruptcy of numerous Wall Street and other US banks, produce the greatest impacts as the financial securitisation GIN worked through the investor communities of different parts of the world's financial innovation system? For this we may turn for guidance to Lewis (2011) who investigated the countries who had lost most from bailing out their sub-prime CDO financial innovators. They can be listed and then dealt with in turn as follows. First is Ireland, where the relation between bank indebtedness and national GDP at the peak of their financial crisis was of the order of 1, 267 %. Second is Iceland, where it was some 1,000 % at the peak of their crisis in 2009. Switzerland and the UK were second and third in late 2011 at 422 % and 408 % respectively. Greece, where total debt was only 252 % of GDP in late 2011 but government borrowing was expected to reach 166 % near the end of 2011. Table 5.1 gives foreign and public debt magnitudes end-2011. Of interest is which countries and their banking systems most engaged with the opportunities misleadingly promised by the evolution of securitisation such that it massively affected the viability of those states and banking systems. In the space available it is only possible to be indicative by referring to the cases of the first three 'worst offenders' with commentary in passing of relevance to absorptive capacity to financial innovation by other countries of interest such as Germany and the US. Briefly we may say something about the last two before summarising the key points of relevance in the case of the financial services (securitisation) GIN as it affected Ireland, Iceland and the UK.

Clearly, from what has been noted already in this sub-section, the securitisation innovator was Wall Street, aided by regulatory loosening, notably the repeal of the Glass–Steagall Act that had maintained 'firewalls' between risky investment banking and normal deposit-based banking. This repeal meant that depositors' money was available for speculation by investment bankers. As we have seen US banks, followed by UK and other banks securitised these as well as the loans and mortgages they had issued or bought from community 'savings and loans banks. A further reason, beyond deregulation, why this was done on such a massive scale is that interest rates were lowered by the Federal Reserve after the US terrorist attacks in 2001 and vast surpluses, notably from China were fuelling low interest rates by means of their purchase of US debt. In Lewis' (2011) account, the connections in the GIN ran as follows:

> ....From 2002 there had been something like a false boom in much of the rich, developed world. What appeared to be economic growth was activity fuelled by people borrowing money they probably couldn't afford to repay......Critically, the big banks that had extended much of this credit were no longer treated as private enterprises but as extensions of their local governments, sure to be bailed out in a crisis. The public debt of rich countries already stood at what appeared to be dangerously high levels and, in response to the crisis, was rapidly growing. But the public debt......included the debts inside each country's banking system, which, in another crisis, would be transferred to the government (Lewis, 2011, xi–xii).

Because in the US this process on the one hand threatened to bring down the global banking system and potentially bankrupt the US Treasury, Lehman Brothers was allowed to go bankrupt 'to encourage the others' but AIG, the huge insurer of enormous quantities of toxic debt, was saved by the Bush administration, thus proving the 'too big to fail' thesis to be true. Various bailout mechanisms were then in a period of institutional panic put in place to facilitate bailouts in the US and other countries, notably the UK, Ireland and Iceland.

Germany was not an innovator and not even a very good learner about the benefits but also the pitfalls of the innovations that had occurred in financial securitisation. According to Lewis, German banks were even more reckless in their appetite for the new derivatives and known to be so by Wall Street and London. As he puts it: '....other countries used foreign money to fuel various forms of insanity. The Germans, through their bankers, used their own money to enable foreigners to behave insanely' (Lewis, 2011, p. 145). Thus they lent to US sub-prime borrowers, to Irish real estate speculators, and to Icelandic banking raiders, building up losses of $21 billion to Icelandic banks, $100 billion in Irish banks and $60 billion in US sub-prime bonds. Because there had been so little *innovation* in German banking they were wholly ill-prepared. They were particularly ill-prepared for the evolution of the global financial system into a means for the strong repeatedly to exploit the weak. The securitisation model that had emerged had, as we have seen, extremely smart traders devising fiendishly complex bets they then scoured the world to find ill-informed customers to accept the bets. These often turned out to be the German Landesbanks like WestLB or Rhineland's IKB, each of which, like Commerzbank, had to be rescued either by other banks or the federal government.

This gives a little further perspective upon the plight of the small economies that also fell foul of securitisation innovation in Iceland and Ireland, and the larger debacle in the UK. Ireland represents the most palpably massive debt to GDP ratio. Based on the data in Table 5.1, Ireland's ratio stands at a frightening 1,093 having been 1,267 % in 2009. Ireland's crisis is like Iceland's mostly a banking driven one.

With the Irish government having forecast a contraction in GDP of 8.3 %, the debt-to-GDP ratio will continue to increase, even without additional foreign investment. However, Irish taxpayers are only responsible for a portion of the debt responsibilities. But even if the banking sector is removed from the total external debt number, Ireland would still have a significant debt to GDP ratio. In Iceland's case, and according to the country's central bank, Iceland's external debt was measured at $104 billion in mid-2009. With a GDP of $10.4 billion, that amounted to a debt-to-GDP ratio of 1,000 %. The Icelandic economy was the hardest hit of any in the financial crisis, and although the country's external debt was not solely to blame, it had a major hand in the country's downward economic spiral. When this is combined with a dramatic drop in the value of its currency the result was a near-government bankruptcy.

Meanwhile, the UK's 'Great Recession' started in early 2008 and ended in the summer of 2009, based on the technical measure of 'recession' as two consecutive quarters of GDP decline. The UK economy shrank by 7.1 % in this period. The economy then slowed after a short revival in the first half of 2011. Government tax rises in VAT—income tax went up temporarily for the richer middle classes, declining again in April 2012—have been made and wider austerity cuts are still kicking in. The final negative factor—and possibly the biggest—is that consumers, companies and the state are all locked in a race to pay down their debts, a toxic combination for the economy in the short and medium term. Much of this contraction arose from the combination of the government having to nationalise or semi-nationalise a number of large banks like RBS and Lloyds alongside the smaller Northern Rock. Much of the debt built up by these banks was tied to toxic sub-prime US investments or over-ambitious corporate and property investments based on inflows of cheap capital. Government policy of running a lightly regulated finance industry and spending tax receipts on the health and education sectors also came unstuck when the downturn began in 2007.

6 Concluding Remarks

This chapter has sought to show that, unlike a widespread perception that the services industries are somewhat overshadowed by manufacturing in relation to innovation, an increasing amount of contemporary innovation actually occurs in services. This occurs in relation to technological change, as it does in manufacturing, but it also emerges in relation to application of recombined or 'modular' knowledge bundles, which occurs but is more rarely written about from the perspective of manufacturing industry. In conducting this analysis the perspective of evolutionary complexity theory (ECT) was deployed. Certain core concepts in this approach, notably 'emergence' proved extremely useful in untangling the key variables that require analytical focus in untangling evolutionary change processes in complex industries operating at global scale. Thus the transition from hierarchical, linear GPNs to distributed, nonlinear GINs could be theorised and given a convincing evidential base according to this methodology.

One of the contributions of the chapter was to show that service innovation, especially in more advanced economic platforms like ICT services and biotechnology takes one or a combination of three forms—architectural, meaning a major reconfiguration of the key elements of the innovation network; modular, meaning recombination of separate but related elements to contribute to the implementation of innovation; and exploration innovation where the result of knowledge exploration, or research, can be the catalyst for innovation on a large, including global scale as with the other modes. A large measure of the weakness of service innovation analysis in the past has been its vertical, sectoral and piecemeal observational method. Clearly, service innovations are horizontal, combining related modules and, accordingly, integrated with hardware innovations in many, but not all, instances rather as Raymond (1999) and Metcalfe and Miles (2000) once observed.

It was further shown that these ECT categories were useful in framing the nature of innovation in a number of major service innovation areas like retail, logistics, R&D services and public services. Accordingly the underlying model was retained as a guide to the understanding of service innovation on a global scale in ICT related to modern smartphone and tablet services such as software and systems design, the development of social networking and 'apps' and the manner in which innovation was shown to be decentralised rather than especially hierarchical in this platform. Elements of architectural, modular and exploration service innovation could also be observed. Finally, this was contrasted with the far more hierarchical financial services global innovation network or GIN for securitisation of financial assets where the elements of architectural, modular and exploration innovation in different aspects of service evolution were again pronounced. However, the very hierarchical nature of the GIN and the relatively poor absorptive capacity and considerable myopia of actors in the GIN meant basically untested, untrialled innovation somewhat typical of services industry in general, led to financial catastrophe from which the advanced world has yet to emerge.

References

Aoyama, Y., & Horner, R. (2011). Service innovation. In P. Cooke, B. Asheim, R. Boschma, R. Martin, D. Schwartz, & F. Tödtling (Eds.), *The handbook of regional innovation and growth*. Cheltenham: Edward Elgar.

Beinhocker, E. (2006). *The origin of wealth: Evolution, complexity and the radical remaking of economics*. Boston: Harvard Business Press.

Buncombe, A. (2012). ARM thrives as it inhabits Apple's core products, *Independent*, 46, February 1.

Castellaci, F. (2010). The internationalization of firms in the service industries: Channels, determinants and sectoral patterns. *Technological Forecasting and Social Change, 77*(3), 500–513.

Chen, S., & Wen, P. (2011). Industrial Upgrading and Global Innovation Networks: Taiwan Case, paper to conference on: *The Economic Geography of ICT in Asia Pacific*, University of Queensland, Brisbane, April 28–30
Chesbrough, H. (2003). *Open innovation*. Boston: Harvard Business School Press.
Christensen, C. (1997). *The innovator's dilemma*. Boston: Harvard Business School Press.
Cohan, W. (2010). *House of cards*. New York: Random House.
Cooke, P. (2007). *Growth cultures: The global bioeconomy and its bioregions*. London: Routledge.
Cooke, P. (2011). Regional services innovation. In P. Cooke, B. Asheim, R. Boschma, R. Martin, D. Schwartz, & F. Tödtling (Eds.), *The handbook of regional innovation & growth*. Cheltenham: Edward Elgar.
Cooke, P. (2012). *Complex adaptive innovation systems: Relatedness & transversality in the evolving region*. London: Routledge.
Cooke, P., & Ehret, O. (2009). Proximity and procurement: A study of agglomeration in the Welsh aerospace industry. *European Planning Studies, 17*, 549–566.
Cooke, P., & Morgan, K. (1998). *The associational economy*. Oxford: Oxford University Press.
Ernst, D. (2009). *A new geography of knowledge in the electronics industry? Asia's role in global innovation networks* (Policy studies 54). Hawaii: East–west Centre.
Fishman, C. (2007). *The Wal-Mart effect*. London: Penguin.
Gereffi, G., Humphrey, J., & Sturgeon, T. (2005). The governance of global value chains. *Review of International Political Economy, 12*, 78–104.
Grove, A. (1996) *Only the paranoid survive*, New York: Doubleday.
Henderson, R., & Clark, K. (1990). Architectural innovation: the reconfiguration of existing product technologies and the failure of established firms. *Administrative Science Quarterly, 35*, 9–30.
Henderson, J., Dicken, P., Hess, M., Coe, N., & Yeung, H. (2002). Global production networks and the analysis of economic development. *Review of International Political Economy, 9*, 436–464.
Immelt, J., Govindarajan, V., & Trimble, C. (2009). How GE is disrupting itself. *Harvard Business Review, 85*, 56–65.
Isaacson, W. (2011). *Steve jobs*. London: Little Brown.
Lester, R., & Piore, M. (2004). *Innovation: The missing dimension*. Cambridge: Harvard University Press.
Lewis, M. (2011). *Boomerang: The meltdown tour*. London: Allen Lane.
MacDonald, C. (2008). *Green Inc.*, Guilford, Lyons-Globe Picquot.
March, J. (1991). Exploration and exploitation in organisational learning. *Organization Science, 2*, 71–87.
Martin, R. (2009). *The design of business*. Boston: Harvard Business Press.
Martinez-Noya, A., Garcia-Canal, E., & Guillen, M. (2012). International R&D service outsourcing by technology-intensive firms: Whether and where? *Journal of International Management, 18*, 18–37. 10.1016/j.intman.2011.06.004.
McKinnon, A., & Piecyk, M. (2010). *Measuring and managing Co_2 emissions*. Edinburgh: European Chemical Industry Council.
Metcalfe, S., & Miles, I. (Eds.). (2000). *Innovation systems in the service economy*. Dordrecht: Kluwer.
Miles, I. (2000). Services innovation: Coming of age in the knowledge-based economy. *International Journal of Innovation Management, 4*, 371–389.
Mudambi, R. (2008). Location, control and innovation in knowledge-intensive industries. *Journal of Economic Geography, 8*, 699–725.
Page, S. (2006). *The difference*. Princeton: Princeton University Press.
Page, S. (2011). *Diversity and complexity*. Princeton: Princeton University Press.
Patterson, S. (2010). *The quants*. New York: Crown Business.
Pavitt, K. (1984). Sectoral patterns of technical change: Towards a taxonomy and a theory. *Research Policy, 13*, 343–373.

Porter, M. (1980). *Competitive strategy*. New York: The Free Press.
Prahalad, C. (2005). *The fortune at the bottom of the pyramid: Eradicating poverty through profits*. London: Pearson.
Raymond, E. (1999). *The cathedral and the bazaar*. Sebastopol, CA: O'Reilly.
Reinhart, C., & Rogoff, K. (2010). *From financial crash to debt crisis*, NBER Working Paper 15795, March.
Schmitz, H., & Strambach, S. (2009). The organisational decomposition of innovation and global distribution of innovative activities: Insights and research agenda. *International Journal of Technological Learning, Innovation and Development, 2*, 231–249.
Smolina, K., Wright, F., Rayner, M., & Goldacre, M. (2012). Determinants of the decline in mortality from acute myocardial infarction in England between 2002 and 2010: Linked national database study. *British Medical Journal, 344*, 2–9.
Sturgeon, T. (2002). Modular production networks: A new American model of industrial organization. *Industrial and Corporate Change, 11*, 451–496.
Tett, G. (2009). *Fool's gold*. London: Little Brown.
Yeoh, B., & Willis, K. (Eds.). (2004). *State-nation-transnation: Perspectives on transnationalism in the Asia-Pacific*. London: Routledge.
Yeung, H. (2011) Industrial clusters, regional systems and global production networks: strategic coupling through innovative learning, paper to conference on: *Clusters in Global Value Chains & Production Networks: What is the Role of Innovation Systems*? San Sebastian, Deusto University/Orkestra, October 20–21

Chapter 6
Reconfiguring the Global Service Economy? The Case of the BRICs

Peter Daniels

> Access to an efficient business services sector is especially important for those sectors which are facing the double challenge of globalisation and the need to innovate in an increasingly knowledge-intensive economy
>
> (OECD, 2007a, p. 19)

1 Background

The global economic recession triggered by the global financial crisis that set in from 2008 has presented numerous challenges to, amongst others, governments, central banks, financiers and businesses of all kinds and sizes. Five years later, the uncertain prospects for the recovery of the global economy linger but this has also been accompanied by an opportunity for those economies that already had a favourable growth trajectory before the crisis to consolidate their position thereafter. Whereas an earlier global financial crisis in 1998 affected the emerging economies of Asia, the crisis that started in 2008 had its roots firmly implanted in the advanced economies of North America and Europe; leaving the emerging and developing economies as the leaders of global economic growth. Between 2007 and 2011 they accounted for 77 % of incremental global GDP (PPP) (IMF, 2011).

It is therefore not surprising that the notion of a shifting economic centre of gravity in the global economy has attracted interest in recent years (Grether & Mathys, 2010; Quah, 2011). Amongst the reasons cited by Quah for undertaking such analyses are issues linked to the rapid emergence of the BRIC (Brazil, Russia, India, China) economies in recent years and the strengthening of Asia's economies generally relative to the major established players in Europe or North America (Mahbubani, 2008). Measures such as GDP growth or changes in personal income

P. Daniels (✉)
School of Geography, Earth and Environmental Sciences, University of Birmingham, Birmingham, UK
p.w.daniels@bham.ac.uk

J.R. Cuadrado-Roura (ed.), *Service Industries and Regions*,
Advances in Spatial Science, DOI 10.1007/978-3-642-35801-2_6,

are used to model the shift in the global economic centre of gravity but there may also be merit in exploring trends specific to economic activities, such as service industries, that were major participants in, or beneficiaries of, the rapid global economic growth prior to 2008. Some of these economic activities have been more resilient than other sectors during the subsequent economic malaise (Das, Banga, & Kumar, 2011; UNCTAD/UNDP, 2010).

After a brief outline of the emergence of the BRIC economies, the stages of economic development approach is used to account for the growing importance of service industries in shaping their development trajectories. It is suggested that knowledge intensive services (KIS) are a useful measure of whether the changing position of the BRIC economies relative to the advanced economies (where KIS are an important contributor to value-added) is real rather than aspirational. If the former is the case, it suggests that the centre of gravity of the global service economy is starting to shift (or to be reconfigured) away from its established heartlands in Europe and North America.

2 The Rise of the BRIC Economies

It is now more than 10 years since the BRIC acronym has become common currency (O'Neill, 2001). Together, Brazil, Russia, India and China account for some 29 % of world's surface area. This is not insignificant in so far as the vast territories involved, especially Russia and China, increases the probability of known and undiscovered mineral resources, the amount of fertile land suitable for agriculture, and biodiversity. In addition to these activities' share of the current composition of GDP, the presence of mineral resources or significant agricultural output may shape the degree of engagement of BRIC economies in international commodity trade while a rich biodiversity can determine the opportunities for the development of significant biotechnology or pharmaceuticals industries. The future development of the BRIC economies is also linked to changes in levels of urbanization. Only two BRICs, Brazil (86 %) and Russia (73 %), had urban populations above the world average (55 %) by 2008 (CIA, 2011, see also OECD, 2010). India (29 %) and China (51 %) lagged behind but had projected estimated urban population growth rates (2005–2010) of 2.4 % and 2.7 % respectively, well ahead of Brazil (1.8 %) or the US (1.3 %) and the UK (0.5 %). In the case of Russia, however, the urban population is actually expected to contract by 0.2 % over the same period.

In his overview of the state of world economy at the end of 2001 and forecasts for the next 10 years, O'Neill (2001) highlighted a shift in the role of the BRIC economies relative to the long-established G7[1] group of countries; real GDP growth in 2001 and 2002 in the BRICs would exceed that of the G7, they accounted for

[1] The G7 comprises: Canada, France, Germany, Italy, Japan, United Kingdom, and United States.

some 8 % of global GDP, and four forecasting scenarios using various nominal GDP assumptions for the BRICs and the G7 all pointed to a rise in the BRICs share of global GDP (in current US$) perhaps to as much as 14 %. Using purchasing power parity (PPP) the forecasted shift is less dramatic but the BRIC share was still anticipated to rise from just over 23 % to 27 %. On both measures, China accounts for the largest share of the increased weight of the BRICs in the global economy. Indeed, China became the world's second largest economy (it replaced Japan) in 2010, overshadowing Brazil, India and Russia, even though in 2010 the former was ranked 8th and the latter ranked 10th (IMF, 2010). Building on the initial forecasts in O'Neill's paper, Lawson and Purushothaman (2003) calculated admittedly 'optimistic' (p.1) but also 'startling' (p.1) statistics for BRIC GDP growth and income per capita that would see India's economy larger than Japan's by 2032 and China eclipsing the US economy by 2041 after leapfrogging every other G7 economy by 2016.

Such forecasts are of course just that; much can change over the lengthy timescales that are often used, such as the on-going effects of the global economic turbulence that took hold in 2008 on a wide range of economic indicators. Thus, even though the high growth rates of the BRIC economies continue to be driven by, amongst other factors, low wages, copyright infringements, limited regulation, or the absence of trade unions, they are ultimately dependent on demand from markets in the advanced economies. Such demand is currently weak and is compromising the growth trajectories of the BRIC economies. Nonetheless, the absolute growth of GDP up to 2016 confirms the scale of China's expansion while the other BRICs (and the UK) remain relatively minor players by comparison (Fig. 6.1). More significantly perhaps, the BRIC economies have very different social, cultural and political as well as economic milieu so that their ability to sustain the development trajectories suggested by economic modelling is far from straightforward.

A comparison of some of the basic economic differences between the BRIC economies and two leading advanced economies (UK and US) is shown in Table 6.1. The fact that China and India together account for some 40 % of world population is well known, together with the significantly higher annual real GDP growth rates amongst the BRICs (although Russia lags behind the rest of the group) compared with those for the US and the UK. Of particular interest in the context of this paper is the composition of national GDP by industry and its relationship with labour force composition by industry occupation. As a broad rule of thumb the share of an industry in GDP is reflected in a similar share of that industry in the labour force. Thus, more than 77 % of UK GDP is derived from services and these also account for more than 80 % of the labour force composition (see Table 6.1). The picture is much the same for the US and for two of the BRIC economies, Brazil and Russia. Indeed the share of services in GDP and occupations is not widely different for those two BRICs and the UK/US. In the case of China and India, however, not only do they have a much lower proportion of GDP accounted for by services, especially China (43 %) but they both also have only some one-third of the labour force (not allowing for the role of the formal economy) in service occupations.

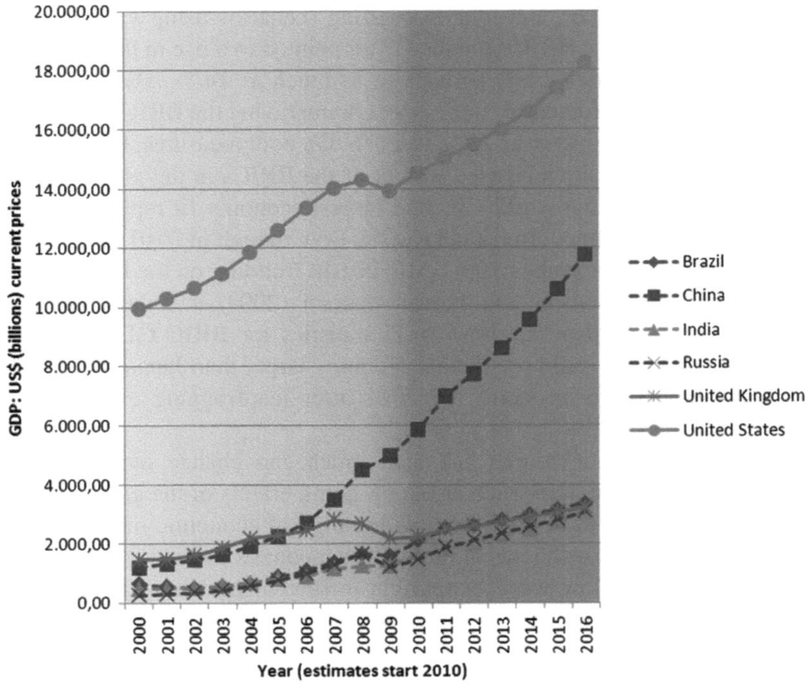

Fig. 6.1 Actual and forecast trends in total GDP, BRICs and UK, US—2000–2016. *Source*: IMF, World Economic Outlook Database, September 2011 (accessed 25 October 2011)

3 Theoretical Perspective

In view of their different political systems and social structures, it is not surprising that the BRICs have followed different paths to reach their current level of development. This has also influenced the relative importance of service activities in the development to date of the BRIC economies. In the case of Brazil, by the 1990s it was clear that the drag imposed on development during the 1980s as a result of an earlier industrial import substitution strategy might be reversed using a different approach; a recognition that the country needed to engage more proactively with a globalizing economy saw the adoption of privatisation, deregulation and a lowering of the obstacles to two-way cross-border trade and FDI transactions. China also relied on very large investments and capital accumulation via state-owned enterprises but after 1978 the rigours of a centrally-planned economic system that did not favour service industries were gradually relaxed in favour of economic internationalization, heavily focused on high volume, export-oriented, relatively low cost manufacturing. The elements of a socialist market economy have also been introduced, including the creation of designated economic zones, especially in the eastern provinces adjacent to the coast, where FDI was encouraged

Table 6.1 Comparison of BRIC economies: basic indicators, 2010

	Population	GDP per capita	GDP, real growth rate	GDP composition, by sector (%)			Labour force, by occupation (%)		
	(mill.)	(US$, current prices)	%[a]	Agriculture	Industry	Services[a]	Agriculture	Industry	Services
Brazil	193.3	10,816	7.5	5.8	26.8	67.4	20.0	14.0	66.0[b]
China	1,341.40	4,382	10.3	10.2	46.9	43.0	38.1	27.8	34.1[b]
India	1,190.50	1,371	10.4	15.3	47.0	55.2	52.0	14.0	34.0[b]
Russia	142.9	10,356	4.0	4.0	36.8	59.1	10.0	31.9	58.1
UK	62.2	36,164	1.3	0.7	21.8	77.5	1.4	18.2	80.4[b]
US	310	46,860	2.8	1.1	22.1	76.8	0.7	20.3	79.0[b]

Sources: CIA, World Factbook 2011 (https://www.cia.gov/library/publications/the-world-factbook/geos/br.html, accessed 3 November 2011); International Monetary Fund, World Economic Outlook Database (http://www.imf.org/external/pubs/ft/weo/2011/02/weodata/index.aspx, accessed 25 October 2011)

[a]2010 estimate

[b]Estimated for different years: Brazil (2003), China (2008), India (2009), UK (2006), US (2008)

using less restrictive labour and tax regimes. Further integration with the world economy was heralded by China's admission to the WTO in 2005 and this has included obligations to improve trade flows in services.

India also relied on an import substitution strategy (1947–1991) followed by economic liberalization (post-1991). Intervention by the IMF in the 1990s prevented India from defaulting on its financial obligations but also required liberalization of the national economy towards a market-based system that involved a reduction in the size of the public sector or the removal of import restrictions. This strategy has paid dividends in the form of a much enhanced engagement with the global economy, especially in service process outsourcing and information services, and growth rates rivalling those of China; but India remains a lower middle-income country according to the World Bank (2011).

As in China, the former USSRs initial reliance on a centrally-planned economy led ultimately to IMF involvement which, again, made the introduction of regulatory liberalisation and planning for more broad-based economic growth a condition of its financial contribution to the post-Soviet adjustment process. Recovery has been steady over the last decade, although the reliance on oil and gas exports continues along with the attendant risks of the impact of fluctuations in energy prices.

Since the BRICS are considered to be emerging market economies there is an implicit assumption incorporated in the growth forecasts for the next 10–40 years that they are on a path to a more advanced level of economic development. The contemporary terminology is different but each of the BRICs is, to varying degrees, positioned along the stages of growth model (Rostow, 1953, 1959): somewhere between take-off and the drive to maturity. There is scope for debate about where in 2012 each country is positioned on this continuum but it seems reasonable to suggest that, as yet, not one of the BRICs has made the final step to the age of high mass consumption (or become an advanced market economy). There is therefore an interesting question about the economic conditions required to ensure the progress of the BRICs towards the stage of high mass consumption. This will be necessary for consolidating the case, for example, that they could ultimately join a reconfigured G7 or a slightly larger group (see for example O'Neill, 2001). While Rostow's conceptualisation of what constitutes the drivers from one stage to the other, and the period of time over which it happens, is far from watertight (see for example Cypher & Dietz, 1997; Kuznetz, 1971) there is perhaps some agreement that a small number of leading sectors are required to underpin the take-off into sustained economic growth. The leading sectors will not be the same for each country but the list could look as follows; timber, manufacturing, especially automobiles, petrochemicals, steel, textiles (Brazil), mining and extractive industries, especially oil, natural gas, timber, all forms of machine building (Russia), IT and business processing services (India), mining and ore processing, iron, steel, aluminium, and other metals, coal; consumer products such as footwear, toys, and electronics; food processing; automobiles, rail cars and locomotives, ships; telecommunications equipment (China).

A more contemporary stage of development approach is exemplified by Schwab (2011) using the Global Competitiveness Index (GCI) devised by the World Economic Forum.[2] Following the economic theory of a stages of development approach, the first stage is factor-driven with their factor endowments and primarily unskilled labour the basis of competition between national economies. Competitiveness is closely linked to competent public and private institutions (pillar 1), a well-developed infrastructure (pillar 2), a stable macroeconomic environment (pillar 3), and workforce trained to at least a basic level of education (pillar 4). The second, efficiency-driven, stage of development requires countries to adopt more efficient production processes and improvements in product quality in response to wage costs that tend to be rising faster than prices. National competitiveness is now shaped by levels of higher education and training (pillar 5), efficient goods markets (pillar 6), soundly operating labour markets (pillar 7), developed financial markets (pillar 8), the ability to harness the benefits of existing technologies (pillar 9), and a large domestic or foreign market (pillar 10). An innovation-driven economy marks the third and final stage; where wages are rising by so much that the associated standard of living can be maintained only if firms are able to compete by offering new, unique, and sophisticated products. At this stage, companies must compete by producing customised and different goods using the most sophisticated production processes (pillar 11) and by innovating new ones (pillar 12). Between 2005 and 2011 the BRICS have moved further apart on the GCI with China moving closer to the US and the UK (World Economic Forum, 2011). In relation to stage of development, however, China and India remained well behind Brazil and the Russian Federation in 2011 although all the BRICs have moved slightly nearer to the US since 2005 (World Economic Forum, 2011).

Each of the BRIC economies relies on different leading sectors as well as a multitude of variations in the political and institutional objectives, priorities and frameworks that guide on-going development. Nonetheless, as Rostow (1959, p. 8) puts it: 'During the drive to maturity the industrial process is differentiated, with new leading sectors gathering momentum to supplant the older leading sectors of the take-off, where deceleration has increasingly slowed the pace of expansion' and 'economic maturity is defined as the period when a society has effectively applied the range of (then) modern technology to the bulk of its resources.' The emphasis on the effective application of technology is perfectly understandable in the context of the period during which Rostow and other proponents of the stages of economic growth model were working. The relevance of technology for lubricating the step from take-off to maturity and particularly the final step to the age of high mass consumption has not diminished; however it may now be more apposite, in place of or in addition to technology (which of course itself embodies knowledge), to advocate the importance of knowledge in its widest sense as the totality of what has been perceived, discovered, or learned. This is especially the case when economies are moving to the stage

[2] Competitiveness is defined as 'the set of institutions, policies and factors that determine the level of productivity of a country' (Schwab, 2011, 4).

of high mass consumption when 'radical improvements in housing and durable consumers' goods and services become the economy's leading sectors.' (Rostow, 1959, p. 9). The idea that services are a leading sector symbolising the transition to the final stage in an economic growth model is also reflected in the proposition that the industry structure of mature economies moves towards a relatively consistent structure of the labour force into primary, secondary and tertiary employment (white-collar and semi-skilled workers) together with an increasing share of the population in urban areas (Clark, 1940).

4 Emerging Role for Knowledge-Intensive Services?

Although there is room for debate, the fact that all the BRICs are positioned somewhere on the stage of development continuum, leaves questions about whether, and how, they will sustain their places into the second decade of the twenty-first century. While acknowledging macro-economic influences, not least the effects of the global financial turbulence since 2008 that continues to threaten the very existence of some national economies (Greece, Spain, Portugal and Ireland for example) and has modified global patterns of trade and investment, it seems likely that more 'local' (nation-specific) constraints, enabling, and challenging factors will ultimately determine the course of events for the BRICs. Such factors have been enumerated, for example, by Kedia et al. (2006, p. 71); they include constraints on growth such as bureaucracy, potential political instability, corruption, ineffective reforms and excessive regulation; growth enablers include factors such as planned economic reforms, deregulation, innovation, lowered/reduced trade barriers, high economic growth, and improved resource utilization. The growth challenges include: the containment of population growth, sustaining and growing FDI inflows (and outflows), improving confidence perception, and sustaining high rates of growth in GDP or GVA.

It could be argued that one of the inputs that will help the BRICs to ameliorate or modify the effects of these factors will be channelled through knowledge-intensive services (KIS). It will be necessary for the BRICs to proactively move towards a knowledge-based (or perhaps more accurately a knowledge-led) economy (OECD, 1996; see also Dasgupta & Singh, 2005, 2006). The leading advanced economies have successfully utilised knowledge as a driver of productivity and growth across all economic sectors with their economic performance focused on harnessing information, technology and learning as a way of transforming businesses into leaner, more innovative, and competitive entities. This, in turn, has required workers to adapt and reconfigure their skills to meet requirements that are more about finding, using and manipulating knowledge than about operating machines. It has also required a better understanding of how knowledge is distributed or diffused, often using technology and informal/formal networks, to the benefit of innovation and economic wellbeing. Such adjustments have not taken place totally of their own volition; they have been enabled and supported amongst other things by a suite of fast growing knowledge-intensive services (KIS).[3] Since the tasks undertaken by KIS are often business-specific, strategically important, and commercially sensitive some of the growth has taken place within firms (as continues to be case for state-owned enterprises (SOEs) in China, for example) but more typically it has been through arms-length provision by specialist firms able to benefit from economies of scope and scale.

KIS therefore perform an important role in improving the efficiency of business processes across all industries and also act as conduits for the transfer of, or joint creation of, new knowledge to clients (Miles, 2005). They therefore also contribute to innovation processes; indeed business service firms themselves, for example, generally demonstrate higher levels of innovation than firms across the economy as whole (OECD 2007a, 2007b, Pro Inno Europe, 2010). The different roles performed by KIS in the innovation process have been categorized by the OECD (2007a, p. 13): renewal services such as R&D or strategic management consulting that are directly related to innovation; routine services such as payroll management improve the maintenance and management of systems within organisations; compliance services such as corporate legal firms environmental consultants enable organisations to conform with the requirements laid down by various national or sectoral regulations; and network services ensure efficient distribution and exchange of knowledge and resources, whether via formal production networks or informal personal networks. While comparable statistics are difficult to assemble, in the case of R&D for example, apart from Russia, the BRICs have significantly fewer researchers and technicians per million persons than the US or the US (Table 6.2).

The need to nurture KIS in the BRIC economies can be illustrated by examining the forward linkages of manufacturing compared with business services (as a surrogate for KIS) (Fig. 6.2). In addition to being an important supplier of inputs to other sectors along the value chain, business services demonstrate linkages across a wider range of industries than manufacturing (OECD, 2007a, p. 9). The data is a decade old and may not necessarily be representative of the scale of current forward linkages, but by using input–output to compare the relationship of individual industries with the rest of the economy (from the demand side) it is possible to demonstrate some clear differences between the BRIC economies and advanced economies such as the US and the UK. The forward linkage 'gap' between business services in the three BRICs for which data are available and other OECD countries has likely narrowed somewhat since 2000 but the evidence suggests there are still

[3] Using ISIC Rev. 3, KIS comprise Category K (Real Estate, Renting and Business Activities) includes three business services Divisions: computer and related services (72), research and development (73) and other business services (74). Within these groups knowledge-intensive business services include IT-consulting (72), R&D services (73), legal (74), accounting (74), marketing and advertising (74), business consulting and human resource development (74). Parts of Category J (Financial intermediation) such as Division 67 (Activities auxiliary to financial intermediation) could also be classified as KIS). (see for example, Miles, 2005; Rubalcaba-Bermejo & Kox, 2007)

Table 6.2 Researchers and technicians in R&D (per million persons): BRICs, UK and US, 2007

	Brazil	China	India	Russia	UK	US
Researchers in R&D[a]	656.86	1,070.94	136.94[c]	3,304.72	4,180.67	4,663.28[d]
Technicians in R&D[b]	..	..	93.58[c]	516.11	878.53	..

Source: World Bank, WDI Database, accessed 1 December 2011

[a]Researchers in R&D are professionals engaged in the conception or creation of new knowledge, products, processes, methods, or systems and in the management of the projects concerned. Postgraduate PhD students (ISCED97 level 6) engaged in R&D are included

[b]Technicians in R&D and equivalent staff are people whose main tasks require technical knowledge and experience in engineering, physical and life sciences (technicians), or social sciences and humanities (equivalent staff). They participate in R&D by performing scientific and technical tasks involving the application of concepts and operational methods, normally under the supervision of researchers

[c]2005

[d]2006

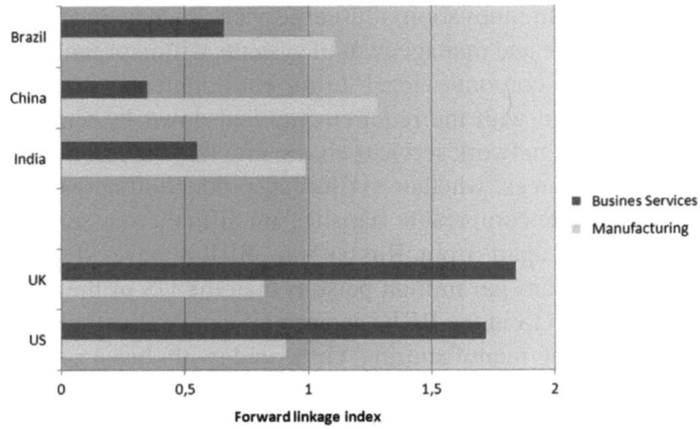

Fig. 6.2 Forward linkages: manufacturing compared with business services, BRICs (excluding Russian Federation), UK and US, 2000. *Source*: OECD (2007a, 2007b, 10), compiled from Fig. 6.2. Forwarded linkages calculated from OECD input–output tables (India, 1999; all other countries, 2000)

challenges ahead. The forward linkage values shown in Fig. 6.2 provide an indication of what happens to a given industry, in this case business services or manufacturing, if the final demand of every other industry was to increase by one unit. Any sector with an index value greater than 1.0 equates with a key role in the economy. Not only are business services well below this threshold for Brazil, India and China, especially the latter, the equivalent values for the UK and the US are up to three times higher and well above 1.0. As expected, the position is reversed for manufacturing although India is more like the US and the UK than the other BRICs.

In relation to services as a whole, their value-added as a proportion of the GDP of the BRIC economies has improved significantly since the late 1980s (Fig. 6.3). By the late 1990s Brazil was on a par with the US (67 %) and well ahead of China and India even though the services share of GDP has flattened out over the last

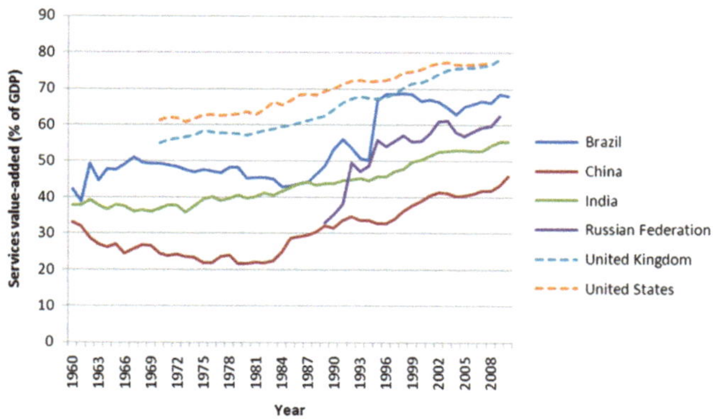

Fig. 6.3 Services value-added (% of GDP), BRICS, UK and US, 1960–2010. *Source*: World Bank, WDI database, accessed 30 November 2011

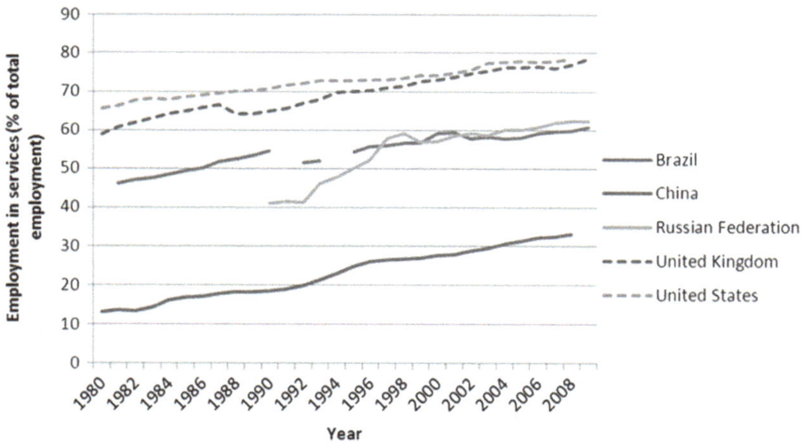

Fig. 6.4 Employment in services (% of total employment): BRICs, UK and US, 1980–2009. *Source*: World Bank, WDI Database, accessed 1 December 2011

decade as the share continued to increase in the UK and the US. This is not as clearly mirrored by the share of employment in services when compared with the US and the UK; in China there has been a steady upward trend but the proportion employed in services (using the World Bank definition) is less than 35 % (Fig. 6.4). The picture is better for Brazil and the Russian Federation although they lag the US and the UK shares of total employment in services by some 20 %. In the context of the argument that KIS are crucial for the progression of the BRIC economies along the stage of development continuum, it is the trend of their share in total employment which is more important. It is difficult to compile accurate statistics and the nearest indicator is the share of total employment in financial intermediation (ISIC-Rev 3 Class J) and in real estate, renting and business activities (Class K) (Fig. 6.5).

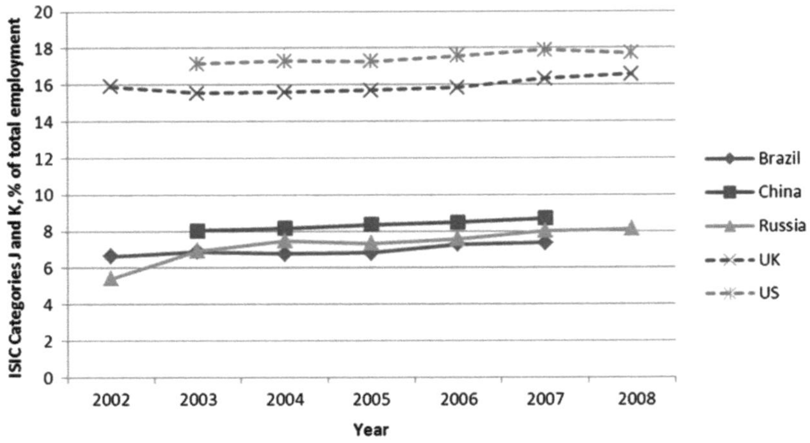

Fig. 6.5 Share (%) of total employment in Financial intermediation (J) and Real Estate, Renting and Business Activities (K), BRICs (excluding India), UK and US, 2002–2008. *Source*: ILO, http://laborsta.ilo.org/, accessed 9 November 2011

As a broad generalization, the share of these sectors in total employment in the BRICs (6–8 %) is approximately half the level typical of economies such as the US and the UK (16–18 %). A slow upward trend in the share of these activities has occurred but there is no indication that it is closing the gap between the BRICs and the benchmark economies.

The evidence derived from WTO data on exports and imports of other commercial services via the BRIC economies also shows an increased share of total world trade in these services and a growing positive balance towards exports between 2005 and 2010 (Table 6.3). The increased share of world exports of commercial services is largest for China (2.5–4.6 %) although India, which already had a larger share in 2005, remained only marginally ahead (4.9 %) of 2010. The exports of these services from Brazil and the Russian Federation more than doubled in value over the 5-year period but their shares remained lower and changed little. In 2005, and perhaps reflecting the limited development of domestic commercial services, all the BRIC economies except India, which had a positive balance, relied on a higher level of imports by value than exports. Russia and Brazil continued to have a negative balance in 2010 as India and China had achieved a significant improvement in the ratio of exports to imports of commercial services by 2010. It is not possible to say, however, whether this reflects transactions by the foreign affiliates of overseas firms in these two countries or direct or foreign affiliate exports by domestic commercial services firms.

Perhaps the closest group to KIS within the other commercial services category is 'other business services.' It is interesting to examine how the position of the BRICs for this narrower group has evolved, in relation to its value and share of the world total, between 2005 and 2010 (Table 6.4). An idea of the scale of the gap between the economies with well-established competitive advantage in services

Table 6.3 World share of exports and imports of other commercial services[a], BRICS, 2005 and 2010

	2005		2010	
Country	Value $bn	Share (%)[b]	Value $bn	Share (%)
Exports				
Brazil	6.9	0.6	19.4	1
China	29.2	2.5	90.2	4.6
India	43.8	3.8	95.9	4.9
Russian Federation	9.3	0.8	20.1	1
Total (BRICs)	89.2	7.7	225.6	11.5
Imports				
Brazil	12.5	1.2	32	1.9
China	33	3.2	74	4.3
India	26.6	2.6	59.1	3.5
Russian Federation	15.6	1.5	31.7	1.9
Total (BRICs)	87.7	8.5	196.8	11.6

Source: Compiled from WTO (2006, 2011)
[a]Communication services (telecommunications, postal and courier services); construction services; insurance services; financial services; computer and information services (including news agency services); royalties and licence fees; other business services; personal, cultural, and recreational services, including audiovisual services
[b]Share of total world exports/imports of other commercial services

such as legal, accounting, management or advertising services is shown in Table 6.4 by including exports and imports to economies outside the EU25 (extra-EU) and from and to the US. Exports of such services are dominated by these and other advanced economies; in 2005 and 2010, for example, the US accounted for a larger or equivalent share of the world total for other business services than the BRIC economies combined even though the change in the value of exports from all the BRICs except India approximately doubled (compared with a more modest growth by some one-third for the Extra Eu-25 and the US combined). The situation in relation to imports is not dissimilar to that for other commercial services as a whole (see Table 6.3), with India the only BRIC in 2005 importing fewer other business services by value than it exported. Russia and Brazil again comprise a group with a continuing negative balance of trade in 2010. Table 6.4 also summarizes the annual percentage change in the value of exports and exports for the sub-groups within other business services. There are wide variations between the sub-groups that can be related to the historical, business, regulatory and other factors that have encouraged or constrained the growth of these, mostly specialised and professional, services in the respective BRIC economies. Generalizations are therefore difficult (and there is no data available for China for many of the sub-groups, and the base values for 2005 are also low). Nonetheless, annual changes in the value of imports of advertising, market research, and public opinion polling (d); architectural, engineering, and other technical services (f) have been greater than the equivalent performance for exports of these services. But the annual export growth of legal, accounting, management consulting and public relations services has been higher

Table 6.4 Trade in other business services by category, BRICs and selected economies, 2005 and 2010

	2005		2010		Annual percentage change, by category, 2005–2010							
	Value ($m)	Share (%)[1]	Value ($m)	Share (%)	a[2]	b	c	d	e	f	g	h
Exporters												
Extra-EU (25)	147,764	25	204,226	22	7.3	13.8	13.2	−5.9	3.7	13.7	6.8	6.6
United States	62,719	11	94,195	10	12.2	−3.6	45.7	106.2	16.2	11.1	166.2	−8.0
China	23,283	4	45,623	5	9.2	..	49.8	23.4	..	..	..	..
India	20,523	3	28,540	3	10.5	44.9	4.1	7.8	5.4	−12.1	105.2	16.6
Brazil	6,720	1	13,867	1	27.7	−6.2	71.6	34.3	152.0	20.8	..	10.0
Russian Federation	5,309	1	11,182	1	..	5.4	32.7	9.0	−4.7	36.0	28.3	42.1
Importers												
Extra-EU (25)	109,717	n.a.	160,550	n.a.	5.0	8.2	10.4	10.4	14.1	3.6	3.9	10.2
United States	37,881	n.a.	61,870	n.a.	274.0	−3.7	47.2	26.0	27.1	118.8	1.1	−6.5
China	16,287	n.a.	34,144	n.a.	20.0	..	23.4	34.3	..	..	..	..
India	16,020	n.a.	21,036	n.a.	9.0	21.3	34.2	6.3	19.4	28.0	85.0	−4.6
Brazil	7,480	n.a.	15,348	n.a.	−1.2	24.8	66.2	35.7	62.1	19.9	..	14.7
Russian Federation	6,459	n.a.	13,742	n.a.	3.2	29.3	24.9	13.6	0.0	49.1	17.1	7.5

Source: Compiled from WTO (2006, 2010)

[1]Share of total world exports of other business services. Equivalent values for share of total world imports not available (n.a.)

[2](a) Trade related services; (b) Operational leasing (rentals); (c) Legal, accounting, management consulting, public relations services; (d) Advertising, market research and public opinion polling; (e) Research and development services; (f) Architectural, engineering, and other technical services; (g) Agricultural, mining and on-site processing; (h) Personal, cultural, and recreational services, including audiovisual services

Table 6.5 Producer services as inputs (%), BRICs, US and UK: mid-1990s–mid-2000s

	Average(%) of producer services in national output (*SI/TO*)			Services input as % of total intermediate inputs (*SI/TI*)			Producer services ratio (*SI/SO*)		
Country	Mid-1990s	Early-2000s	Mid-2000s	Mid-1990s	Early-2000s	Mid-2000s	Mid-1990s	Early-2000s	Mid-2000s
Brazil	15.33	14.79	17.35	32.34	29.81	35.45	33.71	33.37	35.12
China	11.74	12.16	14.12	19.02	18.93	19.96	54.24	53.86	51.99
India	13.25	13.09	13.98	28.00	26.90	27.78	39.75	37.20	37.27
Russia	14.39	15.16	–	27.92	30.72	–	32.28	33.44	–
Average: non-OECD economies	**15.35**	**14.85**	**16.48**	**29.80**	**28.16**	**28.96**	**38.94**	**39.97**	**40.81**
UK	24.41	29.36	28.21	45.29	53.93	55.29	41.95	45.54	41.56
US	20.83	26.21	26.67	42.89	48.96	49.74	34.56	39.23	38.88
Average: OECD economies	**21.32**	**23.49**	**24.52**	**40.14**	**43.11**	**44.68**	**40.31**	**43.12**	**43.17**
Overall average	**20.03**	**22.21**	**23.52**	**37.90**	**40.93**	**42.72**	**40.01**	**42.12**	**42.55**

Source: Extracted from Cheng (2011). Calculations using OECD, STAN Input–output Tables, available at http://www.oecd-ilibrary.org/statistics

than for imports. This provides a very crude indication that the BRICs are building the infrastructure and providing an appropriate environment for supporting their development and links with external markets.

Another way of demonstrating the 'KIS gap' is to examine input–output data which is published by the OECD for selected economies for the mid-1990s to the mid-2000s (Table 6.5) (Cheng, 2011). The producer services input in total national output (SI/TO) for the BRICs is generally lower than the average for the non-OECD economies although there has been a very modest increase since the mid-1990s it is considerably smaller than that for the OECD economies which saw a change from some 21 % to 24.5 %. The services share of total intermediate inputs (SI/TI) also increased very slowly in the BRICs with China consistently below (20 %) the other economies in the group as well as the non-OECD average. Again, the OECD economies outperformed the BRICs both in relation to size of the overall share of services in total intermediate inputs (40–45 %) and the rate of change over the three periods used for the analysis. Finally, the gap between the BRICs and the advanced economies in relation to the share of service inputs from service outputs (SI/SO), which can be considered as the producer services ratio, is relatively narrow and shows a similar upward trend. The values for China have decreased from the first to the third period but they are ahead of the other BRICs as well as the OECD economies. This counter-intuitive characteristic reflects the fact that more than 50 % of producer services in China are provided by three more traditional and labour-intensive sub-sectors: wholesale and retail trade and repairs, hotels and restaurants, and land transport and transport via pipelines. This is a very different pattern to economies such as the US where activities such as finance and R&D dominate the provision of producer services.

5 Capturing a Greater Share of the Production Value Chain in the BRICs

There are numerous challenges for the BRICs to address through the process of national enhancement of the economic environment and infrastructure for supporting the domestic growth and diversification of. KIS as well as KIS-related inward investment. In the meantime, the range and quality of KIS can also be enhanced in the BRICs through the process of capturing a greater share of the production value chain. While China or India, for example, now produce numerous well-trained electronic or software engineers these invariably do not have the industry-specific skills engendered through experience. Thus Dedrick and Kraemer (2006, p. 40) cite an interviewee who notes that "China's engineers 'work perfectly at doing what they have been told, but cannot think about what needs to be done; they lack both creativity and motivation. They are good at legacy systems, but not new things; they can't handle 'what if' situations.'" The example of the notebook PC industry in China is used to examine whether the initial predominance of notebook mass production will gradually be complemented or even replaced with development activities (prototype, design production) or even design (concept, product planning, design review) higher up the production chain. After Taiwan removed restrictions on its notebook makers undertaking final assembly in China (in 2001) there was a wholesale relocation of original design manufacturers (ODMs) to Shanghai/Suzhou, following the already-established ranks of Taiwanese component suppliers. Japanese and other overseas ODMs have also established operations in the same region (as well as elsewhere in China).

With so much notebook and PC production concentrated in a low cost location and well over half of it relying on interaction between ODMs and end-product manufacturers (rather than exclusively in-house development and production—such as Lenovo) there is considerable scope for pulling in other knowledge oriented functions, especially in an industry where proximity between production and pre-production engineering and testing etc. facilitates prompt testing and problem solving on the factory floor (Dedrick & Kraemer, 2006; Lüthje, 2004; Yang, 2006). Dedrick and Kraemer (2006, p. 40) conclude that in the near-term the notebook PC cluster in China will look something like: (a) component-level R&D, concept design, and product planning in the US and Japan, (b) applied R&D and development of new platforms in Taiwan and Japan; and (c) product development for mature products and all production and sustaining engineering, in China. Crucial to the on-going evolution of this division of labour in favour of knowledge-based services and manufacturing will be the degree to which the reliance on overseas markets is replaced by local final demand since this will increase the case for early concept design and product planning (which is significantly shaped by near market requirements) to be undertaken in China rather than in Japan or the US.

It is also the case, however, that at the same time as the BRICs are strengthening their portfolio of KIS activities in the interests of moving up the value chain, the developed countries have recognised that in the face of growing competition from the BRICs and other emerging economies they must engage in further enhancement of intangible assets, technology and knowledge. For example, the OECD (2007b) has shown that investment in knowledge increased in all the member countries during the early 2000s with a corresponding shift, consolidating a trend evident since the 1990s, towards KIS and high level technology-intensive services. A comparison of trade flow data for the G7 countries and the BRIC economies (OECD, 2007b, Fig. 10) reveals how the trade balance for trade in high- and medium-high technology industries in 2003 grew faster than for total manufacturing and, while only a few OECD countries are specialised in high technology manufacturing industries, as a group they have a strong comparative advantage in these industries. The BRIC economies, by contrast, showed comparative disadvantage in high and medium-high technology and only China had (very significant) comparative advantage in medium-low technology manufacturing; the other BRICs only showed a comparative advantage in low technology manufacturing. It would be interesting to undertake a similar analysis using Other Commercial Services in the WTO databases, for example, as a surrogate for KIS but the available data are incomplete.

6 Role for In Globalization of Innovation and Knowledge

But perhaps we should not get carried away by the actual and forecast evidence derived from case examples such as that of the PC notebook industry. Encouraged by the reduced costs of international communications and travel, more rigorous protection (although not everywhere, not least China) of intellectual property rights, increased levels of cross-border mergers and acquisitions, there is little doubt that the globalization of innovation as represented by the spread of the R&D networks of MNEs has been accelerating over the last decade or so. Before then the BRICs very much occupied a secondary position to the triad comprising North America, Europe and Japan in the geography of business innovation. But Bruche (2009a) suggests that, around the turn of the century, FDI in R&D began to shift towards two of the BRICS, China and India, with a simultaneous shift by vendors and contract research organizations in both countries from supplying routine services to knowledge processing and R&D offshoring (see also Chen, 2011). Between 2001 and 2009 the number of R&D MNE centres in China and India has expanded from less than 100 to more than 1,000 (920 MNEs) in the former and around 800 (670 MNEs) in the latter (Zinnov, cited in Bruche, 2009b, p. 2). Global multinationals increased their total R&D sites by 6 % between 2004 and 2007; 83 % were in China and India. Their R&D staff increased by 22 % over the same period; 91 % were located in China and India (Jaruzelski & Dehoff, 2008). Factors such as market size and a pool of low-cost skilled workers have helped India and China, but Russia and Brazil (and some other emerging economies in Eastern Europe) now offer alternative destinations for MNE R&D as wage costs and greater international demand for R&D personnel are eroding their early comparative advantages.

It is also the case that MNE R&D is directed at a very narrow range of sectors, firstly, ICTs (software and engineering R&D in India), secondly, health-related industries such as pharmaceuticals and biotechnology, and, thirdly, the automobile industry (Jaruzelski & Dehoff, 2008). These are however high up the list of international R&D generators. R&D FDI is still heavily dominated by intra-Triad flows with on-going prominence of home country R&D. The clusters that have grown in India and China are also confined to a very few locations: Bangalore, Pune, and the national Capital Region in India; Shanghai and Beijing in China. Furthermore, Bruche (2009b, p. 4) posits that most of the MNE R&D in China and India is directed at routine adaption of existing designs or processes and perhaps 'providing modular contributions transformed into innovative products and processes in the triad's higher order R&D centers.' Perhaps any catch-up effects can arise from knowledge spill overs from inward invested R&D to domestic firms but 'the R&D investment levels even in more advanced Chinese and Indian companies are low and local challengers may even suffer from an in-situ brain drain to MNEs able to offer more stimulating and rewarding work to talented R&D professionals' (Bruche, 2009b, p. 4).

7 Conclusion

While their economic histories vary, there are a number of common challenges that confront all of the BRIC economies as they strive to move along the stages of development continuum. There are population issues that will have economic consequences for work, families, health, education, technology or outmigration arising from the aging of the population in China and in Russia (see for example Bloom, Canning, & Fink, 2011; McKibbin, 2006; Tyers & Shi, 2007) or the continuing growth in India to a level exceeding China within 10–20 years. As their engagement with world markets is expected to continue to grow apace (even if retarded somewhat by the condition of the world economy) it is likely that in the case of China and India (less so Brazil and Russia) that an already large and growing schism between the poorly educated and low income majority and the rest of the population will widen (causing social and political instability that will divert resources away from the economic development agenda and hinder knowledge transfer and inward investment). Perhaps most important in the context of this paper and the argument that KIS are integral to the future success of the BRICs, is that the financial systems on all four lack transparency (especially China and Russia) or are relatively underdeveloped, risking not only the successful integration of the BRICs into the world economy or making them vulnerable during international financial crises, but also the endogenous growth and diversification of KIS, as well as SMEs and other businesses more generally.

Yet, while these and other frailties such as poverty, creaking or non-existent infrastructure, and governments that fail to deliver even basic services, undoubtedly exist and the crude empirical evidence suggests that both the production of, and access to KIS, in the BRICs continues to lag compared to the advanced economies, there are indications that this does not mean that they are 'hollow' i.e. still in low cost, mass manufacturing mode. While the balance of trade in knowledge-related services is still broadly negative for the BRICs the gap is getting smaller (Goswami, Mattoo, & Sáez, 2011; Havlik, Pindyuk, & Stollinger, 2009). There are examples of companies that are designing and producing cheaper products (cars, computers and mobile phones) and services (such as very low cost mobile telephone networks) than their advanced economy equivalents (Jain, 2006; Pearce, 2011). This has required reinvented systems of production or overcoming obstacles in services such as logistics and distribution to enable very large, spatially extensive, and diverse domestic markets to be reached i.e. outside the big cities. The potential for exporting may be restricted by the quality of many of these goods and services, which may not yet meet the expectations of consumers in the advanced economies, but they will be more affordable for consumers in the BRICs and will therefore facilitate import substitution. This will make the world's largest multinationals that are already investing or exporting to the BRICs because they recognise their significant potential for product demand to work even harder to access those markets with goods and services that can compete with the increasingly innovative domestic firms. As knowledge-intensive companies such as IT specialists and consultancies or business process providers have increased the number of people they employ in the BRICs, this can only have the effect of encouraging demand for KIS more generally. Depending on the country and the regulatory environment it seems likely that, in the short term, this will be fulfilled via international joint ventures, licensing agreements, franchising, and cross-border movements of professional workers. In the medium to longer term, however, competitive domestic KIS providers will emerge to complement or even replace the expertise of international providers and this will be a crucial transition as the BRIC economies seek to move towards the final stage in their economic development. Whether this will contribute to a shift in the centre of gravity of the global service economy, as opposed to its reconfiguration, does however remains a moot point.

References

Bloom, D. E., Canning, D., & Fink, G. (2011). *Implications of population aging for economic growth*. Program on the Global Demography of Aging. Boston: Harvard Initiative for Global Health.

Bruche, G. (2009a). The emergence of China and India as new competitors in MNCs' innovation networks. *Competition and Change, 13*(3), 267–288.

Bruche, G. (2009b). A new geography of innovation – China and India rising. *Columbia FDI perspectives*. New York: Vale Columbia Center on Sustainable International Investment.

Chen, C. (2011). *Foreign direct investment in China*. Cheltenham: Elgar.

Cheng, D. (2011). Producer services and the evolution of the Chinese economy. *Symposium on producer services and China's economic and urban development*. Centre of Urban Studies and Urban Planning, University of Hong Kong.

CIA. (2011). *The world factbook 2011*. Washington, DC: Central Intelligence Agency.

Clark, C. (1940). *The conditions of economic progress*. London: Macmillan.
Cypher, J. M., & Dietz, J. L. (1997). *The process of economic development*. London: Routledge.
Das, A., Banga, R., & Kumar, D. (2011). *Global economic crisis: Impact and restructuring of the service sector in India*. Working Paper No. 311. Tokyo: Asian Development Bank Institute.
Dasgupta, S., & Singh, A. (2005). Will services be the engine of Indian economic growth? *Development and Change, 36*(6), 1035–1057.
Dasgupta, S., & Singh, A. (2006). Manufacturing, services and premature de-industrialization in developing countries: A Kaldorian empirical analysis. Working Paper No 327. Cambridge: University of Cambridge Centre for Business Research.
Dedrick, J., & Kraemer, K. L. (2006). Is production pulling knowledge work to China? A study of the notebook PC industry. *Computer, 39*, 36–42.
Goswami, A. G., Mattoo, A., & Sáez, S. (Eds.). (2011). *Exporting services: A developing country perspective*. Washington, DC: The World Bank.
Grether, J.-M., & Mathys, N. (2010). Is the world's economic centre of gravity already in Asia? *Area, 42*(1), 47–50.
Havlik, P., Pindyuk, O., & Stollinger, R. (2009). *Trade in goods and services between the EU and the BRICs*. Vienna: Vienna Institute for International Economic Studies.
IMF. (2010). *World economic outlook: Rebalancing growth*. Washington, DC: International Monetary Fund.
IMF. (2011). *GDP by country, 2010, World Economic Outlook Database*. Washington, DC: International Monetary Fund. http://www.imf.org/external/pubs/ft/weo/2011/02/weodata/index.aspx. Accessed 27 March 2012.
Jain, S. C. (Ed.). (2006). *Emerging economies and the transformation of international business: Brazil, Russia, India and China (BRICs)*. Cheltenham: Elgar.
Jaruzelski, B., & Dehoff, K. (2008). Beyond borders: The global innovation 1000. *Strategy+business*. Fiorham Park, NJ: Booz & Co. *53*, 54–67
Kedia, B. L., Lahiri, S., & Mukherjee, D. (2006). BRIC economies: Earlier growth constraints, contemporary transformations and future potential, and key challenges. In S. C. Jain (Ed.), *Emerging economies and the transformation of international business: Brazil, Russia, India and China (BRICs)* (pp. 46–72). Cheltenham: Elgar.
Kuznetz, S. (1971). *Economic growth of nations: Total output and production structure*. Cambridge, MA: Harvard University Press.
Lawson, S., & Purushothaman, R. (2003). Dreaming with BRICs: The path to 2050. *CEO confidential*. London: Goldman Sachs. 2003/12.
Lüthje, B. (2004). Global production networks and industrial upgrading in China: The case of electronics contract manufacturing. *International Conference on Multinationals in China – Competition and Cooperation*. Sun Yat-sen University, Guangzhou.
Mahbubani, K. (2008). *The New Asian Hemisphere: The irresistible shift of global power to the East*. New York: Public Affairs.
McKibbin, W. J. (2006). The global macroeconomic consequences of a demographic transition. *Working paper 6/2006*. Canberra: Australian National University Centre for Macroeconomic Analysis.
Miles, I. (2005). Knowledge-intensive business services: prospects and policies. *Foresight, 7*(6), 39–63.
OECD. (1996). *The knowledge-based economy*. Paris: OECD.
OECD. (2007a). *Globalisation and structural adjustment: Summary report of the study on globalisation and innovation in the business services sector*. Paris: OECD.
OECD. (2007b). *Moving up the value chain: Staying competitive in the global economy*. Paris: OECD.
OECD. (2010). *Perspectives on global development: Shifting wealth*. Paris: OECD.
O'Neill, J. (2001). Building better economic BRICs. *Global Economics Paper No. 66*. London: Goldman Sachs.
Pearce, R. (Ed.). (2011). *China and the multinationals*. Cheltenham: Elgar.

Pro Inno Europe. (2010). European Innovation Scoreboard (EIS) 2009. *Pro Inno Europe Paper No 15*. Brussels: European Commission (Enterprise and Industry).
Quah, D. (2011). The global economy's shifting centre of gravity. *Global Policy, 2*(1), 3–9.
Rostow, W. W. (1953). *The process of economic growth*. Oxford: Oxford University Press.
Rostow, W. W. (1959). The stages of economic growth. *The Journal of Economic History, 12*(1), 1–16.
Rubalcaba-Bermejo, L., & Kox, H. (2007). *Business services in European economic growth*. Basingstoke: Palgrave Macmillan.
Schwab, K. (2011). *The global competitiveness report 2011–2012*. Geneva: World Economic Forum.
Tyers, R., & Shi, Q. (2007). Global demographic change: Policy responses and their economic implications. *The World Economy, 30*(4), 537–566.
UNCTAD/UNDP. (2010). *Creative economy report 2010: Creative economy – A feasible development option*. Geneva: UNCTAD/UNDP.
World Bank. (2011). *Open data: Country profiles*. Retrieved 10 December 2011.
World Economic Forum. (2011). *Global competitiveness report 2011–12*. Accessed 30 November 2011 at http://google.co./publication/explore.
World Trade Organization. (2006). *International trade statistics 2006*. Geneva: WTO.
World Trade Organization. (2011). *International trade statistics 2011*. Geneva: WTO.
Yang, Y. (2006). *The Taiwanese notebook computer production network in China: Implication for upgrading of the Chinese electronics industry*. Hong Kong: University of Hong Kong Center on China's Transnational Relations.

Part II
Regional Aspects of FDI in Services, Export Decisions, Productivity and Employment

Chapter 7
FDI in Services in European Regions: An Overview

Laura Resmini

1 Introduction

In the last two decades, services have emerged as the largest and most dynamic sector in the world economy, providing about two thirds of global value added and a similar share of employment in most developed and emerging countries (World Bank, 2010). The growth in services has been accompanied by a rising share of services in international transactions: trade in services has grown faster than trade in goods and now it represents about 12 % of world GDP. Moreover, there has been a marked shift of foreign direct investment (FDI) from the manufacturing sector towards the services sector worldwide. The share of services in total FDI stocks has increased to about 63 % in 2010, as compared to 49 % in 1990(Unctad, 2011).

The European Union (EU) has played and still plays a dominant role in international transactions regarding services. It is the largest exporter in the world for services and the largest market for FDI in services (Unctad, 2011; World Trade Organization, 2010). The implementation of the Single Market Programme provided impetus for the expansion of FDI in services sectors and for a EU-wide restructuring of several service industries, accelerating intra-EU services FDI.[1] Moreover, the East enlargements of 2004 and 2007 created new opportunities for FDI in services given the opening up of new markets and the liberalisation of important services, such as telecommunications, banking and transportations (Lejour, 2007). A cornerstone in the EU is the principle that goods, services, capital and people can move freely across

[1] Services are often considered as strategic or sensitive industries and therefore subject to prudential regulations because of their tendency to natural monopolies and market failures, as well as for national security or economic nationalism considerations. See Golup (2009) for a discussion on this issue.

L. Resmini (✉)
University of Milano Bicocca, Milano, Italy

ISLA, Bocconi University, Milano, Italy
e-mail: laura.resmini@unimib.it

J.R. Cuadrado-Roura (ed.), *Service Industries and Regions*,
Advances in Spatial Science, DOI 10.1007/978-3-642-35801-2_7,

member states. However, while the internal market for goods seems to respond to such a principle since the implementation of the Single Market programme in 1992, the internal market in services does not. Service producers face several impediments in exporting their services to or in setting up a subsidiary in another EU member state mainly because of differences in national regulations. In order to boost the internal market in services, in 2004 the European Commission launched a *Proposal for a Directive of the European Parliament and of the Council on Services in the Internal market* (EC, 2004). The Directive, which has become effective from 2010 onwards,[2] aims at removing regulation-based impediments to trade and foreign investments in services by applying the same principle—i.e. the country of origin principle—to most services sectors.[3] According to this principle, EU member states apply mutual recognition of national regulatory regimes.

The rapid international expansion of services, and mainly FDI in services, is supported by two other recent phenomena. On the one hand, it reflects the increase in outsourcing: while consumer services satisfy the final demand, business services, such as telecommunication, transport, and consulting are increasingly used as intermediate inputs in the manufacturing sector. By having the choice of producing internally the services needed to complete and support their manufacturing operations or outsourcing them to external contractors, an increasing number of manufacturing firms have taken the last solution, thus contributing to the diffusion and further development of the services sectors. On the other hand, the technological progress, especially in information and telecommunication technologies and the progressive liberalisation of many services sectors (telecommunication, transportation, finance, etc.) have increased the tradability of several services, making their production increasingly subject to the international division of labour. In this context, multinational enterprises have become the dominant means of delivering abroad those services that, being neither tradable nor storable, have to be produced where they are consumed (Markusen, 2007).

Despite the growing importance of services, there exists very limited literature on FDI in services and its implications for host economies. The existing theoretical literature (Deardorff, 1985; Markusen, 2007) points out that FDI in services are more complex than FDI in goods, since services differ from goods because of their intrinsic characteristics, such as intangibility, non-storability, non-transferability and heterogeneity. Such peculiarities have implications on how investments can occur and suggest that FDI in services may happen for different reasons than FDI in goods and require separate interpretation. The empirical literature, however, has studied FDI in services by using the same conceptual framework as FDI in manufacturing. Not surprisingly, it has not found substantial differences both in

[2] The Service Directive was adopted by the European Parliament and the Council on 12 December 2006. It should have been fully transposed by Member States into their national systems by 28 December 2009.

[3] Financial, healthcare, transport and electronic communication services are not covered by the directive.

the forces driving FDI in services and manufacturing and in their potential impact on host economies (Casi & Resmini, 2010; Davis & Guillin, 2011; Kolstad & Villanger, 2008; Nefussi & Schwellnus, 2007; Riedl, 2010; Unctad, 2004).

The present paper follows this recent stream of literature and provides an analysis of the location of FDI in services in Europe at regional and sectorial level. More in details, this paper aims at investigating the incidence, the sectorial distribution and the determinants of FDI in services across EU regions. This comprehensive view of spatial and sectorial patterns of FDI in services not only contributes to enrich the present knowledge about FDI in services, but also may help in designing more effective FDI promotion policies. As for FDI in goods, in fact, host economies can benefit from FDI in services through employment creation, capital accumulation, transfer of technology, more efficient services and increased competition (Arnold, Javorcik, & AMattoo, 2007; Golub, Jones, & Kierzkowski, 2007). Moreover, FDI in services can also improve manufacturing firms' efficiency by increasing the availability of high quality production-related services, while international outsourcing of services (offshoring) may help firms to restructure and move into more high value-added activities (Gorg & Hanley, 2011; Markusen, Rutherford, & Tarr, 2005; Olsen, 2006).

The originality of this study comes from its regional focus and use of sectorally disaggregated data for FDI. The data are derived from the Amadeus database produced by Bureau Van Dijk (BVD), which contains firm-level qualitative and quantitative information for all European countries, covering all sectors of the economic activities. Data on foreign affiliates have been aggregated at regional and sectorial level over three different periods of time, i.e. 1997–1999, 2001–2003 and 2005–2007. In so doing, it is possible to control potential factors that can affect FDI in specific years and capture most of the changes occurred in the EU integration process with a potential impact on FDI, i.e. the introduction of the euro and the East enlargements (Liebscher, Christl, Mooslechner, & Ritzberger-Grünwald, 2007; Oxelheim & Ghauri, 2004).More recent data have been excluded since the global financial crisis has not only affected but also altered FDI inflows worldwide (Unctad, 2009).[4]

Generally speaking, FDI data are usually reported in terms of stocks and flows. Inward FDI stocks refer to the total accumulated value of foreign-owned assets at a given point of time, whereas FDI flows refer to foreign capital received over a given period of time. Official statistics often report total FDI stocks and flows at country level; sectorial data on FDI are sometimes available at country level, but they are barely comparable at international level. No official FDI data set is available at sub-national level, either for total FDI or for sectorally disaggregated flows and stocks. Therefore, in order to analyse the impact of foreign investments at sub-national level and compare the experience of different regions within and across countries, alternative sources of data should be explored. The most promising are those based on

[4] As it is standard in the literature, a firm is considered as foreign-owned if at least 10 % of its value is owned by an ultimate owner who is established outside the country where the firm is located.

firm-level data, since they enable the users to have an overview of the foreign presence in different geographical units by simply counting foreign firms operating in that area in a given period of time. While newly created foreign firms can be easily identified, the computation of FDI stocks implies the identification of firms active over the whole period of time, net of new entrants and exiting firms in each considered period. Given the complexity of such a computation, this paper considers flows rather than stocks of FDI. The number of new foreign affiliates, disaggregated by the 269 NUTS2 EU regions and by 10 one digit NACE Rev. 1 services industries has been used as a proxy for FDI flows.

This approach offers some advantages and also a few disadvantages. As for the former, the regional distribution of foreign firms is directly observed and not estimated by national data. Therefore, the analysis does not suffer from potential distortions in the geographic distribution of FDI due to the "regionalisation" process of national data. Potential disadvantages may arise, first of all, from the fact that data come from firms' balance sheets; therefore, they may include either plant or firm level information. Despite that, previous studies based on the same source for FDI data have shown that possible biases deriving from using corporate balance sheet information do not distort significantly the results (EC, 2005; Pusterla & Resmini, 2007). Secondly, the use of the number of foreign affiliates located in a given region instead of the total amount of FDI flows received by the same region implies the assumption that FDI flows increase with the number of foreign affiliates set up in a given location, though it is not necessarily true, given that foreign investments involve both the initial transaction and all subsequent capital transactions between the parent firm and the affiliated enterprises.[5] Finally, one cannot distinguish between small and large foreign investments, which are supposed to have a different impact on local economies. Since this paper investigates the processes of foreign-firm locations and not their impact on the host economies, these potential shortcomings should not affect the results.[6]

The remainder of the paper is structured as follows: Sect. 2 provides a brief overview of FDI in services at aggregate level. Section 3 analyses regional and sectorial patterns of FDI in services and their dynamics by using simple descriptive statistics. Section 4 explores the determinants of FDI patterns of location at regional and sectoral levels through econometric techniques and Sect. 5 concludes by summarising main results and discussing some policy implications.

---

[5] Despite that, it has been demonstrated that, on average, there is a positive and strong correlation between the number of foreign affiliates and the value of foreign direct investments received by a location, at least at country level. See Capello, Fratesi, and Resmini (2011) and Pusterla and Resmini (2007).

[6] Other empirical studies on FDI, especially those which consider finer levels of disaggregation both at geographic and sectorial level use the number of foreign firms as a proxy for the value of FDI flows or stocks. Needless to say, regression analyses have been adjusted in order to take into account the discrete nature of data. See, among others, Capello et al. (2011), EC (2005), Basile, Castellani, and Zanfei (2009); Pusterla and Resmini (2007); Guimaraes, Figueiredo, and Woodward (2000).

Table 7.1 Foreign firms in the service sector

	EU27			EU15			EU12		
	97-99	01-03	05-07	97-99	01-03	05-07	97-99	01-03	05-07
Number of foreign firms	9,696	76,552	88,803	9,162	61,340	57,657	534	15,212	31,146
Share on total foreign firms	57	76	80	58	77	83	38	72	76
% Variation (previous period)		690	16		570	−6		2,749	105
Variation on total FDI		486	10		405	−13		1,392	96
Number of intra EU foreign firms	5,742	49,005	57,484	5,290	42,483	37,319	452	6,522	20,165
Share on total foreign firms (services only)	59	64	65	58	69	65	85	43	65
% Variation (previous period)		753	17		703	−12		1,343	209
Number of extra EU foreign firms	3,954	27,547	31,319	3872	18,857	20,338	82	8,690	10,981
Share on total foreign firms (services only)	41	36	35	42	31	35	15	57	35
% Variation (previous period)		597	14		387	8		105	26

Source: Own calculation from FDI Region database

2 FDI in Services in the EU

Over the period of 1997–2007 inward service FDI flows to the EU have substantially increased (Table 7.1): in the last considered period, the number of newly established foreign affiliates is about tenfold larger than it was at the end of the 1990s, bringing the share of FDI in services on total FDI from 57 % to 80 %. The largest increases have been recorded in the early 2000s. The rise in the share of services in economic activities, the externalisation of services to independent providers, the growing service intensity of the production of goods have certainly created new opportunities for such an impressive increase in the internationalisation of services. However, in the EU FDI in services has taken advantage from other two important millstones in the history of the EU, i.e. the deregulation of service markets and the liberalisation of FDI policies within the Single Market Programme and the East enlargement, which made the EU the most open area in the world for FDI in services (Golup, 2009).[7]

These deregulation efforts have enhanced the internationalisation of the services sectors since the harmonisation of national regulations has allowed European firms to reap economies of scales by expanding in other EU member states. Hence, intra-EU FDI increased tenfold, leading its share on total FDI in services to about 65 % in the late 2000s. By contrast, the share of extra-EU FDI on total FDI in services dropped to 35 % from 41 % at the end of the 1990s.

[7] In 2004 the European Commission proposed the so-called Services Directive with the aim of better integrating services markets by reducing differences in national regulations. See Kox and Lejour (2006) on the potential impact of such a Directive on intra-EU FDI.

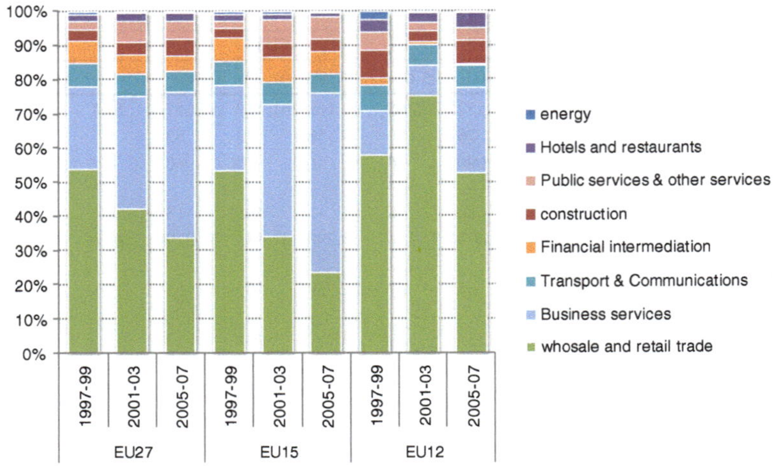

Fig. 7.1 The distribution of foreign firms by sector affiliation

Finally it is worth noticing that, though FDI inflows in services have grown, on average, more rapidly than FDI in other sectors, Western and Eastern Europe follow different patterns: the latter records impressive increases in the whole considered period, while old EU-15 members experiment a drop of about 6 % in the number of newly created foreign firms in 2005–2007 period. This reduction has been driven by intra-EU foreign firms, which may have found more profitable to invest in new EU member states in order to exploit new faster growing markets.

By examining the distribution of FDI flows among the main services sectors (Fig. 7.1), one can note that they are highly concentrated. About 70 % of foreign firms, in fact, concentrates in only two sub-sectors, i.e. wholesale and retail trade and business services. The two branches, however, have followed an opposite trend over time. At the end of the 1990s, trade activities collected about 54 % of newly created foreign affiliates, while the business services sector accounted for about 24 % of total FDI flows. By the years 2005–2007, these percentages have almost reversed, with distribution representing 34 and business services 43 % of FDI inflows. Although these patterns are common to all Europe, they were more pronounced in old EU-15 member states than in EU-12 member states, where FDI in wholesale and retail trade still represents more than 50 % of total FDI inflows in services. No other service sector accounts for more than 10 % of total foreign firms in services. Transport and communication industries collect about 6 % of newly created foreign firms, a share that maintains a constant level in all the considered periods, though the number of foreign affiliates decreased in the EU-15 and increased in Central and Eastern Europe. Financial intermediation services account for about 4 % of total FDI in services. This share, however, is three percentage points less than the share recorded at the end of the 1990s. As expected, most foreign affiliates locate in the EU-15, since it hosts some the most important financial centres in the world. Other less internationalised services include both sensitive sectors, such as energy (electricity, gas and water distribution) and public administration and other social and community related services, as well as less restricted activities, such as construction and the hotel and restaurant sector, which includes the tourism industry.

These disparities in the distribution of FDI flows across services can be only partially explained by differences in the size of each specific service sector. More precisely, they depend on sector-specific transaction costs, the different role played by scale economies and network factors in each service sector, as well as regulatory factors, which may vary not only across sectors but also across nations, to the extent that the Service Directive remains to be transposed (Lejour, 2007).

From this analysis emerges a clear divide between Western and Eastern Europe: foreign firms providing financial and business services concentrate mainly in the former, while new EU member states of Central and Eastern Europe attract more foreign firms providing consumer services, such as wholesale and retail trade. The internationalisation of these services activities is in fact based on their need to exploit economies of scales and scope, as well as access to global dynamic markets and supply capabilities.

3 Spatial Patterns of FDI in Services

The geographical perspective points up to a different picture: FDI inflows do not appear particularly skewed, both at aggregate and a disaggregate level and over time. This not surprising result reflects on the one hand the non tradability of most services and, on the other hand, the fact that the production and the consumption of several service products cannot be separated either in place or in time.

In order to assess the spatial concentration of FDI inflows across EU regions and over time, the following indicator has been computed (Overman, Redding, & Venables, 2003): $\mathrm{LQ}_{ij} = \frac{\mathrm{FF}_{ij} / \sum_i \mathrm{FF}_{ij}}{\sum_j \mathrm{FF}_{ij} / \sum_i \sum_j \mathrm{FF}_{ij}}$. It indicates the share of newly created foreign firms (FF) in sector j on the total number of newly created foreign firms in region i, relative to the share of sector j in all of Europe. LQ_{ij} allows comparisons to be made across regions: $\mathrm{LQ}_{ij} > 1$indicates that region i has attracted a share of foreign firms in sector j larger than the same share measured on a European level. The opposite is the case when $\mathrm{LQ}_{ij} < 1$.

Figure 7.2 shows the spatial distribution of FDI inflows over time by box plot. It indicates that the location of foreign firms has become less dispersed over time, as suggested by the reduction in both the range and the interquartile range.[8] In the latest

[8] In descriptive statistics the range is the smallest interval that includes all the observations and it is calculated as the difference between the maximum and the minimum value of the distribution. The interquartile range, instead, is the difference between the upper and the lower quartile. Both are measures of statistical dispersion.

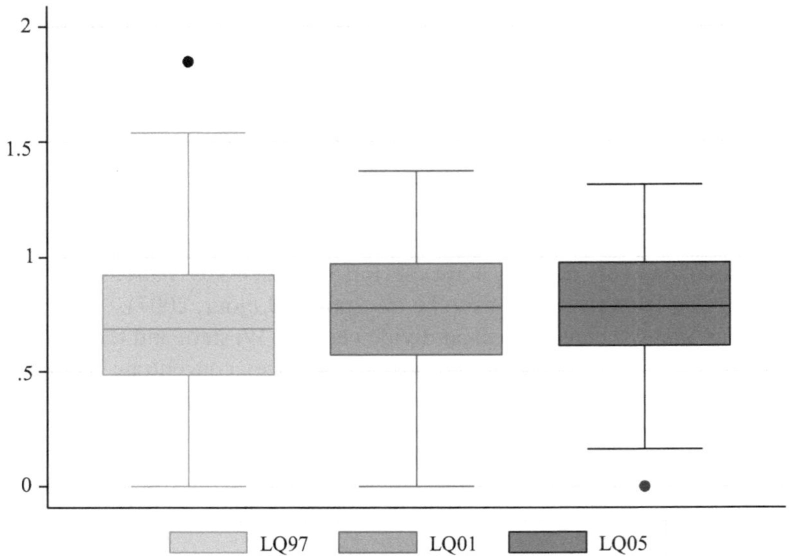

Fig. 7.2 The distribution of newly created foreign firms by regions and over time. *Source:* own calculation

observed period, only one fourth of the EU regions shows a concentration of foreign firms larger than the EU average. However, if one considers that the maximum value of the LQ index is 1.25, it is clear that none of the EU regions substantially deviate from the EU average. Therefore, we can conclude that in 2005–2007, the newly created foreign firms in services were more evenly distributed across EU regions than they were at the beginning of the considered period.

This aggregate picture hides different spatial and time profiles at sectorial level. The discussion on this issue is based on the probability transition matrix (Overman & Puga, 2002; Puga, 2002) that tracks changes over time in the relative position of regions within a given distribution. The transition matrixes in Table 7.2 report changes between the first and the last period considered in the distribution of foreign firms relative to the EU average in eight service sectors.

The transition matrix provides several pieces of information. The given rows reflect initial values, while the columns refer to final values. The main diagonal gives the most important piece of information: it shows the fraction of regions that were in the same range of distribution in the years 1997–1999 and 2005–2007. Initial and final values refer to the standardised LQ indexes, i.e. the relative concentration of FDI inflows in each service sector across EU regions.[9]

[9] Since the LQ index cannot be compared on both sides of 1, it is often made symmetric. The symmetric index, i.e. SLQ = [(LQ − 1)/(LQ + 1)], varies between −1 and +1. Positive values indicate concentration of foreign firms above the EU average while negative values imply dispersion. Values above ±0.50 indicate strong concentration/dispersion, while values ranging between −0.50 and +0.50 suggest weak dispersion/concentration. If SLQ is equal to zero this means that the share of foreign firms in sector *j* and region *i* equals the same share at the EU level.

Table 7.2 Probability transition matrices by sector, 1997–1999 and 2005–2007

Sector E	0	1	2	3	Total	Sector F	0	1	2	3	Total
0	120	17	21	51	209	0	67	26	58	12	163
	57.42	8.13	10.05	24.4	100		41.1	15.95	35.58	7.36	100
1	1	2	2	4	9	1	7	12	5	0	24
	11.11	22.22	22.22	44.44	100		29.17	50	20.83	0	100
2	3	2	1	2	8	2	9	12	20	0	41
	37.5	25	12.5	25	100		21.95	29.27	48.78	0	100
3	14	2	4	14	34	3	5	6	18	3	32
	41.18	5.88	11.76	41.18	100		15.63	18.75	56.25	9.38	100
Total	138	23	28	71	260	Total	88	56	101	15	260
	53.08	8.85	10.77	27.31	100		33.85	21.54	38.85	5.77	100
Sector G	**0**	**1**	**2**	**3**	**Total**	**Sector H**	**0**	**1**	**2**	**3**	**Total**
0	15	6	24	1	46	0	121	36	29	16	202
	32.61	13.04	52.17	2.17	100		59.9	17.82	14.36	7.92	100
1	8	29	31	1	69	1	3	6	0	0	9
	11.59	42.03	44.93	1.45	100		33.33	66.67	0	0	100
2	6	56	81	2	145	2	8	16	2	1	27
	4.14	38.62	55.86	1.38	100		29.63	59.26	7.41	3.7	100
3	–	–	–	–	–	3	6	6	6	4	22
	–	–	–	–	–		27.27	27.27	27.27	18.18	100
Total	29	91	136	4	260	Total	138	64	37	21	260
	11.15	35	52.31	1.54	100		53.08	24.62	14.23	8.08	100
Sector I	**0**	**1**	**2**	**3**	**Total**	**Sector J**	**0**	**1**	**2**	**3**	**Total**
0	55	27	45	10	137	0	128	33	23	6	190
	40.15	19.71	32.85	7.3	100		67.37	17.37	12.11	3.16	100
1	5	20	16	0	41	1	6	11	4	0	21
	12.2	48.78	39.02	0	100		28.57	52.38	19.05	0	100
2	4	22	28	4	58	2	9	17	7	6	39
	6.9	37.93	48.28	6.9	100		23.08	43.59	17.95	15.38	100
3	3	5	13	3	24	3	4	2	1	3	10
	12.5	20.83	54.17	12.5	100		40	20	10	30	100
Total	67	74	102	17	260	Total	147	63	35	15	260
	25.77	28.46	39.23	6.54	100		56.54	24.23	13.46	5.77	100
Sector K	**0**	**1**	**2**	**3**	**Total**	**Sector L-P**	**0**	**1**	**2**	**3**	**Total**
0	35	48	24	–	107	0	89	44	52	9	194
	32.71	44.86	22.43	–	100		45.88	22.68	26.8	4.64	100
1	2	41	43	–	86	1	0	11	5	0	16
	2.33	47.67	50	–	100		0	68.75	31.25	0	100
2	5	31	25	–	61	2	7	8	10	0	25
	8.2	50.82	40.98	–	100		28	32	40	0	100
3	2	4	0	–	6	3	6	10	9	0	25
	33.33	66.67	0	–	100		24	40	36	0	100
Total	44	124	92	–	260	Total	102	73	76	9	260
	16.92	47.69	35.38	–	100		39.23	28.08	29.23	3.46	100

Grey lines report frequencies, white lines report transition probabilities. Legend: 0 = −1<=LQ=<−0.5; 1=−0.5<LQ=<0; 2=0<LQ=<0.5; 3=0.5<LQ=<1. E = energy; F = constructions; G = wholesale and retail trade; H = hotels and restaurants; I = transportations and telecommunication; J = financial services; K = business services; L-P = public and other community, social and personal related services.

By taking a broader perspective, in the years 2005–2007, the percentage of regions with a concentration of FDI inflows well above the EU average (SLQ $>$ 0.5) is very low in all sectors, as indicated by the last row of each transition matrix. The only exception is represented by public utilities (sector E) where about 27 % of EU regions has a concentration of foreign firms well above the EU average. Contrarily, more than 50 % of EU regions shows a concentration of foreign firms below the EU average (SLQ > -0.5) in most consumer services, such as hotels and restaurants (H) and public administration and other community related services (L-P), as well as in financial intermediation (J).

When considering changes in status, the matrixes suggest that persistency is more likely in regions with an initial concentration of FDI below rather than above the EU average, as indicated by the main diagonal of each matrix. The sector with the largest share of regions with a symmetric LQ close to one in both the considered periods of time is financial intermediation (J). It is worth noticing that in two sectors, i.e. business services (K) and public administration and other community related services (L-P), no region showed a concentration of foreign firms well above the EU average in the years 2005–2007. This suggests that foreign firms widespread over time across regions looking for new markets.

4 Drivers and Determinants

In order to estimate the determinants of location choices of foreign service suppliers, a discrete choice model has been applied to a large set of foreign firms investing in Europe. In particular, a negative binomial model has been used since it allows to take into account the overdispersion that usually characterises counted data.[10] Given the panel structure of the data, a fixed effect model has been used, according to which partial regression coefficients are considered constant across cross-sectional units (European regions) while intercepts can vary across them.[11]

The dependent variable is the number of FDI in each sector j, time t and region r, considered as a whole, and disaggregated between intra- and extra-EU foreign firms and across services sectors. As for the explanatory variables, their choice has been inspired by the existing empirical and theoretical literature. Traditionally, most empirical investigations regarding the determinants of FDI have focused on FDI flows aggregated across economic sectors or across manufacturing sectors. However, as most of services are non-tradable, the determinants of inward FDI flows

[10] Over dispersion occurs when the conditional variance exceeds the conditional mean, which may be small because of the presence of many zeros in the data. Negative binomial regression analysis allows to deal with these complications. See Camerun and Trivedi (1998; 2009) for an in depth discussion of count data analysis.

[11] A random effect model has also been estimated. It is not supported by the data, as suggested by the Hausman test statistics reported at the bottom of the tables summarising main results.

may differ from those in the manufacturing sector. The specific nature of services suggests to place attention on investor motivations for becoming multinationals.

Since services are neither storable nor tradable and should be produced where they are consumed, FDI in services is expected to be primarily driven by market-seeking motivations (Nefussi & Schwellnus, 2007; Unctad, 2004). However, it is not easy to identify the boundaries of the market of interest. Services targeted to final consumers need large local markets, while services complementary to business production may be less sensitive to it. Moreover, the recent developments in Information and telecommunication technologies have improved services tradability, thus increasing the size of the potential market that producers may serve from a specific location. In order to account for all these possibilities the regression analysis includes two different measures for market size, i.e. GDP per capita (*GDPpc*) and market potential (*MKT POT*) in the formulation suggested by Head and Mayer (2004). The expected relationship between FDI flows and these variables is positive: the higher the local demand for services—proxied by GDP per capita—and market potential, the larger are potential FDI inflows.

Agglomeration forces can also be crucial in location decisions, as suggested by previous studies on foreign firm-location choice, though non-specifically targeted to FDI in services (Crozet, Mayer, & Mucchielli, 2004; Head, Ries, & Swenson, 1995, 1999; Pusterla & Resmini, 2007). Generally speaking, in order to enjoy agglomeration economies, foreign investors prefer to set up their subsidiaries where the same or related industries are already located. In the services sectors, however, agglomeration effects may not only have a different intensity according to the type of activity carried out by the foreign producers, but may also respond to functional rather than to cost reasons. From this point of view, foreign firms in services might prefer to locate in urban or densely populated areas, where there are already a large number of firms operating in different economic sectors, rather than close to other domestic or foreign firms operating in the same economic sector. In order to test these hypotheses and compare their relative importance, two different proxies for agglomeration economies have been included in the analysis: the first is a dummy variable (AGGLOM) identifying those regions with a city with more than 300,000 inhabitants and a population density of at least 300 inhabitants per squared kilometre or a population density of 150–300 inhabitant/km^2;the second, instead, refers to the relative specialization of each region in the manufacturing sector (*MAN*).[12] The expected sign of the estimated coefficient for agglomerated regions is positive, implying that the choice of location of foreign firms is mainly driven by functional reasons, while the expected sign of the estimated coefficient of the region's specialisation in manufacturing business is ambiguous. A positive sign suggests first that FDI in services are driven by the demand of the manufacturing firms and, secondly, that foreign service producers exploit inter-sectorial spill-overs, while a negative sign indicates that FDI in services prefer to locate close to

[12] Regional relative specialization in manufacturing has been computed by comparing the share of the manufacturing sector on total regional value added to the same share computed at the EU level.

other firms operating in the services sectors in order to save information costs and exploit intra-sectorial knowledge spill-overs from existing producers.

Since it has already be proven that FDI inflows in services are sensitive to the degree of openness of services sectors (Kox & Lejour, 2006; Lejour, 2007), the possibility that FDI inflows in services may be hampered by sector-specific restrictions is accounted for by including a dummy variable (*RESTR*) taking the value of 1 if a service sector is more restricted than the average for all sectors included in the sample.[13] Restricted sectors are energy, transportation and telecommunication services and financial intermediation (Golup, 2009).[14] Another sector-specific dummy has been included in order to understand whether and to what extent FDI drivers vary across groups of homogenous services, i.e. producer and consumer services (PROD).[15]

Finally, since larger regions attract more FDI than smaller regions, the size of the region—measured in squared kilometres—(*AREA*) has been added to the regressors' set in order to correct for possible distortions due to differences in size. The dummy EU15, instead, helps in identifying differences in foreign firm location patterns in Western and Central and Eastern Europe.

The regression equation, thus, takes the following form:

$$\begin{aligned} \mathrm{FDI}_{jrt} = {} & \alpha_r + \beta_1 \log \mathrm{GDPpc}_{rt} + \beta_2 \log \mathrm{MKT\ POT}_{rt} + \beta_3 \log \mathrm{AREA}_r \\ & + \beta_4 \log \mathrm{MAN}_{rt} + \beta_5 \mathrm{AGGLOM}_r + \beta_6 \mathrm{RESTR}_j + \beta_7 \mathrm{PROD\ SER}_j \\ & + \beta_8 \mathrm{EU15}_r + \alpha_t + \alpha_J + \varepsilon_{jrt} \end{aligned} \tag{7.1}$$

Table 7.3 shows the main results for foreign firms as a whole and intra- and extra-EU foreign firms. As the table indicates, higher levels of regional GDP per capita have a positive effect on FDI inflows, regardless of the origin of foreign firms, thus confirming the horizontal nature of FDI in services and the importance of local markets in the location processes of foreign firms. Contrarily, agglomeration forces do not seem to be important in these processes. However, the negative sign of the coefficient of the specialisation in the manufacturing sector variable suggests that foreign firms prefer to locate close to other service producers in order to enjoy intra-sectorial knowledge spillovers, while the positive sign of the proxy for the settlement structure indicates that location externalities arise for functional reasons. As expected, stricter regulations negatively affect FDI flows, while some sector specificities may also have an impact on the estimated relationships, as

[13] On how to measure openness to FDI in the service sectors, see Golup (2009).

[14] Needless to say, restrictions refer to specific branches of the above mentioned sectors such as air transportation or banking. However, the lack of more disaggregated data does not allow using a finer classification.

[15] Producer services include energy (E), constructions (F), transport and communication services (I), financial intermediations (J) and Business services (K), while wholesale and retail trade (G), hotels and restaurants (H), as well as public administration, education and health and social work service activities (L-P) encompass to the consumer service group.

Table 7.3 The determinants of FDI in services

	FDI	Sig.	extra-EU FDI	Sig.	intra-EU FDI	Sig.
GDP per capita	1.01	***	0.89	***	1.08	***
	(0.078)		(0.006)		(0.081)	
Market potential	0.22	***	0.11		0.26	***
	(0.055)		(0.086)		(0.057)	
Area	0.20	***	0.10	**	0.24	***
	(0.031)		(0.046)		(0.032)	
Manufacturing sector	−0.07		−0.13		−0.09	*
	(0.047)		(0.083)		(0.048)	
Agglomerated regions	0.02		0.01		0.03	
	(0.049)		(0.070)		(0.050)	
EU-15	−0.76	***	−0.51	***	−0.92	***
	(0.08)		(0.123)		(0.080)	
Restricted sectors	−0.74	***	−0.88	***	−0.75	***
	(0.040)		(0.057)		(0.041)	
Producer services	1.04	***	0.88	***	1.07	***
	(0.036)		(0.048)		(0.037)	
Constant	−14.44	***	−12.20	***	−15.55	***
	(0.651)		(0.901)		(0.680)	
Hausman test	379.85	***	379.5	***	441.67	***
Log likelihood	−17,697.30		−9,381.44		−16,173.98	
No. of obs.	8,305		7,733		8,305	

*, **, *** indicate significance at 10, 5 and 1 percent level, respectively. Standard errors in parenthesis
All regressions include time and sector-specific dummies

potentially indicated by the positive and significant sign of the coefficient of the dummy identifying producer services, which is robust to all specifications. Finally, it is worth noticing that regions belonging to the old EU-15 member states are able to attract, *ceteris paribus*, less newly created foreign firms than regions belonging to new EU-12 member states. No relevant differences seem to emerge between intra- and extra-EU foreign firms if one does not consider that intra-EU foreign service producers are not attracted by regions specialised in manufacturing, as indicated by the negative though weakly significant sign of the corresponding variable. Therefore, this distinction has been abandoned in the following analysis.

In order to further explore the role of sector specific effects in the location process of foreign firms, the sample has been split into two different sub-samples: producer and consumer services. Here, some interesting differences emerge, as indicated by the results shown in Table 7.4. In particular, FDI in producer services are more oriented to the local market than FDI in consumer services, as indicated by the coefficient of the market potential variable which is positive in both specifications but significant only for consumer services. Moreover, FDI in consumer services are particularly sensitive to within-services agglomeration externalities, while patterns of location of foreign firms in producer services are not affected by agglomeration

Table 7.4 The determinants of FDI inflows by groups of homogenous sectors

	Consumer services	Sign.	Producer services	Sig.
GDP pro capite	1.49	***	2.10	***
	(0.125)		(0.125)	
Market potential	0.50	***	0.11	
	(0.090)		(0.090)	
area	0.40	***	0.29	***
	(0.053)		(0.050)	
Manufacturing	−0.19	***	−0.08	
	(0.072)		(0.075)	
Agglomerated regions	0.06		−0.02	
	(0.083)		(0.074)	
EU15	−1.10	***	−1.73	***
	(0.134)		(0.129)	
Restricted sectors	...		−2.15	***
			(0.065)	
Constant	−17.91	***	−21.45	***
	(0.981)		(0.940)	
Hausman test	28.27	**	99.44	***
Log likelihood	−6,559.34		−7,987.30	
No. of obs.	4,518		3,745	

*, **, *** indicate significance at 10, 5 and 1 per cent level, respectively. Standard errors in parenthesis. All regressions include time and, when appropriate, sector-specific dummies

forces, regardless of the motivations at the base of their surge. Sector-specific regulations exert a negative impact on FDI inflows in producer-service sector.

Table 7.5 summarizes the determinants of FDI inflows in individual service industries. Although some estimates cannot be explained either from a theoretical or an economic point of view, sufficient evidence that the effects of FDI determinants may vary across service sectors has been found.[16] More specifically, consumer services, such as wholesale and retail trade and hotels and restaurants, seem to be more sensitive to local and potential market conditions than producer services, such as transportation and communication and business services. Moreover, only one service sector, i.e. transportation and communication services, seems to respond to agglomeration forces. However, foreign firms operating in this sector are attracted by within-sector externalities rather than inter-sectoral externalities, while regions densely populated and with larger cities attract, *ceteris paribus*, less FDI inflows than less urbanized regions, as indicated by the negative and significant coefficient of the agglomerated region dummy variable. This result suggests that in regions hosting larger cities or densely populated domestic and foreign competition may be stronger for these service activities than for other services. It is worth noticing that

[16] In particular, results cannot explain FDI inflows in the energy sector, the most restricted one at least according to the existing literature (Golup, 2009) and in public services and other social and personal services. Sector peculiarities in terms of specific regulations and the public nature of most of these activities may explain these poor results.

Table 7.5 The determinants of FDI by individual services sectors

	E	Sig.	F	Sig.	G	Sig.	H	Sig.	I	Sig.	J	Sig.	K	Sig.	L-P	Sig.
GDP per capita	0.31		2.13	***	1.30	***	1.98	***	2.06	***	1.58	***	2.77	***	1.95	***
	(0.653)		(0.318)		(0.232)		(0.379)		(0.357)		(0.423)		(0.290)		(0.224)	
Market potential	−0.45		0.64	**	0.54	***	0.64	**	0.55	*	0.77	**	0.03		−0.08	
	(0.476)		(0.261)		(0.189)		(0.284)		(0.298)		(0.345)		(0.212)		(0.160)	
Area	−0.78	***	0.64	***	0.38	***	0.54	***	0.55	***	0.43	**	0.23	**	0.16	*
	(0.298)		(0.148)		(0.104)		(0.193)		(0.154)		(0.185)		(0.114)		(0.090)	
Manufacturing	0.32		−0.22		−0.20		−0.28		−0.46	**	0.06		−0.20		−0.03	
	(0.389)		(0.203)		(0.144)		(0.209)		(0.221)		(0.286)		(0.186)		(0.144)	
Aggl. regions	0.27		0.17		−0.10		−0.12		−0.39	*	−0.22		−0.17		0.01	
	(0.430)		(0.224)		(0.167)		(0.313)		(0.221)		(0.280)		(0.171)		(0.129)	
EU15	−0.44		−1.72	***	−0.78	***	−1.93	***	−1.52	***	−0.73		−2.06	***	−1.82	***
	(0.652)		(0.331)		(0.254)		(0.404)		(0.364)		(0.454)		(0.314)		(0.244)	
Constant	4.65		−27.34	***	−16.81	***	−24.88	***	−25.27	***	−21.04	***	−27.78	***	−20.44	***
	(5.073)		(2.326)		(1.773)		(2.882)		(2.309)		(2.797)		(2.081)		(1.682)	
Hausman test	16.77	**	58.31	***	49.46	***	22.34	***	62.21	***	29.33	***	92.65	***	162.73	***
log likelihood	−423.03		−895.21		−1855.20		−597.75		−1027.97		−661.76		−1582.06		−2686.93	
n. of obs.	534		678		753		581		687		534		724		2,592	

*, **, *** indicate significance at 10, 5 and 1 percent level, respectively. Standard errors in parenthesis
All regressions include time dummies

agglomeration externalities between manufacturing and services sectors positively affect FDI inflows only in energy and financial services, though it is not significant.

Quite surprisingly, foreign firms in consumer services are not attracted by regions hosting larger cities, as indicated by the negative though not significant coefficient of the corresponding dummy variable. As before, this may be explained by the fact that in these regions, foreign firms may be more exposed to competition than in less urbanised regions.[17]

5 Summary and Conclusions

This paper has investigated the spatial and sectorial distribution of inward FDI flows in the European services industry. It has demonstrated that over time the distribution of FDI flows in Europe has changed, both from a geographical and a sectorial point of view. In particular, relatively more FDI is directed towards services sectors and more FDI has been flowed into the new member states of Central and Eastern Europe.

This paper has also explored the determinants of FDI inflows in the services sectors by using a cross-regional panel of three periods of time, which altogether cover more or less the 2000s decade. The empirical analysis indicates that FDI in services are mainly driven by market reasons, though some minor differences in terms of size and significance of the estimated coefficients do exist across different service industries. In particular, local demand seems to be more important than market potential, indicating that fragmented service markets across Europe still exist. The findings also confirm the hypotheses that high levels of regulation reduce FDI inflows, as well as the East enlargements of the EU created further opportunities for FDI in the services sectors. Agglomeration effects, instead, are very weak: within-service externalities drive FDI inflows in consumer service sectors, while agglomeration for functional reasons do not exert any impact on inflows of FDI. Only FDI in transportation and communication services seems to be affected by the settlement structure of EU regions: regions hosting larger cities are less attractive than less urbanized regions, because of stronger competition effects. No other sector-specific differences emerge from the analysis, even when one compares intra-EU direct investments with extra-EU FDI.

Despite its limitations, the present research contributes to the existing literature in several ways. First of all, it represents a first comprehensive view of the composition and spatial distribution of foreign investment flows in the European services industry. Secondly, it offers more insights into the determinants of FDI in services. Last, but not least, it allows to assess the "missed" opportunities for intra-EU FDI due to the existence of two related phenomena, i.e. residual restrictions to the free circulation of

[17] Possible collinear effects with the GDP per capita variable cannot be excluded, though not detected by the correlation matrix.

some services and persisting differences in national regulations that prevent the full functioning of the single market. Thus, the main policy implication that can be drawn from this study is the need for a more open market for services both at a European and worldwide level. To this respect, the full implementation of the Service directive may represent an important step forward in this direction, though important services activities are not included in it. Given the special nature of some services, however, this process should be carefully monitored and driven by competent regulatory agencies at both national and international level, in order to minimize negative effects of FDI and maximize the positive ones. The need to balance costs and benefits of foreign investments is not a new issue in the policy debate, but it is much more sensitive in the case of services, especially those concerning socially or culturally sensitive areas, because of their greater human intensity and because a number of services take the form of public goods.

References

Arnold, J., Javorcik, B., & Mattoo, A. (2007). Does service liberalisation benefit manufacturing firms? Evidence from the Czech Republic. World Bank working paper series n. 4109. Washington, DC: The World Bank.

Basile, R., Castellani, D., & Zanfei A. (2009). National boundaries and the location of multinational firms in Europe. *Papers in Regional Science, 88*(4), 733–748.

Camerun, A., & Trivedi, P. (1998). *Regression analysis of count data.* Cambridge: Cambridge University Press.

Camerun, A., & Trivedi, P. (2009). *Microeconometrics using stata.* College Station, TX: Stata Press Publication.

Capello, R., Fratesi, U., & Resmini, L. (2011). *Globalization and regional growth in Europe: Past trends and future scenarios. Advances in spatial science.* Berlin: Springer.

Casi, L., & Resmini, L. (2010). Evidence on the determinants of foreign direct investment: The case of EU regions. *Eastern Journal of European Studies, 1*, 93–118.

Crozet, M., Mayer, T., & Mucchielli, J. L. (2004). How do firms agglomerate? A study of FDI in France. *Regional Science and Urban Economics, 34*, 27–54.

Davis, R., & Guillin, A. (2011). *How far away is an Intangible? FDI, Space and Services.* UCD Centre for Economic Research, working paper n. 11/20.

Deardorff, A. (1985). Comparative advantage and international trade and investment in services. In R. Stern (Ed.), *Trade and investment in services: Canada/US perspectives* (pp. 39–71). Toronto: Ontario Economic Council.

EC. (2004). *Proposal for a Directive of the European Parliament and of the Council on Services in the Internal Market*, SEC(2004), 21, Brussels.

EC. (2005). The location of multinationals across EU regions. In *The EU economy 2005 review*, pp. 237–249.

Golub, S., Jones, R., & Kierzkowski, H. (2007). Globalization and country-specific service links. *Journal of Economic Policy Reform, 10*, 63–88.

Golup, S. (2009). Openness to foreign direct investment in services: An international comparative analysis. *The World Economy, 32*, 1246–1268.

Gorg, H., & Hanley, A. (2011). Services outsourcing and innovation: An empirical investigation. *Economic Inquiry, 49*, 321–333.

Guimaraes, P., Figueiredo, O., & Woodward, D. (2000). Agglomeration and the location of foreign direct investment in Portugal. *Journal of Urban Economics, 47*, 115–135.

Head, K., & Mayer, T. (2004). Market potential and the location of Japanese Investment in the European Union. *The Review of Economics and Statistics, 86*, 959–972.
Head, K., Ries, J., & Swenson, D. (1995). Agglomeration benefits and location choice: Evidence from Japanese manufacturing investments in the United States. *Journal of International Economics, 38*, 223–247.
Head, K., Ries, J., & Swenson, D. (1999). Attracting foreign manufacturing: Investment promotion and agglomeration. *Regional Science and Urban Economics, 29*, 197–218.
Kolstad, I., & Villanger, E. (2008). Determinants of foreign direct investment in services. *European Journal of Political Economy, 24*, 518–533.
Kox, H., & Lejour, A. (2006). The effects of the services directive on intra-EU trade and FDI. *Revue économique, 57*, 747–769.
Lejour, A. (2007). FDI in services: Recent developments and prospects in Europe. In K. Liebscher, J. Christl, P. Mooslechner, & D. Ritzberger-Grunwald (Eds.), *Foreign direct investment in Europe: A changing landscape* (pp. 175–189). Cheltenham, UK: Edward Elgar.
Liebscher, K., Christl, J., Mooslechner, P., & Ritzberger-Grünwald, D. (2007). *Foreign direct investment in Europe: A changing landscape*. Cheltenham, UK and Northampton, MA, USA: Edward Elgar.
Markusen, J. (2007). *Trade and foreign direct investment in business services: A modelling approach*. Mimeo.
Markusen, J., Rutherford, T., & Tarr, D. (2005). Trade and direct investment in producer services and the domestic market for expertise. *Canadian Journal of Economics/Revue canadienne-d'Economique, 38*(3), 758–777.
Nefussi, B., & Schwellnus, C. (2007). *Does FDI in manufacturing cause FDI in business services? Evidence from French firm-level data*, CEPII working paper n. 21.
Olsen K. B. (2006). *Productivity impacts of offshoring and outsourcing: a review*. OECD Science, Technology and Industry Working Papers n. 1.
Overman, H. G., & Puga, D. (2002). Unemployment clusters across European regions and countries. *Economic Policy, 17*, 115–148.
Overman, H. G., Redding, S., & Venables, A. (2003). The economic of geography of trade, production and income: A survey of empirics. In K. Choi & J. Harrigan (Eds.), *Handbook of international trade*. Oxford, UK: Blackwell Publishing Ltd. Ch. 12.
Oxelheim, L., & Ghauri, P. (2004). *European Union and the race for foreign direct investment in Europe*. Bingley, UK: Emerald Group Publishing.
Puga, D. (2002). European regional policies in light of recent location theories. *Journal of Economic Geography, 2*, 373–406.
Pusterla, F., & Resmini, L. (2007). Where do foreign firms locate in transition countries. *The Annals of Regional Science, 41*, 835–856.
Riedl, A. (2010). Location factors of FDI and the growing services economy. Evidence for transition countries. *The Economics of Transition, 18*, 741–761.
UNCTAD. (2004). *World investment report 2004. The shift towards services*, chapter III, Geneva: United Nation.
UNCTAD. (2009). *Assessing the impact of the current financial and economic crisis on global FDI flows*. Geneva: United Nations.
UNCTAD. (2011). *World investment report 2011. Non-equity modes of international production and development*. Geneva: United Nation.
World Bank. (2010). *World development indicators*. www.worldbank.org. Accessed on February 2012.
World Trade Organization. (2010). *World trade report 2010*, Geneva.

Chapter 8
Export Decisions of Services Firms Between Agglomeration Effects and Market-Entry Costs

Henk L.M. Kox

1 Introduction

Last decade has seen a fast expansion of theoretical and empirical studies on international trade with firms that are heterogeneous with respect to some productivity performance parameter. The growth of this literature is partly generated by the recent availability of international transaction data at the firm level. These data have enabled researchers to test a host of predictions made by the new trade models with heterogeneous firms. New in these models and studies is that they explain why not all firms participate in exports, and why exporters differ from non-exporters. Basically, the models assume that exporters face sunk entry costs in foreign markets and that only firms with superior productivity are profitable enough to absorb these sunk costs. The self-selection by firms into exports explains why exporters are systematically more productive than non-exporters and why many firms refrain from exporting. The predictions by now have been corroborated by firm-level studies for many countries.[1]

More recently, scientists in the regional economics specialisation have drawn attention to the fact that the heterogeneous-firms trade economists perhaps jump to conclusions too easily: they should consider the impact of agglomeration effects. Regional economists have, in the footsteps of Adam Smith and Alfred Marshall, firmly established that firms and workers in large cities are on average more productive than comparable firms and workers in non-urban areas. If high-productivity exporters are predominantly located in urban areas, the trade researchers may in fact have (re)discovered agglomeration effects. The main problem with the estimation of agglomeration effects is that it is difficult to distinguish between two competing

[1] Comprehensive surveys of the results have been provided by Wagner (2007, 2012), Greenaway and Kneller (2007), Mayer and Ottaviano (2007), and ISGEP (2008).

H.L.M. Kox (✉)
CPB Netherlands Bureau for Economic Policy Analysis, The Hague, Netherlands
e-mail: h.l.m.kox@cpb.nl

J.R. Cuadrado-Roura (ed.), *Service Industries and Regions*,
Advances in Spatial Science, DOI 10.1007/978-3-642-35801-2_8,

explanations for the positive correlation between agglomeration and productivity. First, productivity is high because of agglomeration effects: local spatial externalities (e.g. thick labour markets, shared suppliers, concentration of high-skilled workers, learning potential). Second, agglomeration is an endogenous consequence (rather than a cause) of high productivity, if workers and firms are drawn to regions where successful high-productivity firms are established. Econometric research of the latter causality chain typically requires long time series, because the geographical shift of firms and workers is a process that materialises in the long term.

Combes, Duranton, Gobillon, Puga, and Roux (2008, 2012) using French data find that agglomeration effects contribute positively to above-average productivity of firms in larger cities. They claim that the process is generated by the fact that Darwinian competitive selection among firms is tougher in large cities than elsewhere in a country.[2] If we assume that the results by Combes et al. (2012) are correct, this must have consequences for the export decision of firms. We may infer the prediction that for large-city firms the step towards exports will be smaller than for firms elsewhere in a country where both the local Darwinian selection among firms and the agglomeration advantages are weaker. The present paper will investigate this implication.

The paper contributes in three ways to the self-selection versus agglomeration discussion. Firstly, it establishes the role of productivity based self-selection for export decisions of services firms. Secondly, the paper investigates whether productivity-based self selection of exporters is stronger in non-urban areas than in large-city areas. A third contribution to the literature is the finding that productivity self selection for exporters also depends on the market structure, more in particular the degree of product homogeneity.

The findings of the paper can be summarised as follows. The export decision of services firms goes along with positive, performance-based self selection, even after controlling for a range of other factors that may affect performance. This self-selection is strongest in markets with more or less homogeneous products. The magnitude of productivity-based self selection by exporters (in response to sunk foreign market-entry costs) tends to be over-estimated if one does not control for productivity sorting in large-city areas. Taking account of agglomeration effects should become standard procedure in empirical tests of heterogeneous-firms trade models. Finally, productivity-based self selection for exporters is stronger in rural areas than in urban areas, because rural services firms lack the agglomeration-caused productivity benefits.

The structure of the paper is as follows. Section 2 positions the research questions in the context of the international literature from which researchable hypotheses will be developed. Section 3 describes the data and the empirical strategy. Section 4 first presents some descriptive results from the data. Section 5 investigates the static aspects of productivity-based self selection and agglomeration for exporting firms.

---

[2] Allowing only the most productive ones to survive and thus generating higher average productivity compared to rural areas (like in Melitz & Ottaviano, 2008).

I distinguish between markets with relatively homogeneous and markets with relatively differentiated products. Section 6 concentrates on dynamic, time-dependent aspects of the self-selection process and the way this is affected by agglomeration factors. Section 7 discusses the main results and concludes.

2 Agglomeration, Productivity Sorting and Exporting

Bernard and Jensen (1995) were the first authors to question the causality direction between productivity differences and export participation. Until then, received theory had it that firm productivity increased by the export experience itself, through the contact with competitors in international markets. The "self-selection school" came into being with almost simultaneous papers by Bernard, Eaton, Jensen, and Kortum (2003) and Melitz (2003). Both assume that firms are heterogeneous in terms of productivity. Melitz (2003) so far has turned out to be the most fruitful approach. It assumes monopolist competition and one-off sunk entry costs for firms entering export markets.[3] Only firms with productivity above a certain threshold level expect to overcome the sunk market-entry costs and decide to start exporting. Lowering of sunk market barriers creates market opportunities for exporters that have lower productivity than the firms that were exporting before the sunk market barriers decreased. Nonetheless, the increased import penetration drives out the least-productive domestic firms. The resulting market reallocation causes average productivity in the importing country to go up. The Melitz model has been followed by several papers offering variants and elaborations of the original model.[4] Helpman, Melitz, and Yeaple (2004) extend the analysis to foreign market entry through direct investment, i.e. local production by multinational enterprises (MNE). They claim that the sunk costs required to become a MNE are even higher than those for an exporter. Based on this, their model predicts that productivity self-selection for firms that start multinational activities will even be stronger than for export starters.

The heterogeneous-firms trade model by Melitz (2003) is intrinsically complex. In technical terms, it requires a joint solution of the trade equilibrium and the entry conditions for all firms to identify the marginal domestic and the marginal exporting firms. Some of the later papers try to simplify the model without loss of generality and to make it more easily testable (e.g. Baldwin, 2005; Lawless, 2009). Both Balistreri, Hillberry, and Rutherford (2008) and Chaney (2008) achieve a simplification by decomposing the model into a general equilibrium module and a

[3] Examples of such costs are language barriers, complying with foreign regulations, adapting the product to the foreign market, the fixed costs of reaching foreign customers. Early trade models with sunk costs are Dixit (1989), Baldwin (1988) and Baldwin and Krugman (1989).

[4] See Redding (2010) for a review. Prominent theoretical papers are Yeaple (2007), Eaton, Kortum, and Kramarz (2011), Melitz and Ottaviano (2008), and Chaney (2008).

partial-equilibrium module. The general-equilibrium part evaluates the relative prices, expenditure structures, comparative advantage and the terms of trade. The partial equilibrium module determines the industrial-organisation aspects: market entry, production and employment and export decisions. An attractive property of the Chaney model is that it predicts that firms in homogeneous markets (high substitution elasticity between products) need a bigger performance premium to enter an export market. Conversely, exporter premia are predicted to be lower in markets with more product differentiation. I will test this prediction later in the paper.

The literature on the impact of agglomeration on the productivity distribution of firms and workers is vast. It is by now well-established that this impact is statistically significant and positive (e.g. Andersson & Lööf, 2009; Ciccone, 2002; Ciccone & Hall, 1996; Duranton & Puga, 2004; Melo, Graham, & Noland, 2009; Rosenthal & Strange, 2004). A positive "urban productivity premium" has been widely researched and it could be registered in many countries. Agglomeration impacts may work along different lines. Three main micro-foundations for urban agglomeration effects are: (1) sharing of resources and suppliers, (2) local matching of quality-based search for specific inputs and labour skills and (3) local learning spill-overs (e.g. Duranton & Puga, 2004; Moretti, 2004). Larger regions may be able to sustain a larger variety of input suppliers and a deeper division of labour that makes workers more productive. Urban agglomerations offer opportunities for the generation, the diffusion, and the accumulation of knowledge (Marshallian externalities). Furthermore, the increasing-returns-to-scale effects of a larger market when many firms and workers crowd together, come on top of this (Krugman, 1991).

The interaction of agglomeration and productivity self selection of exporters is an aspect in the trade literature that so far has received only limited attention. Early papers on trade and agglomeration are Krugman (1991) and Ottaviano, Tabuchi, and Thisse (2002), but at that time the seminal papers on trade with heterogeneous firms still had to appear. The first papers to link productivity self selection and agglomeration are Baldwin and Okubo (2006), Melitz and Ottaviano (2008) and the papers by Combes (2008, 2012). Baldwin and Okubo (2006) focus on the fact that firms in large markets benefit more from forward and backward linkages. Melitz and Ottaviani (2008) analyse the relation between toughness of competitive selection and country size, which subsequently feeds back into impacts on the selection of exporters in the large-country markets.

Combes et al. (2012) concentrate on firm selection in large cities, with selection defined as "the inability of weak firms to survive when faced with tougher competition in large markets". Their model is a generalisation of Melitz and Ottaviano (2008). It predicts that: (1) stronger selection in large cities left-truncates the productivity distribution; (2) stronger agglomeration right-shifts the productivity distribution and dilates/widens it. An important innovation in their paper is that the authors propose a number of solutions to overcome the econometric difficulty of identifying agglomeration economies and to separate the questions whether average productivity is higher in denser cities, because:

- Shared resources in large agglomerations generate productivity benefits for all local firms.
- There is greater selection that eliminates the least productive firms.

Earlier papers like Melitz and Ottaviano (2008) or Syverson (2004) predict that in larger and more dense markets a tougher competition causes left truncations of the productivity distribution and therefore a raised average productivity. Combes et al. (2012) argue that it is insufficient to only use summary statistics about the productivity distribution (variance, median, bottom decile), because both self selection and agglomeration can be the cause of an increase in median productivity or in the performance of the bottom productivity decile. So an identification problem remains. They propose that a proper test should look at all quantiles of the productivity distribution without imposing particular distributional assumptions or restrictions. Using French data they found that self selection alone cannot fully explain the left truncation of the distribution and the right-shift of median productivity, hence agglomeration must play an additional role. Moreover, the agglomeration advantages of large cities accrue to all local firms and therefore show up in the form of changes over the *entire* productivity distribution.

Combes et al. (2012) is by now a well-travelled paper that has been around for some time (e.g. Combes et al., 2008), so we probably may quite safely assume that their findings stood the test of all types of professional critique. Hence, it is time to think about its consequences. It means that the standard empirical tests of the heterogeneous-firms trade models are probably flawed for two reasons. Firstly, because agglomeration effects generally are not taken into consideration in this empirical literature. Secondly, because the standard procedure of testing these models has often been a test on summary statistics about the productivity differences between exporters and non-exporters. A typical test like the existence of a "productivity premium" for exporters (cf. Wagner, 2007, 2012) is unable to distinguish between agglomeration and self selection effects. This leads to biased measurement as is illustrated in Fig. 8.1. The national population of firms in a particular services industry is split up in two part: firms in large-city agglomerations (thin, dashed line) and firms in small and rural agglomerations (solid line). The productivity distribution of the large-city firms has a right-shift due to positive agglomeration benefits that accrue to all large-city firms. Moreover, due to a more fierce competitive selection in the large cities, we also see a left-truncation of the productivity distribution. As a consequence, the least-productive large-city firms (B) are more productive than the least-productive firms (C) in the rest of the country. Now assume that all exporters in the country face the same sunk foreign market-entry costs and face the same price for their products. The minimal productivity required to compensate for the sunk foreign entry costs is depicted by A, the cut-off productivity for all exporters. Because the distribution area dominated by AB is smaller than the area dominated by AC, the productivity premium in large cities is considerably smaller than that in the rural areas. The standard procedure is, however, to measure with national data with neglect of agglomeration differences. If the two sub-sample distributions of Fig. 8.1 are added up to one national distribution, the distribution

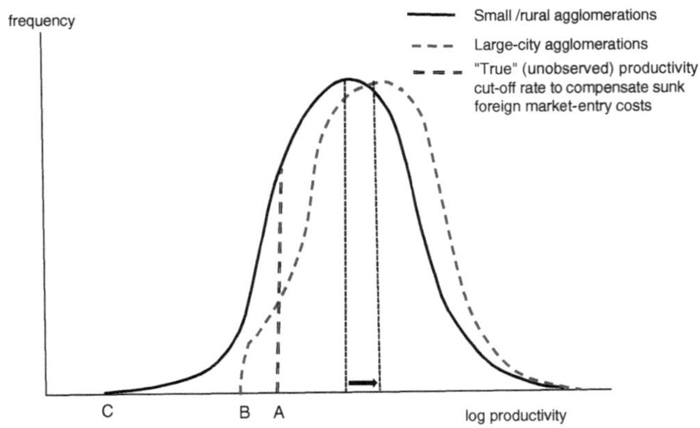

Fig. 8.1 Agglomeration-caused measurement bias of productivity premium for exporters

tail AC will become "fatter" due to the fact that the AB firms of the large-cities are added. The result is that the average productivity premium for large-city exporters will be *over*-estimated, while the average productivity premium for rural exporters is *under*-estimated. It depends on the relative size of both sub-sample how large the overall bias is.

In this paper I address these measurement challenges for Dutch services exporters when testing performance-based sorting into export by explicitly dealing with agglomeration differences. Moreover, instead of estimating productivity premia (a potentially misleading summary statistic) I use a probit estimation procedure that evaluates the entire productivity distribution.[5]

3 Data and Empirical Strategy

The exports of firm *i* can be decomposed into a decision on export participation, a decision on country choice, a decision on export products, a decision on export quantities and a decision on the export price. Some of these decisions may be taken simultaneously, but for analytic reasons it is useful to dissect the export decision in a number of sub-decisions. Figure 8.2 shows that the decision structure has four levels. At each level we may distinguish a participation decision (extensive margin) and a decision on the quantity and volume (intensive margin). Using the symbols of the figure, the export of firm *i* decomposes into three extensive margin decisions (A_i, C_i, E_i) and the price and volume decision (G_i, P_i). Sunk, fixed and variable trade barriers may affect firm export behaviour at each of these levels. This paper concentrates on the level-1 decision: the role of productivity self selection and agglomeration on the export participation level of individual firms.

[5] Although productivity premiums will be shown as a form of robustness test.

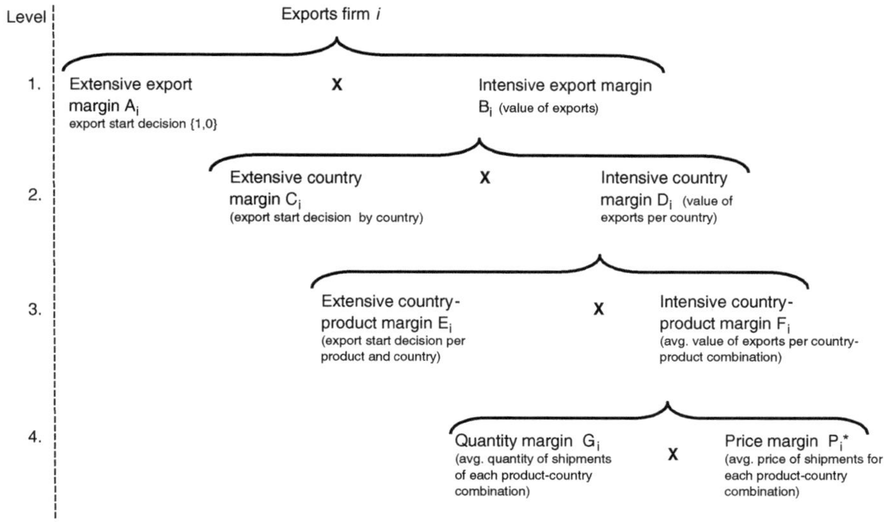

Fig. 8.2 The structure of export decisions (export margins)

When testing the impacts of self selection and agglomeration on export decisions one has to deal with endogeneity and causality issues. Agglomeration effects in large-city areas may have a positive impact on services firms, but at the same time the awareness of such impacts works like a magnet on high-productivity services firms to relocate to these agglomerations. Several endogeneity loops may play a role in this process:

- Mobile workers consume where they work, generating a large local demand pool and thus activating the increasing-returns-to-scale effects on productivity of firms that produce consumption items for workers (Krugman, 1991). Since services are often produced and consumed locally, this endogeneity bias can be strong when we want to assess agglomeration impact on services productivity.
- High-educated workers are often endogenously drawn to regions or cities where successful high-productivity firms are already established. This creates a relatively thick local labour market for high-educated workers, with positive productivity spill-overs for firms in this region.
- Firms may require the outputs of their sector as intermediate inputs (Krugman & Venables, 1995). This generates another endogeneity loop between local demand and increasing-returns-to-scale effects on productivity of local firms. Input–output studies show that business services and other intermediary services have high within-industry input intensities (Kox, 2002). Hence, the endogeneity bias could be particularly strong in services.

The endogeneity biases are probably strongest for the headquarters and top hierarchical layers of services firms. The latter typically have a strong demand for high-skilled workers, and also more demand for inputs of articulated producer services. If one wishes to assess the specific impact of agglomeration on productivity, it is probably better to choose a plant-level or establishment-level approach. For instance, Moretti (2004) and Henderson (2003) estimate *plant-level* production functions that are extended with variables reflecting the local environment.[6] If one would choose the location of headquarters as reference, the measured productivity effects of agglomeration could more easily be "polluted" and over-rated by the endogenous sorting into cities where other successful high-productivity firms are already established. I will therefore study agglomeration effects using establishment data, the lowest level of observation of economic units in the business demography.

Establishment-level data for Dutch services are produced and have kindly been provided by Statistics Netherlands. These data called Productiestatistieken (PS) are collected through annual surveys. In the definition by Statistics Netherlands, an establishment is characterised by relative independence in production or distribution, and it offers its products to an external market. Because also export data are available at the level of establishments, this seems an excellent basis for assessing how agglomeration effects and productivity-based self selection affect the export-participation decision. Establishments with 50 or more employees are represented each year, while smaller firms are represented on the basis of a rotating and stratified annual sample.[7] The probability that a small establishment (with less than 50 employees) is in the sample during a number of consecutive years is therefore small.[8] Due to statistical breaks in the time series we only use data for the period 1999–2005. On average we have 7,500 annual observations for services establishments. The services data include mainly construction, transport and business services. We have enriched the establishment data with characteristics of the associated firms using the General Business Register (ABR) and the Financial Statistics for Firms (SFO). The ABR data allow linking firm-level and establishment-level data, providing inter alia data on location, municipality and ownership by multinational firms.

[6] Moretti (2004) focuses on the education-level of the employees in the region, whereas Henderson (2003) focuses on the number of other firms in the same industry in the region as a source of spill-over effects.

[7] The sampling data are provided in the form of an expansion factor that says for each sampled establishment how much other establishments it represents in its stratum. This expansion factor is used as a weight in regressions.

[8] The problem of a long under-represented tail has been reduced by adopting a cut-off size of at least ten employed workers for services establishments. Export participation and association with multinational firm are less important for these small establishments. Data entries holding imputations by Statistics Netherlands were removed from our sample, keeping only questionnaire-based establishment data.

Rather than inquiring into specific types of external economies, I refer to agglomeration economies as a comprehensive term for urban increasing returns (cf. Fujita & Thisse, 2002). To capture the potential impact of urban increasing returns on the productivity of establishments I use a set of three variables: (a) the population size of the municipality; (b) a rural area dummy; and (c) an agglomeration density measure for the postal code at which the establishment is registered. The agglomeration density measure is used to derive five urbanisation classes.[9] The "rural area dummy" has been set to 1 if the number of addresses per postal code amounted to <1,501 km^2, and zero otherwise. I conclude that agglomeration has a positive effect on productivity and export participation decisions of establishment if the "rural area dummy" has a negative sign, and when the municipal population size and/or the agglomeration density measure has a positive sign. Agglomeration-relevant variables about the location of establishments have been added from the Basic Geographic Register (GBR).

I have constructed human capital indicators and productivity indicators from the raw data. Using the establishment-level data at the lowest level of detail, it was possible to construct an indicator for human-capital intensity per worker, using the following exploitation sheet items: expenditure on R&D, patents and licenses, internal education programs, costs of knowledge-intensive intermediary services (consultants, accountants), travel and communication costs, ICT expenditure, and also establishment-level earnings from patents, licenses, and intra-company services charged to affiliated companies. The sum of these items is expressed per full-time employee. The resulting indicator appeared to be correlated strongly with wages per worker, which was available for a much larger sample of establishments. For pragmatic reasons, I therefore used wages per worker as indicator of human capital use. For firm performance three indicators have been used: (a) labour productivity defined as value added per full-time worker; (b) sales per worker; and (c) profitability, defined as gross value added minus wages and minus depreciation.

As suggested by Vogel and Wagner (2011), I have removed extreme outliers from the sample to achieve robust results. Compared to many other studies using firm- or establishment-level data, this paper controls for a rich set of attributes that are likely to influence productivity. In addition to capital intensity, human-capital intensity, a control dummy is added for a firm's affiliation with (domestic) multinational corporation, its median size class during the observation period, its 4-digit industry affiliation.

[9] The agglomeration density variable measures the average number of addresses per square kilometre within a circle of a 1-km ray, measured at the beginning of each year. The five urbanisation classes are: (1) very strong urbanisation (≥2,500 addresses per km^2); (2) strong urbanisation (1,500–2,500 addresses per km^2); (3) moderate urbanisation (1,000–1,500 addresses per km^2); (4) weak urbanisation (500–1,000 addresses per km^2); (5) non-urban area (<500 addresses per km^2). In the regressions I have taken the log of the urbanisation-class indicator.

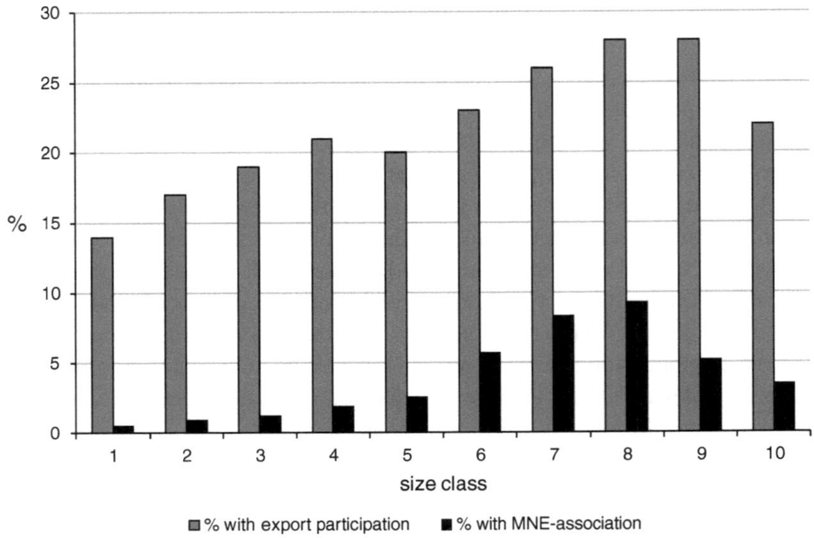

Fig. 8.3 Services establishments with exports and multinational affiliations by size class, 2005

4 Some Descriptive Results

This section shows some descriptives on the extensive and intensive export margin, productivity characteristics of establishments with international activities, the relation of productivity with agglomeration characteristics. Figure 8.3 differentiates the export participation rate by size class.[10]

Export participation rates amount to around 20 % on average, peaking at 28 % participation in size class 8 (500–999 employees). This class also has the highest incidence of establishments that are associated with Dutch multinational firms.[11] The graph shows that some of the largest services providers are more oriented to the domestic market than the intermediate-sized establishments in size classes 7 and 8 (those that employ between 250 and 999 employees).

Table 8.1 shows that services establishments with international activities differ significantly from establishments that are only active in the national market. I distinguish between establishments in four internationalisation groups: (a) locals with only domestic sales, (b) exporters without multinational affiliation, (c) establishments with

[10] The size classes are based on the number of employed persons and cover the following intervals: (1) 10–19 employed persons; (2) 20–39 employed persons; (3) 40–59 employed persons; (4) 60–80 employed persons; (5) 80–124 employed persons; (6) 125–249 employed persons; (7) 250–499 employed persons; (8) 500–999 employed persons; (9) 1,000–1,999 employed persons; (10) >2,000 employed persons. This size class definition is consistently applied in the remainder of the paper.

[11] The data do not allow a reliable identification of establishments that are associated with foreign multinational firms.

Table 8.1 Performance statistics by internationalisation group, establishments, pooled data, 1999–2005

Internationalisation group	No. of observations	Value added per worker in 1,000 euro	Sales per worker in 1,000 euro	Average wage per worker in 1,000 euro	Gross profit[a] in 1,000 euro
Services total[b] of which	52,144	55.9	91.3	20.0	35.9
(a) Domestic-oriented	36,789	52.2	87.1	19.0	33.3
(b) Export only	10,500	63.2	87.6	22.4	40.9
(c) Non-exporter, Dutch MNE affiliated	3,013	71.8	158.0	23.5	48.3
(d) Exporter, Dutch MNE affiliated	1,462	73.0	116.3	25.2	47.8

Source: Own calculations based on PS, ABR and SFO databases

Notes:

[a]The gross profit is calculated as valued added minus wages

[b]Services total includes firms (not shown) for which foreign MNE affiliation could not be identified in a reliable way

Dutch MNE affiliation but without exports, (d) exporting establishments with Dutch multinational affiliation. A first comparison of productivity performance indicates that—when disregarding other firm characteristics—there is a strictly monotonic performance hierarchy between the internationalisation groups. Dutch affiliated multinational firms are substantially more productive than establishments that only export. The productivity advantage of services exporters increases by type of firm. However, this hierarchy does not hold for sales per worker as the labour productivity indicator: non-exporting MNEs are more productive than exporting ones. Exporting firms (irrespective of MNE affiliation or not) are more productive than non-exporting firms. Table 8.1 also shows that the internationalisation groups differ by average wages and gross profits per worker. MNEs tend to pay higher wages than non-multinationals, while profits are distinctively higher for MNEs. Irrespective of MNE affiliation, exporters pay higher wages and earn bigger profits.

Figure 8.4 shows the labour productivity (lnvalp) distribution for exporting (right panel) and non-exporting establishments (left panel). The distributions for both sub-samples differ in two important ways. Firstly, for exporters we see a clear left truncation of the distribution, which can be regarded as evidence that some cut-off productivity threshold plays a role. Secondly, there is a right-shift of the mean and the median productivity of exporters compared to non-exporters.

The discussion that I want to address is how the agglomeration factors fit into this picture. For this reason I have split up both samples in rural versus non-rural (large cities) establishments. Figure 8.5 shows the distributions for the four sub-samples. The figure suggests that—in line with our hypothesis—the productivity self selection of exporters in the rural areas (panel 1,1) is stronger than for exporters in urban areas (panel 1,0), both in terms of the degree of left truncation and in terms

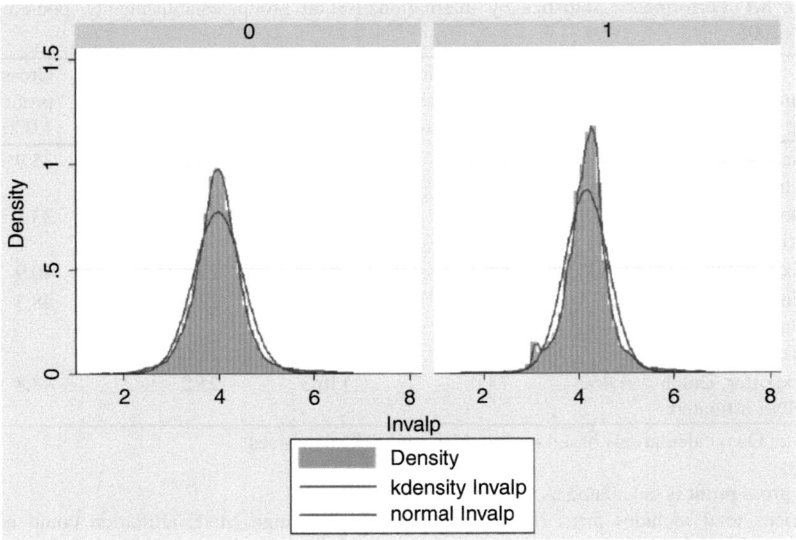

Fig. 8.4 Distribution characteristics of labour productivity for exporting (*right*) and non-exporting (*left*) services establishments, 2005

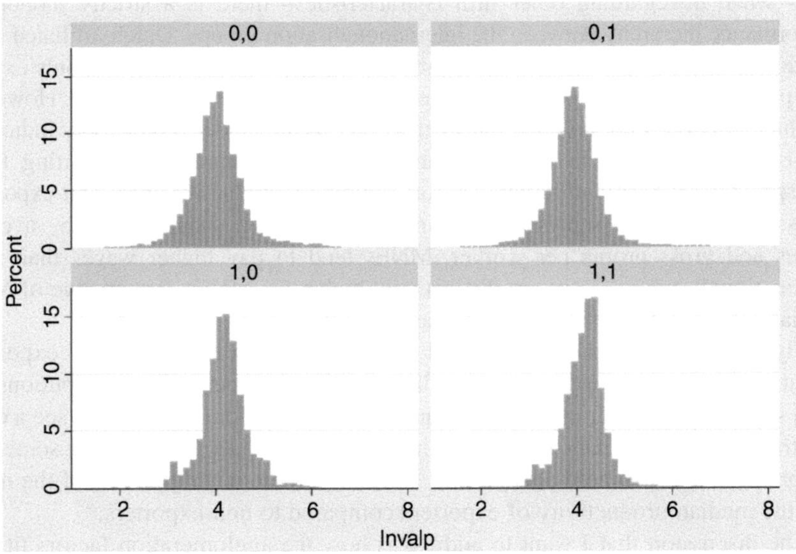

Fig. 8.5 Distribution characteristics of labour productivity for exporting rural (1,1), non-exporting rural (0,1), exporting urban (1,0) and non-exporting urban (0,0) services establishments, 2005

of right-shift of the median, compared to the corresponding non-exporters.[12] For urban exporters, we also see a dilation of the distribution compared to rural exporters; following Combes et al. (2012), this can be interpreted as an impact of agglomeration. Also the median shift is much stronger in rural areas than in urban areas. This also can be interpreted as an effect of agglomeration, according to Combes et al. (2012).

The descriptives so far indicate that both performance-based self selection and agglomeration effects play a role. The next two sections investigate whether these indications hold when adding further control variables that may affect the productivity performance of firms.

5 Productivity Based Self-selection for Services Exporters

The dominant empirical tests of the predictions of the heterogeneous-firms trade models use panel data regressions that test to what extent exporting firms have a significant productivity performance premium compared to non-exporters, after controlling for other export-invariant factors as well (cf. Wagner, 2007, 2012).[13] A positive exporter premium is indeed to be expected if positive self selection drives the choice behaviour on the extensive export margin. However, the productivity premium may also result from exporters being mainly located in urban areas. In this paper I use probit regressions as the main econometric test for factors that affect export participation decisions. The standard panel data regressions that estimate the productivity premium are applied as a robustness check.

5.1 Testing the Probability of Becoming an Exporter

The main prediction of the heterogeneous-firms trade model is that firms opt for exporting if their productivity is sufficient to absorb the fixed entry costs in the export market. Actual export behaviour can be adequately described by a latent

[12] The distinction between rural and non-rural areas is made on the basis of the rural area dummy. The "rural area dummy" has been set to 1 if the number of addresses per postal code amounted to $<1{,}501$ per km^2, and zero otherwise.

[13] Given that for this regression international transaction data were not available, the following assumptions are imposed to test the predictions of the heterogeneous-firms trade model: (a) establishments in each sector (4-digit) have the same information about market size, variable and sunk trade barriers, covering all relevant countries; (b) establishments share a common country set as potential export markets and (giving assumption a) have an identical ranking within their set of preferred export countries; (c) establishments in a (4-digit) sector enter potential export countries according to an identical country sequence based on market size, distance and sunk market-entry costs. In other research (Smeets, Creusen, Lejour, & Kox, 2010) we have estimated the country-specific sunk market entry costs for Dutch manufacturing firms.

variable model in which the export preference of firm i in year t, y_{it}^{*} precedes actual export decisions. The heterogeneous-firms trade model can then be reinterpreted in the following way. The decision to export y_{it}* depends on a set of observable firm characteristics x_{it} and on an unobserved characteristic ε_{it} (e.g. the sunk entry costs firms expect to face in the export market). The main observable firm characteristics in x_{it} are performance characteristics (i.e. productivity, profitability). The assumed distribution of the unobserved characteristics ε_{it} determines the eventual export decision. I assume that the firm's preference for exporting $y_{it} \in \{1, 0\}$ depends on a linear additive relationship between the vector of observed x_{it} characteristics and the unobserved ε_{it} characteristic that determines net export benefits:

$$y_{it}^{*} = \beta x_{it} + \varepsilon_{it} \tag{8.1}$$

If the latent decision variable y_{it}^{*} exceeds a certain threshold level, the firm exports.[14] Consequently, if $ES_{it} \in \{1, 0\}$ is firm i's export status in year t, $ES_{it} = 1$ is only observed if $y_{it}^{*} > 0$ and $ES_{it} = 0$ otherwise. The probability of exporting can be expressed as:

$$\begin{aligned} P\{E_{it} = 1\} &= P\{y_{it}^{*} > 0\} = P\{\beta x_{it} + \varepsilon_{it} > 0\} = P\{-\varepsilon_{it} \leq \beta x_{it}\} \\ &= F(\beta x_{it}) \end{aligned} \tag{8.2}$$

where $F(.)$ denotes the distribution function of $-\ \varepsilon_{it}$. This yields a binary choice model that depends on the distribution of ε_{it}. As the scale of the firm preference y_{it}^{*} is not identified, a normalisation on the distribution of ε_{it} is required.[15] Using a standard normal distribution, the binomial probit model for the export decision is given by:

$$y_{it} = \beta x_{it} + \varepsilon_{it} \tag{8.3}$$

where $\varepsilon_{it} \sim NID(0,1)$ and $y_{it} = 1$ if $y_{it}^{*} > 0$; and $y_{it} = 0$ if $y_{it}^{*} \leq 0$. The probit estimation results allow calculating the vector of impact elasticities $\partial \ln P\{ES_{it} = 1 \mid \mathbf{x}_{it}\} / \partial\ ln\ \mathbf{x}_{it}$ (the percentage change in the probability of a positive export preference after a 1 % change in the log of firm characteristic x_{it}).[16]

To start with I test the most simple version of the probit model, concentrating on the impact of single performance variables x_{it} on the probability that an establishment engages in exports; all other possible impacts on the export decision are left

[14] The threshold value can be set at zero without loss of generality.

[15] Usually this means that its variance is fixed at a given value (Verbeek 2004). Since F(βx_it) is also bounded between 0 and 1, it is plausible to choose a standard normal distribution φ(βx_it). There is no reason to expect that the standard normal distribution does not apply.

[16] Impact elasticities instead of marginal effects are presented, because the intuitive interpretation of elasticities is easier. I evaluated point elasticities at the mean and at the median values of ln βx_it. Since differences between both were very small, I only report point elasticities at the mean.

Table 8.2 Impact of establishment performance on export probability: probit regressions, pooled data, 1999–2005

Performance indicator (log)	Elasticity on export probability[a]	Standard errors	z-value	No. of observations
Value added per worker	2.39	0.783	3.06***	37,717
*Urban sub-sample	2.74	0.932	2.94***	28,132
*Rural sub-sample	1.69	0.499	3.38***	11,585
Profitability	2.97	0.836	3.55***	39,562
*Urban sub-sample	0.0023	0.001	1.94**	28,132
*Rural sub-sample	−0.005	0.006	−0.81	11,585

Source: Own calculations based on PS database

Confidence levels: *** Indicates significance at 1 % level, ** at 5% level

Notes:

[a]Post-estimation calculations of point elasticities evaluated at the means of the independent variables, using the estimated probit model with clustered standard errors by 2-digit industry. The "rural area" dummy has been used to split the total sample into two sub-samples

unspecified for the moment. Table 8.2 shows the point elasticities of two performance variables. Both have a statistically significant and positive impact on the export participation. The same regression has been done for two sub-samples, one for rural establishments and one for urban establishments. The split is made on the basis of the "rural area" dummy. It turns out that in the urban areas a 1 % higher labour productivity yields a 2.74 % increase in the probability that the establishment exports, while this is only 1.69 % in rural areas.

In a next step I extend the probit model by adding control variables that may also affect the export participation decisions of establishments: size of the establishment, industry-specific effects, input characteristics, affiliation with a multinational firm, and time shocks. To prevent multicollinearity between input characteristics and the performance indicator x_{it} the input characteristics of the establishment are lagged θ years. The probit model now becomes:

$$P\{ES_{it} = 1\} = F(\beta x_{it} + \gamma \mathbf{G}_{\mathrm{it}-\theta} + \lambda \mathbf{R}_{\mathrm{i}} + +\eta \mathbf{T}_{\mathrm{t}}) \tag{8.4}$$

where x_{it} is the performance indicator (e.g. value-added per worker, profitability), $\mathbf{G}_{\mathrm{it}-\theta}$ is a vector of firm characteristics lagged θ years. $\mathbf{R}_{\mathrm{i}}$ is a vector of time-invariant environment variables of the firm, and $\mathbf{T}_{\mathrm{t}}$ is a vector of year dummies to control for time shocks. The $\mathbf{G}_{\mathrm{it}-\theta}$ vector includes a dummy for affiliation with a multinational company (in year $t - 1$), a lagged indicator of the firm's human capital (using the average wage per worker in year $t - 3$ as a proxy), and the establishment's median size class over the entire interval that it is in our data panel.[17] The $\mathbf{R}_{\mathrm{i}}$ vector includes 2-digit industry dummies and the agglomeration variables (rural area dummy, agglomeration density measure for the postal code at

[17] The size class is measured on a 10-point Likert scale {1, ... ,10} that increases in employment size. We took the median size category for the firm over the full observation period. The result is expressed as a natural logarithm.

Table 8.3 Impact elasticities of performance self-selection and agglomeration variables on probability of export participation, based on probit regression[a], pooled data, 1999–2005

	Performance variable: value added per worker		Performance variable: profitability	
Explanatory variables	Without agglom. variables	With agglom. variables	Without agglom. variables	With agglom. variables
Performance indicator (log)	1.421***	1.324**	0.949***	0.880*
Agglom. density measure (log)		0.479**		0.481**
Rural dummy		−0.187***		−0.190***
Control variables				
Human capital indicator (log)	1.486	1.645	1.886***	2.028*
Median size class (log)	0.277***	0.307***	0.053	0.307***
MNE dummy	0.007	0.009	0.009	0.011
Industry dummies (2-digit)	Yes	Yes	Yes	Yes
Year dummies	Yes	Yes	Yes	Yes
No. of observations	13,200	9,141	13,200	9,130
Predicted ES_{it} after probit[b]	0.23	0.22	0.23	0.22
Log pseudo likelihood	−7,024	−4,751	−7,036	−4,760

Source: Own calculations based on PS database, combined with data from SFO and ABR
Significance levels are coded as: *at 10 % level; **at 5 % level; ***significance at 1 % confidence level
Notes:
[a]Post-estimation calculations of the point elasticities evaluated at the means of the independent variables, using the estimated probit model with clustered standard errors by 2-digit industry
[b]This indicator gives the joint prediction power of the probit estimate, predicting the probability that $ES_{it} = 1$ (actual probabilities by size class are shown in Fig. 8.3)

which the establishment is registered).[18] The estimation accounts corrects for clustered standard errors by 2-digit industry.

Table 8.3 presents the results for this extended probit model, covering pooled observations for the period 1999–2005. The table shows the results for two x_{it} performance variables, with and without the agglomeration variables included in the $\mathbf{R}_i$ vector. The performance indicators in all cases have a statistically significant marginal effect, even after controlling for a host of environment variables. Without agglomeration variables, a 1 % higher labour productivity would increase the probability of being an exporter by 1.4 % in services; a 1 % higher profitability would raise this same probability by 0.95 %. These results are consistent with the performance self selection predicted by the heterogeneous-firms trade models.

[18] The population size and the agglomeration density scale are expressed in logs.

However, inclusion of the local agglomeration variables lowers these effects to 1.3 % and 0.88 %, while at the same time the statistical significance of the self-selection variables diminishes. The confidence level for labour productivity self selection drops from the 1 % level to the 5 % level, and for profitability self selection from the 1 % level to the 10 % level. We may conclude from this that measured exporter self selection is in fact partly driven by agglomeration effects.[19]

5.2 *Testing Productivity Premia for Exporters and MNEs*

As a robustness check on the previous result I run a standard econometric test to find positive export and MNE productivity premia. This is done by running panel data regressions where we construct dummy variables for the export status (ES_{it}) and the MNE-affiliation, and include these dummies as explanatory variables for the establishment's performance levels, in particular labour productivity. Productivity is the crucial performance variable in heterogeneous-firms trade models (e.g. Baldwin, 2005; Chaney, 2008; Melitz, 2003) and also in models that explain agglomeration economies (e.g. Baldwin & Okubo, 2006; Melitz & Ottaviano, 2008). Export starters need a higher productivity rate allowing them to absorb the fixed or sunk entry costs in the foreign market, i.e. they should have a significant positive performance premium compared to non-exporters, all other things equal. This holds a fortiori for firms that engage in multinational activities (Helpman et al., 2004). I test whether there is robust evidence for the existence of such productivity performance premia after correction for agglomeration and control variables and fixed effects at the establishment level. The tested model reads:

$$ln\ \ xt_{it} = \alpha + \beta\ ES_{it} + \psi\ MNE_{it} + \gamma\ \mathbf{G}_{it} + \eta\mathbf{T}_t + \varphi\ \mathbf{Z}_i + \varepsilon_{it} \tag{8.5}$$

where xt_{it} is the performance variable for firm i in period t, ES_{it} is a {0,1} dummy for the firm's exporter status, MNE_{it} is a {0,1} dummy for the establishment's affiliation with a multinational company, $\mathbf{G}_{it}$ is a vector of environment control variables (industry dummies, period-median size class, agglomeration variables), $\mathbf{T}_t$ is a vector of year dummies to control for time shocks, $\mathbf{Z}_i$ is a vector holding establishment-level fixed effects, and ε_{it} is the error term. Regression equation (8.5) is tested with a weighted least squares (WLS) estimator, applying sample-to-population expansion factors as weights. The WLS estimator accounts for non-response and for under-representation of small firms with less than 50 employees. Table 8.4 reports the results, before and after including control variables for local agglomeration effects.

[19] Andersson and Lööf (2009) conclude that firms located in larger regions are more productive, even when controlling for size, human capital, physical capital, ownership structure, import and export, industry classification and time trend. Second, they find that results from dynamic panel estimations suggest a learning effect in that agglomeration enhances firms' productivity. Third, the role of agglomeration phenomena does not seem to have a clear coupling to firm size.

Table 8.4 Labour productivity premia for exporters and MNE-affiliated establishments, with and without agglomeration variables, using establishment-level fixed effects, pooled data, 1999–2005

Internationalisation group	Without agglomeration variables	With agglomeration variables
Exporters versus non-exporters	2.2 %***	1.9 %***
MNE versus local firms	−2.6 %	−3.0 %
No. of observations	47,934	35,448
No. of establishments	20,559	15,300

Sources: Own calculations based on the SFO and PS database
Note: Panel regressions with establishment-level fixed effects, dummies for years, size, and 2-digit industry. Productivity indicator is in logs. Premia calculated as: 100*[exp(β)−1], where β is the estimated coefficient. The MNE identifier solely refers to Dutch-owned MNE
Significance levels are coded as: *at 10 % level; **at 5 % level; Confidence levels: *** Indicates significance at 1 % level

After accounting for fixed effects at the establishment level, exporters have 2.2 % exporter premium without agglomeration variables, and 1.9 % when agglomeration is taken into account. This suggests that about one-seventh of the productivity premium is caused by agglomeration effects.[20] The MNE premium completely evaporates when fixed effects (unobserved heterogeneity at the establishment level, like management capacity) are taken into account.

5.3 Testing the Impact of Market Structure

Chaney (2008) extends the heterogeneous firms trade model of Melitz (2003) to include the role of market structure. It can be derived from the Chaney model that a firm in homogeneous markets (high substitution elasticity between products) needs a bigger performance premium to enter an export market. Conversely, exporter premia are predicted to be lower in markets with more product differentiation. We test this hypothesis by splitting the samples in two parts on the basis of competition characteristics. The heterogeneous products group is made up of industries with strong product differentiation and low substitution elasticity. The homogeneous products group is characterised by weak product differentiation and high substitution elasticity.

The distinctive criterion for product homogeneity is based on the idea that in an industry with homogeneous products, competition will have mainly the character of price and cost competition. Inefficient firms with low productivity will then either shrink or drop out and more efficient firms will survive and grow. As a result of these movements, the dispersion of productivities in such homogeneous-products industries will be lower than average for all industries. Conversely, in industries with more differentiated products the competitive process is driven less by price and

[20] The caveat raised at the end of Sect. 2 about the unreliability of summary statistics for measuring self selection (in the presence of agglomeration factors) remains valid.

cost competition, and we expect more than average dispersion of productivities. Using these insights I calculated the dispersion of firm productivities in each 4-digit industry. The "homogeneous products" dummy was set to 1 if the variation coefficient of value added per worker over the entire observation period was less than 75 % of the average for Dutch manufacturing and services, and set to 0 otherwise.

According to this criterion, about two-thirds of manufacturing and services establishments were found to operate in homogeneous products industries. With the split samples I again estimated the full probit regression model of equation (8.4). For brevity, Table 8.5 only gives the estimated parameters and the number of observations.

The results are consistent with the predictions of the Chaney (2008) model: we find that self selection based on productivity is stronger in markets with homogeneous products than in markets with differentiated, heterogeneous products. This suggests that self selection in industries with strong cost and price competition is based on productivity as major selection parameter. For heterogeneous services, the elasticity of labour productivity on the probability of being an exporter is less than one-third compared to the homogeneous services. That effect does not change after inclusion of agglomeration variables. Establishment size is important at the highest confidence level in all regressions, suggesting that fixed entry costs affect scale economies. Multinational affiliation and human capital intensity appear to be mostly insignificant as predictors of export starting. Overall, these results can be read as implying that the type of sunk entry costs in foreign markets differs between homogeneous and heterogeneous products. Agglomeration variables only have a significant impact on the selection process in markets with more or less homogeneous products.

6 Dynamic Self-selection into Foreign Markets

So far, I calculated static productivity premia for the pooled dataset. This means that the data include all-time exporters, new exporters and non-exporters. Such regressions can only yield a rough approximation of the self-selection process and how it is affected by agglomeration effects. In order to understand dynamic self selection it is necessary to deal with time-dependent anticipation behaviour. According to the heterogeneous-firms theory, a firm self-selects into export participation on the basis of its relative performance in the domestic market *before* export starts. If export starts in year t we should find a positive *ex ante* performance premium in year $t - \theta$. I test this using a modified version of the probit analysis of (8.4). The all-time exporters are removed from the dataset and the focus is now exclusively on how future new exporters compare with domestic establishments that do not export. The equation array being regressed is:

$$P\{ES_{it} = 1\} = F(\beta x_{it-\theta} + \gamma \mathbf{G}_{it-\theta} + \lambda \mathbf{R}_i + + \eta \mathbf{T}_t) \tag{8.6}$$

New exporters are identified as establishments that started exporting during our data period (1999–2006) and that did not have exports in the θ years before year t when it starts exporting. It is known from other trade literature that export

Table 8.5 Impact elasticities of performance self-selection and agglomeration variables on probability of export participation after including market structure, post-probit elasticities[a], pooled data, 1999–2005

	Markets with homogeneous products		Markets with heterogeneous products	
Explanatory variables	Without agglom. variables	With agglom. variables	Without agglom. variables	With agglom. variables
Labour productivity (log)	3.549***	3.854***	1.051***	1.010***
Agglom. density measure (log)		0.581***		0.054
Rural area dummy		−0.092***		−0.185
Control variables				
Human capital indicator (log)	1.851	1.890	0.781	0.796*
Median size class (log)	0.293***	0.358***	0.202***	0.158**
MNE dummy	0.030	0.036	−0011	−0.013
Industry dummies (2-digit)	Yes	Yes	Yes	Yes
Year dummies	Yes	Yes	Yes	Yes
No. of observations	7,161	5,971	6,057	4,937
Predicted ES_{it} after probit[b]	0.20	0.25	0.20	0.19
Log pseudo likelihood	−3,860	−3,142	−2,975	−2,363

Source: Own calculations based on PS database, combined with data from SFO and ABR
Significance levels are coded as: *at 10 % level; **at 5 % level; ***significance at 1 % confidence level
Notes:
[a]Post-estimation calculations of the point elasticities evaluated at the means of the independent variables, using the estimated probit model with clustered standard errors by 2-digit industry
[b]This indicator gives the joint prediction power of the probit estimate, predicting the probability that $ES_{it} = 1$ (actual probabilities are shown in Fig. 8.3)

decisions are at least partly governed by trial-and-error behaviour.[21] In order to keep the trial-and-error exporters out of the export starter samples, I add the restriction that the export starters remain active as exporter at least 2 years (t, t + 1) even though this diminishes the size of the sample.

The characteristics of the new exporters are compared with non-exporting firms that neither had exports in the year t nor in period $t - \theta$. This sample selection implies that we lose many observations, leaving only small samples of export

[21] Cf. Békés and Muraközy (2008) for Hungary. Besedes and Prusa (2006) established for the USA that trade relationships typically start small and that almost half of the "small relationships" end within a year, while larger initial purchases result in longer, stable relationships. From this they advance a matching model of international trade in which uncertainty and the costs of searching reliable trade partners play important roles. The search cost idea can be reconciled with the Melitz model, because search costs are in fact country-specific sunk entry costs. Albornoz, Calvo Pardo, Corcos, and Ornelas (2009) go one step further in evaluating the role of uncertainty and learning. In their view a strategy of sequential exporting to different country markets is a rational firm strategy to discover their own competitive advantage.

Table 8.6 Impact elasticities[a] of *ex-ante* performance on probability that an establishment exports in period t − θ: probit regressions, new exporters, 1999–2006

	Value added per worker		Profitability	
Explanatory variables	Without agglom. variables	With agglom. variables	Without agglom. variables	With agglom. variables
Performance indicator, t−3 years	0.452	0.564	−0.612	−0.506
Number of observations	4,607	3,034	4,590	3,204
Number of export starters	297	297	297	297
Predicted ES_{it} after probit[b]	0.05	0.01	0.05	0.04
Performance indicator, t−2 years	0.705	0.639	0.281	0.370
Number of observations	8,611	5,470	8,572	5,640
Number of export starters	724	724	724	724
Predicted ES_{it} after probit[b]	0.06	0.02	0.06	0.06
Performance indicator, t−1 year	1.692**	1.758*	1.288***	1.308*
Number of observations	9,718	6,267	9,675	6,247
Number of export starters	627	624	627	627
Predicted ES_{it} after probit[b]	0.05	0.053	0.05	0.02

Source: Own calculations using the PS database

Includes control variables: human capital indicator, size class, and industry (2-digit) and year dummies

Significance levels are coded as: *at 10 % level; **at 5 % level; ***significance at 1 % level

[a]Post-estimation calculations of the marginal effects at the means of the independent variables, using the estimated probit model with clustered standard errors by 2-digit industry

[b]This indicator gives the joint prediction power of the probit estimate, predicting the probability that $ES_{it} = 1$

starters each year. The number of observations decreases with the length of the lead period θ. We experimented with lead periods of 1–3 years. The small numbers of export starters make it harder to establish statistically significant effects.

Table 8.6 gives the results for the *ex-ante* probit self-selection model. For $\theta = 3$ and $\theta = 2$ no significant *ex ante* self selection behaviour could be established. There was, however, significant *ex ante* self selection with respect to labour productivity and profitability in $t - 1$ as predicted by the heterogeneous-firms trade model. In contrast to the services results, Kox & Rojas-Romagosa (2010) found evidence for ex ante self-selection behaviour for $\theta = 3$ and $\theta = 2$. A possible explanation for the difference between manufacturing and services may be that fixed-equipment investments for export capacity are smaller in services, and that the lead times for starting exports are shorter in services. The *ex ante* effect did not disappear after inclusion of agglomeration variables, on the contrary: if anything, the self-selection effect became stronger after accounting for agglomeration factors, even though the statistical significance dropped. The *ex ante* self selection was corroborated by a robustness test for *ex ante* productivity premia (not shown). Summarizing, the results support the self-selection hypothesis that export starters have a significant productivity advantage—with respect to non-exporters—before they begin to export.

7 Discussion and Conclusion

This paper assessed the role of agglomeration effects on the export decision of services firms. Our findings support the predictions done by heterogeneous-firms trade models about positive productivity sorting among new exporters, in anticipation of sunk market-entry costs in foreign markets. We found significant *ex ante* productivity-based self selection by future exporters 1 year before actual exports started. The productivity self selection is three times stronger in services markets that are characterised by relatively homogeneous products, compared to services industries where products tend to be more differentiated. A possible explanation for this difference is that price- and cost competition is stronger in markets with homogeneous services products, whereas in markets with differentiated products each supplier is in a sense a monopolist in his own product niche and thus has more leeway for passing on higher costs to foreign consumers.

The empirical findings do not corroborate the correctness of the claim by Helpman et al. (2004) that firms with multinational affiliations will have a stronger productivity self selection than firms that only export. The "MNE productivity premium" disappeared entirely when we corrected for fixed unobserved effects at the establishment level. As a caveat in this regard it should be repeated that the data only allowed identifying links with Dutch multinational enterprises, but not with foreign-owned multinationals.

Agglomeration effects in large-city areas were found to contribute to above-average productivity for exporting firms. Without the explicit check on agglomeration effects the above-average productivity bonus for exporters would incorrectly have been attributed to productivity selection for anticipated sunk foreign market-entry costs. Or alternatively stated, the impact of sunk foreign market-entry costs would have been over-rated. Based on our results it is fair to conclude that many earlier empirical evaluations of the heterogeneous-firms trade models have probably upwardly biased estimates of the effects of sunk foreign market-entry costs.

Time seems ripe for theoretical effort to unify the models of productivity sorting in large agglomerations and productivity sorting in trade models. In both cases this sorting behaviour is driven by fixed/sunk costs and increasing returns to scale. The

Melitz-based trade models have structured the productivity sorting as a lottery result: upon accepting sunk entry costs the firm's true productivity is revealed to itself as a draw from a known productivity distribution.[22] If the drawn productivity is sufficiently high the firm stays, otherwise it exits immediately. The same mechanism is used to model the productivity hierarchy between non-exporters and exporters: if the productivity draw allows it, the firm exports, otherwise it stays only active in the domestic market. From this it is only a small step to also integrate the productivity selection that goes along with agglomeration. In order to be allowed to operate in large cities (with their cultural amenities, with large local supply of consumption varieties, thick labour markets for high-educated workers, intricate division of labour, and the local supply of advanced intermediary deliveries and services), firms must annually accept to absorb a given amount of sunk urban costs.[23] Their own productivity draw determines whether the firm stays in the large-city area, moves to areas without the sunk urban costs, or completely exits. The firm's own productivity sorting then yields the left-truncation of the urban productivity distribution. In this way a unified model of trade and agglomeration is achieved.

References

Albornoz, F., Calvo Pardo, H., Corcos, G., & Ornelas, E. (2009, September). *Sequential exporting*. Paper presented at European Trade Study Group (Rome, 2009), University of Birmingham University of Southampton, NHH Oslo and London School of Economics.

Andersson, M., & Lööf, H. (2009). *Agglomeration and Productivity- evidence from firm-level data* (CESIS Electronic Working Paper Series No. 170). Centre of Excellence for Science and Innovation Studies, Royal Institute of Technology, Stockholm.

Baldwin, R., (1988). Hysteresis in import prices: the beachhead effect. *American Economic Review, 78*(4), 773–785.

Baldwin, R., (2005). *Heterogeneous firms and trade: Testable and untestable properties of the Melitz model* (Working Paper 11471). Cambridge, MA: NBER.

Baldwin, R., & Krugman, P. (1989). Persistent trade effects of large exchange rate shocks. *Quarterly Journal of Economics, 104*, 635–654.

Baldwin, R., & Okubo, T. (2006). Heterogeneous firms, agglomeration and economic geography: Spatial selection and sorting. *Journal of Economic Geography, 6*(3), 323–346.

Balistreri, E., Hillberry, R., &, Rutherford, T. (2008). *Structural estimation and solution of international trade models with heterogeneous firms* (Research Paper #1056). University Of Melbourne, Department Of Economics, Melbourne.

Békés, G., &, Muraközy, B. (2008, May). *Permanent and temporary trade*. Paper presented at the CAED conference (Budapest, 2008), Hungarian Academy of Sciences, Budapest.

Bernard, A., Eaton, J., Jensen, J., & Kortum, S. (2003). Plants and productivity in international trade. *American Economic Review, 93*, 1268–1290.

Bernard, A., & Jensen, J. (1995). *Exporters, jobs, and wages in U.S. Manufacturing, 1976–1987*. Brookings Papers on Economic Activity, Microeconomics, Washington, DC.

[22] The mechanism is taken from Jovanovic (1982) and Hopenhayn (1992).

[23] Think of taxes that are linked to urban land and property prices, such taxes embody the agglomeration mark-up linked with being in an attractive business and living agglomeration.

Besedes, T., & Prusa, T. (2006). Ins, outs, and the duration of trade. *Canadian Journal of Economics, 39*(1), 266–295.

Chaney, T. (2008). Distorted gravity: The intensive and extensive margins of international trade. *American Economic Review, 98*(4), 1707–1721.

Ciccone, A. (2002). Agglomeration effects in Europe. *European Economic Review, 46*, 213–227.

Ciccone, A., & Hall, R. (1996). Productivity and the density of economic activity. *American Economic Review, 86*(1), 54–70.

Combes, P., Duranton, G., Gobillon, L., Puga, G., & Roux, S. (2008). *The productivity advantages of large markets: Distinguishing agglomeration from firm selection.* Mimeo: University of Aix-Marseille.

Combes, P., Duranton, G., Gobillon, L., Puga, G., & Roux, S. (2012). The productivity advantages of large markets: Distinguishing agglomeration from firm selection. *Econometrica, 80*(6), 2543–2594.

Dixit, A. (1989). Entry and exit decisions under uncertainty. *Journal of Political Economy, 97*, 620–638.

Duranton, G., & Puga, D. (2004). Micro-foundations of urban agglomeration economies. In V. Henderson & J. F. Thisse (Eds.), *Handbook of regional and urban economics* (Vol. 4, pp. 2063–2117). Amsterdam: North-Holland.

Eaton, J., Kortum, S., & Kramarz, F. (2011). An anatomy of international trade: Evidence from French firms. *Econometrica, 79*(5), 1453–1498.

Fujita, M., & Thisse, J.-F. (2002). *Economics of agglomeration: Cities, industrial location and regional growth.* Cambridge: Cambridge University Press.

Greenaway, D., & Kneller, R. (2007). Firm heterogeneity, exporting and foreign direct investment. *The Economic Journal, 117*, F134–F161.

Helpman, E., Melitz, M., & Yeaple, S. (2004). Export versus FDI with heterogeneous firms. *American Economic Review, 94*(1), 300–316.

Henderson, V. (2003). Marshall's scale economies. *Journal of Urban Economics, 53*, 1–28.

Hopenhayn, H. (1992). Entry, exit and firm dynamics in long-run equilibrium. *Econometrica, 60*, 1127–1150.

ISGEP. (2008). Understanding cross-country differences in exporter premia: Comparable evidence for 14 countries, International Study Group on Exports and Productivity. *Review of World Economics, 144*(4), 596–635.

Jovanovic, B. (1982). Selection and the evolution of industry. *Econometrica, 50*, 649–670.

Kox, H. (2002). *Growth challenges for the Dutch business services industry: International comparison and policy issues* (CPB Special Publication #40). CPB/Koninklijke De Swart, The Hague.

Kox, H., & Rojas-Romagosa, H. (2010). Exports and productivity selection effects for Dutch firms. *De Economist (Springer), 158*(3), 295–322.

Krugman, P. (1991). Increasing returns and economic geography. *Journal of Political Economy, 99*(3), 483–499.

Krugman, P & Venables, T. (1995). Globalization and the Inequality of Nations. *Quarterly Journal of Economics*, 110, 857–880.

Lawless, M., (2009). *Deconstructing gravity: Trade costs and extensive and intensive margins.* Paper submitted to EIIE/FREIT, Central Bank and Financial Services Authority of Ireland, Dublin.

Mayer, T., & Ottaviano, G. (2007). *The happy few: The internationalisation of European firms.* Bruegel, Brussels: Bruegel Blueprint III.

Melitz, M. (2003). The impact of trade on aggregate industry productivity and intra-industry reallocations. *Econometrica, 71*(6), 1695–1725.

Melitz, M., & Ottaviano, G. (2008). Market size, trade and productivity. *Review of Economic Studies, 75*(1), 295–316.

Melo, P., Graham, D., & Noland, R. (2009). A meta-analysis of estimates of the urban agglomeration economies. *Regional Science and Urban Economics, 39*(3), 332–342.

Moretti, E. (2004). Workers' education spillovers, and productivity: Evidence from plant-level production functions. *American Economic Review, 94*, 656–690.

Ottaviano, G., Tabuchi, T., & Thisse, J.-F. (2002). Agglomeration and trade revisited. *International Economic Review, 43*(2), 409–435.

Redding, S. (2010). *Theories of heterogeneous firms and trade* (NBER Working Paper #16562). Cambridge, MA: National Bureau of Economics Research.

Rosenthal, S., & Strange, W. (2004). Evidence on the nature and sources of agglomeration economies. In *Handbook of regional and urban economics* (Vol. 4, pp. 2119–2171). Amsterdam: North-Holland.

Smeets, R., Creusen, H., Lejour, A., & Kox, H. (2010). *Export margins and export barriers: Covering market entry costs of exporters in the Netherlands* (CPB Document 208). CPB, The Hague.

Syverson, C. (2004). Market structure and productivity: A concrete example. *Journal of Political Economy, 112*(6), 1181–1222.

Verbeek, M., (2004). A guide to modern econometrics, 2nd edition, John Wiley & Sons, Oxford.

Vogel, A., & Wagner, J. (2011). Robust estimates of exporter productivity premia in German business services enterprises. *Economic and Business Review, 13*(1–2), 7–26.

Wagner, J. (2007). Exports and productivity: A survey of the evidence from firm level data. *The World Economy, 30*(10), 60–82.

Wagner, J. (2012). International trade and firm performance: A survey of empirical studies since 2006. *Review of World Economics, 148*, 235–267.

Yeaple, S. (2007). A simple model of firm heterogeneity, international trade, and wages. *Journal of International Economics, 65*, 1–20.

Chapter 9
Do Services Play a Role in Regional Productivity Growth Across Europe?

Andrés Maroto-Sánchez and Juan R. Cuadrado-Roura

1 Introduction

Over recent decades, increasing attention has been paid to the relationship between the economic structure of economies and their productivity growth, particularly considering the role being played by service industries. Those pioneer contributions to this topic, during the 1970s and 1980s, focused on two processes. On one hand, "*deindustrialization*" which started with the economic crisis of the 1970s, trying to explain the continuous growth of service sector in the developed economies compared to the manufacturing decline.[1] On the other, the progress towards a service society or an increasingly *tertiarized society*.[2] The majority of these works underlined that changes involved in a transfer of labour from sectors with low productivity to other more dynamic sectors was one of the main reasons for the overall productivity growth in an economy.

Nevertheless, a wave of economic literature, from the foremost contribution by Fourastié (1949) and, particularly, since the seminal work by Baumol (1967) and their well-known "cost disease", has supported the thesis that the continuous increase of services in the economic structures as part of the development processes, together with

[1] See, among others, Blackaby (1978) or Gemmell (1982).

[2] See, among others, Bell (1974) or Fuchs (1968).

A. Maroto-Sánchez (✉)
Autonomous University of Madrid (UAM), Department of Economic Analysis: Economic TheoryC/ Francisco Tomas y Valiente, 5 28049 Cantoblanco (Madrid), Spain
e-mail: andres.maroto@uam.es

J.R. Cuadrado-Roura
University of Alcala, Madrid, Department of Applied Analysis, Rafael Calvo, 17 – 4 . PD 28010, Madrid, Spain
e-mail: jr.cuadrado@uah.es

J.R. Cuadrado-Roura (ed.), *Service Industries and Regions*,
Advances in Spatial Science, DOI 10.1007/978-3-642-35801-2_9,

the low productivity in these types of activities as compared with the manufacturing industries, entail a clear threat for future growth, while its rates should be pushed down.[3]

The relationship between the growth of services and labour productivity, comparing different samples of OECD countries and time periods, has been revised in the recent literature. This chapter tries to translate these issues to the regional sphere. Productive specialisation can be one of the main causes of the differences between regional[4] behaviour and that of the countries. The evolution of those regions with a higher specialisation in dynamic activities will be far higher than the average of their corresponding countries.

On doing so, two hypotheses were considered. The first discusses what role structural changes play in overall economic productivity and particularly focuses on the growth of services activities. The idea underlying this hypothesis is whether the transfer of labour from less to more productive sectors does or does not propel an increase in the overall productivity of the economy. The second hypothesis tries to verify whether any differences are noted in productivity depending on the different branches of the services sector. Some recent studies have demonstrated this hypothesis confirming that some tertiary branches of the most advanced countries show equal or better productivity levels than those of the manufacturing branches, and therefore demonstrating that they contribute to the overall productivity growth of their respective economies. The paper aims to assess whether a regional analysis allows us to draw similar or identical conclusions to those obtained from those studies based on national data. To be precise, this is not the only concern in this work. It is also being considered the possibility that differences arise and in that case and in that case we should be able to explain them. For this purpose, regions taken as a reference for the analysis are NUTS-2[5] from a sample of 16 European countries (EU-15 with the exception of Luxembourg, plus Norway and Switzerland) in the period between 1980 and 2008.[6]

The structure of the analysis is the following. Firstly, we set out some theoretical thoughts regarding the relationships between structural changes, services and productivity (Sect. 2). Then, we offer an overview of the results obtained from the application of shift-share techniques both at national and regional level (Sect. 3). Following on from this we will contrast the previous results with estimated econometric data panel models highlighting coincidences and

[3] See Maroto (2012) for a survey on the relationships between services and productivity. See Towse (1997) for more information on Baumol's "cost disease".

[4] See, among others, Amiti (1999), Combes and Overman (2003), Ezcurra et al. (2006) or OECD (2009)

[5] In the case of Germany and the United Kingdom, we have used NUTS-1 because the dimension of NUTS-2 is too small to make a realistic and accurate comparison. Additionally, Azores Islands (POR), Ceuta and Melilla (SP) and the overseas French territories have been excluded. In the case of Greece, all islands are considered as a single region.

[6] Although the dataset provided by Cambridge Econometrics show estimations for later years, these are only forecasting data. For this reason, in this chapter we have decided to handle the data until 2008.

discrepancies (Sect. 4). And, finally, the paper ends with some final remarks on the most significant results and a summary of the questions that have been posed.

2 Structural Change, Service Industries and Productivity Growth in Recent Literature

As mentioned, increasing attention has been recently paid by different authors to the relationship between the economic structure of a country and its overall productivity growth. Along the second half of the twentieth century, those pioneer papers on this subject[7] have been followed by others focused on the manufacturing sector.[8] Nevertheless, the influence of services sector has not been analyzed empirically as much as would have been expected given its dominant role in highly developed countries.

A controversial topic in last decades has been, precisely, the extraordinary increase in the weight of services in advanced economies, as well as its challenges and policy implications. An important aspect is whether this increasing weight of the service industries does have or not an impact on the performance of the overall productivity. Except for some papers (i.e.: Dutt & Lee, 1993; Maroto & Cuadrado, 2007, 2009), this factor has not been dealt empirically in the depth required and only a very few papers have analyzed this problem at regional level and practically all them referred to a single country. This paper aims to contribute to fill the gap and to feed the debate around productivity in service sector from a regional perspective.

Baumol (1967) and himself with the collaboration of Blackman and Wolff (1989) produced some suggestive ideas on the relationship between the progressive growth of services in advanced economies and their low productivity. Nusbaumer (1987) and De Bandt (1991) have also agreed on Baumol's approach. Using the labour force in order to explain the differences in productivity among industries, such theories concluded that economic growth and overall productivity growth of "service" economies would show a trend to a slowdown. Empirical evidence commonly shows that there is a negative relationship between the overall labour productivity growth and the weight of the services sector in advanced economies. Figure 9.1 shows aggregate evidence on this for a wide group of OECD countries. It can be seen that there is a negative relationship between the overall labour productivity growth rate of the economy and the weight of the services sector.[9] Data show that the economies having higher productivity growth are also those in which the

[7] See, for example, Salter (1960), Denison (1967), and Chung and Denison (1976).

[8] See, among others, Young (1995), Fagerberg (2000), Timmer and Szirmai (2000), Carree (2003), or Krüger (2008).

[9] Specifically, the correlation coefficient in the case of employment is −0.5223, significant to 1 % (p-value = 0.0040). Results are robust if the weight of service sector is measured in terms of value added. Then, the correlation coefficient is −0.5838, also significant to 1 % (p-value = 0.0015).

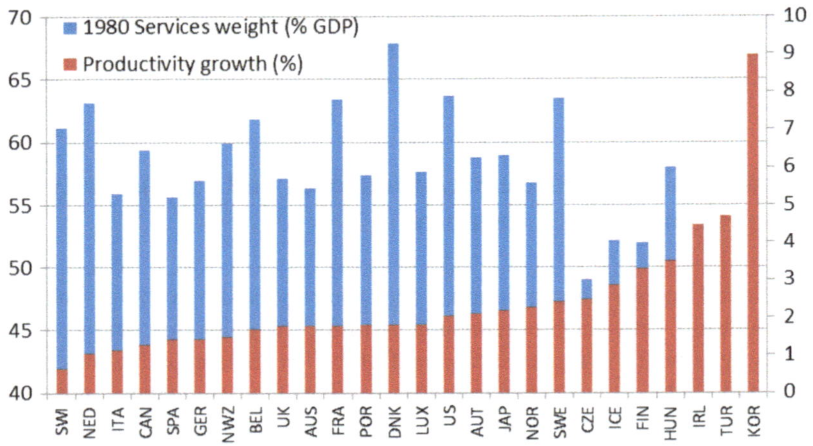

Fig. 9.1 Relationship between service sector weight (and productivity growth 1980–2008. OECD countries sample. *Source*: Own elaboration. Data Conference Board (2011) and OECD (2009)

services sector had a lower percentage of the total at the beginning of the 1980s, as occurs in the case of Korea, Ireland, Iceland, Finland and some of the New Members of the EU. On the contrary, countries showing a high percentage of services (in total production and employment) register lower productivity growth rates, as it is the case of Italy, Germany, Canada, the Netherlands, Spain or the United Kingdom. The only exceptions in this Fig. 9.1 are countries such as some Nordic (Sweden or Denmark) or the United States.

The latter affirmation was based on the hypothesis of a lower productivity growth within the services sector. But, in recent years this thesis has been smothered or even refuted by empirical evidence in some papers. Even Baumol (2002) rectified his previous position by admitting that it is necessary to differentiate between types of services and stressing the role of innovation and knowledge in the evolution of services. Triplett and Bosworth (2006) have also criticized the traditional theories on the services sector and even believe they have found the "cure" for Baumol's cost disease. Generally speaking, criticism and revision are based on the following components (Maroto, 2012): (1) the need to take into account the indirect effects of some service activities on the productivity growth within other industries; (2) biases in the definition and measurement; or (3) the possibility of using indirect indicators of productivity as consequence of the conceptual and statistical debates generated over the last 10 years.

Some empirical studies have proved that the traditional affirmation that services contribute to the stagnation of overall productivity growth in the long term might actually be questioned. The data at international level highlight the patterns of dynamic productivity in some branches of services, mainly those related to ICT, both in Europe (O'Mahony & van Ark, 2003; van Ark & Piatkowski, 2004) and in the US (Bosworth & Triplett, 2007; Stiroh, 2001; Triplett & Bosworth, 2004). The high growth rates have been almost continuous over the last decades, fact which suggests that these service industries do not seem to be asymptotically "*stagnant*". On the contrary, the dynamism observed in some advanced economies from the middle of the 1990s may indicate an environment for potential improvements in the future.

Additionally, the theories which currently explain the reason for the growth of services and which condition their productivity are not limited exclusively to the labour factor, but are related to multiple factors, such as those linked to the nature of the services, the organization and segmentation of their markets, or the special substitution relationships between labour and capital (Rubalcaba, 2007). Finally, others authors have highlighted the interrelationship between globalization, trade and growth of services (Cuadrado et al., 2002).

Empirical evidence summarized in Table 9.1 shows that productivity growth in relation to the evolution of employment and production is not homogeneous in all service branches. Communications and some branches of transport show high productivity growth rates, although without regard for strong employment reduction processes. On the other hand, part of the transport services, the financial activities, wholesale trade and renting services are characterized by an intensive use of factors boosting productivity, such as innovation or human capital. All of them show also positive employment growth.

3 Structural Change, Service Sector and Productivity Growth: A Decomposition Analysis

3.1 Data and Methodological Approach

European Regional Database provided by Cambridge Econometrics will be used in order to develop our analysis. It offers indicators on gross value added, employment and other relevant economic variables both for countries and regions at a sector level since the beginning of the 1980s. Despite the narrow industrial disaggregation of this source, we have chosen it due to the homogeneity with the other sections in the paper. The sample of countries used includes all of the EU-15 with the exception of Luxembourg, plus Norway and Switzerland. The time span used is the one available in the chosen source, which ranges from 1980 to 2008. Finally, the selected breakdown by economic sectors is as follows: agriculture (01–05 level of the ISIC), manufacturing and mining (10–39), construction (45), market services (50–74), and non-market services (75–99). As service sector constitutes the focus of our analysis, market services have been broken down into five branches: distribution (50–52), hotels and restaurants (55), transport and communications (60–64), financial and insurance services (65–67), and other market services, including real state and business services (70–74).

Table 9.1 Typologies of European services, 1980–2008

		Productivity growth			
		Above the aggregate average		Below the aggregate average	
Employment growth	*Above the aggregate average*	*Dynamic services*		*Labour intensive services*	
		GVA above the average	*GVA below the average*	*GVA above the average*	*GVA below the average*
		Air transport (62) and auxiliary transport activities (63) Financial services (65) and auxiliary finance services (67) Wholesale (51) Renting of equipment and machinery (71)	–	Computer services (72) R&D services (73) Other business services (74) Real state (70) Domestic service personnel (95) Health and social services (85) Other community and personal services (90–93)	Commerce (50) and retailing (52) Hotels and restaurants (55) Education (80) Insurance services (66)
	Below the aggregate average	*Services reestructuring through employment*		*Backward Services*	
		GVA above the average	*GVA below the average*	*GVA above the average*	*GVA below the average*
		Post and communications (64)	Land (60) and sea transport (61)	–	Public Administration and Defence (75)

Source: Own elaboration

To analyze what is the impact of structural changes on the productivity growth we will use the data above described, pointing out the heterogeneity of the different branches within service sector. To do it, a *shift-share* type analysis is used. This technique provides a convenient tool to research how aggregate growth is mechanically linked to differential growth of labour productivity and the reallocation of labour between industries. It breaks down overall productivity growth into two effects: structural changes (net or static effect and dynamic effect) and the within-sector productivity growth. Formally, the method applied here may be derived as follows:

$$\dot{\pi}_r = \frac{\pi_{r,t}}{\pi_{r,t-n}} = \frac{\sum_{i=1}^{N} \pi_{ir,t}\left(s_{i,t} - s_{ir,t-n}\right) + \sum_{i=1}^{N} \left(\pi_{ir,t} - \pi_{ir,t-n}\right)\left(s_{ir,t} - s_{ir,t-n}\right) + \sum_{i=1}^{N} \left(\pi_{ir,t} - \pi_{ir,t-n}\right) s_{ir,t}}{\pi_{r,t-n}} \tag{9.1}$$

where: π is the labour productivity; *t-n* is the initial year; *t* is the final year; *i* corresponds to each economic sector; *r* to regions, and *s* is the sector weight in terms of employment $s_i = \frac{L_i}{L}$.

According to the methodology, the overall growth of labour productivity can be broken down into three differentiated effects. The first is the contribution from changes in the allocation of labour between industries. The second one measures the interaction between changes in productivity in individual industries and changes in the allocation of resources. Finally, the third effect would be the contribution of productivity growth within individual industries (weighted by the share of these in total employment).

Decomposition techniques do not just allow us to analyse structural changes over time and their effects on productivity, but also structural changes in space and their effects. For this purpose, we have used a decomposition technique (shift-share) in order to analyse the regional productivity growth (and the variables it depends on: production and employment) by using two effects of a multiplicative nature[10]: the country effect (CE) and the net effect (NE) of the region. The latter can also be broken down into the product of the proportional effect (PE), based on the productive structure of the region, and the differential effect (DE), which represents the rest of the identifying variables of the region itself. Although Eq. (9.1) will be used for both the analysis by countries and by regions, the technique described below will only be used for the regional analysis.

[10] Instead of the additive nature which is usually used in this kind of techniques. The reasons are: on the one hand, the elimination of effects of scale originated from the use of several variables with different units. On the other hand, the possibility of combining the three variables under consideration: productivity, added value and employment, in just one indicator, in line with what was previously shown graphically in Fig. 9.1.

Each index or effect can be greater than one (if the region has grown above national average) or lower than one (otherwise). The mathematical expression in the analysis of production and employment would be as follows:

$$\dot{\xi}_r = \frac{\xi_t}{\xi_{t-n}} = \frac{\sum_{i=1}^{N} \xi_{ir,t-n} \left(\frac{\sum_i \sum_r \xi_{ir,t}}{\sum_i \sum_r \xi_{ir,t-n}} \right)}{\sum_{i=1}^{N} \xi_{ir,t-n}} \cdot \frac{\sum_{i=1}^{N} \xi_{ir,t}}{\sum_{i=1}^{N} \xi_{ir,t-n} \left(\frac{\sum_i \sum_r \xi_{ir,t}}{\sum_i \sum_r \xi_{ir,t-n}} \right)} = CE_\xi \cdot NE_\xi \quad (9.2)$$

$$NE_r = \frac{\sum_{i=1}^{N} \xi_{ir,t}}{\sum_{i=1}^{N} \xi_{ir,t-n} \left(\frac{\sum_i \sum_r \xi_{ir,t}}{\sum_i \sum_r \xi_{ir,t-n}} \right)}$$

$$= \frac{\sum_{i=1}^{N} \xi_{ir,t-n} \left(\frac{\sum_r \xi_{ir,t}}{\sum_r \xi_{ir,t-n}} \right)}{\sum_{i=1}^{N} \xi_{ir,t-n} \left(\frac{\sum_i \sum_r \xi_{ir,t}}{\sum_i \sum_r \xi_{ir,t-n}} \right)} \cdot \frac{\sum_{i=1}^{N} \xi_{ir,t}}{\sum_{i=1}^{N} \xi_{ir,t-n} \left(\frac{\sum_r \xi_{ir,t}}{\sum_r \xi_{ir,t-n}} \right)} = PE_\xi \cdot DE_\xi \quad (9.3)$$

where ξ represents the analysis variable (gross value added, Y, or employment, L), i represents the N productive sectors, r corresponds to the regions considered, and t and t-n are the two points of time chosen in the analysis (1980 and 2008).

Regional productivity growth can be obtained from the previous equations as the quotient between the growth of gross value added and regional employment. The aforementioned productivity growth π can be broken down again into its country, proportional and differential effects, on the basis of the following equation:

$$\dot{\pi}_r = \frac{\pi_{r,t}}{\pi_{r,t-n}} = \frac{\dot{Y}_r}{\dot{L}_r} = \frac{CE_{Y,r} \cdot NE_{Y,r}}{CE_{L,r} \cdot NE_{L,r}} = \frac{CE_{Y,r} \cdot PE_{Y,r} \cdot DE_{Y,r}}{CE_{L,r} \cdot PE_{L,r} \cdot DE_{L,r}}$$
$$= CE_{\pi,r} \cdot PE_{\pi,r} \cdot DE_{\pi,r} \quad (9.4)$$

In accordance with formulas (9.2) and (9.3), a region r can be classified according to six different typologies or categories, three with a *NE* greater than one and three with a *NE* lower than one:

1. NE, PE, DE > 1: *Dynamic* regions.
2. NE, PE > 1, but DE < 1: Regions *specialised in dynamic sectors.*
3. NE, DE > 1, but PE < 1: Regions with *advantages of location.*
4. NE, PE, DE < 1: *Backward* regions.
5. NE, PE < 1, although DE > 1: Regions *specialised in backward sectors.*
6. NE, DE < 1, although PE > 1: Regions with *disadvantages of location.*

Table 9.2 Decomposition analysis of productivity growth, 1980–2008

	Labour productivity growth	Net or static effect	Interaction or dynamic effect	Within effect
Euro zone				
Total	**1.33**	**0.67**	**−0.54**	**1.19**
	=	=	=	=
Manufacturing	(2.85)	+0.04	−0.02	+0.07
Services	(0.60)	−0.01	+0.00	+0.00
Rest	(1.88)	+0.64	−0.52	+1.12
Sample of 17 OECD countries				
Total	**1.50**	**0.51**	**−0.47**	**1.46**
	=	=	=	=
Manufacturing	(3.06)	+0.07	−0.05	+0.21
Services	(0.83)	−0.01	+0.00	+0.00
Rest	(2.12)	+0.45	−0.43	+1.25

Note: "Rest" refers to those main sectors not included under "manufacturing" or "services", i.e. "agriculture" and "construction"
Source: Based on Cambridge Econometrics

3.2 National Results

According to Eq. (9.1), results of national calculations for the period 1980–2008 are shown in Table 9.2, both for the countries belonging to the Euro-zone and to the sample of 176 OECD economies, broken down into individual contributions by the three main economic sectors. Table 9.3 shows analogous results broken down by specific service industries. In line with the Eq. (9.1) on the breakdown of the overall productivity, the sum of the static and dynamic effects, as well as the within-industry growth, is equal to the average growth rate of labour productivity in the according aggregate (first cell in each sub-table). This is how the data sums up horizontally. Vertically, for each of the three components, the contributions made by each sector also sum up to the corresponding number in the first line of each sub-table. As additional information, the number in brackets show the average growth of labour productivity within individual sectors or service industries (Table 9.3), and don't sum up neither in the horizontal nor in the vertical dimensions. They facilitate us the work of identifying whether there are any regular patterns of differential productivity growth between industries.

Supported by data from Table 9.2, some stylized facts can be underlined. First of all, the structural components emerge to be generally dominated by the within effects of productivity growth, which is consistent with the results obtained by some authors and referring to other economic areas.[11] This means that, in aggregated terms, the

[11] See: Peneder (2002, 2003) for 28 countries of the OECD; Havlik (2005) for the new Eastern European countries belonging to the EU; Fagerberg (2000) for the manufacturing sectors in 39 countries based on the UNIDO; Timmer and Szirmai (2000) for the manufacturing sectors of four

Table 9.3 Decomposition analysis

	Labour productivity growth	Net or static effect	Interaction or dynamic effect	Within effect
Euro zone				
Services	**0.60**=	**0.55**	**−0.07**	**0.12**
Distribution	(0.94)	−0.08	−0.02	+0.03
Hotels and restaurants	(0.09)	−0.05	+0.00	+0.00
Transport and communications	(1.09)	+0.08	+0.02	+0.04
Financial and insurance	(1.16)	+0.09	+0.03	+0.04
Other market services	(−0.73)	+0.62	−0.09	+0.00
Non market services	(0.24)	−0.11	−0.01	+0.00
Sample of 17 OECD countries				
Services	**0.83**	**0.57**	**0.00**	**0.27**
	=	=	=	=
Distribution	(1.15)	+0.24	+0.02	+0.06
Hotels and restaurants	(0.25)	+0.63	+0.01	+0.00
Transport and communications	(1.44)	−0.50	−0.04	+0.10
Financial and insurance	(1.86)	+0.03	+0.00	+0.11
Other market services	(−0.14)	−0.47	+0.00	+0.00
Non market services	(0.29)	+0.64	+0.01	+0.00

Service industries, 1980–2008
Source: Based on Cambridge Econometrics

reallocation of labour among those sectors with low and high productivity has only had a weak net effect on overall growth. This fact is even more noteworthy since the mid-1990s, a period in which productivity growth rates of the European countries in relation to other areas, such as the US, began to fall notably. Secondly, it can be seen that there are not significant differences between the two areas analyzed. Euro-zone performance differs somewhat from the case of the broader sample, where the productivity growth rate is a little bit higher (due to the higher productivity growth rates experienced in most of Northern European countries) and the structural effects, both static and dynamic, are barely lower than in Euro-zone countries. Thirdly, the data obtained show the simultaneous operation of opposing mechanisms captured under the static and the dynamic shift effects. The structural burden of resource reallocation seems to be robust in the European case, where the dynamic effect is negative for the broad 3-sector break down. Finally, if we analyze the performance by sectors, most of the effects on the overall productivity come from non-tertiary activities. This suggests that, despite the progress obtained as regards productivity by the services sector, those non tertiary activities are still providing the major contribution to the growth of the overall productivity of the advanced economies.

Asian countries; Maroto and Cuadrado (2007, 2009) for Spanish economy, and EU-15 and US, respectively; and van Ark (1995) for a group of 8 countries of the EU and the USA.

This aggregated approach could conceal important structural aspects in each individual sector. This perspective is particularly interesting in the case of the service sector, where the overall contribution to productivity is divided practically between two of the components analyzed here: the within growth and the static effect. In other words, services contribute to GDP per capita via two different channels. Firstly, through their within growth of the GDP per hour worked, just as in any other sector and secondly, and this is an exclusive factor of services sector, through the growth of the weight their activities suppose in terms of employment. This is consistent with the traditional hypothesis on growing percentages in the demand for the services sector due to its greater income-elasticity.[12]

If we deep into the service sector (Table 9.2), calculations show that productivity growth of the service sector in the sample of 17 advanced countries (0.83 per 100) is rather higher than the growth in the Euro-zone (0.60 per 100) and both rather distant from the one in the US (1.3 per 100). But, disaggregating the heterogeneous branches of services, there are some, particularly transport, communications and financial services, which show high within growth (last column), similar to those within sectors traditionally characterized by higher productivity levels. Moreover, most of the productivity growth comes from the reallocation of resources and not from the within growth. Consequently, the traditional view of the (aggregated) service sector being scarcely productive might be refuted when certain tertiary activities are studied, consistent with the findings of some of the more current empirical studies. Again, the case of the Euro-zone differs to some extent from the broader sample of 16 countries. Additionally, detailed analysis of these data shows, as in Table 9.1, that *structural burden* hypothesis is clearly confirmed for the service sector in the Euro-zone, although the effect in the EU16 is null. Alternatively, the *structural bonus* hypothesis (positive static effect) can also be observed—with few exceptions—in most service industries.

The results presented are consistent with those found by other authors for previous periods (Bonatti & Felice, 2008; Fagerberg, 2000; Maroto & Cuadrado, 2007, 2009; Peneder, 2002, 2003; van Ark, 1995). The *structural change*[13] has a positive effect, although this is relatively weak, on the overall productivity growth. No clear or univocal tendency to the reallocation of labour to those sectors with higher productivity levels has been found. However, the robust existence of a so-called structural burden can be observed due to the fact that, in the sectors with faster productivity growth, the expansion of production is not generally accompanied by growth in employment. Thus, it is possible to speak about a stylized fact. In contrast with periods previous to the economic crises of the 1970s, the results of the period analyzed here show that the structural changes do

[12] See: Schettkat and Yocarini (2006) for a review of the literature on the shift to services employment. Fourastie (1949) and Fuchs (1968) have been pioneers introducing this theory. Some empirical applications on this hypothesis are: Peneder et al. (2003); or Gregory et al. (2007).

[13] This combined effect of the static and dynamic components is named "*structural effect*" or simply the "*effect of structural change*" by some authors (Maddison, 1996), and analyzed together although the analysis is deeper if both effects are distinguished.

not notably boost productivity growth. The novelty of our results emerges, neither from the methodological approach used nor from the main conclusions arisen, but from the disaggregated focus of the service industries, clearly characterized by a heterogeneous composition of activities. This will extend findings of previous papers on the service sector, the most important agent in advanced economies.

3.3 Regional Results

The previous section revealed the relationships between structural changes and, particularly, the growth of the services sector and the evolution of aggregate productivity in the European countries. However, the objective of this section is to demonstrate the degree of influence of productive specialisation on the evolution of regional productivity in Europe, paying special attention to the role played by the growth of services.

Productive specialisation can be one of the main causes of the differences between regional[14] behaviour and that of the countries. The evolution of those regions with a higher specialisation in dynamic activities will be far higher than the average of their corresponding countries. The main objective of this section is to analyse the importance of these factors, where the contribution of services activities to growth is particularly significant. The main conclusion drawn is that services play a role in the growth of productivity in the European regions under consideration. In order to reach this conclusion, the decomposition techniques described in Eqs. (9.2)–(9.4) are used.

The starting hypothesis of this work was the existence of a positive relationship between the weight of the services sector and the evolution of productivity in the European economies. The previous section revealed the first evidences at a state level. This section tries to draw the same conclusion at a regional level. By using the concept of Camagni and Cappellin (1985), which we previously applied in order to create Fig. 9.1 in Sect. 1, we try to analyse the evolution of labour productivity in a certain European region, together with the evolution of production (added value) and employment of such a region, taking the average behaviour of the country where each region belongs to as a reference. For this purpose, we consider the regional net effects—once isolated from the country effect.

The aforementioned allows us to simplify the information included in the previous tables by classifying the European regions into four different groups or typologies: (1) dynamic regions with net effects greater than one regarding productivity and employment; (2) regions under reconstruction through employment (dynamism regarding productivity arises mainly due to net effects lower than one regarding employment); (3) creation of employment-intensive regions (the net

[14] See, among others, Daniels (2004), Combes and Overman (2003), Midelfart-Knarvik et al. (2003), Ezcurra et al. (2006), Amiti (1999), Haaland et al. (1998), Hallet (2000), and Molle (1996).

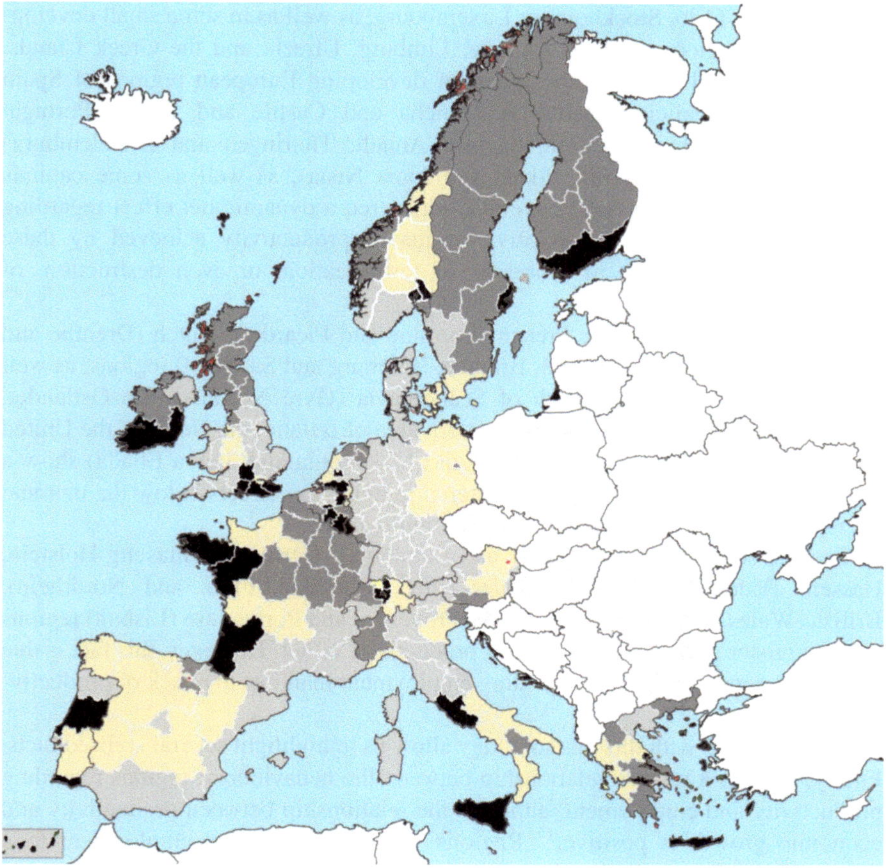

Map 9.1 Regional clustering according productivity and employment, 1980–2008. *Note: Black coloured* regions identify dynamic regions; beige coloured regions recognize restructuring regions; *light grey coloured* categorize labour intensive regions; and, finally, *dark grey coloured* regions classify backward regions. *Source*: Own elaboration. Data: Cambridge Econometrics

effect greater than one regarding employment leads to a lower growth regarding productivity); and, finally, (4) backward regions (which show a lower growth regarding both productivity and employment).

Map 9.1 shows this classification for the regions analysed during the 1980–2008 period. It is difficult to draw general conclusions from the data obtained due to the high level of heterogeneity between the 170 regions included in the sample since behaviours and explaining factors of a different nature and origin are intertwined. However, European regions can be classified, in a broad outline, according to their productivity growth and their capacity to simultaneously create employment or not.

Thus, dynamic regions (those with good results regarding both productivity and employment) are concentrated in some capitals and financial centres, such as

Zürich, Lazio, Oslo, Stockholm or Luxembourg, as well as in some small developing outlying regions, such as Algarve, Limburg, Utrecht and the Greek Islands. Some regions belonging to the group of developing European regions of Spain (Extremadura, Galicia, Castile-La Mancha and Castile and Leon), Portugal (Alentejo), Germany (Sachsen, Sachsen-Anhalt, Thuringen and Brandemburg) and Greece (Ipeiros, Dytiki Ellada and Ionia Nisia), as well as some capitals (Paris, London and Brussels) have also registered a dynamic net effect regarding productivity. However, the positive results of productivity achieved by these regions are mainly due to processes of low creation, or even destruction, of employment.

On the other hand, some French (Lorraine and Picardie), Dutch (Drenthe and Groningen) and German (Berlin, Bremen, Hamburg and Saarland) regions, as well as some others from the North of Scandinavia (Ovre Norrland, Sor-Ostlandet, Nord-Norge and Smäland) and the western area of Ireland, the North of the United Kingdom and some Greek regions (Dytiki Makedonia and Sterea Ellada) show a deterioration, because they have registered regional net effects below the national average regarding both productivity and employment.

Finally, some Spanish (Madrid and Catalonia), German (Schlasung-Holstein, Hassen, Baden-Württenberg, Niedersachen, Rheinland-Pfatz and Nordrhein), British (Wales and Yorkshire), French (Provence) and Portuguese (Lisbon) regions do not register good results regarding productivity either. However, this fact is due more to dynamic net effects regarding employment than just to a lack of productivity or efficiency.

Data obtained with this methodology allow us to highlight several stylised facts. Firstly, there is a reverse relationship between the behaviour of regions regarding productivity and employment, although the relationship between productivity and economic growth is positive.[15] Regions which have created employment to a greater extent in recent decades are generally associated with lower growths of productivity. More specifically, the correlation coefficient between the growth of employment and the growth of productivity in the sample of analysed regions is -0.226 (with a p-value of 0.000), while the correlation coefficient between the growth of productivity and the growth of added value is 0.570 (which is also statistically significant for any significance level). When only the services sector is taken as a reference of employment and production compared to the productivity of the region in question, the results obtained are similar.

However, the relationship between the evolution of productivity and the growth of the weight of the services sector, regarding production and employment, is significantly positive (with a correlation coefficient of 0.151), though not very high. Data seem to show a slight relationship between the weight of services and the growth of productivity in the regions under analysis during the period from 1980 to 2008. This is an important conclusion as it coincides with what was obtained in the previous Sect. 3.1 regarding the national analysis, but even more so because it

[15] See, among others, Salter (1960) and Baumol et al. (1989).

can be used as a link and foundation for the econometric analysis constituting the core of Sect. 4. The objective of that Sect. 4 is, precisely, to statistically contrast the existence of the aforementioned relationship between the growth of the weight of services in European regions and the evolution of their productivity.

4 Tertiarization and Productivity Growth: An Econometric Analysis

The results obtained until now should not be taken as an implication that the structural changes or growth of services do not play an important role in the evolution of overall productivity. What they do show is that structural changes, on average, do not involve significant growth in that area. One economic sector that deserves an in-depth analysis in this respect is the service sector. Based on this fact, we will analyze the impact of the growth of services on overall productivity growth in the sample of seventeen European countries since 1980. From a merely accounting point of view (as in Table 9.2), some service industries are characterized by both high productivity levels and high growth rates. Nevertheless, the methodology developed previously does not obtain the indirect effects that the tertiarization of the economies have on other sectors (outsourcing, off-shoring, etc.), and maintains the intrinsic difficulties concerning definition and measurement.

4.1 Data and Methodology

To develop this analysis the *European Regional Database* provided by Cambridge Econometrics will be used again in order to homogenize our results with those in the previous section. However, as previously mentioned, this source only provides information on production, employment and physical capital. In order to complement those items and to explore some additional explanatory factors the *Regional Database* provided by the OECD will be used as well. The only disadvantage of using both sources is the different time range. While the data base provided by Cambridge Econometrics begins at 1980, the starting date in the one provided by the OECD is 1995, reducing the size of the sample.

The aim of this section will be to explore to what extent an increase in the share of resources assigned to the service industries is relevant to the productivity growth of an economy at a regional level. To achieve this, a panel data model was used, carrying out regressions of the overall productivity growth over the change in the weight of services. Additionally, two other explanatory variables are included: the initial level of productivity (introduced to achieve *catching-up* or technological convergence) and the initial weight of the service sector (which distinguishes between those countries (regions) which, while undergoing equal growth in the percentages of employment, differ significantly in their levels or weight). As overall productivity growth is also influenced by other variables, besides structural change, a matrix of auxiliary conditioning variables has also been included in the regressions. This matrix includes the investment effort (measured as the ratio between the gross stock of physical capital over GDP), the demographic composition changes (as the relationship between active and total population), the level of human capital (approximated through the percentage of employees with secondary and higher education in the total employment), and the degree of trade openness of the country which each region belongs to.

The final specification of the model to be used is the following:

$$\Delta\pi_i = \alpha + \beta\pi_{i,t-n} + \gamma\Delta s_i + \delta s_{i,t-n} + \phi Z_{i,t} + v_i + \varepsilon_{i,t}$$

where $i = 1.2,\ldots,K$ are the regions in the sample (with $K = 170$), n is the length of the period considered (with $n = 28$), s_i is the weight of the service sector (over total employment) in the country i, and $\Delta\pi_i$ represents the labour productivity growth rate. Z_i is the matrix of auxiliary variables. v_i is the random effects component, and ε_{it} the residue of the model. The idea of fixed effects is discarded despite its generalised use in panel data models, as this does not admit within-group constant variables, such as the case of the initial weight of the service sector or the initial productivity level in our analysis.

4.2 *National Results*

Table 9.4 summarizes the main results of the model used.[16] A simpler model relates the growth of overall productivity only to services growth (column 3.1). Then we have added the initial level of productivity (3.2) and the initial level of tertiarization (3.3). Finally, the matrix of auxiliary variables was included in our model (3.4). The main result is that the increase in the weight of services, from 1980 to 2008, had a positive[17] effect on overall productivity growth. However, this positive effect is limited. An absolute increase of 1 per 100 in the weight of the service sector in terms of employment would be associated to an increase of 0.3 points in the rate of absolute overall productivity growth (during the whole period). The estimations are

[16] A standard OLS regression model in a cross-section (for example, in Fagerberg, 2000) has also been implemented. Conclusions, although calculations are not included in the text, do not differ from the conclusions drawn in the paper based on a panel-data regression model.

[17] The positive relationship between service growth (regressor) and labour productivity (dependent variable) might be endogenous, so results could be influenced by reverse causation matters. In order to solve this, Granger causality tests were implemented (Granger, 1969). According to our data, the growth of services could explain productivity growth (with the usual number of lags up to 14, null hypothesis that growth of services does not cause productivity growth will be rejected with any usual level of statistical confidence). Nevertheless, reverse causality will not be accepted (null hypothesis that productivity growth does not cause growth of services will not be rejected with any usual level of statistical confidence). Summarizing, likely reverse causation matters seem to be solved in the model regressed here.

Table 9.4 Structural change and productivity growth in European countries, 1980–2008[a]

	3.1	3.2	3.3	3.4
Service sector growth	1.68***	1.68***	0.70***	0.30***
Initial productivity level		−1.28e^{-7}	−1.23e^{-5}***	−1.27e^{-5}***
Initial services weight			2.22***	2.27***
Physical capital				0.26***
Demographic composition				0.46
Adjusted R^2	0.70	0.71	0.73	0.75
Num. observations	476	476	476	476
Market services	1.22***	1.22***	1.22***	1.20***
Non market services	0.45***	0.45***	0.47***	0.48***
Initial productivity level		−4.72e^{-7}	−5.65e^{-6}	−5.83e^{-6}
Initial market services weight			1.31**	1.33***
Initial non market services weight			1.46***	1.50***
Physical capital				0.22***
Demographic composition				0.28
Adjusted R^2	0.72	0.72	0.72	0.73
Num. observations	476	476	476	476

Constant coefficient is not shown, although it was included in the model
Note: Specifications 3.1, 3.2 and 3.3 have been run on the reduced sample which is used in specification 3.4 and the results are robust with those presented in this table
Source: Own elaboration. *Data* Cambridge Econometrics
Statistical significance level at *10 %; **5 %; ***1 %
[a]Data panel estimation, random effects

highly significant (at 1 %) and stable throughout the different specifications of the model.

Convergence or *catching-up* effect (approximate for the level of labour productivity in 1980) is also statistically significant, with a negative coefficient, as predicted by the traditional theories, although this is relatively low. Those countries which started with higher levels have seen how their overall growth rates were below those which were further behind at the end of the 1970s. Additionally, the weight of services at the beginning of the period is also statistically significant and demonstrates a positive sign. This fact may support the hypothesis that those countries which were more tertiarized from the beginning had a more dynamic overall productivity growth rate than those which started with a lower weight of services.

One of the features that characterizes the service sector is a marked heterogeneity (as observed, among other results, in the calculations shown in Fig. 9.1 and Table 9.2), as well as its atomization and diversification of supply due to the fact that market activities and other non-market services coexist in this sector. Consequently, it is reasonable to suppose that the likely impact on overall productivity growth might differ depending on the different kind of services involved. In order to differentiate the results obtained so far depending on service clusters, bottom-block in Table 9.4 shows the results of our model. The innovation is the way in which we distinguish between market and non-market services.

The results highlight that, following the logic stated above, the market services have a higher (and statistically significant) coefficient than that observed in the case

Table 9.5 Structural change and productivity growth in European regions, 1980–2008[a]

	3.1	3.2	3.3	3.4[b]
Service sector growth	1.10***	1.06***	1.09***	0.89**
Initial productivity level		$-3.82e^{-6}$***	$-9.42e^{-6}$***	$-1.51e^{-5}$***
Initial services weight			1.17***	0.42***
Physical capital				0.14***
Human capital				0.01***
Demographic composition				0.25**
Openness				0.10***
Adjusted R^2	0.50	0.53	0.57	0.67
Num. Observations	4,688	4,688	4,688	1,140

Constant coefficient is not shown, although it was included in the model
Note: Specifications 3.1, 3.2 and 3.3 have been run on the reduced sample which is used in specification 3.4 and the results are robust with those presented in this table.
Source: Own elaboration. *Data* Cambridge Econometrics and OECD Regional Database
Statistical significance level at *10 %; **5 %; ***1 %
[a]Data panel estimation, random effects
[b]1995–2008

of the non-market services. Thus, an increase of 1 % in the weight of market services would suppose an increase in the absolute overall productivity growth amounting to 1.2 % points, whilst the same increase in those services outside the market involves a relatively lower change amounting to 0.45 % points. Additionally, the performance of the other variables included in our model follows the same behaviour patterns as when the service sector as a whole was analyzed in up-block in Table 9.3.

4.3 Regional Results

Table 9.5 summarizes the main results of the model with a panel[18] of regional data belonging to the 17 European countries of our sample. A simpler model relates the growth of overall productivity only to services growth (column 3.1). Then we have added the initial level of productivity (3.2) and the initial level of tertiarization (3.3). Finally, the matrix of auxiliary variables was included in our model (3.4). The main result is that the increase in the weight of regional service sector, from 1980 to 2008, had a positive effect on overall productivity growth. An absolute increase of 1 per 100 in the weight of the service sector in terms of regional employment would

[18] A standard OLS regression model in a cross-section (for example, in Fagerberg, 2000, or Maroto & Cuadrado, 2009) has also been implemented. Additionally, estimations with subsamples and different time spans have been developed. Conclusions, although calculations are not included in the text, do not differ from the conclusions drawn in the paper based on a panel-data regression model.

be associated to an increase of 1.1 points in the rate of absolute regional productivity growth (during the whole period). The estimations are highly significant (at 1 %) and stable throughout the different specifications of the model. The explanatory capacity of the model, through its adjusted R-squared, is also relatively acceptable. Moreover, regional results not only argue with previous country ones, but the positive coefficient is even a little bit higher.

The positive relationship between service growth (regressor) and labour productivity (dependent variable) might be endogenous, so results could be influenced by reverse causation matters. In order to solve this, Granger causality tests[19] were implemented (Granger, 1969). According to our data, the growth of services could explain productivity growth (with the usual number of lags up to 14, null hypothesis that growth of services does not cause productivity growth will be rejected with any usual level of statistical confidence). Nevertheless, reverse causality will not be accepted (null hypothesis that productivity growth does not cause growth of services will not be rejected with any usual level of statistical confidence). Summarizing, likely reverse causation matters seem to be solved in the model regressed here.

Related to the other explanatory variables of the model, convergence or *catching-up* effect is also statistically tested in the model, although its role is quite low. Those regions which started with higher levels have seen how their overall growth rates were below those which were further behind at the end of the 70s. Additionally, the weight of services at the beginning of the period is also statistically significant and demonstrates a positive sign.

With respect to the auxiliary matrix, and taking into account its incorporation into the model as a complement to the central analysis, all ancillary variables are statistically significant and have a positive coefficient. Both physical and human capital, measured in this analysis as levels, in line with various papers which stress the role of these two factors in economic growth and in the good performance of the productivity growth, have a positive impact on the growth of overall productivity. This is greater in the case of physical capital. Those regions with a greater quantity of qualified working population and more extended capitalization processes are those which have presented a more dynamic growth in productivity. Additionally, demographic issues and the degree of openness of the countries where regions are located also boost productivity growth. Finally, results of the last column in Table 9.3 show that the positive effect of structural changes, and particularly of the services sector growth, is lower when other auxiliary variables are included in the model. This does imply a lower effect of tertiarization on the productivity growth since the mid-1990s. While this effect accounted for 1.1 in the 1980–2008, the relative coefficient was only up to 0.6 when we analyze only the 1995–2008 period. This result follows some of the most recent works in the literature. The role

[19] A time series X is said to Granger-cause Y if it can be shown, usually through a series of F-tests on lagged values of X (and with lagged values of Y also known), that those X values provide statistically significant information about future values of Y.

Table 9.6 Structural change and productivity growth, 1980–2008[a]: market services versus non-market services

	4.1	4.2	4.3	4.5[b]
Market services	0.61***	0.60***	0.60***	0.16***
Non market services	0.43***	0.42***	0.45***	0.11***
Initial productivity level		$-3.68e^{-6}$***	$-8.71e^{-6}$***	$-1.47e^{-5}$***
Initial market services weight			1.02***	0.49***
Initial non market services weight			1.26***	0.70***
Physical capital				0.58***
Human capital				0.02***
Demographic composition				0.63***
Openness				0.13***
Adjusted R^2	0.50	0.52	0.54	0.61
Num. observations	4,688	4,688	4,688	1,140

Constant coefficient not shown, although it was included in the model
Note: Specifications 3.1, 3.2 and 3.3 have been run on the reduced sample which is used in specification 3.4 and the results are robust with those presented in this table
Source: Own elaboration. *Data* Cambridge Econometrics and OECD Regional Database
Statistical significance level at *10 %; **5 %; ***1 %
[a]Data panel estimation, random effects
[b]1995–2008

of structural changes over the productivity growth in advanced economies has lost its major role for the within productivity effects since the 1980s (Cuadrado et al., 1999). However, the responsibility of tertiarization, and specially the growth of some professional and dynamic market services since the mid-1990s, has played an important role in the productivity growth of these economies.[20]

Following the schedule applied in the previous section and looking for differentiating the results obtained so far depending on market and non-market services, Table 9.6 shows the results of our model. The results highlight that, following the logic stated above, the market services have a higher (and statistically significant) coefficient. In those non-market services, the behaviour is quite the opposite. Thus, an increase of 1 % in the weight of market services would suppose an increase in the absolute overall productivity growth amounting to 0.61 % points, whilst the same increase in those services outside the market involves a relatively lower change amounting to 0.43 % points. Additionally, the performance of the other variables included in our model follows the same behaviour patterns as when the service sector as a whole was analyzed in Table 9.3.

[20] See, among others, Bosworth and Triplett (2007) and Triplett and Bosworth (2004) for the United States; Crespi et al. (2006) for the United Kingdom; McLachlan et al. (2002) for Australia; Maroto and Cuadrado (2009) for a simple of OECD countries; and Maroto and Rubalcaba (2008) for the European Union.

5 Final Remarks and Open Research Issues

As established in the introduction, the two starting hypotheses of this paper were related to the impact of the growth of services on the evolution of productivity. The first entailed the verification of the role played by structural changes, and particularly the growth of services, on the evolution of economic productivity. The second determined whether the variety of services branches demonstrated different behaviours in this field, in contrast to what has been considered by some more traditional approaches. Furthermore, the preparation of this paper has been inspired by two facts. On the one hand, the results obtained in a recent article (Maroto & Cuadrado, 2009), which showed that structural change has played an important role in the evolution of productivity in a wide sample of developed countries. And, on the other hand, to verify if this is also the same at a regional level, due to services playing an increasingly important role, although there are notable differences among regions.

The analysis by countries, which has been replicated taking 17 European economies as a reference and using data for a substantial period of time (1980–2008), does not produce different results from those obtained in the previous study based on a sample of OECD countries from 1980 to 2005. Conventional theory regarding the relationships between the services sector and labour productivity, according to which the expansion of the former would cause a lower growth of such productivity, cannot be supported in absolute terms. Some services branches register an increase in productivity which is comparable to, or even higher than the one corresponding to manufacturing, although those services branches characterised by a high and irreplaceable use of labour register comparatively low productivity levels.

At a regional level, the results obtained from the sample of 170 European regions during the same period (1980–2008) lead us to conclude that structural change still plays a significant role in the improvement of productivity of each region as a whole. However, as verified at a national level, most of the growth of productivity was due to the improvement within each activity branch and not just to the reallocation of resources between the various sectors.

The shift-share analysis used allowed us to break down the productivity growth in the regions into two components of a multiplicative nature: the country effect and the net effect of the region itself. The latter can also be broken down into the product of the proportional effect and the differential effect. The calculations made have shown that regions can be classified into different categories according to the results of the net, proportional and differential effects. Data obtained have been simplified in order to form four categories or groups of European regions, as illustrated in Map 9.1. Despite this synthesis effort, there is a great heterogeneity in the evolution of the different regions, because of the influence of many behaviours and different factors. However, the analysis reveals that the most dynamic regions are concentrated in various large capital cities and European financial centres, as well as in some outlying regions and regions of a lower weight, some of these related to the growth in tourism. Other comparatively backward regions, where structural change has boosted the increase of productivity to a greater extent than in the most developed regions, must be included.

The econometric analysis carried out has added some interesting results related to the role played by services. It has been demonstrated that the growth of services and productivity is positive and significant. Moreover, it has been verified that there is a process of convergence regarding productivity between those regions registering higher productivity levels at the beginning and the more backward regions. It is also confirmed that those regions specialising in services to a greater extent also register more positive dynamics regarding productivity growth. And, finally, as was expected, those services branches subject to market conditions have a greater impact on the variation of productivity, and this is contrary to the case of non-market services.

This analysis leaves an open door for further exploration of some analytical possibilities. Firstly, the differentiated behaviour of regions must be analysed in more depth and more detailed explanations must be pursued. Furthermore, it seems necessary to verify if the training levels of population—human capital—have an influence on productivity and to what extent. And, finally, a method to delve deeper into the issues considered could be to focus on significant countries or, as an alternative, to make a detailed analysis of those regions included in some of the aforementioned categories.

References

Amiti, M. (1999). Specialization patterns in Europe. *Review of World Economics, 135*(4), 573–593.

Baumol, W. (1967). Macroeconomics of unbalanced growth. The anatomy of urban crisis. *American Economic Review, 57*(3), 416–426.

Baumol, W. (2002). Services as leaders and the leader of the services. In J. Gadrey & F. Gallouj (Eds.), *Productivity, innovation and knowledge in services* (pp. 147–163). Cheltenham, UK: Edward Elgar.

Baumol, W., Blackman, S. A., & Wolff, E. N. (1989). *Productivity and American leadership: The long view*. Cambridge, MA: MIT Press.

Bell, D. (1974). *The coming of post industrial society*. London: Heinemann.

Blackaby, F. (1978). *Deindustrialisation*. London: Heineman.

Bonatti, L., & Felice, G. (2008). Endogenous growth and changing sectoral composition in advanced economies. *Structural Change and Economic Dynamics, 19*, 109–131.

Bosworth, B., & Triplett, J. (2007). The early 21st century US productivity expansion is still in Services. *International Productivity Monitor, 14*(Spring), 3–19.

Camagni, R., & Capellin, R. (1985). *La productivité sectorielle et la politique régionale*. Bruselas: Comisión Europea.

Carree, M. A. (2003). Technological progress, structural change and productivity growth. A comment. *Structural Change and Economic Dynamics, 14*, 109–115.

Chung, W., & Denison, E. (1976). *How Japanese economy grew so fast: The sources of post-war expansion*. Washington DC: Brookings Institution.

Combes, P., & Overman, H. (2003). The spatial distribution of economic activities in the EU. In V. Henderson & J. Thysse (Eds.), *Handbook of urban and regional economics*. Amsterdam: North-Holland.

Crespi, G., Criscuole, C., Haskel, J., & Hawkes, D. (2006). Measuring and understanding productivity in UK market services. *Oxford Review of Economic Policy, 22*(2), 186–202.

Cuadrado, J. R., Garcia, B., & Raymond, J. L. (1999). Regional convergence in productivity and productive structure: The Spanish case. *International Regional Science Review, 22*(1), 36–54.

Cuadrado, J. R., Rubalcaba, L., & Bryson, J. (2002). *Trading services in the global economy*. Cheltenham, UK: Edward Elgar.

Daniels, P. W. (2004). Reflections on the "old" economy, "new" economy, and services. *Growth and Change, 35*(2), 115–138.

De Bandt, J. (1991). *Les services*. Paris: Productivité et prix. Economica.

Denison, E. (1967). *Why growth rates differ: Post-war experience in nine western countries*. Brookings Institution: Washington D.C.

Dutt, M. J., & Lee, K. H. (1993). The service sector and economic growth: Some cross-section evidence. *International Review of Applied Economics, 7*(3), 311–329.

Ezcurra, R., Pascual, P., & Rapun, M. (2006). The dynamics of industrial concentration in the regions of EU. *Growth and Change, 37*(2), 200–270.

Fagerberg, J. (2000). Technological progress, structural change and productivity growth: A comparative study. *Structural Change and Economic Dynamics, 11*, 393–411.

Fourastié, P. (1949). *Le grand espoir du XXeme siecle. Progrés technique, progrés économique, progrès social*. Paris: Presses Universitaires de France.

Fuchs, V. (1968). *The service economy*. New York: Columbia University Press.

Gemmell, N. (1982). Economic development and structural change: The role of the service sector. *Journal of Development Studies, 19*(1), 37–66.

Granger, C. W. (1969). Investigating causal relations by econometric models and cross-spectral methods. *Econometrica, 37*, 424–438.

Gregory, M., Salverdad, W., & Schettkat, R. (2007). *Services and employment explaining the US-European gap*. New York: Princeton University Press.

Haaland, J., Kind, H., Midelfart-Knervik, K., & Tortensson, J. (1998). *What determines the economic geography of Europe*? Discussion Paper 19/98. Norwegian School of Economics and Business Administration.

Hallet, M. (2000). *Regional specialization and concentration in the EU* (pp. 1–29). Economic Papers. European Commission 141.

Havlik, P. (2005, January 2005). *Structural change, productivity and employment in the New EU Member States* (WIIW Research Reports, 313).

Krüger, J. (2008). Productivity dynamics and structural change in the US manufacturing sector. *Industrial and Corporate Change, 17*(4), 875–902.

Maddison, A. (1996). Macroeconomic accounts for European countries. In B. van Ark & N. Crafts (Eds.), *Qualitative aspects of post war European economic growth*. Cambridge, MA: CEPT/ Cambridge University Press.

Maroto, A. (2012). Productivity in the services sector. Conventional and current explanations. *Service Industries Journal, 32*(5), 719–746.

Maroto, A., & Cuadrado, J. R. (2007). *Productivity and tertiarization in industrialized countries. A comparative analysis* (Efficiency Working Series, 15-2007). University of Oviedo.

Maroto, A., & Cuadrado, J. R. (2009). Is growth of services an obstacle to productivity growth?'. *Structural Change and Economic Dynamics, 20*(4), 254–265.

Maroto, A., & Rubalcaba, L. (2008). Service productivity revisited. *Service Industries Journal, 28* (3), 337–353.

McLachlan, R., Clark, C., & Monday, I. (2002). *Australia's service sector: A study in diversity*. Productivity Commission Staff Research Paper, AUSINFO: Canberra.

Midelfart-Knarvik, K. H., Overman, H. G., Redding, S. J., & Venables, A. J. (2003). Monetary union and the economic geography of Europe. *Journal of Common Market Studies, 41*(5), 847–868.

Molle, W. (1996). The regional economic structure of the European Union: an analysis of long term developments. In K. Peschel (Ed.), *Regional growth and regional policy within the framework of European integration.* Heidelberg: Physica-Verlag.

Nusbaumer, J. (1987). *The services economy: Lever to growth.* Boston: Kluwer Academic Publishers.

O'Mahony, M., & van Ark, B. (2003). *EU productivity & competitiveness an industry perspective. Can Europe resume the catching-up process?* Brussels: EC Enterprise Publications.

OECD. (2009). *How regions grow.* Paris: OECD.

Peneder, M. (2002). *Industrial structure and aggregate growth* (WIFO Working Papers, 182). Wienn.

Peneder, M. (2003). Industrial structure and aggregate growth. *Structural Change and Economic Dynamics, 14*, 427–448.

Peneder, M., Kaniovski, S., & Dachs, B. (2003). What follows tertiarisation? Structural change and the role of knowledge -based services. *Service Industries Journal, 23*(2), 47–66.

Rubalcaba, L. (2007). *Services in the European economy: Challenges and implications for economic policy.* Cheltenham, UK: Edward Elgar.

Salter, W. (1960). *Productivity and technical change.* Cambridge: Cambridge University Press.

Schettkat, R., & Yocarini, L. (2006). The shift to service employment: A review of the literature. *Structural Change and Economic Dynamics, 17*, 127–147.

Stiroh, K. (2001). *Information technology & the US productivity revival. What do the industry data say.* New York: Federal Reserve Bank of New York.

Timmer, M., & Szirmai, A. (2000). Productivity growth in Asian manufacturing: The structural bonus hypothesis examined. *Structural Change and Economic Dynamics, 11*, 371–392.

Towse, R. (1997). *Baumol's cost disease: The arts and other victims.* Cheltenham: Edward Elgar.

Triplett, J., & Bosworth, K. (2004). *Productivity in the US service sector. New sources of economic growth.* Brookings Institution Press: Washington D.C.

Triplett, J., Bosworth, K., Triplett, J., & Bosworth, K. (2006). Baumol's disease has been cured. IT and multifactor productivity in US service industries. In D. W. Jansen (Ed.), *The new economy and beyond: Past, present and future.* Cheltenham, UK: Edward Elgar.

van Ark, B. (1995). *Sectoral growth and structural change in post-war Europe* (Research Memorandum GD-23). Groninge: GGDC.

van Ark, B., & Piatkowski, M. (2004). *Productivity innovation and ICT in old and new Europe.* GGDC Research Memorandum, 69. Groningen: Groningen Growth and Development Centre.

Young, A. (1995). The tyranny of numbers: Confronting the statistical realities of the East Asian growth experience. *Quarterly Journal of Economics, 110*, 641–680.

Chapter 10
Service Employment and Unemployment in the Great Recession: Trends in OECD Countries and U.S. States

William B. Beyers

1 Introduction and Overview

The recent global recession—The Great Recession—appears to have been strongly linked to events in the United States service economy—in particular the real estate lending sector, that was in turn strongly tied to a variety of financial institutions. Whether these relations were the causal agent can be debated, but it is clear that globalization of the financial sector helped spread problems in U.S. real estate markets to financial institutions in many other countries. The up-front problems in the financing of the U.S. housing market led to downstream impacts that were much broader sectorally and globally. Thus, this recession may have been different in its root causes than previous recessions, and the downstream effects may also be different than other recent recessions. This paper explores the employment impacts of The Great Recession, with data for U.S. states and OECD countries.

In the United States the residential real estate sector was historically dominated by mortgage instruments for single structures, which were held by the lender for these mortgages, typically a local bank. However, in the last 50 years mortgage debt has gradually become securitized like other assets, and packaged in instruments that were traded nationally and internationally. As is painfully clear now, some of these instruments were not protected by financial regulations primarily designed during or after the Great Depression of the 1930s. Banking and financial entities engaged in developing creative financial instruments that turned out to be unsustainable. Traditional purchasers of such debt—such as insurance companies using resources from their insured—have found themselves without income needed to repay their creditors, leading to dramatic financial collapses such as AIG. Banks have been caught without sufficient resources to repay creditors, and have collapsed or have been bought up by surviving institutions (such as WAMU's absorption by Chase).

W.B. Beyers (✉)
Department of Geography, University of Washington, Seattle, USA 98195-3550
e-mail: beyers@uw.edu

J.R. Cuadrado-Roura (ed.), *Service Industries and Regions*, Advances in Spatial Science, DOI 10.1007/978-3-642-35801-2_10,

Much hand-wringing has taken place about the nature of new institutions needed to provide oversight so that these problems do not recur. This real-estate "bubble" was fueled by public policies to promote home ownership, rapid growth rates in particular regions where prices escalated speculatively to levels unsustainable when compared to household income, and by an investment community eager to expand its scope.

The initial impact in the United States was a sharp realization that entities holding this housing debt were in a financially untenable position, leading the federal government to intervene with the "TARP" funds, aimed at propping up these financial institutions until they could regain "normal" operating positions. This meant disruption in housing markets that is ongoing, as people lose their homes when they are unable to restructure their debt to affordable levels, or when they simply default on ownership. It is unclear whether current interventions will be sufficient to stem these housing market effects, and whether the financial entities will recover to reasonable positions of regulated solvency.

The secondary consequences of these disruptions in financial markets have been sharp, ongoing, and have spread far beyond the United States. This paper explores these recent events, using December 2007 as the benchmark date for the start of The Great Recession. It relies on data from the U.S. government, as well as OECD sources, to focus on the structural impacts of the downturn. The next section provides some background literature on the role of services in periods of recession. This is followed by analyses of recent historical trends in the OECD and U.S. economies. This analysis leads to the articulation of a model for this recession, and some comparison of its contours up to the year 2010 in comparison to other recent downturns. The paper concludes with some remarks regarding needed research.

2 Background Literature

The literature on the relationship between business cycles and the shift to a service dominated economy is not large. Several analyses deserve mentioning, including the work of Moore, Rubalcaba-Bermejo, and Cuadrado-Roura.

It is well known that business cycles have historically been associated with strong swings in investment levels, that in turn have led to strong fluctuations in the demand for construction and the goods that enter into the investment process. One of the consequences of a shift to the service economy has been a dramatic change in the composition of business investment. The share of investment associated with structures and equipment has fallen. Figure 10.1 shows for the U.S. economy the changing shares of Gross Domestic Product from 1970 to the 1st quarter of 2012. The long term trend has been a reduction in the relative importance of government, a gradual drift upward in the share of consumption, rising shares of exports and imports, and a rise in the share of investment. However, it is evident in Fig. 10.1 that The Great Recession has decreased the contribution of investment, and has

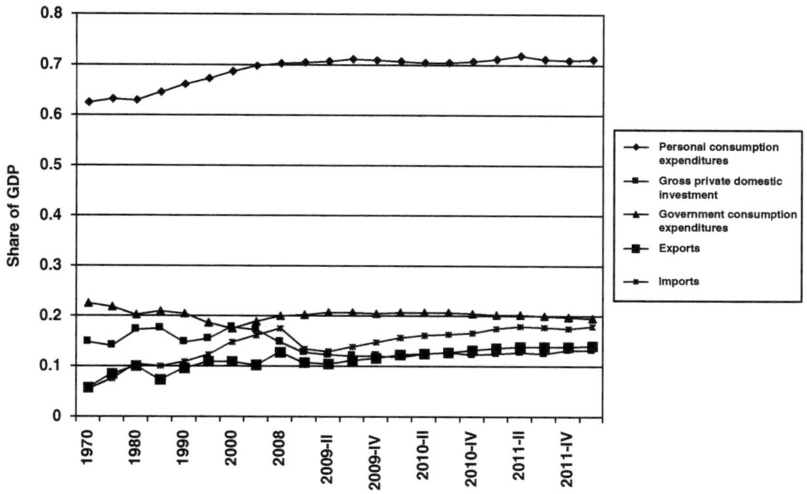

Fig. 10.1 Share of Gross Domestic Product United States. *Source*: Calculated by author from U.S. Bureau of Economic Analysis

increased the share associated with government. Exports and imports have also decreased as a share of gross domestic product.

The collapse of the residential investment sector in The Great Recession is clearly evident in Fig. 10.2. This figure shows that the share of residential investment hovered around 30 % in the United States for the 1996–2007 time period. However, this share plummeted to 20 % in The Great Recession, and it has not recovered even though technically the Great Recession ended in 2010.

The Great Recession also impacted the mix of nonresidential investment. Figure 10.3 shows the long run trend of about three-quarters of this investment being associated with software and equipment, and about one-quarter associated with structures. In large measure the changing relative ratio has been related to the growth of the service economy, where compared to manufacturing, the share of investment in structures is lower than in equipment and software. The sharp break in 2007, related to The Great Recession, put the ratio of equipment and software back to the level of the late 1990s, but it is evident in Fig. 10.3 that after 2010 the share of equipment and software returned to pre-recession levels.

With the overall shift in production towards a larger service economy, it has been argued that business cycle tendencies will be dampened due to the relatively stable demand for services. Cuadrado-Roura makes this argument:

> ....the consumption of non-durable goods and a good number of services shows slower relative fluctuations. One of the main explanations is that consumption of durable goods can be postponed—for obvious reasons, while stocking services presents evident difficulties and, as a result, makes their consumption more stable. (Cuadrado-Roura, 2001, p. 105).

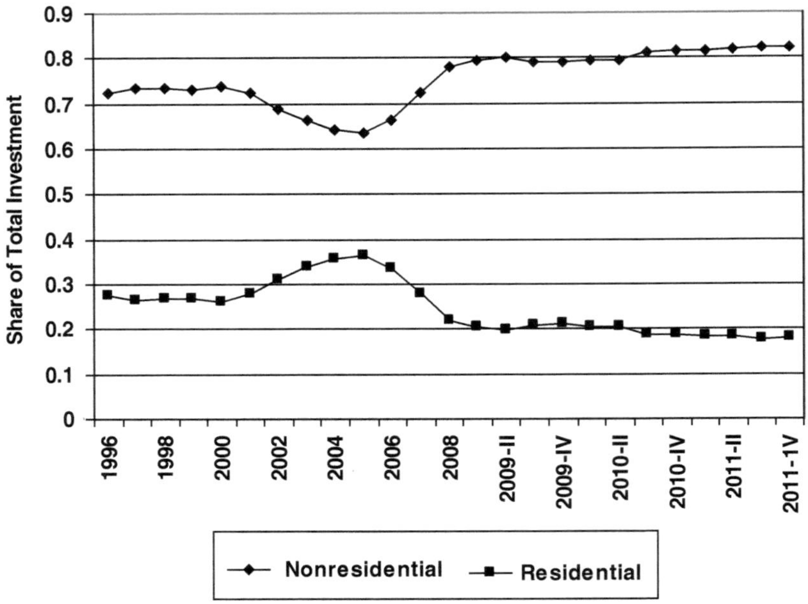

Fig. 10.2 Residential and Nonresidential Shares of Private Investment United States. *Source*: Calculated by author from U.S. Bureau of Economic Analysis

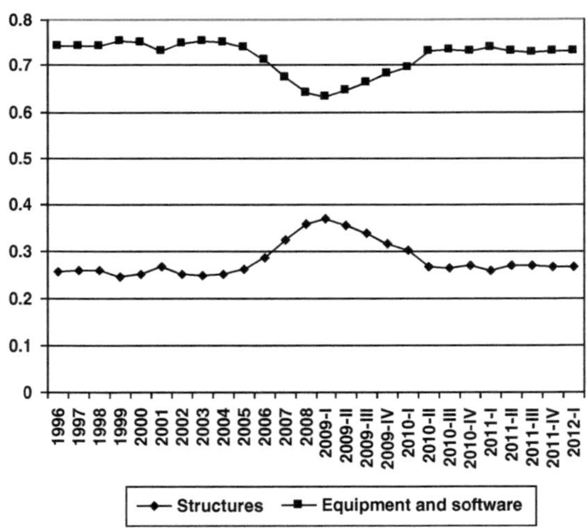

Fig. 10.3 Share of nonresidential investment in structures vs. equipment and software. *Source*: Calculated by author from U.S. Bureau of Economic Analysis

Cuadrado-Roura notes that there has been a systematic change in the nature of business cycles in modern times, with longer cycles and longer periods of expansion, and shorter periods of contraction (Cuadrado-Roura, 2001, p. 105). He notes

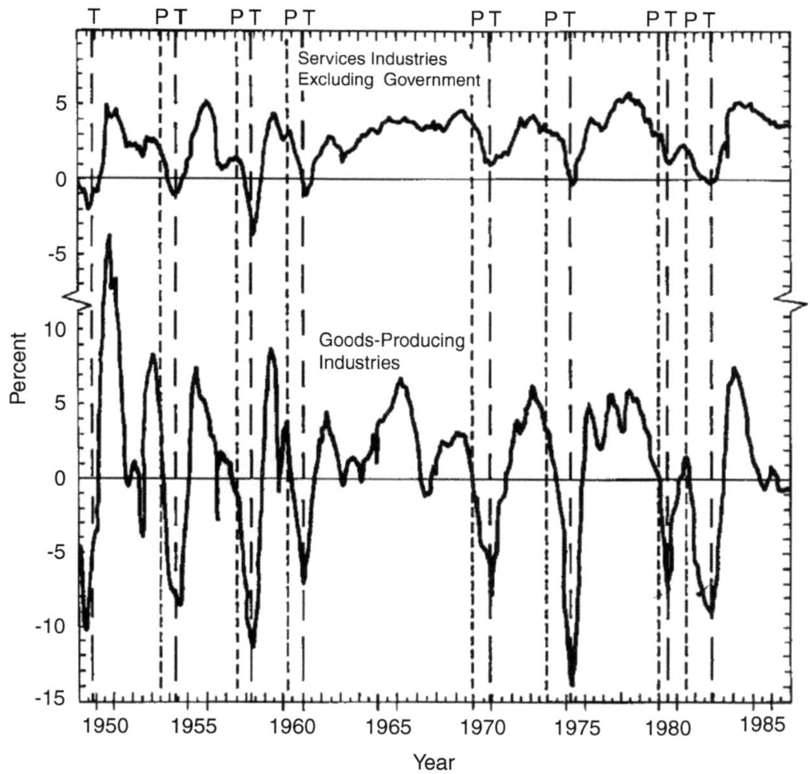

Fig. 10.4 Growth rate in employment in the service industries and goods-producing industries. *Source*: Moore (1987) (as reproduced in Guile & Quinn, 1988)

that part of this due to the growth of the demand for private services, but also is related to the expansion of the demand for public services that are also less subject to demand fluctuations. In a case study of the Spanish economy, he shows how these trends have played out in this style over the 1940–1990 time period.

An earlier analysis of these relationships was reported by Moore. Using data for the U.S. economy over the 1949–1986 time period, he shows that the levels of change in employment in goods producing industries was much greater than in service producing industries. Figure 10.4 is taken from Moore's paper, and this graph shows not only less fluctuation in employment in the services compared to goods-producing industries, but also after the recession of the early 1960s no job losses on the downturn in the service sector. Moore projects likely changes in employment in two hypothetical recessions, occurring in 1990 and 1995. In these projections, he shows private services and government expanding, while the goods producing sectors are predicted to be the source of all job losses.

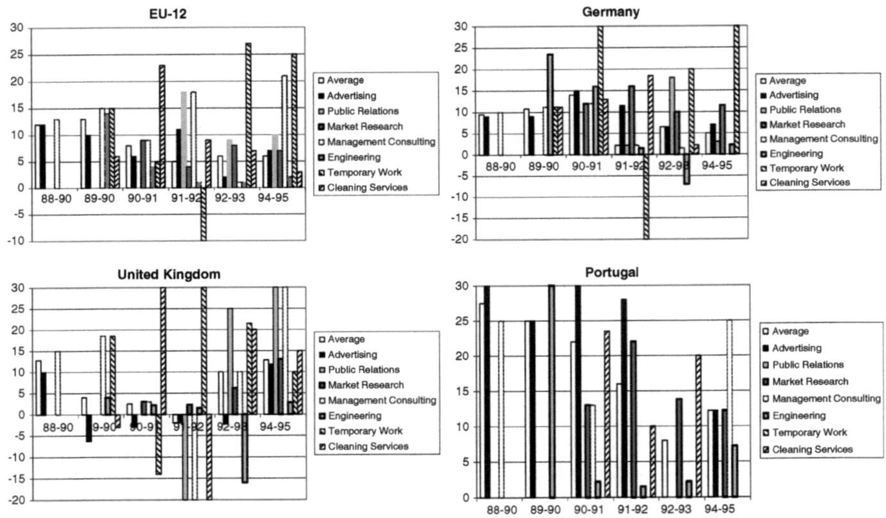

Fig. 10.5 Examples of short-term evolution of some activities of business in Europe (annual growth 1989–1995). *Source*: Rubalcaba-Bermejo (1999, Graph 3.11, p. 175)

Moore discusses the demand for services in relation to these cyclical patterns. He writes:

> The purchasing power created by additional jobs in services must help to maintain demand for consumer goods and hence delay downturns and hasten upturns in the goods-producing sector itself. This development in turn would contribute to the prolongation of business cycle expansions and shortening of recessions. Although we have not tried to measure this directly, the near equality shown. . .between the lengths of business cycle phases and those in total non farm employment carries with it the implication that the rapid growth in service industries employment has had a favorable effect on the relative length of the prosperous and depressed phases of the business cycle (Moore, 1987).

Moore and Cuadrado-Roura do not discuss the changing nature of the service economy, and particularly the emergence of the large producer services sector, whose demand is not primarily with consumers, but rather across the spectrum of industries. Their arguments are phrased largely in the context of consumer oriented services. Analyses of the markets of producer services find that their clients are spread across the economy, and it would seem likely that the demand for these services would be impacted by client sectors hit hard by business cycles. This topic will be revisited later in this paper.

Rubalcaba-Bermejo has also examined the role of services in business cycles in Europe (Rubalcaba-Bermejo, 1999). His analysis concentrated on the role of business services, with statistical analyses focused on the 1989–1995 time period. Figure 10.5 shows results of his analyses for the EUR-12, Germany, the United Kingdom, and Portugal. This graph clearly shows cyclicality in the growth rates for business services, with a significant drop associated with the 1991–1992 recession.

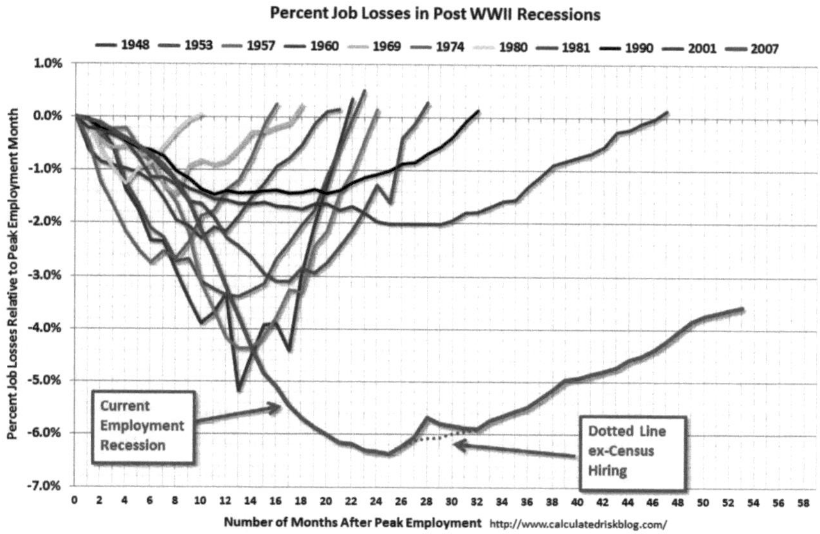

Fig. 10.6 Patterns of job losses in U.S. recessions since 1948. *Source*: McBride (2012)

For the EUR-12 temporary work shows the strongest degree of cyclicality, and it is clear that there are major differences in these patterns across countries in Europe.

Rubalcaba-Bermejo concludes as follows:

> Our results emphasize, still more, if that is possible, the heterogeneity of situations according to country and activity. Additionally, they verify that business cycles affect business services without qualification. Nevertheless, it also seems evident that a structural growth component exists that tends to take precedence over the pro-cyclical one. For this reason, business services' slump during recessions is by any reckoning less than that which takes place in the vast majority of other economic sectors. The strong structural growth of business services explains the causes of this peculiarity. This makes business services a potential instrument for policies that aim to guarantee an equilibrated and stable economic path (Rubalcaba-Bermejo, 1999, p. 176).

An analysis by McBride provides useful insights into the recent downturn in the United States, compared to other business cycles since 1948. Figure 10.6 shows job losses in the recent downturn compared to the last 10 recessions, and it is very clear that the recent downturn was much steeper than the average recession in the United States, and job recovery as of February 2012 had not returned to the late 2007 level when this recession began.

3 Analysis

A data base was developed for the U.S. and the OECD to explore changes in employment in The Great Recession. This analysis was benchmarked against either December 2007, the fourth quarter of 2007, or the year 2007. In the U.S. economy,

Table 10.1 Patterns of employment and unemployment change OECD countries 2007–2010

	Unemployment up	Unemployment down
Employment up	14	1
Employment down	15	0

December 2007 marked the peak level of employment, and has nominally been regarded as the date of the beginning of The Great Recession. However, the seeds of the downturn were evident earlier, with the emergence of problems in the financial sector related largely to real estate financing as discussed earlier. The goal of the analysis was to explore the role of services in the current economic downturn. The outward presumption was that this downturn was different than other recent downturns, as it was essentially precipitated by problems in a service sector. However, the mechanisms by which employment was impacted could be such that while certain financial services were associated with the needs of governments to intervene, the actual impacts followed other channels, linked to consumption, investment, and intermediate demand for services. If that were the case, then this recession should have structural similarities to other recent recessions.

Two levels of analysis are presented. First, recent data for OECD countries are examined. Second, an analysis of change in the United States as a whole is reported, followed by an analysis of impacts on states of the United States of the Great Recession.

3.1 OECD Countries

OECD data on employment and unemployment conditions in many member countries were accessed. These data reported the level of employment in 2007 and in 2010, unemployment for 2007 and 2011, the 2007 unemployment rate, the change in the unemployment rate from 2007 to 2011, the percent of employment that was considered long-term in 2008, and the growth rate of personal consumption from the first quarter of 2008 to the first quarter of 2009. Table 10.1 summarizes the employment and unemployment situation in OECD countries from 2007 to 2010. Almost all countries reported a rise in the level of unemployment, with half of these experiencing a rise in overall employment, and half recording a drop in total employment levels. In most of these cases, the absolute gain in employment was small. In Germany the level of unemployment fell, while total employment rose. No country had declines in employment and unemployment.

Table 10.2 reports unemployment rates in OECD countries averaged 5.5 % in 2007, and rose to an average of 8.2 % in 2011. Particularly striking increases in unemployment rates were observed in Greece, Ireland, and Spain. Total employment in OECD countries increased by 5.8 million jobs between 2007 and 2010, while unemployment increased by 13.9 million, much higher than the rate of job-creation.

Table 10.2 Employment trends in OECD countries (absolute values in thousands)

	2007 employment	2010 employment	2007 unemployment	2010 unemployment	2007 unemployment rate (%)	change unemployment rate 2007–2011	2007 (%) Services	Change (%) services 2007–2010	Long term unemployment (%) 2010
Australia	10,539	11,247	4,821	611	4.4	0.84	75.3	0.5	18.5
Austria	4,028	4,085	186	200	4.4	−0.33	66.7	3.0	25.2
Belgium	4,380	4,489	353	390	7.5	−0.33	73.7	1.6	48.8
Canada	16,866	17,041	1,079	1,430	6.0	1.43	76.3	1.3	12.0
Chile	6,449	ND	496	633	7.1	−0.1	ND	ND	ND
Czech Republic	4,908	4,870	276	385	5.3	1.3	56.1	2.7	49.3
Denmark	2,779	2,659	115	196	4.0	3.77	74.2	5.2	19.1
Finland	2,482	2,438	183	215	6.9	0.7	69.7	2.6	23.6
Germany	38,210	38,549	3,602	3,061	8.6	−2.97	67.9	1.9	47.4
Greece	4,510	4,427	407	514	8.3	12.2	66.2	1.3	45.0
Hungary	3,890	3,756	312	442	7.4	3.53	62.4	2.2	50.6
Iceland	177	167	4	12	2.3	4.41	73.4	2.7	21.3
Ireland	2,090	1,851	101	264	4.6	10.13	67.6	7.3	49.0
Italy	22,970	22,612	1,461	2,145	6.0	3.15	66.6	0.7	48.5
Japan	64,120	61,950	2,568	3,307	3.9	0.57	68.4	1.0	37.6
Korea	23,433	23,829	7,830	817	3.2	−0.12	66.9	1.6	0.3
Luxembourg	201	359	8	12	3.9	1.07	76.3	0.8	29.3
Mexico	42,552	44,380	1,495	2,498	3.4	1.55	61.1	2.1	2.4
Netherlands	7,264	8,370	342	413	4.5	0.37	77.7	3.6	27.6
New Zealand	2,174	2,180	83	159	3.7	2.7	70.8	1.5	9.0
Norway	2,427	2,494	63	75	2.5	0.87	75.9	1.8	9.5
Poland	15,240	15,961	1,619	1,471	9.6	0.37	54.5	3.4	25.5
Portugal	5,135	4,953	449	563	8.0	6.07	57.6	3.7	52.3
Slovak Republic	2,357	2,318	292	375	11.0	3	56.4	10.0	59.3
Spain	20,356	18,352	1,834	4,327	8.3	14.67	66.2	6.3	45.1

(continued)

Table 10.2 (continued)

	2007 employment	2010 employment	2007 unemployment	2010 unemployment	2007 unemployment rate (%)	change unemployment rate 2007–2011	2007 (%) Services	Change (%) services 2007–2010	Long term unemployment (%) 2010
Sweden	4,446	4,546	285	400	6.0	1.47	76.1	1.9	16.6
Switzerland	4,413	4,568	158	213	3.5	0.35	72.4	1.4	34.3
Turkey	20,750	22,594	2,368	3,270	10.2	−1.9	49.8	−1.2	28.6
United Kingdom	29,230	28,763	1,653	2,412	5.4	2.9	76.6	3.0	32.6
United States	146,047	139,064	7,078	13,251	4.6	4.1	78.8	2.5	29.0
Total	514,430	520,177	30,135	44,061	5.5	2.71			32.4

ND no data

Table 10.3 Selected correlations for OECD countries

	2007 unemployment rate	Change in unemployment 2007–2011	Percent services 2007	Change percent services 2007–2010
2007 unemployment rate				
Change in unemployment 2007–2011	−0.601**			
Percent services 2007	−0.618**	0.529**		
Change percent services 2007–2010	0.252	−0.187	−0.16	

**Significant at 0.01 level

Table 10.2 also reports variations in the percentage of employment in OECD countries in services. High income countries such as the U.S., Sweden, and the U.K. had relatively high employment percentages in services in 2007, and this percentage rose during the Great Recession. However, Table 10.2 also reports gains in the percentage of service employment in almost all other OECD countries, including strong gains in a number of lower income countries.

Major differences can be seen in the importance of long term unemployment rates across the OECD countries, as reported in Table 10.2. This is defined as persons seeking work for at least 6 months. The United States and the Nordic countries show relatively low long term unemployment rates compared to many European countries. Long term unemployment rates of one-third to one-half of current unemployment are common across much of continental Europe.

Current data were not available when this paper was written for OECD countries for detailed industry employment categories. Data were reported for agriculture, manufacturing and construction, and services. These data were used for the period from 2007 through 2010, to identify the share of employment associated with each of these broad industry aggregates. Per capita gross national income was not significantly correlated with measures of unemployment. However, as can be seen in Table 10.3, there was a statistically significant inverse correlation between the level of unemployment in 2007 and the share of employment in services. Expressed alternatively, countries with high levels of employment in services had relatively low levels of unemployment in 2007. However, there was no statistically significant relationship between the change in unemployment rates and the share of services employment—increases in the share of services employment were not positively associated with increases in unemployment rates. However, there was a direct and statistically significant correlation between the change in unemployment levels and the change in the percentage of services employment. This result is undoubtedly related to declines in manufacturing and construction employment—which raised the share of services employment—and the overall rise in unemployment levels across almost all OECD countries.

Table 10.4 Goods and Services Employment change in recent U.S. business cycles (thousands of jobs)

	Total change	Change goods	Change services	Change goods (%)	Change services (%)
January 1999 to February 2001	5,050	69	4,981	0.3	4.8
February 2001 to August 2003	−2,708	−2,767	59	−11.3	0.1
August 2003 to December 2007	8,330	335	7,995	1.5	7.4
December 2007 to February 2010	−8,564	−4,214	−4,350	−19.1	−3.7
February 2010 to May 2012	3,421	478	2,943	2.7	2.6

Source: Calculated by author from U.S. Bureau of Labor Statistics (BLS) data

3.1.1 Section Summary

OECD countries experienced significant rises in unemployment levels, and had varying changes in employment levels between 2007 and 2011. Data were not available for detailed service industries to explore the role of particular lines of services to the current downturn. However, the data are compelling regarding the structure of job losses in these countries—it has been concentrated in manufacturing—with job losses in services being quite small for this group of countries. These data also suggest that the drop in manufacturing was related to a drop in consumption spending. Very recent data on investment were not found, and they too could be correlated with the decline in manufacturing. High levels of structural unemployment in many European OECD countries appear to be strongly related to recent increases in unemployment. These high levels of structural unemployment may be pose difficulties for turning around levels of personal consumption expenditures, which appear to be related to the strong downturns in manufacturing. Persistent high levels of unemployment in Greece and Spain have been at the core of recent concerns about Euro-zone recovery from the Great Recession.

3.2 U.S. Trends

The United States had a significant increase in unemployment between December 2007 and early 2010. The national rate of unemployment rose from 5 % in December 2007 to 9.9 % in April 2010, and it had fallen to 8.2 % in May 2012. Literature on business cycles for the United States discussed earlier in this paper argued that services have not been major contributors to business cycle job losses. Table 10.4 reports job changes through four recent business cycles in the U.S, and into the recovery that began in early 2010. These data indicate no job losses in services in the brief recession in 2001–2003, and rapid growth rates in services employment gains in the two expansions reported in this table. However, in the Great Recession BLS data show a larger decline for services than reported in the

Table 10.5 Change in employment in the United States December 2007 through May 2012 (nonfarm payrolls, seasonally adjusted, thousands of jobs)

	December 2007	February 2010	Change	Change (%)	July 2010	Change February–July	Change (%)	May 2012	Change Feb 2010–May 2012	
Total	138,078	12,9588	−8,490	−6.1	130,242	654	0.50	133,009	3,421	2.6 %
Private	115,745	10,7131	−8,614	−7.4	107,737	606	0.57	111,040	3,909	3.6 %
Goods producing	21,976	17,829	−4,147	−18.9	1,8023	194	1.09	1,8307	478	2.7 %
Natural resources and mining	739	690	−49	−6.6	733	43	6.23	838	148	21.4 %
Construction	7,465	5,577	−1,888	−25.3	5,573	−4	−0.07	5,516	−61	−1.1 %
Manufacturing	13,772	11,562	−2,210	−16.0	1,1717	155	1.34	1,1953	391	3.4 %
Durable Mfg	8,739	7,065	−1,674	−19.2	7,210	145	2.05	7,479	414	5.9 %
Nondurable Mfg	5,033	4,497	−536	−10.6	4,507	10	0.22	4,474	−23	−0.5 %
Service providing	116,102	111,759	−4343	−3.7	112,219	460	0.41	1,14702	2,943	2.6 %
Private service providing	93,768	89,302	−4,466	−4.8	8,9714	412	0.46	9,2733	3,431	3.8 %
Wholesale	6072.9	5559.9	−513	−8.4	5585.8	25.9	0.47	5622.6	62.7	1.1 %
Retail trade	15487.8	14417.4	−1070.4	−6.9	14433.7	16.3	0.11	14755.6	338.2	2.3 %
Transportation and warehousing	4539.9	4133.3	−406.6	−9.0	4189.5	56.2	1.36	4378.4	245.1	5.9 %
Utilities	557.1	558	0.9	0.2	552.2	−5.8	−1.04	561.2	3.2	0.6 %
Information	3,018	2,738	−280	−9.3	2,712	−26	−0.95	2,628	−110	−4.0 %
Financial activities	8,252	7,620	−632	−7.7	7573	−47	−0.62	7724	104	1.4 %
Finance	3756.7	3376.1	−380.6	−10.1	3379.2	3.1	0.09	3479.2	103.1	3.1 %
Insurance	2466.5	2287.6	−178.9	−7.3	2260.8	−26.8	−1.17	2282.3	−5.3	−0.2 %
Real estate	2193.9	1956.1	−237.8	−10.8	1932.9	−23.2	−1.19	1962.5	6.4	0.3 %
Professional and Business services	18,131	16,551	−1580	−8.7	16674	123	0.74	17815	1,264	7.6 %
Legal	1173.9	1105.7	−68.2	−5.8	1100.3	−5.4	−0.49	1119.9	14.2	1.3 %
Accounting	993.3	915.1	−78.2	−7.9	893.2	−21.9	−2.39	951.1	36	3.9 %
Architecture and engineering	1460.4	1281.9	−178.5	−12.2	1272.3	−9.6	−0.75	1323.7	41.8	3.3 %
Computer services	1391.4	1438.3	46.9	3.4	1451.7	13.4	0.93	1586.4	148.1	10.3 %
Consulting	994.3	984.9	−9.4	−0.9	990.4	5.5	0.56	1126.6	141.7	14.4 %

(continued)

Table 10.5 (continued)

	December 2007	February 2010	Change	Change (%)	July 2010	Change February–July	Change (%)	May 2012	Change Feb 2010–May 2012	
Management of companies	1847.8	1818.6	−29.2	−1.6	1824.2	5.6	0.31	1943.7	125.1	6.9 %
Administrative and waste serv.	8462.8	7316.5	−1146.3	−13.5	7444.4	127.9	1.75	7963.7	647.2	8.8 %
Employment services	3566.9	2669.8	−897.1	−25.2	2764.5	94.7	3.55	3148.8	479	17.9 %
Education	2984.5	3119.2	134.7	4.5	3146.5	27.3	0.88	3,339	219.8	7.0 %
Health care and social assistance	15583.2	16277.4	694.2	4.5	16411.5	134.1	0.82	16968.3	690.9	4.2 %
Liesure and hospitality	13,635	13,019	−616	−4.5	1,3097	78	0.60	13576	557	4.3 %
Arts, entertainment and recreation	2010.3	1893.2	−117.1	−5.8	1914.8	21.6	1.14	1905.3	12.1	0.6 %
Accommodation	1858.1	1726.6	−131.5	−7.1	1768.7	42.1	2.44	1815.5	88.9	5.1 %
Food services and drinking places	9766.6	9399.2	−367.4	−3.8	9413.3	14.1	0.15	9854.7	455.5	4.8 %
Other services	5,507	5,308	−199	−3.6	5,339	31	0.58	5,365	57	1.1 %
Government	22,333	22,457	124	0.6	2,2505	48	0.21	2,1969	−488	−2.2 %
Federal	2,735	2,863	128	4.7	3,017	154	5.38	2,819	−44	−1.5 %
State	5,153	5,171	18	0.3	5,134	−37	−0.72	5,073	−98	−1.9 %
Local	14,445	14,423	−22	−0.2	1,4354	−69	−0.48	1,4077	−346	−2.4 %

Source: U.S. Bureau of Labor Statistics, Table B-3 Employment and Earnings

OECD statistics (in fact about double the OECD estimate). Over half of the job losses in the United States in The Great Recession were in services, but this is only a 3.7 % decline in services employment, compared to a 19 % decline in employment in goods production. However, the recovery has been very different than in other recent business cycles. Services employment growth has been slower goods employment growth; details of this trend are presented in Table 10.5

Table 10.5 reports detailed changes in industry employment in the United States from December 2007 through May 2012, marking the peak level of unemployment in early 2010. The data source here is slightly different than in Table 10.4. Table 10.5 shows much stronger percentage losses of employment in goods producing sectors than in the services. And, like Table 10.4, about half of the job losses in the United States between December 2007 and February 2010 were in services. Sectors such as wholesaling and transportation, related to the channel of distribution for goods, show relatively large percentage employment reductions. The information and professional and business services sectors show a larger percentage rate of employment loss than the service sector as a whole. Detail in the business and professional services sector reveal a sharp drop in employment services (temporary help). The downturn in investment is undoubtedly related to declines in architecture and engineering. Employment in education and health care continued to show gains, counter to the broad-based downturn in services employment.

Since 2010 the data in Table 10.5 report some sectors with relatively rapid growth that are symptomatic of hiring patterns during a recovery (such as employment services), but the sluggish housing market in the United States has still been pulling down construction employment. Across a broad range of services, Table 10.5 reports slow employment growth, and in contrast to many recoveries, government has had employment contraction. Fiscal difficulties of state and local governments have precipitated layoffs of these government workers, raising questions about the traditional stabilizing impact of government through business cycles. The health care sector continues to exhibit strong growth, as is also the case through a number of lines of business and professional services.

Figure 10.7 shows detail on monthly employment in the financial services sector, an area clearly associated with The Great Recession in the United States. This figure shows employment levels indexed against January 2006, nearly 2 years before the employment peak associated with the statistical beginning of The Great Recession (December 2007). Banking, and real estate rental and leasing, had peak employment in December 2006; real estate employment peaked in May 2007, while insurance employment peaked in December 2007. Thus, many key financial sectors were on the downturn well before the economy as a whole, and have fallen much more than the national average. Security brokerages and trusts continued employment expansion until April 2008, but since that date have dropped strongly, with May 2012 employment at about 90 % of pre-recession levels. The rental and leasing sector shows continued declines in employment, while banking, securities & trusts, and banking have had stagnant employment levels since late 2010.

While the financial services sector may have been a leader in drawing the U.S. and the global economy into recession, Table 10.5 makes it clear that the downstream

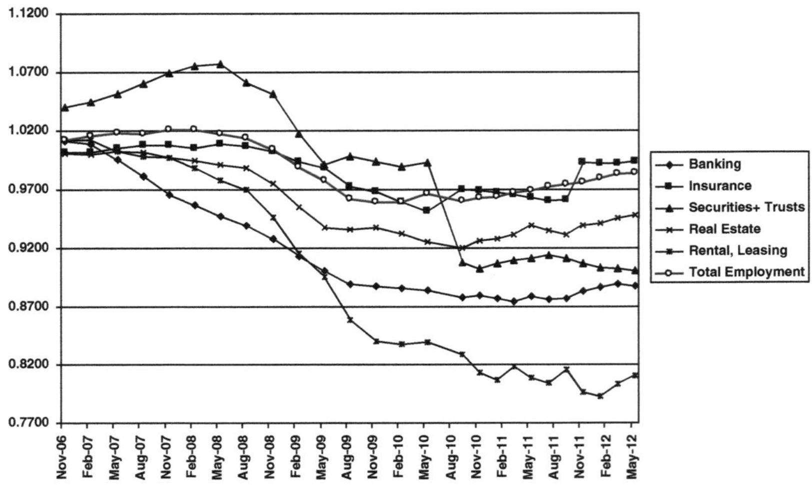

Fig. 10.7 Employment indices: financial sectors and total employment (January 2006 = 1.0). *Source*: Calculated by author from BLS data

impacts have been much stronger in goods production, as investors and consumers reduced their spending on commodities. A model of these relationships will be presented later in this paper.

3.3 State Trends in the United States

Originally the author had hoped to develop a database of unemployment by sector at the state level, but these data were not available. Data were available on employment by state for December 2007 and February 2010. Location quotients were calculated for states by sector. Data on unemployment by state were also accessed from the BLS. Analyses were conducted to evaluate the relationship between industry concentration, as measured by location quotients, and unemployment levels. Figures 10.8 and 10.9 show the level of unemployment by state in December 2007 and April 2010. Nationally, BLS data indicate a rise in U.S. unemployment from 5 % to 9.9 % over this time period. Figure 10.9 shows unemployment at or above the national average concentrated in many states in the old Industrial Belt, California, Oregon, Nevada, Mississippi, South Carolina, and several states bordering the old Industrial Belt. Figure 10.9 shows April 2010 unemployment rates, and it shows a pattern not unlike that in Fig. 10.8. In fact, the unemployment rate in December 2007 was an excellent predictor of unemployment rates in April 2010, as shown in Table 10.6. This table indicates a very strong correlation between unemployment levels in December 2007 and April 2010, and a slightly weaker relationship between the change in unemployment rates and unemployment in December 2007. Although data are now available up to 2012 on state

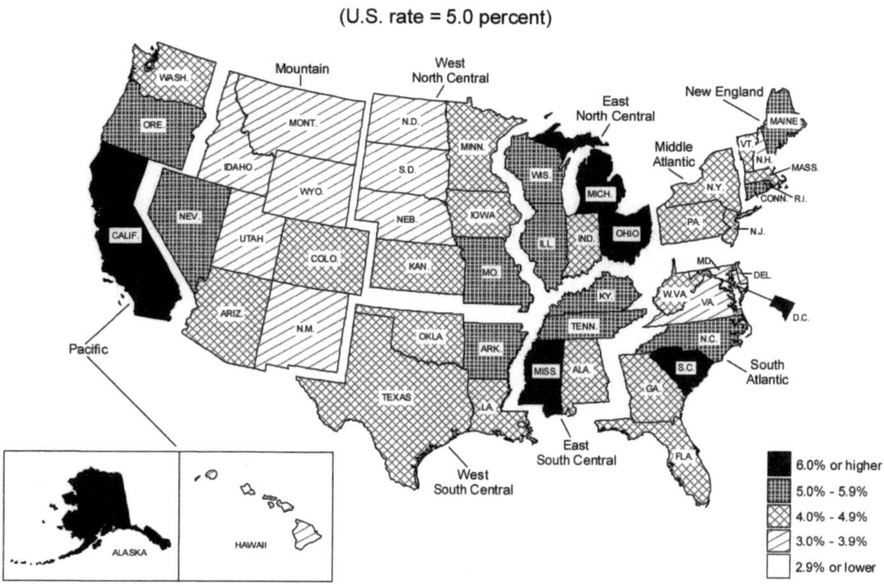

Fig. 10.8 Unemployment rates by state, seasonally adjusted, December 2007. *Source*: U.S. Bureau of Labor Statistics

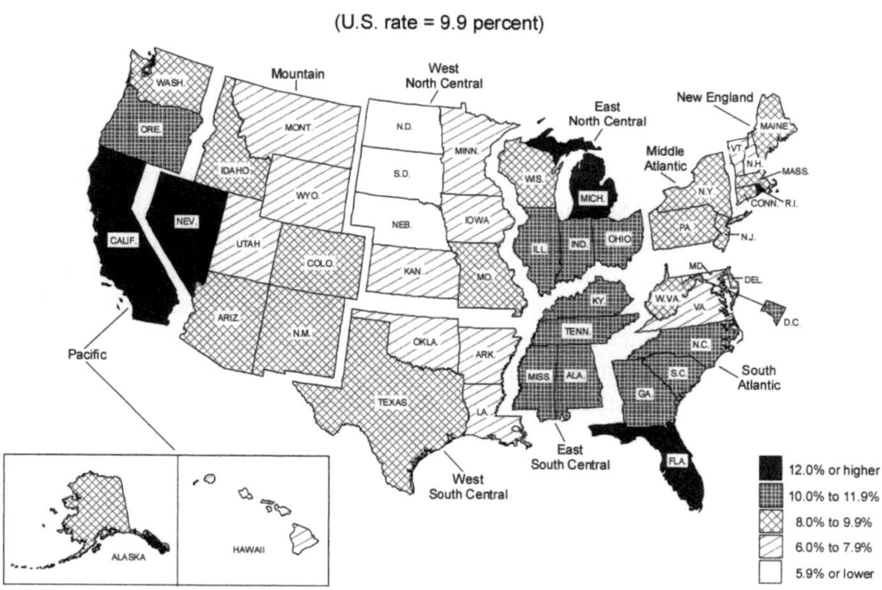

Fig. 10.9 Unemployment rates by state, seasonally adjusted, April 2010. *Source*: U.S. Bureau of Labor Statistics

Table 10.6 Correlations of unemployment statistics United States

	Unemployment April 2010	Change in unemployment 2007–2010	Unemployment December 2007
Unemployment April 2010			
Change in unemployment 2007–2010	0.887**		
Unemployment December 2007	0.706**	0.298*	

Source: Calculated by author from BLS statistics
*Sig. at 0.05 level; **sig. at 0.01 level

Table 10.7 Significant correlations between state location quotients, April 2010 unemployment rates, and change in unemployment rates from December 2007 to April 2010

Construction location quotients and manufacturing location quotient	−0.378**
Manufacturing location quotient and business service location quotient	−0.384**
Business service location quotient and unemployment rate	0.374**
Business service location quotient and change in unemployment rates	0.454**
Transportation and trade location quotients and government location quotient	−0.541**
Leisure location quotient and education and health location quotient	−0.411**
Government location quotient and change in unemployment rate	−0.303*

*Sig. at 0.05 level; **sig. at 0.01 level

level employment and unemployment, the analysis reported in this chapter was not extended to 2012 due to the very strong correlations found in these relationships.

Figure 10.7 clearly shows high levels of unemployment in 2007 in part of the old U.S. Industrial Belt, but also in California, Mississippi, and South Carolina. Low unemployment rates are found in the upper Midwest in a number of rural states strongly focused on farming, mining, and forest products. By the peak of unemployment in the United States, in 2010, this pattern changes somewhat. Unemployment rates are still high in many states in the old Industrial Belt, but now Florida and Nevada are added to the high unemployment states. These are two states that had been highly dependent upon housing industry growth and immigration. Their high unemployment rates are most likely related to the housing industry collapse that has still not revived in the United States.

Analyses were conducted of the concentration of employment in major industries and unemployment rates in December 2007 and April 2010. Table 10.7 reports significant results from these analyses. Interestingly, there were few significant correlations between the concentrations of particular industries and the state unemployment rates in April 2010 or their change from December 2007. Not reported in this table are correlations with the December 2007 unemployment rates and the location quotients for major industries. There was a statistically significant negative correlation between the unemployment rate and construction activity in December 2007. Thus, states with strong concentrations of construction activity in December 2007 tended to have low unemployment rates. However, by April 2010 this relationship disappeared. Only the location quotients in the business services sector show a significant positive correlation with unemployment rates in April 2010, and with their change from December 2007 to April 2010. Thus, states with a high concentration of employment in the business and professional services tended to have relatively high unemployment rates in April 2010, and had relatively high increases in their unemployment rate from December 2007. Government shows the reverse relationship—states with strong concentrations of government employment tended to have lower changes in their unemployment rate from December 2007 to April 2010. This latter finding is not unexpected, given the counter-cyclical role of government in business cycles. It should be noted that subsequent to this analysis there has been a reduction in employment in many state and local governments due to tax revenue reductions.

Several industries in Table 10.7 exhibit statistically significant correlations between their location quotients, but do not necessarily exhibit significant links to unemployment rates or changes in unemployment rates. The negative correlation between construction and manufacturing location quotients reflects the fact that construction activity was most strongly concentrated in states such as Florida, Arizona, and Nevada that had very strong housing market activity before the Great Recession, while manufacturing was concentrated in slower growth states (with more modest construction activity). The states with high location quotients in leisure (such as Nevada with its high level of gambling activity) tended to have weaker concentrations in education and health care, while states such as Massachusetts with strong concentrations in health care tended to have weak concentrations in leisure activities. The negative correlation coefficient between transportation and trade, and government is more puzzling. Transportation and trade are relatively evenly distributed, while government exhibits strong concentrations in the District of Columbia, Maryland, and Virginia.

It is surprising that states with strong concentrations of manufacturing employment did not exhibit a larger than average level of unemployment or increase in the level of unemployment, given the large job losses that have occurred in the manufacturing sector. This result means that job losses associated with manufacturing were widely distributed across the states, and/or non-manufacturing job losses contributed to the overall pattern of job change in ways unrelated to the concentration of manufacturing employment.

States with concentrations of manufacturing industry tended to have weaker concentrations of employment in the business and professional services sector, and in leisure, government, and other employment. In contrast, the states with strong concentrations of manufacturing also tended to have strong concentrations in employment in trade and transportation services. States with strong concentrations of construction activity tended to have strong concentrations in trade and transportation as well as leisure—likely a reflection of growth in states with large retirement populations or concentrations of gambling and amusements (such as Nevada and Florida).

What lessons are there from this analysis in the United States of The Great Recession? First, the problems that started in the housing sector with lending practices that were unsustainable financially for both homeowners and lenders were evident early on in parts of the financial services industry. Second, as employers saw demand begin to erode for a broad array of goods and services they responded with downturns in the levels of production and employment. Third, this downturn was not confined to goods production, but has been felt equally by the services sector. Fourth, while the downturn in jobs was roughly equal between goods and services production, the percentage impact within services has been much lower than in the goods producing (and distributing) sectors. Thus, the efforts of the Obama administration to stimulate the economy have been largely aimed at reviving consumer demand and investment, so as to rekindle employment and production.

4 A Model for This Downturn

Figure 10.10 presents a familiar diagram linking consumption to production, and the flow of funds, goods, and services. This model should be seen as set in time and space, and in todays interdependent world, each arrow is implicitly tracking interregional and international flows of goods and services and funds. Thus, what started as a downturn due to inappropriate financial practices in the United States housing market quickly spread to the global economy due to these linkages. Each of the arrows in this model should be seen as a flow whose magnitude is influenced not just by simple linear equation relationships, but by policies and behaviors influenced by expectations and experiences—such as becoming unemployed.

In 2010 savings rates in countries like the U.S. rose to levels not seen in decades, as people were unable to find lending opportunities or feared the loss of future income. Not captured in the arrows in this model are the impacts of public policies that regulate or influence the costs of the elements of this model, through fiscal stimulus packages, tax policies, and other incentives to impact demand (such as the infamous "cash for clunkers" program that had wild success in jump-starting the demand for fuel efficient automobiles in the United States recently). When confidence rebuilds, either through public policy or shifts in consumer behavior, the magnitudes of the flows in this system will begin to expand, and the employment cycle will move in the direction of job creation. Clearly, this is a process of adjustment that has affected almost all of the OECD countries, and global patterns of exports and imports.

In this model production has been broken into two components: (1) goods production and consumer services, and (2) producer services. Linkages are shown between producer services and goods production and consumer services, reflecting the inputs of producer services to all segments of the economy. The double headed arrows suggest internal connectivity within these two broad divisions of the economy. Arrows are also shown between the capital markets (investment) and producer services and other productive sectors. As reported earlier in this paper (Table 10.5), the sharp downturn in architecture and engineering employment can be traced to the decline in construction activity, which is in turn strongly linked to the investment process. There is a link also shown between producer services and

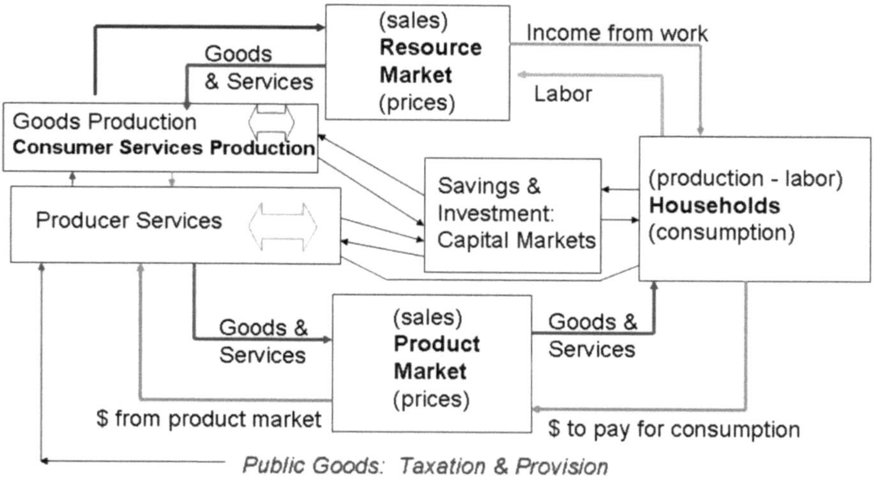

Fig. 10.10 The circular flow of production and consumption

households, to make it clear that there are some demands by households for producer services (such as legal services). Given the long-run growth of the producer services, it is useful to isolate or identify this sector specifically, as the correlation analysis presented in this paper suggests differentiated responses by the service sector to business cycles, such as the current cycle strongly associated with turbulence in housing markets and related financial activities.

We can calibrate the flows in this model with hindsight, through systems of national accounts, and measures of international trade. The changing role of services in the recovery that is anticipated has been debated. Kirkegaard's analysis of recent patterns of structural adjustment found relatively few sectors exhibiting structural gains, such as health care and food services and drinking places (Kirkegaard, 2009). However, students of business cycles have tended to focus very much on the role of consumers and consumer demand, and have tended to overlook the bases of demand for producer services. These demands are contained in the boxes outside the household box in Fig. 10.10, and as Rubalcaba-Bermejo reminds us, the systematic expansion of the producer services has come about through a variety of forces expanding the relative size of business services (Rubalcaba-Bermejo, 1999). This ongoing expansion of the division of labor is likely to continue to occur, and it is interesting to note the rich array of structural gains documented by Kirkegaard in these sectors in the recent business cycle (Kirkegaard, 2009). It would be very interesting to have data allowing us to develop for an international or interregional system the linkages captured in Fig. 10.10 over the course of the recent downturn, but more importantly, in the recovery phase that has begun.

5 Concluding Comments

This paper has used recent statistical information to provide an overview of employment trends in OECD countries and within the United States in The Great Recession. It has highlighted the role of services in this business cycle. With the long-run shift of employment to service industries, the impact of that shift on business cycles has not gone unnoticed. The general argument has been made that business cycles are less dramatic in a service-dominated era due to less wild swings in the demand for services. Data presented for OECD countries confirms this trend in the current recession. The model presented above in Sect. 4 of this paper anchors that view of demand with consumers, but as the relative importance of producer services has expanded, it is important to examine how this sector is affected by business cycles.

The data for the United States show the information, financial services, and business and professional services sectors had relatively large employment downturns in The Great Recession. The largest of these impacts are related to relatively contingent sectors, such as temporary help, or to sectors whose demand is linked to construction such as architecture and engineering. Just as linkages within the manufacturing sector have ripple effects when output of products such as automobiles drop precipitously, we should expect that services linked to the production of goods such as housing will be affected when housing markets drop. Analysis of market structure for advanced services from sources such as input–output models may be helpful in identifying potential demand impacts of business cycles on advanced services.

Sector specific data for employment in OECD countries were not available for the same dates as those available for the U.S. when this paper was written. The types of analysis done for the U.S. with regard to the current business cycle could be repeated for OECD countries and for Europe when such data become available. Differences in national production systems will likely temper such analyses. Likewise, the industrial composition of unemployment both at the OECD level and within countries such as the United States at a state level remains unstudied. If we had these data, it would be possible to construct accounts for models similar to Fig. 10.10, focusing on *changes* in business activity, as opposed to cross-sectional models that can be derived from national accounts.

References

Cuadrado-Roura, J. R. (2001). Business cycle and service industries: General trends and the Spanish case. *Service Industries Journal, 21*(1), 103–122.

Guile, B. R., & Quinn, J. B. (Eds.). (1988). *Technology in services: Policies for growth, trade, and employment*. Washington, DC: National Academy Press.

Kirkegaard, J. F. (2009). *Structural and cyclical trends in net employment in US business cycles, 1949–2009: Implications for the next recovery*. Washington, DC: Peterson Institute of International Economics.

McBride, B. (2012). Percentage job losses: Great recession and great depression. http://www.calculatedriskblog.com/2012/02/percent-job-losses-great-recesson-and.html. Accessed 1 July 2012.

Moore, G. H. (1987). The service industries and the business cycle. *Business Economics, 22*(2), 12–17.

Rubalcaba-Bermejo, L. (1999). *Business services in European industry: Growth employment and competitiveness*. Brussels: European Commission, DGIII-Industry.

Part III
Location of Service Industries: Cities, Internationalization, the Case of Some Specific Activities and Regional Development Policies

Chapter 11
The Location of Service Industries

Juan R. Cuadrado-Roura

1 Introduction

Heterogeneity is one of the most outstanding features of service activities. This sector is formally and statistically made up of a wide group of industries, although, as it is well known important differences do exist among service activities and firms. Airlines, hypermarkets, taxi service or industrial cleaning companies have in common almost exclusively that they "produce" certain specific "services" required by people and/or firms. Yet, the differences regarding the type and size of the companies producing such services or regarding in terms of the characteristics of the services rendered are absolutely evident. This heterogeneity of service industries also implies, among other things, that the location of these companies is characterised by a highly notable diversity, which undoubtedly hinders the possibility of suggesting or making generalisations.

As far as the manufacturing sector is concerned, a high number of theories have been developed about the location of firms. The reason is that the place where these enterprises are located often depends on specific factors such as the availability of basic resources or the accessibility to these in any place or region, the transport costs, the labour supply, and the agglomeration economies. These theories have contributed to explain the location of either an iron and steel company or that of a canning factory or, at least, to approach which may be the most appropriate location from the perspective of economic rationality. In fact, "industrial location" theories have a long tradition since the end of the nineteenth century until the first half of the twentieth century, leaded by Nordic and German authors (from Von Thünen, 1875, and Launhardt, 1882, to Engländer, 1924; Lösch, 1940; Palander, 1935; Predöhl, 1925; Weber, 1909; and others), later joined by North-American

---

J.R. Cuadrado-Roura (✉)
Department of Applied Economics, University of Alcala, Madrid, Spain
e-mail: jr.cuadrado@uah.es

J.R. Cuadrado-Roura (ed.), *Service Industries and Regions*,
Advances in Spatial Science, DOI 10.1007/978-3-642-35801-2_11,

authors including Hoover (1936 and 1948); Isard (1956); and Moses (1955); among others. From this basis, the most modern developments on industrial location arose, although some later changes have forced these theories to be revised. Among such more recent changes, it is worth to mention those derived from the development of ICTs and the reduction of transport costs. Moreover, the subjective factors in the decisions related to location and the influence of institutional aspects have also required increasing attention.[1]

Any case it is unquestionable that conventional theories on the location of productive activities have focused mainly on manufacturing and rarely on services. However, some outstanding contributions exist in this latter case, such as the theories on market areas and central places, or those related to the location of companies within cities, which will be referred to below. Yet, the diversity of services itself makes these theories quite insufficient. On the other hand, the empirical studies on the location of service companies were not abundant either, although they have increased considerably in recent years, particularly regarding some activities such as commerce and business services.

To the intrinsic difficulties already existing to "make generalisations" on the main factors that can have an influence upon the location of services, we should add some changes and processes over the last 15 years which are having important effects on the production and supply of services. Among these processes, one stands out: the development of *ICTs*, which are promoting the emergence of new services and new forms of providing existing services. Also, the *higher accessibility to transport* by people has extended substantially the possible areas for rendering a great amount of services. Both of these changes have favoured what some authors call the "reduction of geography and distance", an undoubtedly relevant but controversial issue (Rietveld & Vickerman, 2004). On the other hand, the significance of the services *internationalisation* cannot be disregarded as this has promoted not only the growth of trading of services, but also international off-shoring and outsourcing processes, which are gaining more and more importance from the viewpoints of production and demand.

The basic objective of this chapter is to provide a general outlook of some of the *theories explaining the location of services*, and, above all, to set out and try to define which are *the most decisive factors for the location* of different types of services. Some old and new theories regarding the location of service companies are put forward quite synthetically in Sect. 2. Section 3 focuses on the location of some specific activities on the basis of the most recent researches. Retailing, the hotel and catering sector, business services, and the special case of capital cities as magnets for locating companies headquarters are dealt with in this section. Finally, the chapter concludes with some final remarks.

---

[1] See, for example, Souza and Stutz (1994), McCann (2001), Maggioni (2002), Hayter (1997), and Garrido (2007).

2 A Reference to Some Old and New Theoretical Contributions Regarding the Location of Services

Despite the fact that the most traditional location theories cannot readily accommodate the case of service activities, this should not prevent us from making a brief reference to some contributions which have been and still are useful and interesting. We will also comment upon some factors being recently taken into consideration in order to explain some current trends upon the location of service companies.

As in the case of any location problem, the analysis of service industries location depends on the level of analysis applied: local, regional, national, or international. In all these cases, but mainly in the first three, the location of service companies is highly correlated to the spatial distribution of cities, towns, and villages and, obviously, to their economic, social, and political importance. However, once this idea admitted, the analysis of the location of service industries can be evidently dealt with by taking into account two main levels:

(a) The **inter-urban level**, by considering the territory in its widest extent and verifying the influence of the distribution of population and services.
(b) The **intra-urban level**, by focusing on how service activities are located within cities, particularly in the case of large cities and metropolitan areas.

2.1 Theories and Approaches Contributing to Understanding the Inter-urban Location of Services

Many authors have explored the relationship between the types of services and their location regarding the territorial distribution of population. This analysis has also been object of a large number of applied researches mainly based on the *central place theory*. However, the gravitational analysis and about urban hierarchies, made from the contributions by Reilly (1929), Converse (1938), and Zipf (1949), among others, are also present in such researches, although less directly.

Two of the most widespread theoretical approaches, the *central place theory* and the *export-base theory*, emerged before the Second World War and were more widely diffused since the 1950s. Together with some brief comments on both contributions, the attention paid to *urban hierarchies* and the *New Economic Geography* approach will be commented upon further on in this chapter.

1. **The "central place" approach: main contributions and limitations**
   As it is well known the *central place theory* is based on the original contribution by the geographer Christaller (1933). However, Lösch (1940) incorporated this theory into a more comprehensive approach of spatial equilibrium, which included the location of many manufacturing industries. The core issue set out by both theories is the relationships between the different types of services or industries and space. According to Christaller, cities, towns and villages are

"central places" that provide certain goods and services to the population and its environment. The goods and services provided differ from one city or town to another according to their importance or extent (the distance accepted by a service consumer or user taking into consideration the transport costs that he/she will need to bear) and to the spatial threshold required by the emergence of the supply of a service or good to those who demand them (minimum number of potential clients that would allow the producer/distributor to obtain profits for providing the service or good in question).

This approach, as it is known, is based on the need of a certain proximity between the service activities and their consumers and/or users. The differences among the various services regarding their range, from the most common to the most specialised ones, lead to them being located according to a hierarchy of "central places" in different geographical areas. Those services with the lowest or most common level and with a more limited client area will be located in small cities or villages. The service activities of a higher category and, therefore, likely to aspire to a widest area, will tend to concentrate in bigger cities and be provided in a much larger environment or area. Furthermore, the theory holds that producers of goods and services of the same level tend to be located following a certain grouping arrangements in order to take advantage from scale and agglomeration economies. Therefore, a series of services are rendered in similar and smaller central places as these are regularly demanded by their populations (groceries, chemist shops, hairdressing saloons, etc.). At a higher level, we find those companies providing services or products whose demand is linked to a more differentiated population (fashion and specialised perfume stores, specialised repair services, etc.). And also, there are other levels where the demand of the services offered is even more selective requiring, thus, a broader area of possible clients. This leads to their location in bigger towns. Obviously, the top central places not only offer services of a higher level, but also those of a lower level, although these are redistributed over different neighbourhoods within the city.

On the basis of some assumptions (e.g., the uniform distribution of population, the inexistence of obstacles to relocation, transport facilities in all directions, etc.), this approach allows us to establish, as known, a theoretical distribution of the "central places" arranged in accordance with a hierarchy and covering certain market areas which are limited, in turn, by other adjoining central places. Consumers will go to the central places according to their proximity and to the range of services offered, minimising distances and, consequently, travelling costs. In case that there is just one good/service, the distribution of those providing it would theoretically tend to get organised in an hexagonal market system, whose surface is determined by a frontier where another provider operates. If there are several goods/services, the providers established in the same area will tend to regroup in order to take advantage of scale and agglomeration economies.

The empirical analyses based on this model have contributed a certain level of evidence to the relationship between the supply/demand of services and the territory. However, more realistic changes have been necessary to be incorporated to the genuine ideas of this model. Technical improvements in transport and its lower costs modify any regular distribution and bring about the creation of corridors or axes comprised by various towns. This could result in a substantial extension of the frontiers of the higher-level centres and, of course, in the fact that these are not regular, as established by the theory. Moreover, the political-administrative regulations that exist or existed in some countries for the location of specific services (e.g., chemist shops; petrol stations) could also modify the availability of the services in the hierarchy of central places.

Despite the merits of the central place theory, this has certainly revealed its limitations in terms of how services can be located. Firstly, because it undoubtedly simplifies the consumer's behaviour, particularly nowadays when transport facilities have changed as well as residential preferences; and secondly, because it is inadequate to understand the location of certain service industries, such as those linked to tourism, business services, and a great amount of public services whose location is more discretionary.

2. **The export-base model** (Innis, 1920; North, 1955) has been also used to explain the role and expansion of services in a given region or metropolitan area, although not to explain inter-urban distribution.

   The approach of this model differs substantially from the previous one, as its main objective is to explain the growth or non-growth of a city or region. Ultimately, this is based on the hypothesis that economic development of a certain region or spatial area (either a city and its metropolitan surroundings, or a region) depends on its economic "base", which is fundamentally comprised of productive activities exporting to the rest of the country or abroad. In short, the model establishes, as known, that the activity in a region or town can be divided into two components: everything produced to satisfy local demand (known as local or residential activity) and all which is earmarked for the export market.

According to the first versions of this theory, exporting productive activities corresponded almost exclusively to the primary and industrial sectors, but also to some service activities (e.g. tourism-related services), which could also be object of exports. Nevertheless, a great amount of non-basic or residential activities corresponds to those services locally demanded by the population. The expansion of those services will depend on the extent of the exporting capacity of the territory or city analysed. This exporting capacity generates income and monetary flows, which are transferred at the same time to the demand of local goods and services.

The historical moment when the aforementioned model was conceived did not possibly allow understanding that services (and not only tourism-related services) may also be a part of the export base. Nowadays, there exist numerous examples of cities and regions where services are produced and rendered to other territories, i.e., these are exported, besides meeting local demand. Examples can be found in trade, transport, finance, and business service sectors, among others.

3. **Shifting to a new urban and service location hierarchy: metropolisation** One of the most salient changes in recent decades has been a clear shift in urban hierarchies. Understanding this shift also requires modifying the "regional" and "national" dimension of services and cities. As noted by Camagni (1992, 2005), it makes less and less sense to define an extensive urban hierarchy within a country and to explore the range of services at a host of different levels since only three types of cities can currently be recognised, i.e., global cities; a network of national specialised cities; and a network of regional cities. Other authors (De Roo, 1993) have proposed to replace the hierarchy of central places and of their catchment area, according to Christaller's approach, with a *hubs and spokes model* closely linked to the flows of goods, people, information, and services. Thus, large cities would operate as *hubs* or nodes that receive and channel the main flows, while the other cities would be linked to the most important city as a *hub.* In recent years, a great amount of works have dealt with these issues and there have emerged a number of empirical studies aimed at determining the "new" hierarchies of cities, this being the subject of various publications of interest (e.g., GIP-Reclus, 1989; Rozenblat & Cicille, 2003).

In this context, there has been a conspicuous shift in the idea that services are or should be necessarily related to the "proximity" of those who use them (citizens or customers, whether individuals or companies), since closeness is no longer a clear constraint for many services. This is true both for services to families and for certain business services including technical, legal, commercial, and auditing services. On the one hand, changes and improvements in the field of transport (Glaeser & Kohlhase, 2004) and ICTs do not require services to be located according to a clear urban hierarchy. On the other hand, these improvements do not require either that the companies supplying such services be necessarily located in a central site in cities, close to their customers, since these companies can operate from the surrounding towns or from other locations in the metropolitan periphery.

In many European regions, this has spurred a process of "de-concentration" or of "relocation" of some services to smaller towns in the vicinity of large cities, and even to rural areas. In fact, in many services the location of users does not determine that of the producers to such a great extent as before. Service producers may be located quite far away and travel to visit their clients, but this is not always essential.[2] In a similar vein, the *back offices* of some companies (banks, insurance companies and the like) have undergone a process of "relocation" to a place sometimes exceeding 50 and 75 km in distance from where they were located in the past. In these cases, telecommunications and other new technologies allow back offices to stay easily in contact with both the front offices and the customers, if necessary. Yet, this implies a clear reduction in set-up costs and other costs.

[2] It should be noted, though, that the "image" of some companies requires that their office (or part of the central services) be located in the centre of a large city, or in representative locations granting prestige.

The existence of fast means of transportation (highways, high-speed trains, airports, etc.) makes it possible for certain production services to cater to their customers—and to expand their number in territorial terms—without having to be located close to these customers as required in the past and still today by some services (retail businesses, routine services for companies). Furthermore, even in the case of family-oriented services, there has been a process of "relocation", as with new large supermarkets, hypermarkets, shopping centres and malls, some of them relocated in the periphery revitalising towns that were previously experiencing a low level of development. All these changes require theoretical approaches different from the ones formulated in the past, which have delved into theories on the use of land and the formation and spreading of cities.

In short, the process of "metropolisation" observed in all advanced countries proves to be linked to all these changes previously mentioned and is due to two main reasons. Firstly, because both citizens and a significant number of services have moved to an extensive urban periphery, where there are jobs, services for residents, and services and recreational activities for residents and non-residents. This is one of the configurations of the spread effect. Yet, secondly, the original city still retains its role as a leader in the area, as a true magnetic pole that spills into the entire metropolitan area, attracting migrants from smaller towns, new activities, and new ways of access to some services as well as innovations in the provision of services.

Due to all these reasons, the analysis of the catchment areas of cities and large metropolis has relied for a long time on useful and well-known analytical tools either based on classical gravitational models (Converse, 1938; Reilly, 1929), on probabilistic models (Huff, 1963), or potential models which attempt to measure the attraction capacity of a particular location in the territory (Beguin & Thill, 1985; Camagni, 2005).

4. **The contribution of the New Economic Geography (NEG)**
   The sudden emergence of this new current of analysis in the early 1990s has encouraged a reformulation of some of the ideas on the relationships among cities and/or regions, the production structure, and the territorial structure. As noted by Krugman (2000), the NEG aims at building a modelled approach that allows discussing issues such as the economy of New York, Chicago, or any other city within the context of the economy as a whole. This should enable us to assess both the way in which the *centripetal forces* that pull economic activity work simultaneously with the *centrifugal forces* that push it out and the tensions of both forces in relation to the geographical structure of an economy.

The work by Fujita, Krugman, and Venables (1999) is a must-read when assessing the contribution of the founders of the NEG, which has been particularly fruitful regarding the development of the core-periphery model concerning the relations among regions and also among cities of some size and their environment. *Agglomeration economies* play a leading role in the models developed from the NEG. These reveal how a centripetal trend may stem from micro-behaviours (of a manufacturing plant, of a company), from the need to be located near a large market due to transport costs, and from the mobility of producers, who, when moving to a certain location, increase the market size associated to that location, making it even more attractive. This defines a growing agglomeration of producers engaged in various activities.

Lastly, the NEG has incorporated into the field of theoretical Economics models that explain where economic activity is located and why, while, at the same time, it has demonstrated how the core-periphery relationship in a concentration of activity works and the way in which the relationships among regions and cities operate. Services and their dynamics have also been the subject of such models, although not to the same extent in which manufacturing, population movements, and the creation of jobs were taken as a reference. A recent volume of the Journal of Economic Geography (vol. 11 (2), March 2011) incorporates some interesting reflections on the contributions of the NEG, its achievements and limitations. In this regard, see also Cuadrado-Roura (2012).

2.2 *The Intra-urban Location of Services*

The analysis of how and why cities are organised internally, how land is occupied, and which key elements explain this is an essential part of Urban Economics and Geography. Contributions to these issues have been abundant and some of them are particularly useful to understand, at least theoretically, the location of services inside the cities and big metropolis. From the opposite perspective, the location of some services is also often a determining factor in the morphology of metropolitan areas and cities and the development of their centres and sub-centres.[3]

There would be little point in providing a comprehensive overview of the main contributions available on these issues, these appearing in detail in any handbook. Yet, it proves valuable to select some of the approaches which are to some extent transferable to services location decisions "within" a city and its immediate metropolitan area.

1. **The monocentric city and the Central Business Districts**
   Traditionally, *city centres* have concentrated a large share of services, from small businesses and specialised businesses to the so-called "office" services and a large amount of public services (religious, administrative, and cultural). The city centre also played the role of promoting social relations, with activities both during day and at night. This still holds true for many cities although there have been significant changes in large cities, i.e., the emergence of new centres or the relocation to the suburbs. In addition, a great number of cities—particularly in Europe—have implemented actions intended to "recover" the city centre after

[3] Consider, for example, the power of attraction of certain department stores when located in different parts of a city and attract, sooner or later, other commercial activities, offices, public transport, etc. as well as more and better urban transport systems.

being exposed to a sometimes long period of deterioration and neglect by the population, business and leisure activities, "office" services, and traditional light industries. The process of deterioration of the central areas of cities and the efforts to revitalise these areas have been quite significant in most European countries and have, in general, yielded very positive results.

Nowadays and in American literature, though, the "downtown"/"city centre" area has been associated to a larger extent to the concept of "Central Business District", occupied by financial services, professional offices (lawyers, consultants, etc.) and other specialised activities such as media and advertising as well as corporate headquarters, certain cultural and leisure activities, and restaurants among other businesses. This way to organise cities and their various structures and changes have provided the city centre a distinct role which is compatible with the relocation of activities to different areas of the city (Anas, Arnott, & Small, 1998; Derycke, 1992; Glaeser & Gottlieb, 2009).

2. **The contributions of land rent theories**
   These theoretical contributions also allow us to approach, albeit partially, the location of some services in intra-urban areas. Hurd (1903) already made a pioneering contribution on the use of land and buildings in a city. His ideas were later enriched with the incorporation of the principle of accessibility or of spatial competition, which had already been advanced and developed by Von Thünen (1875). A number of years later, Alonso (1964) made a major contribution by defining a modelled approach on the formation of land rent in urban environments. His model would become one of the pillars of urban economics from the seventies. The resulting developments include the fact that the rent curves for a portion of land -a hectare, for instance- differ significantly according to the distance from the centre. Companies' headquarters, professional offices, specialised businesses, and banking services can record higher costs per ha than small industries, workshops, and residential rents. This would explain the concentration of such services in the city centre and in certain business areas, as they are able to face higher rents. In fact, the model allows us to understand the evolution over time from full centrality to new polarised centres.

Although it does not specifically refer to the case of services, along a line just as interesting as the previous discussion, we should mention Lowry's model (1964). The internal logic of this model merges two theoretical hypotheses: the urban economic base theory, which allows to estimate the entity of two variables related to the economic base, the resident population, and employment in services, and the principle of spatial interaction in the form of two models of gravitational potential with a restriction: they are used to locate the population near jobs—residential model—and the occupation in services around—service location model.

3. **Relocation to new areas and intra-urban polarization**
   In last three to four decades, medium-sized European cities (at least exceeding 150,000 inhabitants) and, above all, large cities have experienced a series of changes, which have been analysed and theorised, in connection with the location

of some services. These changes are related to some centrifugal forces that have acted on productive activities and also on the population. Industries took the lead in moving to the peripheral districts or even far away from urban areas. Certain services have also undergone a process of relocation, though, generally, they have stayed within the urban fringe or in some areas of the urban periphery. This is the case of wholesale businesses, repair services (maintenance of machinery, garages, etc.) or warehouses for goods. On the other hand, the services generally described as upper tertiary services, i.e., administration and business management, human resources, political and administrative services, hotel facilities, etc., have only moved to urban fringes not so far from the old city centre. In addition, a not negligible proliferation of new city centres or sub-centres has taken place in conjunction with a growing trend to relocate typically urban leisure areas such as cinemas, restaurants, indoor sport grounds to outlying areas, especially in large cities.

There are various reasons underlying this process. Some are related to the cost of land and the lack of access to enough land offers for such activities. In addition to these, other reasons can be added such as the congestion and degradation problems of some central areas, the need to move away from unsafe areas, as well as a number of municipal regulations that have favoured these relocations encouraged at times by the availability of new transport infrastructures and improved urban connections (underground, urban surface transport, airport). And, of course, a major role in the emergence of new centres or urban poles has also been played by the real estate actions to create new residential areas and/or business parks, which, in turn, attract services and require new infrastructure to be developed. As a result, *cities end up becoming polycentric*. These incorporate new centres where land value increases rapidly and also offer many advantages regarding the supply of premises, the new nature of buildings, and a better image and/or quality of infrastructures. Theories about the cost of land and its development explain only partially the way in which these new urban nodes have evolved, but prices reveal the preferences of space, the opportunities offered by the new area as well as the existing restrictions.

In the field of Urban Economics, some theoretical contributions are useful to understand and explain the location of services within the cities. Literature in this field has increased in the last two decades in conjunction with some contributions arising from the ideas of the *New Economic Geography,* as already mentioned. The ideas on the centre-periphery relationships and the economies of scale and agglomeration provide a useful basis for approaching the concentration of services, although this remains a not well explored area from the perspective of this scientific current.

2.3 New Directions on Location of Services

To understand the location of some services and their working within the economy, some theses and models previously noted are still invoked even though their assumptions have been criticised. This is due particularly to the fact that, while these ideas provide a theoretical basis for reasoning, reality is far richer and more variable.

Changes in population size, in the purchasing power of people as consumers, in the very composition of their spending, or in the supply of new services may modify radically what once might have been a hierarchy of central places. The expansion of services most directly related to people and their presence in a territory is linked, of course, to the *population size* and to its *spending power* and when both increase, the demand for services changes both in terms of quantity and variety. In fact, the improvement in living standards has substantially shortened the gap in the demand for services between remote rural areas and large cities.[4] At the same time, new forms of business organisation (creation of branches and chain stores, franchise systems, outsourcing) have entered the services sphere spreading the supply in population areas historically considered of a lesser significance.

In line with these changes, the location of services from a territorial or inter-urban perspective has become increasingly complex, although in the case of many services to families (health and other personal services, basic retail distribution, etc.) and some unsophisticated business services (accounting, legal services, provision of technical support), closeness to users or customers remains a necessary condition. However, the development of transport (Glaeser & Kohlhase, 2004; Rietveld & Vickerman, 2004) and some well-known technological changes—ICTs in particular—have transformed distances, making it possible for services not to be necessarily located at the centre of a large area, but in the periphery of cities. In this regard, we may speak of a "deconcentration" of services in large metropolitan areas. Furthermore, more and more services can be now obtained from home or from the office without having to travel. These include purchasing materials, travel tickets, books; getting information; or resolving administrative matters.

Furthermore, the studies on *urban hierarchy* have also changed. The place of cities is now in connection more than ever before with an *international context*. The relationships of cities with their catchment areas and with other cities have also changed as a result of *transport* and *connections systems*. In recent decades, these changes have been the subject of numerous studies and contributions conducted within the fields of Urban Economics and Geography. Thus, the analysis of these location processes and the driving factors underlying them has brought about the development of new theories on ***city networks and systems***. These have highlighted the occurrence of three major changes in time: (1) the elimination of a hierarchy of lower-level centres; (2) the preservation and strengthening of the major poles and national centres, particularly in the case of capital cities; and (3) the emergence of a limited number of world cities, following the conception advanced by Sassen (1991) later developed and expanded by other authors (Taylor, 2004, 2012).

[4] Consider, for instance, the demand for nurseries or for some medical services such as dentists.

Somehow, all this relates to the globalisation process and to a process described by some authors as the *metropolisation of the tertiary sector*, as already pointed out, which results in the growing importance of the major cities through an ever-increasing concentration of the supply (and demand) of services (Veltz, 1996; Taylor, Catalano, & Gane, 2003, 2009). This is a specific type of *polarisation* which drives not only the most dynamic sectors and the ones that create the most jobs and require the highest amount of qualified staff (financial, insurance, advice, and consultancy activities; computer services, real estate and advertising) to set themselves up in certain metropolis, but it also encourages in such cities the expansion of employment opportunities in low-skilled services such as cleaning and security services among others as well as wholesale trade and logistics.

As a result of these changes, new approaches to the location of services tend now not to establish generalisable "rules" but to develop studies and research to a much more disaggregated level. In other words, there is a prevailing tendency to give up building "models" on the locational behaviour of the service industries, and to focus, instead, on the study of what happens in the case of some specific activity branches in a well-established geographical context or city.

However, some features of the location of certain groups of services can be regarded as having a fairly general nature, as we will point out in the next section, bringing us much closer to reality.

3 Trends in the Location of Specific Activities

The purpose of this section is to present a synthesis of the prevailing trends in the location of some service activities. Specifically, we will examine the case of retail trade; hotel and catering activities; business services; and, finally, the location of the headquarters of service companies. It would be of interest to provide an account of the existing trends in other service industries, but, given the spatial restrictions of this chapter, we have selected a number of examples that appear to be significant enough. Additionally, numerous empirical analyses and extensive literature support these examples.

3.1 The Case of Retailing Activities

Retailing activities have been given special attention by researchers and, particularly, by geographers, economists, and sociologists. Among other reasons, such interest stems from the fact that these activities have not only provided a continuing high volume of employment in the economy, but is also due to its relation to demography, to patterns of social behaviour, and to the processes of territorial concentration of population. Although the issues to consider could be numerous, attention here will focus on three specific points: the changes that have occurred and that are related to the location of this activity, the trends in urban location, and the case of medium-sized cities and towns.

1. ***Changes affecting retailing from the supply and consumer sides***
   In recent decades, the operation and location of retailing activities have changed in such a way that they have moved away from an approach as simple as the original *central place theory*. These changes are extremely clear both from the supply side and from the side of consumer behaviour and preferences.

From the point of view of supply, at least four major changes have brought about significant changes in retail trade. The first refers to an event that certainly revolutionised a number of years ago how people do business: *the extension of self-service*. Its implementation has not only affected the food trade but also a large number of goods including DIY, clothing, household furniture, sports equipment, etc. Self-service has served as a basis for the development of a host of commercial forms or structures that are now widespread, such as hypermarkets, supermarkets, and discount shops among others, described generically as "businesses without windows". These commercial configurations have led to changes in the location of establishments, which will be discussed later, and they have also encouraged new behaviours on the part of consumers.

Special attention must be accorded to a second change: *the extension of franchising* due to its impact on supply. This system is based on the exploitation of many different brands of products through specialised stores. Among other things, franchises have meant the openness of retailing to internationalisation and to the significance of brands and trademarks in sales; they are a landmark within the globalisation process of commercial services (Rubalcaba & Cuadrado-Roura, 2002).

The third major change relates to *the increasing use of ICTs* in the sale and distribution of food, books, music products, perfumes, etc. Their impact is currently already clear, and the expectations for growth in the coming years are very high. Such expected development is primarily linked to the rapid advances in ICT applications, which offer consumers increasingly versatile and smaller devices featuring a great ability to obtain information and perform operations without having to travel to the store or see a seller.

Finally, also from the supply side although linked to changes in the behaviour of people,[5] we should mention specifically the emergence of a certain type of commercial structures with a noteworthy regional, in addition to a commercial, impact: the proliferation of *commercial malls*. Generally, these are located at a distance from the big city and other surrounding towns and feature a combination of a choice of businesses and leisure activities, while providing good accessibility and parking

[5] In fact, one might consider the issue from a double perspective: whether the evolution of people and their preferences in housing, employment, transport, and ways of life have given rise to the "malls", or if the incorporation of this commercial structure has pulled society into new shopping formulas and individual and family behaviours.

opportunities (Gorter, Nijkamp, & Klamer, 2003), a wide range of offers (Schenk, Löffler, & Rauh, 2007), sales, and a high degree of satisfaction (Léo & Philippe, 2002).

From the perspective of consumer behaviour, some changes have influenced and are still affecting retailing activities and their location. These changes include not only those relating to the characteristics of consumers but also to the very location of shops and retail centres as well as to the reasons to buy and the products that consumers wish to purchase (Van Leeuven & Rietveld, 2011).

The *socioeconomic characteristics* of consumers affect, for instance, the degree of mobility within a territory. Improvements in income levels have gone hand in hand with an increase in the amount of shopping done outside the city or at some poles in the greater metropolitan area. This is related to the availability of individual and collective means of transport which have greatly increased people's mobility (Mouhoud, 2010; Thompson, 1971; Papadopoulos, 1980). Obviously, this increased mobility may become restricted due to age; older people are, therefore, likely to do their shopping at shops near their place of residence (Pinkerton, Hasinger, & O'Brien, 1995; Powe, Hart, & Bek, 2009).

Another fact that has changed the location of retailing as a result of consumer behaviour has been the *relocation of population to the periphery* of large cities, at new residential areas, or also to surrounding cities and towns. These trends have been extensively studied by geographers and town planners. Furthermore, several studies have stressed that the workplace can influence where people choose to do their shopping (e.g., in a large trade centre) if both locations are close and if the times allow for this option (Papadopoulos, 1980). This affects people who commute daily—a process of increasing importance in many European regions considered by some studies as a determining factor in out-shopping (Finlay et al., 2001; Finlay & Sparks, 2008).

Finally, we must not forget that, regarding the consumer behaviour in connection with the location of retailing activities, whether in cities or in the establishments in the periphery, there exist two different attitudes which also play a role in location. The first attitude refers to the visits considered *as useful for the consumer,* as they are prompted directly by the need to buy certain products. The second attitude *refers to the leisure component* of shopping centres and malls as a destination for visitors to just "have a look and walk" ending up ultimately also doing some shopping.

2. ***Trends in the urban location of retailing activities***
   The dominant features in the current location of retail activities in urban areas include, according to most recent studies[6]: (1) a strong development of distribution in peripheral urban areas; (2) the development of secondary poles in large cities; and (3) a number of changes which have taken place in the commercial city centre, backwards for some years and of recovery more recently, at least in Europe.

[6] This applies especially to Europe.

Each of these features deserves further in-depth treatment with the caveat that, on the one hand, the three processes just outlined may not be treated as self-contained areas, that is, no sharp differentiations among them can be established; and, on the other hand, both diversity and heterogeneity are inherent to retail trade. This in particular should also compel us to refine some assertions based on the type of trade taken as a reference (food, fashion, household items, appliances, hardware and DIY, etc.).

There is no doubt that the development of superstores has involved a significant transformation in the geography of European countries by decreasing the amount of shops in the city centre and increasing their presence in the periphery and the outermost fringes of metropolitan areas (Mouhoud, 2010). There has been a tendency to locate hypermarkets, shopping centres, and specialised superstores in peripheral and outlying urban areas.[7] The reasons behind this include not only the availability of land, which was no longer the case in the city centre, but also its lower cost and the advantages of some areas in terms of travel and connections. This turned the areas close to the motorways and carriageways out of large cities into a priority. In this regard, location decisions have generally followed the experience of some large multinational retailing companies, which have reproduced in several countries the practices experienced in other countries (Wrigley, 2003).

However, the dynamics over time have not been the same in all countries. In Europe, the Nordic countries took the lead and the southern countries joined later. Besides, municipal and regional regulations have also influenced the development and location of these large retailing areas. In The Netherlands, the United Kingdom and the Scandinavian countries, the standards governing these new forms of business as well as land use were well established almost from the outset, while, in other countries such as Belgium, France, Spain, and other Mediterranean countries, policies to control these new forms of business have been in place but, not infrequently, they have lacked a clear connection with the potential requirements of land planning.

In their location decisions, these large shopping centres have chosen easily accessible areas having, additionally, an important population potential in their surroundings, both in relation to the city to which they are linked and in terms of existing smaller towns in the hinterland and their future potential growth. There appears in conjunction with this type of structures a new generation of "poles" or "urban centres" located within the expanding conurbation. Some of these new centres began as a second or third "centre" of the city, following the relocation process of the commercial structure of the main centre toward the rural–urban fringes (Mouhoud, 2010). Yet, the development of some of these "centres" is linked to large residential operations performed in the urban fringe, where the new concentration of population has encouraged the establishment of small shops of all kinds, neighbourhood supermarkets, and specialised medium-sized shops selling furniture, appliances, books, stationery, etc. All this results from the evolution towards polycentric urban structures,, a process already mentioned in Sect. 2. The

[7] In recent years, distribution companies—especially food companies—have complemented this trend by opening supermarkets and discount shops in the city centre and in other neighbourhoods.

vitality of these commercial poles has been shown to hinge largely on the socio-economic level of the neighbourhood or development created, on their distance from other sub-centres, and, of course, on good accessibility.

The third feature previously highlighted as part of the evolution of retailing activities is the deterioration suffered by the traditional city centres of many European countries. This has not been so in North America, where, as already pointed out, the "downtown" of large cities has always been much more a business centre (finance, insurance, professional offices, etc.) than an area with residents and retail businesses.

It is worth noting that the deterioration of the traditional centre of many European cities has not happened in absolutely general terms. This was due to several factors and interrelated processes including the deterioration of facilities and housing; the relocation of population with a medium and high income level to other neighbourhoods or areas in the city; the increased insecurity in the traditional centre; the occupation by immigrants; the aging of the existing businesses themselves; and, last but not least, the relocation of the headquarters of companies, among other factors.

However, in the last 25 years a great number of the traditional city centres has been recovered, usually through the implementation of regeneration policies promoted by the authorities. Decisions to pedestrianize certain areas and streets have helped to revitalise commerce and restaurants in such areas,, attracting also new shops and stores and promoting the modernisation of existing ones. At the same time, the execution of moderately expensive infrastructure projects, the enhancing of the numerous monuments, the improvements in urban transport, and the policies supporting house renovations and cleaning of facades through subsidies and tax and credit incentives have brought about quite spectacular changes. In Europe, these policies have also been supported by EU programmes and funding (Urban Programs). Different but significant examples, such as those of Montpellier and Le Havre (FR), Liverpool (UK), Dortmund (D), or Barcelona (SP) are by no means any isolated cases as the list of successful regeneration actions features more than 140 cities in Europe. In all of these cities, business activity has resumed its dynamism with the opening of specialised shops as well the return of the population and festive activities.

3. ***Trade in small and medium-sized towns***

   The picture of the location of retailing activities and of commercial vitality would be incomplete if we left out the role played by medium-sized cities and towns (between 5,000 and 20–25,000 inhabitants) and their supply of retail goods and services to the local households. A recent study conducted by Van Leeuven and Rietveld (2011) on six towns featuring these characteristics in five different European countries (England, France, The Netherlands, Poland, and Portugal) provides particularly relevant data and indications regarding the behaviour of the inhabitants of these medium-sized towns and, to some extent, about the role of retailing activities and other forms of business. This work and other less ambitious works reveal the significance of businesses located in smaller towns with a hinterland of 12–25 km around them, where the business distribution and customer behaviour follows still the central place system.

Highlighted in the aforementioned work is precisely the fact that consumer behaviour is largely influenced by the availability and accessibility of retail businesses and their existence is highly valued by households. However, there are many differences among countries. In England the number of shops in rural areas (towns having fewer than 12,500 inhabitants) is reduced, but these are larger as, on average, they have seven employees. In Portugal, shops within the town are numerous and smaller (average of two employees), and the same happens in Poland. Nevertheless the ratio of the number of inhabitants per employee is not very different.

Surveys show that the shops within the town are still preferred in the small and medium-sized towns examined in Portugal, Poland, and France, and less so in England. Households in the hinterland (7–10 km around the town) have few shops available and tend to buy in the city, as is the case in France, Portugal, and Poland. However, in England, since there exist well-stocked shops in the regional and supra-regional (25 km) surroundings, families shop to a large extent, in comparative terms, outside the region (around 25 %). For its part, the case of The Netherlands is somewhat different since businesses are more scattered and people in the hinterland shop in the hinterland itself rather than in the city. In France, there is a high concentration of people in cities and towns and, by contrast, the number of people in the hinterland is quite low. This reinforces the role of cities from the perspective of purchases of goods and services.

The aforementioned study shows that there exist differences in Europe regarding the structure of retailing activities in small and medium-sized towns caused both by the degree of dispersion of the population and by socio-cultural and historical reasons. Yet, what seems clear is that towns are the place where town and hinterland households do most of their shopping and procure most of their services including pharmacy products and services, food and groceries, and personal services. Shopping beyond the hinterland is done for other varied and occasional purposes (footwear, fashion, gardening, furniture, etc.) not covered in the small town or village.

3.2 *Hotel and Catering Sector*

The hotel, restaurant and catering sector is in some countries an extremely important activity, both from the standpoint of employment and its contribution to added value. In France, this sector represents more than 1.5 million jobs (Cesaer & Cemagref, 2008; Mouhoud, 2010) and, in Spain, 1.2 million full-time-equivalent jobs (Cuadrado-Roura & López, 2011). In many other countries, this activity accounts for 6–8 % of the employed population.

This sector generally appears closely linked to tourist activities and it certainly should be so. However, an analysis through Input–output Tables shows that hotel and catering activities also play a significant role as an input in a number manufacturing and service sectors, having shares of total intermediate demand by industries varying from 2 % to 4 %. The hotel and catering sector is, as it is well known, an important and growing component in final domestic consumption, as shown by the surveys of household expenditure and as stated in the National Accounts of all advanced countries.

The literature on the geography and location of tourism hotel activities is extensive and keeps expanding. Several features stand out in terms of the location of hotel and catering establishments and the like. First, there exists a large amount of this kind of establishments spread over the territory as, in all countries, these are located in every small and medium-sized town and even in small villages (Sund, 2006). However, at the same time, attention must be brought to the fact that there are high concentrations of these establishments at a local, national, and international level. They are particularly present in areas designated as tourist areas per se receiving international and domestic tourists, especially during certain periods of the year (sun and beach tourism, snow tourism). Their presence is also particularly important in large cities, both in Europe and the United States (Arbel & Pizam, 1977; Shoval, 2006; Urtasun, 2006). Their demand stems essentially from two sources: firstly, travel and business meetings, conferences and relations with corporate headquarters (ECM, 2010); and, secondly, the growing importance of short-stay urban tourism both nationally and internationally (Mazanec & Wöber, 2010). This has resulted in recent years in a qualitative increase in supply (5- and 4-star hotels) as well as in a greater demand of value for money regarding hotels, restaurants and the like.

The concentrations of this kind of establishments in the tourist areas is obviously determined by the existence of adequate resources and natural conditions. This characterises the difference between the location of hotels and the like as opposed to the location of other commercial activities, although some of the latter (from small shops to super- and hypermarkets) have also grown exponentially in tourist areas. At any rate, factors such as the climate, the landscape, or any special circumstances for skiing, where applicable, determine the geographic concentration of businesses as the natural environment is an essential requirement. These conditions also determine the type and category of hotels set up in these areas and, of course, also their prices (Rigall et al., 2011). In addition, other factors include the distance from large towns or cities (either as an advantage or a drawback) and the existence of good connections, which are not particularly essential in the early stages of development but in their continuity and expansion.

Attention must be brought also to a new kind of supply in connection with changes in demand, e.g., sports (golf among others) and cultural tourism, stays for health reasons, and senior tourism, all of which do not necessarily lead to locating businesses in traditionally tourist areas but in new areas not yet saturated and with a better environment.

If there is anything that characterises the presence and location of hotels, it is the diversity of tourist trends and spaces, although they are predominantly concentrated in particularly tourist areas and large cities (Cuadrado-Roura & Maroto-Sánchez, 2010). At the same time, two notable features prevail in the hotel sphere and in its location. The first one refers to the existence of *large hotel chains,* most of them international and ranking high in terms of their supply and quality. They seek to be present in the most important tourist areas and cities worldwide (Johnson & Vanetti, 2005; Pan, 2005) and they provide similar and contrasted standards and quality levels above all. The second one concerns the development of *"low cost" hotel chains,* also of an international nature and with a range of facilities not only in big cities but also in medium-sized towns.[8]

3.3 Business Services and KIBS

The subsector of producer services and, particularly, business services, has grown exponentially in all developed countries, both in terms of employment and value added. This is due, in part, to the service outsourcing processes carried out by a large number of companies in recent decades, which brings about a significant "statistical effect". Yet, this growth results also from the increasing demand for services by firms that produce goods and services in order to cope with the changes and new demands in the economies (see Chap. 1 of this book). In this regard, the data provided by the Input–output Tables of all economies reveal the increase occurring in the use of services as a production input in all sectors, where business services constitute their main component.

The report prepared by Ecorys-NEI (2004) showed clearly the consumption of business services by other sectors (Table 11.1), as well as the intensity of use of business services (as a share of total intermediate inputs), both in terms of direct and indirect use.

The internal composition of the business services sector shows that it incorporates a conglomerate of extremely varied industries. The activities within this sector range from unsophisticated services, or routine services as often described in the literature, such as industrial cleaning services, security and surveillance, or temporary work agencies, to services related to IT developments, R&D, innovation, strategic and technical advice, often included in the literature under KIBS. Table 11.1 proposes an arrangement of the various sets of services that make up this industry, although the position of some activities is subject to discussion and the list could be extended (Table 11.2).

1. ***General trends regarding the location of business services***
   Based on the observation of the location of business services at present, we must highlight three features. Firstly, the higher level of spatial concentration that characterises these services (both at an inter- and intra-urban scale) in comparison with other services, such as retailing services. Secondly, the high level of internationalisation of this sector, not only from the perspective of companies

[8] An in-depth analysis of the hotel sector would require a specific and more extensive study. The works cited serve to complete some of the aspects succinctly mentioned here.

Table 11.1 Consumption of business services: sector shares of total intermediate demand for business services inputs (%)

	Agriculture and mining	Manufacturing	Utilities	Construction	Trade and hotels	Transport and storage	Post and telecom	Finance and insurance	Real estate	Business services	Public Sector
EU 9	1.4	28.2	1.7	6.1	10.8	3.9	1.2	8.9	4.9	20.4	12.3
Denmark (1997)	2.7	15.3	0.7	20.1	17.4	4.4	2.0	5.3	2.6	12.9	16.4
Finland (1995)	1.9	48.1	1.9	3.2	8.2	2.6	1.5	4.6	7.2	8.1	12.8
France (1995)	1.0	32.9	2.5	7.6	6.4	2.3	1.0	8.6	2.7	24.2	10.8
Germany (1995)	2.4	32.2	2.0	4.6	9.6	3.2	0.7	7.6	10.0	17.1	10.6
Greece (1994)	1.8	37.8	0.6	9.1	17.9	6.2	0.2	4.6	4.5	3.1	14.2
Italy (1992)	0.3	24.4	0.7	4.5	21.1	6.5	0.8	8.6	4.7	14.2	14.2
Netherlands (1998)	1.7	23.6	0.9	5.5	15.6	5.4	3.7	5.8	1.1	24.9	11.8
Spain (1995)	0.9	32.2	2.1	11.5	12.1	3.3	1.3	6.3	1.0	13.6	15.5
UK (1998)	0.8	17.8	0.9	5.3	10.4	5.6	1.9	14.1	1.8	26.1	15.2

Source: Ecorys-NEI, 2004. I–O Tables OECD

Table 11.2 Business Services as part of Producer Services (Rubalcaba & Kox, 2007)

<table>
<tr><td rowspan="13">Producer services</td><td rowspan="12">Business-related services</td><td rowspan="7">Business services</td><td rowspan="6">Knowledge intensive-business services (KIBS)</td><td>• Software and computer services</td></tr>
<tr><td>• Strategy and management consultancy</td></tr>
<tr><td>• Accountancy, tax and legal advise</td></tr>
<tr><td>• Marketing services, opinion polling</td></tr>
<tr><td>• Technical services, engineering</td></tr>
<tr><td>• Personal training, headhunting</td></tr>
<tr><td>Operational business services</td><td>• Security services
• Facility management, cleaning
• Administration bookkeeping
• Temporary labour recruitment
• Other operational services (e.g. catering, photography, translating, call centers)</td></tr>
<tr><td colspan="3">Distribution and trade services</td></tr>
<tr><td colspan="3">Transport and logistics</td></tr>
<tr><td colspan="3">Banking, insurance, stock exchange</td></tr>
<tr><td colspan="3">Telecommunication, couriers</td></tr>
<tr><td colspan="3">Energy services</td></tr>
<tr><td colspan="4">Consumer services partly used by enterprises (business travel, company health services, social insurance services)</td></tr>
</table>

and groups of companies, but also regarding the type of services supplied and how these are rendered. Finally, the role played by business services in the development of national and regional economies, which accompanies growth and contributes to improve the efficiency of manufacturing industries and virtually all the remaining service activities.

The wide spatial concentration generally characterising the location of the majority of business services can be explained according to three factors: (1) the requirements in terms of staff and labour, and of their qualifications above all; (2) the interrelations among the different business service activities; and (3) the need to be set up near the clients and remain in contact with general or specific sources of information, as the case may be. This attaches importance not only to the labour market, but also to transport, accessibility, communication facilities, and the economic and social environment (Daniels & Bryson, 2005; Ecorys-NEI, 2004; Illeris, 1989; Jouvaud, 1995). In short, all this underlines the significance of external and urban agglomeration economies in this case.

The trend towards spatial concentration has been demonstrated in almost all researches regarding the location of business services, where cities and particularly large metropolitan areas and capital cities lead the sector (Cuadrado-Roura & Maroto-Sánchez, 2010; Cuadrado-Roura & Rubalcaba, 2000; Doloreux, Amara, & Landry, 2008; Muller & Doloreux, 2009; Taylor, 2012; Taylor, Evans, Hoyler, Derudder, & Pain, 2009; Wood, 2002). However, it is worth noting that there are remarkable differences among business services and their higher or lower concentration/dispersion. This is mostly due to the relationships generated by the use of these services: the need or lack of need of a frequent contact between clients and providers; the level of standardisation of the service; the frequency with which it is required; or the type of clients, either large enterprises or SMEs.

Some of these characteristics modify the need for *proximity* between service providers and clients. Thus, proximity may be advisable for highly routine services and specific types of companies served (i.e., SMEs) (for example, accountancy, fiscal advice, etc.). In other cases, however, users can, and prefer to, travel to the company's headquarters, or large business service companies travel to the clients' facilities. In this regard, the development of telecommunications makes it possible for the information required to provide services such as the aforementioned to be at a further distance, with no need for proximity and to travel.

A great amount of recent analyses on the location of business services at an inter-urban level, i.e., from a regional perspective or from the viewpoint of countries reveal that large business service companies (auditing, consulting, major law firms, etc.) locate their headquarters in large metropolitan areas and set up other offices in second-tier cities according to market potentials. On the other hand, those business service companies that rarely require any contact with their clients (e.g., data treatment) are located on the outskirts of metropolitan areas and even long distances from these, with the aim of reducing location costs and having abundant labour available. Conversely, those companies requiring highly-qualified staff are concentrated in areas where this is abundant, easy to entice or obtain, i.e., important cities and large national and international metropolitan areas. Finally, "proximity" and the possibility of frequent contacts and accessibility are taken much more into consideration by the most local business service companies, such as administrative agencies, accounting, tax advice, and law firms, as well as those services not requiring highly-qualified labour (cleaning and surveillance services, for instance), although many of these are integrated into larger companies.

From the *intra-urban* viewpoint, it is worth mentioning a high level of *spatial concentration*[9] or clusterisation of business service companies in those areas

[9] Thus, we can speak of a double concentration: firstly, in large cities and metropolitan areas; and secondly, "within" those cities, in specific areas.

considered to be central for businesses as these areas provide office buildings and a great deal of prestige despite higher costs. All the analyses regarding business services and cities demonstrate that business service companies are located principally in very specific areas of the city in question, either in a main avenue and its side streets, or in the business centre. In the North American literature, this was called Central Business District (CBD). All in all, the more recent development of business parks and the opening of new centres within large cities have made some business services and headquarters to set up in more peripheral areas. Likewise, those services linked to transport centres and wholesale trade services are traditionally established more peripherally within metropolitan areas.

Internationalisation is undoubtedly a feature of current business services. It is part of the on-going globalisation processes and of the creation of large companies and business service groups of international prestige (Cuadrado-Roura et al., 2002). Companies implement a multiple office policy across many cities, and the reason is not only to serve and gain clients, but also to provide a seamless service and protect global brand integrity, while maintaining control over their way of working at the same time (Taylor et al., 2011). All this favours and clearly promotes the expansion of large cities at an international scale, as pointed out by Sassen (1991, 1994). Financial activities sector is clearly one of the major protagonists of international concentration, with London, Frankfurt, New York, and Tokyo at the top of the scale, followed by a good number of cities, mainly capital cities, at a continental level. Therefore, one could speak of a marked international hierarchy of financial centres, leaded by those globally important, followed by the continental centres, and those with an international significance.

2. ***The case of Knowledge-Intensive Business Services (KIBS)***
   The majority of ideas presented in the previous section can apply to the case of KIBS. However, some aspects and differences must be taken into consideration. This type of services has played an increasingly central role in advanced economies for more than three decades. For this reason, the literature on KIBS and their location have increased substantially since the 1980s (Beyers & Alvine, 1985; Coffey & Polèse, 1987; Drennan, 1987) until more recently (Alvesson, 2004; Cuadrado-Roura & Maroto-Sánchez, 2010; Fischer & Frolich, 2001; Gallego & Maroto, 2010; Koch & Stahlecker, 2006; Muller & Doloreux, 2009; Rubalcaba & Cuadrado-Roura, 2001).

The features determining the *location of KIBS* can be summarised on the basis of some recent works.[10] The first aspect to be highlighted is *geographical proximity*, which is linked to the possibilities of personal contact offered and the existence of research and innovation centres. Proximity boosts innovation processes, based to a

[10] Works by Delgado-Marquez & Garcia Velasco and by Cooke, included in this volume, among others.

large extent on the exchange of information and knowledge among the different actors and frequently on learning-by-interacting. Knowledge transfer requires contact, confidence, frequent communication, common understanding, and face-to-face contact, which is favoured by reduced physical distances (Cooke, Heidenreich, & Braczyk, 2004; Morgan, 2004).

Close *interaction* between providers and clients is recognized to be an important factor to develop new and innovative services (Boschma & Weterings, 2005). KIBS companies have their own catalogues and protocols of the services they provide, although adjustments and new solutions for new problems are always required. Solutions are not often suitable for standardisation, and service-providing companies learn in each specific case, innovate, and transfer their experience to other clients and problems. All this relates to very direct client-provider contact, and the location close to the client, or the possibility of having quite direct contacts involves advantages and better expansion prospects for those companies well located at a local level.

The importance of *physical distance* between KIBS clients and providers really depends on many aspects, but mainly on the convenience of adjusting the service to the needs of the client (private client or public agency), *offering a customised service*. If the requirement is high, proximity gains importance, despite that nowadays ICTs make it possible to solve problems and contrast ideas and solution proposals with no need to travel. All in all, the advantage of being near the official organisations—if these are or may become clients—and near the headquarters of client private companies constitutes a positive factor for the majority of KIBS companies. This is the reason why these companies are located in capital and main cities, although with offices in other regional cities.

Two more factors play also a significant role in the location of KIBS and the trend to be set up in the principal cities: easy access to good information and having skilled staff and technological capital available in the vicinity. The access to a good information may include a great amount of elements: knowing what competitors do; obtaining full information (data, people, references); and knowing in time regulatory changes made by the corresponding authorities. The access to skilled staff is also easier in large cities, where academic and research centres, competitive companies, and movements of qualified and experienced staff (either resident or coming from other cities or regions) are concentrated (Cook et al., 2007; Koch & Stahlecker, 2006; Sokol, Van Egeraat, & Williams, 2008). In fact, KIBS companies are attracted by areas or cities where human capital is more abundant, particularly large cities and capital cities. Evidence shows that the higher the intensity of knowledge required by a service activity, the more important the trend towards geographical concentration.[11]

---

[11] See the empirical results included into the contribution of Delgado-Marquez and Garcia-Velasco in this volume.

3.4 A Specific Case: Location of Headquarters of Large Service Firms in Capital Cities

On the basis of recent studies on the location of business decision-making centres, which generally coincide with their headquarters, three major trends in favour of a high concentration of headquarters of large enterprises of any kind in a reduced number of important capital cities can be highlighted:

- At an international level, but also within each country, globalisation process favours large cities and more specifically capital cities, boosting the concentration of the headquarters of large companies in such cities (Kller, 2006). This is particularly so when, as it almost always occurs, these cities have good connections and easy accessibility at an international scale.
- The level of concentration of business decision-making centres is obviously much higher than that recorded by the spatial concentration of productive activity (Klier & Testa, 2002; Ono, 2006). The trend towards concentration is not occurring according to the production centres, which are either still located in those places chosen many years ago, or they relocate to other regions or countries in search of lower costs, tax advantages, and a higher presence in new markets.
- Those companies operating in the most regulated sectors tend to locate their headquarters in the country's capital city, where the public sector is often concentrated and where the corresponding regulatory body is usually established. In Europe, this is clearly the case of London and Paris but also of Stockholm, Copenhagen, Madrid, Brussels, Warsaw, Dublin, Lisbon, Vienna and Prague. Nevertheless, those cities with a high economic importance, sometimes even higher than capital cities, are also very likely to attract this type of firms. Some European examples include Milan, Amsterdam, Frankfurt, Zurich, and Barcelona.

These general trends can be applied to the particular case of headquarters of service companies. The majority of available studies already demonstrated in the past the trend towards the concentration of headquarters in large cities, with a significant preference for capital cities (see Howells & Green, 1986). Therefore, this is not a new phenomenon, although it has intensified in recent years. However, it is worth mentioning that capital cities are not always the priority, as certain historical backgrounds related to the creation and development of some service companies justify the fact that headquarters or at least some of them are still located in other cities that are not necessarily the most important ones. In the Spanish case (Cuadrado-Roura & Maroto-Sánchez, 2010), but also in other countries, there are examples of companies with two headquarters: the one located in the city where the company was originally established or located, and the *de facto* operational central headquarter, set up in the capital city of the country in question. In this regard, some examples of banks include the BBVA Bank (which has its official headquarters in Bilbao, and its operational headquarters in Madrid), and the Santander Bank

Table 11.3 Location of important service companies' headquarters by the number of employees

	40,000–70,000	10,000–40,000	5,000–10,000	2,500–5,000	1,000–2,500	<1,000
Madrid	3	4	10	11	28	116
Barcelona	1	5	1	10	14	93
Valencia		1		4	3	23
Bilbao		2		4	4	11
Zaragoza	3		1			15
Seville				1	3	15

Source: Prepared by the author. AE contrasted data

(having its official headquarters in Santander, a relatively small city, and an important concentration of its central services in Madrid).

A study on large service companies in Spain and their location has yielded interesting results. 584 large firms were analysed, excluding banks and savings banks as they were considered as a particular case. In the distribution of headquarters obtained in this research, Madrid is at the top of the ranking (Table 11.1), with 192 service companies of various sizes, followed by Barcelona, with 125. At a considerable distance, we find Valencia (28) and Bilbao (21), followed by Zaragoza, Seville, and Pontevedra-Vigo (all of them with 19). This high concentration in Madrid is even more pronounced when the companies' turnover is taken into consideration. In this case, the Spanish capital city is far above the rest, with 248 companies, which is twice the number of companies located (Table 11.3) in Barcelona (124).

Several reasons explain the high number of headquarters of large service firms located in Madrid. One of these reasons refers to the fact that Madrid is the country capital, which allows it to take advantage of the so-called "capital city effect", a common attraction effect for the majority of capital cities. The attraction held by capital cities is based not only on the accessibility to the Central Administration and state bodies and agencies (regulatory agencies; contact with the Administration services which provide projects; apply for services; get information regarding projects, needs; etc.), but also on the relationships with politicians, including those associated with the government and the party or parties supporting it, and those belonging to the opposition parties.

On the other hand, state capital cities have also a "representation" and prestigious effect (at least apparently) in the perception of users and clients. And last, but not least, capital cities have, despite some exceptions, an important dimension in terms of population and economic weight, as well as wealth of infrastructures, services, and institutions (from universities to business and professional associations), which are frequently far above the rest of the cities in the country. In the case of Madrid, and in Paris and London to a greater extent, we must consider the facilities for international relations and contacts as positive factors (e.g., international airports with connections to all continents; centralised railway network; motorway network; the existence of foreign multinational companies). All this is reinforced in Madrid due to the relations with Latin America and the Mediterranean countries.

Strategic importance is an issue to be taken into consideration regarding the location of headquarters of large companies in any city. Setting up the headquarters of such companies undoubtedly involves quite positive outcomes for any metropolitan area or city. In practice, this leads authorities to try to attract new headquarters and avoid their relocation to other places.

There are many reasons for a city to wish to host the headquarters of large firms, but three stand out in particular. Firstly, attention must be brought to the fact that such headquarters attract and demand qualified workers and professionals (human capital) to fill the jobs offered. Secondly, their presence in a city contributes to boost its prestige thereby attracting possibly other companies and activities of all kinds. Finally, from the strategic and political viewpoints, the importance of such a city or metropolitan area increases at a national and an international level. It seems that, if there is a limited number of headquarters of important companies, or if these decrease over time, the weight of the region or city tends to be undeniably lower as opposed to others holding a growing attraction for companies to be located in them (Vives, 2002).

The phenomenon towards a growing concentration noted in the French, Spanish and UK cases have also been observed in some analyses referred to other countries. In United States, a research carried out over more than 5,000 listed companies (Shilton & Stanley, 1999) demonstrated that more than 40 % of headquarters were located in just twenty counties, and a special concentration by sectors could be observed, particularly in the case of industry. As for services, companies related to technological developments and business service companies were highly concentrated in the west (California and Washington State), while financial companies and business service companies were particularly concentrated in various eastern cities (New York, Boston, Washington, Philadelphia). Another study referred to North America (Kolko, 1999) revealed that business service companies and management-related companies are set up in large cities. In the case of Europe, the trend towards the concentration of headquarters of service companies can be observed in all countries at a national level and virtually always in favour of the capital city of the country. Moreover, there are some European cities with a worldwide scope which have recorded a remarkable concentration of headquarters of large service companies. London, Paris, and Amsterdam,[12] followed by Munich, Stockholm, Zurich, and Brussels, stand out in particular.

4 Final Remarks

This chapter is divided into two parts: firstly, in Sect. 2, attention was paid to exploring which theoretical approaches have been or can be applied to the location of services; and, secondly, in Sect. 3, the analysis was focused on specific cases of location, taking as a reference some significant service industries on account of their importance in the economies and their dynamics.

[12] See RECLUS-DATAR (1989) and Rozenblat and Cicille (2003).

The main conclusion drawn from the section regarding the available theoretical contributions is that their number is not very high. Moreover, it is worth mentioning that changes that have taken and are taking place in recent years are not easy to be classified on the basis of some traditional approaches. It is clear that Geography and Urban Economy do currently feature a great amount of conceptual and theoretical elements which contribute to understand the services location processes. In particular, the differentiation made between inter- and intra-urban locations seems to be a requirement to clarify any analysis regarding the location of services.

Section 3, has allowed us to analyse three of the most interesting cases: retailing and distribution, hotel and catering activities, and business services and KIBS (one of its most important subsets). As a special case, we have also studied the location of headquarters of large service companies. The leitmotif in the case of business services, KIBS, and headquarters is *spatial concentration* not only at a wider territorial level (countries, regions), but also within metropolitan areas. However, this does not exclude some movements towards dispersion observed in business services and in the new retailing structures. In the case of hotels, a wide *dispersion* is combined with *concentration* in certain geographical areas offering attractive resources for tourism, leisure, and even sport. Such resources–generally natural—require in some way that the hotels are set up in the places where they are located, as the resources cannot be moved. At the same time, some new forms of travelling and transport facilities, the undertaking of business activities, urban tourism, and meetings and congresses are bringing about a high concentration of hotels in large cities.

As a final remark, we must point out that, although the studies on trends in the location of the different types of services have increased in recent years, additional attention and new contributions are required, as on-going changes (internationalisation; integration of ICTs; reduction of transport costs; changes in the population's behaviour, and other) constitute a challenge that requires further research.

References

Alonso, W. (1964). *Location and land use. Toward a general theory of land rent*. Cambridge, MA: Harvard University Press.

Alvesson, M. (2004). *Knowledge work and knowledge-intensive firms*. Oxford: Oxford University Press.

Anas, A., Arnott, R., & Small, K. A. (1998). Urban spatial structure. *Journal of Economic Literature, 36*, 1426–1464.

Arbel, A., & Pizam, A. (1977). Some determinants of urban hotel location. *Journal of Travel Research, XV*(3), 18–22.

Beguin, H., & Thill, J.-C. (1985). Une approche théorique: les aires de rayonnement des villes belges. *Bulletin trimestriel du Crédit Communal de Belgique, 154*, 205–215.

Beyers, W., & Alvine, M. (1985). Export services in postindustrial society. *Papers in Regional Science, 57*, 33–45.

Boschma, R., & Weterings, A. (2005). The effect of regional differences on the performance of software firms in The Netherlands. *Journal of Economic Geography, 5*(5), 567–588.
Camagni, R. (1992). Organisation économique et réseaux des villes. In P.-H. Derycke (Ed.), *Espaces et dynamiques territoriales*. Paris: Economica.
Camagni, R. (2005). *Economia urbana*. Barcelona: A. Bosch Edit.
Cesaer & Cemagref. (2008). *Localisation des services résidentiels. Analyse de la territorialisation de l'économie résidentielle à l'échelle intra-métropolitaine* (Rapport pour la DATAR). Paris.
Christaller, W. (1933/1966). *Die Zentralen Orte in Südeutschland*. Iena: G. Fischer. English version: *Central places in southern Germany*. Englewood Cliffs: Prentice Hall.
Coffey, W., & Polèse, M. (1987). Trade and location of producer services. A Canadian perspective. *Environment and Planning, 19*, 597–611.
Converse, P. D. (1938). *The elements of marketing*. New York: Prentice Hall.
Cook, G. A., et al. (2007). The role of location in knowledge creation and diffusion. Evidence of centripetal and centrifugal forces in the city of London financial services agglomeration. *Environment and Planning, 39*, 1325–1345.
Cooke, P., Heidenreich, M., & Braczyk, H. J. (Eds.). (2004). *Regional innovation systems*. London: Routledge.
Cuadrado-Roura, J. R. (2012). ¿Es tan 'nueva' la Nueva Geografía Económica?. Sus aportaciones, sus límites y su relación con las políticas. IAES WP 1/2012. Madrid. http://www.iaes.es.
Cuadrado-Roura, J. R., & López, J. M. (2011). El turismo un sector clave en la economía española. *Papeles de Economía E, 128*, 2–20.
Cuadrado-Roura, J. R., & Maroto-Sánchez, A. (2010). *Servicios y regiones en España*. Madrid: Funcas.
Cuadrado-Roura, J. R., & Rubalcaba, L. (2000). *Los servicios a empresas en la industria española*. Madrid: Instituto de Estudios Económicos.
Cuadrado-Roura, J. R., Rubalcaba, L., & Bryson, J. (2002). *Trading services in the global economy*. Cheltenham: E. Elgar.
Daniels, P., & Bryson, J. R. (2005). Sustaining business and professional services in a second city region: The case of Birmingham, IK. *Service Industries Journal, 25*(4), 505–524.
De Roo, O. (1993). La metropolité. en: A. Salley (director). *Les Villes lieux d'Europe*. París: Datar.
Doloreux, D., Amara, N., & Landry, R. (2008). Mapping regional and sectorial characteristics of KIBS: Evidence from the province of Quebec. *Growth and Change, 39*(3), 464–496.
Drennan, M. (1987). New York in the world economy. *Survey of Regional Literature, 4*, 7–12.
ECM—European Cities Marketing. (2010). *European cities marketing benchmarking report 2010*. Dijon: ECM.
Ecorys-NEI. (2004). *Business services: Contributing to growth and productivity in the EU* (Technical report). European Commission, Enterprise D.G.
Engländer, O. (1924). *Theorie des Güterverlehrs unde der Frachtsätze*. Jena: G. Fischer.
Finlay, A., & Sparks, L. (2008). Weaving new retail and consumer landscapes in the Scottish borders. *Journal of Rural Studies, 24*(1), 86–97.
Finlay, A. M., et al. (2001). Mobility as a driver of change in rural Britain: An analysis of the links between migration, commuting and travel to shop patterns. *International Journal of Population Geography, 7*, 1–15.
Fischer, H. M., & Frolich, J. (Eds.). (2001). *Knowledge, complexity and innovation systems*. Heidelberg: Springer.
Fujita, M., Krugman, P., & Venables, A. (1999). *The spatial economy*. Cambridge, MA: MIT.
Gallego, J., & Maroto, A. (2010). Análisis de la localización regional de los servicios intensivos en conocimiento a nivel europeo. *Investigaciones Regionales, 17*, 71–92.
Garrido, R. (2007). *Localización y movilidad de empresas en España*. Madrid: OEI.
GIP-Reclus. (1989). *Les Villes Européennes*. Paris: Datar.
Glaeser, E. L., & Gottlieb, J. D. (2009). The wealth of cities: Agglomeration economies and spatial equilibrium in the US. *Journal of Economic Literature, 47*(4), 953–1028.

Glaeser, E. L., & Kohlhase J. E. (2004). Cities, regions and the decline of transport costs. Papers in Regional Science, *83*, 197–228.
Gorter, C., Nijkamp, P., & Klamer, P. (2003). The attraction of out-of-town shopping malls: A case study on run-fun shopping in the Netherlands. *Tijdschrift voor Economische en Sociale Geografie, 94*(2), 219–229.
Hayter, R. (1997). *The dynamics of industrial location*. Chichester: Wiley.
Hoover, E. M. (1936). The measure of industrial localization. *The Review of Economics and Statistics, 18*, 162–171.
Hoover, E. M. (1948). *The location of economic activity*. New York: McGraw-Hill.
Howells, J., & Green A. G. (1986). *Location, technological innovation and structural change in UK services* (FAST Occasional Papers, 101). Brussels: European Commission.
Huff, D. (1963). A probabilistic analysis of shopping centre trade areas. *Land Economics, 53*, 81–90.
Hurd, R. (1903). *Principles of city land values*. New York: The Record and Guide.
Illeris, S. (1989). *Services and regions in Europe*. Aldershot: Gower Publ Company Ltd.
Innis, H. (1920). *The fur trade in Canada*. New Haven: Yale University Press.
Isard, W. (1956). *Location and space-economy*. Cambridge, MA: MIT.
Johnson, C. B., & Vanetti, M. (2005). Las estrategias de localización de las cadenas hoteleras internacionales. *Annals of Tourism Reearch en Español, 7*(2), 351–376.
Jouvaud, M. (1995). *Localisation des services aux entreprises: Combination de facteurs et types d'activités*. Notes du CER, 173, 1995/4, Aix-en-Provence.
Klier, T., & Testa, W. (2002). Location trends of large company headquarters in the 90s. *Economic Perspectives* (QIV), 12–26.
Kller, T. H. (2006). Where the headquarters are: Location patterns of large public companies, 1990-2000. *Economic Development Quaterly, 20*(2), 117–128.
Koch, A., & Stahlecker, T. (2006). Regional innovation systems and the foundation of knowledge-intensive business services. A comparative study in Bremen, Munich and Stuttgart. *European Planning Studies, 14*(2), 123–146.
Kolko, J. (1999). *Can I get some service here?. Information technologies, service industries and the future of cities* (W.P.). Harvard University.
Krugman, P. (2000). Where in the world is the new economic geography. In G. L. Clark, M. P. Feldman, & M. Gertler (Eds.), *The Oxford handbook of economic geography*. Oxford: OUP.
Launhardt, W. (1882). Die Bestimmung des zweckmässigsten Standortes einer gewerbliche Anlage. *Zeitschrift des Vereins Deutcher Ingenieure, 26*, 106–115.
Léo, P.-Y., & Philippe, J. (2002). Retail centres: Location and consumer's satisfaction. *Service Industries Journal, 22*(1), 123–146.
Lösch, A. (1940). *Die raumliche Ordnung des Wirtschaft*. Iena G. Fischer (English version, 1954), *The economics of location*. New Haven: Yale University Press).
Lowry, I.S. (1964). *A model of a metropolis, RM-4035-RC*, Santa Monica: Rand Corporation.
Maggioni, M. A. (2002). *Clustering dynamics and the location of high-tech firms*. Heidelberg: Physica.
Mazanec, J., & Wöber, K. (Eds.). (2010). *Analysing international city tourism*. Heidelber: Springer.
McCann, P. (2001). *Urban and regional economics*. Oxford: Oxford University Press.
Morgan, K. (2004). The exaggerated death of geography: Learning proximity and territorial innovation systems. *Journal of Economic Geography, 4*, 3–21.
Moses, L. N. (1955). The stability of interregional trading patterns and I-O analysis. *The American Economic Review, XLV*(5), 803–832.
Mouhoud, E. M. (Director). (2010). *Économie des services et développement des territoires*. Paris. La Documentation Française, Datar.
Muller, E., & Doloreux, D. (2009). What we should know about knowledge-intensive business services. *Technology in Society, 31*(1), 64–72.

North, D. (1955). Location theory and regional economic growth. *Journal of Political Economy, 63*, 243–258.
Ono, Y. (2006). What do census data tell us about headquarters location? *Economic Development Quarterly, 20*(2), 129–141.
Palander, T. (1935). *Beiträge zur Standortstheorie*. Uppsala: Almqvist & Wiksell.
Pan, C. M. (2005). Market structure and profitability in the international tourist hotel industry". *Tourism Management, 26*(6), 845–850.
Papadopoulos, N. G. (1980). Consumer out-shopping research: Review and extension. *Journal of Retailing, 56*(4), 41–58.
Pinkerton, J. R., Hasinger, E. W., & O'Brien, D. J. (1995). Inshopping by residents of small communities. *Rural Sociology, 60*(3), 467–480 (Quoted by Van Leeuwen, & Rietveld, 2010).
Powe, N. A., Hart, T., & Bek, D. (2009). Market towns in England: Meeting the challenge of maintaining their contemporary relevance. *Planning Practice and Research, 24*(3), 301–319.
Predöhl, A. (1925). Das Standortsproblem in der Virtschaftstheorie. *Welwirtschaftliches Archiv, 21*, 294–331.
Reclus-Datar. (1989). *Les villes européennes*. Paris: DATAR.
Reilly, W. J. (1929). *Methods of the study of retail relationship* (University of Texas Bulletin, n. 2944)
Rietveld, P., & Vickerman, R. (2004). Transport in regional science: The 'death of distance' is premature. *Papers in Regional Science, 83*, 229–249.
Rigall, R., Rigall, R., et al. (2011). The effects of beach characteristics and location with respect to hotel prices. *Tourism Management, 32*(5), 1150–1158.
Rozenblat, C., & Cicille, P. (2003). *Les Villes Européennes. Analyse comparative*. Paris: DATAR.
Rubalcaba, L., & Cuadrado-Roura, J. R. (2001). Spain: KIS. A paradigm of economic change. In P. Wood (Ed.), *Consultancy and innovation. The BS revolution in Europe*. London: Routledge.
Rubalcaba, L., & Kox, H. (Eds.). (2007). *Business services in European economic growth*. Hampshire: Palgrave MacMillan.
Rubalcaba, L., & Cuadrado-Roura, J. R. (2002). Services in the age of globalization: Explanatory interrelations and dimensions. In: J. R. Cuadrado-Roura, L. Rubalcaba, & J. Bryson (Eds.) (pp. 27–57).
Sassen, S. (1991). *The global city. New York, London, Tokyo*. Princeton: Princeton University Press.
Sassen, S. (1994). *Cities in a world economy*. Thousand Oaks: Pine Forge press.
Schenk, T. A., Löffler, G., & Rauh, J. (2007). Agent-based simulation of consumer behaviour in grocery shopping on a regional level. *Journal of Business Research, 60*, 894–903.
Shilton, L., & Stanley, C. (1999). Spatial patterns of headquarters. *Journal of Real State Research, 17*(3), 341–364.
Shoval, N. (2006). The geography of hotels in cities: An empirical validation of a forgotten model. *Tourism Geographies, 8*(1), 56–75.
Sokol, M., Van Egeraat, C., & Williams, B. (2008). Revisiting the informational city: Space of flows, policentrity and the geography of knowledge-intensive business services in the emerging global city-region of Dublin. *Regional Studies, 42*(8), 1133–1146.
Souza, A. R., & Stutz, F. P. (1994). *The world economy. Resources, location, trade and development*. Englewood Cliffs, NJ: McMillan.
Sund, K. J. (2006). The geographical distribution and concentration of hotels in Switzerland. In K. Weiermair, H. Pechlaner, & T. Bieger (Eds.), *Time shift, leisure and tourism* (pp. 317–324). Berlin: E. Schmidt.
Taylor, P. J. (2004). *World city network: A global urban analysis*. London: Routledge.
Taylor, P. J. (2012). *Extraordinary cities: Millenia of moral syndromes, world-systems and city/state relations*. London: Routledge.
Taylor, P. J., Catalano, B., & Gane, N. (2003). A geography of global change: Cities and services, 2000-2001. *Urban Geography, 24*, 431–441.

Taylor, P. J., Evans, D. M., Hoyler, M., Derudder, B., & Pain, K. (2009). The UK space economy as practiced by advanced producer service firms: Identifying two distinctive polycentric city-regional processes in contemporary Britain. *International Journal of Urban and Regional Research, 33*, 700–718.

Taylor, P. J., Ni, P., Derudder, B., Hoyler, M., et al. (Eds.). (2011). *Global urban analysis: A survey of cities in globalization*. Earthscan: London.

Thompson, J. R. (1971). Characteristics and behavior of out-shopping consumers. *Journal of Retailing, 47*(1), 70–80.

Urtasun, A. (2006). Hotel location in tourism cities, Madrid 1936–1998. *Annals of Tourism Research, 33*(2), 382–402.

Van Leeuven, E. S., & Rietveld, P. (2011). Spatial consumer behavior in small and medium-size towns. *Regional Studies, 45*(8), 1107–1119.

Veltz, P. (1996). *Mondialisation, villes et territoires. L'économie de l'archipel*. Paris: P.U.F.

Vives, X. (2002). Centres de decisió empresarial i activitat económica: Els efectes de la globalització (*Papers d'economia industrial*, 17). Barcelona: Generalitat de Catalunya.

Von Thünen, J. H. (1875) (firstly published in 1826). *Der Isolierte Staat in Beziehung auf Landwirtschaft und Nationalökonomie* (3rd ed.). Berlin: Wiegardt, Hempel & Parey.

Weber, A. (1909). *über den Standort der Industrien, I: Reine Theorie des Standorts*. Tubingen (Translated into English as: *Theory of the Location of Industries*, 1929) Chicago: The University of Chicago Press.

Wood, P. (2002). *Consultancy and innovation: The business service revolution in Europe*. London: Routledge.

Wrigley, N. (2003). The globalization of retail capital: Themes for economic geography. In G. Clark, M. P. Feldman, & M. S. Gertler (Eds.), *The Oxford handbook of economic geography*. Oxford: Oxford University Press.

Zipf, G. K. (1949). *Human behaviour and the principle of least effort: An introduction to human ecology*. Cambridge, MA: Addison-Wesley Press.

Chapter 12
European Cities in Globalization: A Comparative Analysis Based on the Location Strategies of Advanced Producer Services

Peter J. Taylor, Michael Hoyler, and Simón Sánchez-Moral

1 Introduction: Comparisons Between Countries and Relations Between Cities

Today there is a key question that lurks behind any consideration of Europe and its cities: is this foundation core zone of the modern world-system showing symptoms of dropping out of the contemporary core zone? It certainly appears that in the period of crises since 2008, Europe has been falling behind other major world-regions. Dubbed the "austerity region" of the world, such an interpretation sees Europe as the first part of the world-economy core to be subject to what are effectively structural adjustment programmes, largely self-imposed but still resulting in a process of peripheralization. Although uneven in impact, this is clearly a result of Europe's states failing to adequately manage and regulate the economic activities within their territories. However it is far too soon to say whether such a monumental global economic shift is happening but we can investigate the current unevenness of economic globalization amongst European states. We compare three of these states that represent different degrees of potential peripheralization: Spain showing the stronger symptoms, Germany with least symptoms, and Britain somewhere in between. Our study is based upon an original analysis of advanced producer services that combines comparisons between countries and relations between cities.

In this paper we take a Jacobs' (1969, 1984, 2000) view of economic development that treats states as economic jurisdictions but not actual functioning economies; cities (and city-regions) are the critical scale of economic process that

P.J. Taylor (✉)
Human Geography, Northumbria University, Newcastle upon Tyne, UK
e-mail: crogfam@yahoo.com

M. Hoyler
Department of Geography, Loughborough University, Loughborough, UK

S. Sánchez-Moral
Department of Regional Geographical Analysis, Complutense University, Madrid, Spain

J.R. Cuadrado-Roura (ed.), *Service Industries and Regions*,
Advances in Spatial Science, DOI 10.1007/978-3-642-35801-2_12,

expands and develops economies. Thus states are "grab bags of very different economies, rich regions and poor ones within the same nation" whereas "cities are unique in their abilities to shape and reshape the economies of other settlements, including those far away from them geographically" (Jacobs, 1984, p. 32). It follows that if Europe is to address its potential peripheralization it will have to do so through economic expansion of its cities; Europe's cities need to become dynamic again to match the cities of the more successful world regions. Our comparison of countries, therefore, is between the leading cities of Spain, Britain and Germany.

By focusing on cities we introduce relational thinking into the analysis. The vibrancy of dynamic cities derives from two distinct but related urban processes. Each generates an externality, a benefit beyond the market, a positive bonus for locating economic activity in the city. First there are dense patterns of intra-city relations that create agglomeration effects and cluster advantages. Second there are strong flows of inter-city relations that create network effects and connectivity advantages. All successful cities combine these agglomeration externalities and network externalities to maximise the bonus of an urban location. There is a massive and sophisticated literature on agglomeration effects (e.g. Fujita & Thisse, 2002), but network externality effects have been much less studied. We concentrate on the latter as part of the Globalization and World Cities (GaWC) research programme where the study of network externalities has been pioneered. Specifically we draw on results from the research report *Global Urban Analysis* (Taylor et al., 2011) that presents findings from analysis of the 2008 survey of leading advanced producer service firms by GaWC in association with the Chinese Academy of Social Sciences.

2 Methodology

Contemporary network externalities are a function of world city network formation. This process has been modelled as an interlocking network. Networks usually consist of two layers, the net level and the node level. Formal city government associations work in this way with the cities (members) as nodes, the city association represents the net level, and the formal relations between members within the association define the network. Such networks can be an important component of global governance but this is not how cities operate as key components of the global economy. In the latter, it is advanced producer service firms that are the network makers; they create the world city network through their everyday practices linking offices across the world. This defines a different type of network, an interlocking network that is unusual in having three layers. In the case of the world city network there is the net level of the global economy, the node level of cities, and an additional sub-nodal level of service firms. The latter are not just an additional level, they define the critical level: this is where the agents of network formation are found. In the global economy, it is firms who are the network makers not the cities themselves. Thus for studying the world city network it is service firms that are investigated in order to understand the city network as the outcome. In other words, it is through studying the locational strategies of firms that it is possible to measure and analyse the world city network.

Why focus on these service firms? In the 1970s two separate industries, computers and communications, merged their technologies to enable work to be coordinated worldwide based upon simultaneous connections. Early on Sassen (1994) spotted two contrasting economic geography effects: first, a dispersal of production to cheaper labour locales, and second, a contrary trend towards concentration of management and business service industries. The latter were required to organize the new worldwide production and were concentrated in cities. As Sassen (1991) originally argued, it is concentration of management alongside financial, professional and creative services that characterises contemporary "global cities". Of course, service firms have always clustered in cities to provide such services to their clients but under conditions of contemporary globalization those specialised services became worldwide with fundamental implications for work practices. Firms need a multiple office policy across many cities to provide a seamless service and protect global brand integrity by keeping all work in-house.

This is how it came to be that from the 1980s onwards there have been hundreds of large service firms with trans-national office networks, many of them global in scope. Each firm had its own locational strategy—which cities to have offices in, what size and functions those offices will be, and how the offices will be organised. It is the work done in these offices that "interlock" various cities in projects that require multiple office inputs. Thus the inter-city relations in these servicing practices are numerous electronic communications—information, instruction, advice, planning, interpretation, strategy, knowledge, etc., some tele-conferencing as required, and probably travel for face-to-face meetings at a minimum for the beginning and end of a given project. These are the working flows that combined across numerous financial, professional and creative projects in multiple firms to constitute the world city network (Taylor, 2001, 2004).

So we have to study service firms to describe and analyse the world city network but, unfortunately, there is no feasible way that data could be collected from firms on these working flows. As well as the obvious confidentiality issues with competing private firms, there is also a feasibility issue: the degree of research collaboration that would be needed from a large number of firms makes such a data collection exercise beyond reasonable social science research logistics. However, this is not a particularly rare situation in measurement practices: where direct measures cannot be obtained, there is the fall back position of carrying out indirect measurement. This requires access to more easily available data plus credible assumptions about how the firms operate.

As mentioned previously, service firms offer a seamless service across their office networks. This means that the geographical distribution of their offices, and their scope and range, are important selling points in attracting new clients. Hence such information is commonly available on service firms' web sites. This has been the main source of data for measuring the world city network: for each firm, offices are assessed individually by asking what is the importance of this office in this city within the firm's overall office network? Answers to this question are termed the service value of a city to a firm. These values are coded and become the quantitative input into the study: the coding ranges from 0 (a firm having no office in a city) to 5 (a city housing the headquarters of a firm); standard or typical offices of a firm score 2, minor and major offices 1 and 3, respectively, leaving 4 for scoring cities housing exceptionally important offices such as regional headquarters. The credible assumption that is made is that the more important an office the more working flows it will generate. Therefore two important offices will generate a much higher level of flow between their respective cities than two minor offices between their respective cities. These data and this assumption are combined to generate estimates of inter-office working flow levels between cities for each firm; they are not actual working flows, but potential working flows, indirect measures derived from the data and the model assumptions. Aggregating all potential working flows for all firms located in a city generates estimates of its working flow relations with other cities; when this is done for all cities it constitutes the world city network.

Network connectivity is the main measure of importance of a city in this model (Taylor, 2001). It is computed from the products of service values for the city with each other city for all firms. Thus assuming m advanced producer service firms and n cities we can define a service value for firm j in city i as v_{ij}. The basic relational unit of measurement is given by

$$r_{ab,j} = v_{aj} \cdot v_{bj} \tag{12.1}$$

which defines the relation between cities a and b in terms of firm j. This is an elemental interlock between two cities for one firm. The aggregate cities interlock between the cities is then given by

$$r_{ab} = \sum r_{ab,j} \tag{12.2}$$

For each city there are n-1 such interlocks and the network connectivity for a city is given by

$$C_a = \sum r_{ai} \quad \text{where } a \neq i \tag{12.3}$$

where C_a is the network connectivity of city a. This relates city a to all other cities within the network through its firms and measures the degree of integration of the city into the world city network.

The data collection to operationalize this model and provide the results reported below was carried out in 2008 (Taylor, et al., 2011). Office networks were coded for 175 advanced producer firms chosen as leading firms in their respective sectors: the top 25 in accountancy, advertising, law and management consultancy, and the top 75 firms for financial services (banking, insurance and diversified finance). For accountancy and advertising, firm choice was based on global ranks by revenue in World Accounting Intelligence (www.worldaccountingintelligence.com) and Advertising Age (www.adage.com). Law firms are those ranked highest by chambersandpartners.com, and management consultancies are leading firms by "prestige" as identified by www.vault.com. The 75 financial firms are those ranked top in the Forbes Global 2000 list (www.forbes.com). Their offices were scrutinized across 525 cities worldwide. The end result is a 525 cities × 175 firms matrix with each cell indicating the importance of a specific city in the office network of a specific firm, 91,875 service values in all.

From this large amount of customized data we compute network connectivities of cities as defined in Eq. (12.3) to show the degree of a city's integration into the world city network. The values computed from Eq. (12.3) are relatively large and therefore to make them easier to interpret we present them as proportions of the highest scoring city. In addition we disaggregate these network connectivities in two ways.

1. By sector. This allows us to see the connectivity of a city generated by firms in just one sector. Thus we produce financial connectivities from the 75 firms in this sector, and accountancy connectivities, advertising connectivities, law connectivities and management consultancy connectivities from the 25 firms in each of the sectors. These connectivities are also presented as proportions of the largest connectivity.
2. By geographical area. This allows us to explore the "hinterworlds" of cities (Taylor, 2004). Here we focus on six connectivities based upon links to Pacific Asian cities, to Northern American cities and to European cities at one scale, and to New York and London, to Beijing, Hong Kong and Shanghai, and to other cities in a city's home country (i.e. Spain, Britain or Germany). These connectivities are shown as standardized variables (mean = 0 and standard deviation = 1) to indicate where a city is relatively over-linked and relatively under-linked.

Results are shown and discussed for Europe as a whole first before focusing on our three chosen countries. For the latter we focus on just the top five cities in terms of network connectivity in each country.

3 European Cities in the World City Network

Although it can be argued that the European Union has provided the general economic framework for the region's cities to prosper, their role as economic units has been largely neglected in European policy circles (van den Berg, Braun, van der Meer, & Mingardo, 2007). European spatial planning (Faludi, 2002) has only very recently begun to address questions of globalization, mainly as international competitiveness. Thus although the EU now stretches across most of Europe from the Atlantic to the Black Sea, it has had little direct effect on European cities except where its major institutions are located, Brussels as "capital of Europe" (Baeten, 2001; de Groof, 2008; Elmhorn, 2001), and to a much lesser extent, Luxembourg, Strasbourg and Frankfurt.

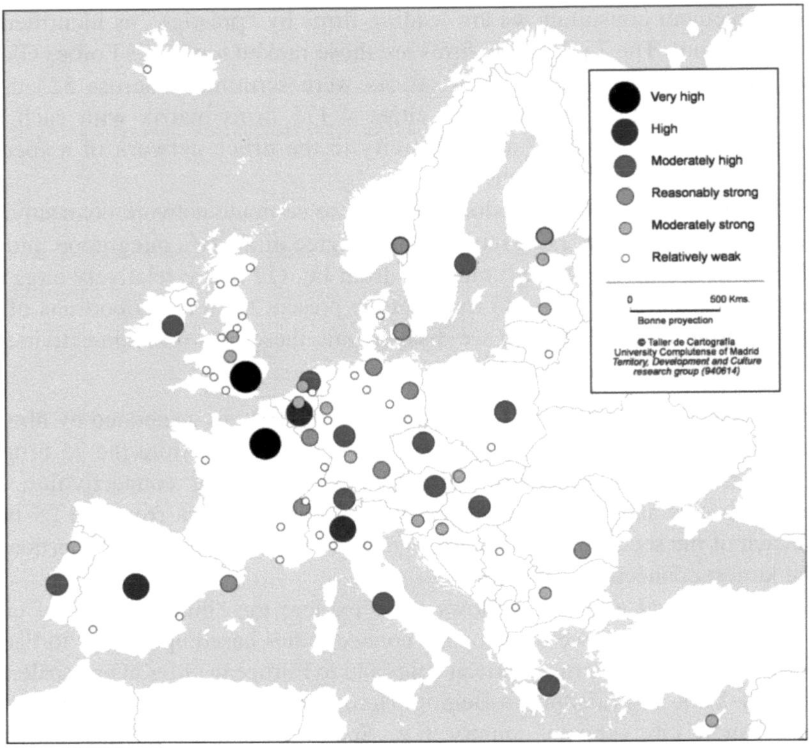

Fig. 12.1 Global network connectivities of European cities

The evidence provided in this section consists of three sets of results. First, the cities with global network connectivities above 0.1 define the 74 cities that are depicted in Fig. 12.1; using the actual connectivity proportions, gaps in these values are employed to identify six strata of cities. In addition, the top 25 ranks, designated the "leading" European cities in the world city network, are listed in the first city ranking in Table 12.1. Second, the city connectivities are disaggregated by service sector producing five new rankings: for financial services the top 25 cities are shown as the second ranking in Table 12.1; for accountancy, advertising, legal services and management consultancy the top 10 cities are shown in Table 12.2. Third, city connectivities are disaggregated by geographical orientation in Table 12.3, showing over- or under-linkage to other European cities, and over- or under-linkage to the other two main globalization arenas, to Northern American cities and to Pacific Asian cities.

Figure 12.1 shows the many European cities integrated into the world city network to varying degrees—this world region has more such cities than any other world region (Taylor et al., 2011). This is partly because Europe remains a region of multiple states and globalizing service firms will want a presence in different states to tap into "national" markets. Very often this is accomplished by

Table 12.1 Top 25 European cities for global network connectivity (GNC) and financial network connectivity (FNC)

Rank	City	GNC	City	FNC
1	London	1	London	1
2	Paris	0.78	Paris	0.79
3	Milan	0.69	Madrid	0.7
4	Madrid	0.65	Milan	0.7
5	Brussels	0.63	Frankfurt	0.61
6	Warsaw	0.56	Zurich	0.6
7	Zurich	0.55	Brussels	0.57
8	Amsterdam	0.55	Amsterdam	0.56
9	Dublin	0.54	Dublin	0.56
10	Rome	0.53	Warsaw	0.5
11	Lisbon	0.52	Stockholm	0.44
12	Frankfurt	0.5	Geneva	0.43
13	Stockholm	0.49	Luxembourg	0.41
14	Prague	0.49	Prague	0.4
15	Vienna	0.48	Athens	0.39
16	Budapest	0.48	Lisbon	0.39
17	Athens	0.48	Rome	0.35
18	Barcelona	0.42	Budapest	0.33
19	Bucharest	0.4	Vienna	0.32
20	Oslo	0.4	Munich	0.29
21	Berlin	0.39	Berlin	0.23
22	Helsinki	0.39	Bucharest	0.22
23	Geneva	0.38	Barcelona	0.22
24	Copenhagen	0.37	Düsseldorf	0.2
25	Hamburg	0.37	Bratislava	0.2

locating an office in a country's capital city, hence the map being dominated by capital cities: thus 14 of 17 cities recorded in the 3 "high" categories in Fig. 12.1 are capitals, 13 out of 24 in the 2 "strong" categories, but only 4 out of 32 in the "weak" category. This domination of capital cities among the more connected cities is specifically shown in the first ranking list of Table 12.1 showing global network connectivities for leading European cities. With its score of 1, this shows London to be the most connected city in the world. With Paris, the two leading European cities are indisputably "global cities" as famously described by Sassen (1991). For the rest of the table we divide the cities into groups at a much finer level than the map strata. Below Paris, Milan, Madrid and Brussels form a distinctive group with global network connectivities around two-thirds of the maximum. Brussels is not a surprise but the two southern European cities are less predictable at this level. The next group contains Warsaw, Zurich, Amsterdam, Dublin and Rome. These cities represent, in order, the post-communist rise of an Eastern European capital city, two important traditional financial centres, one of the major success stories of globalization, and the capital city of one of Europe's largest countries. The following group continues with a mix of capital cities of medium-sized countries (Lisbon, Stockholm, Vienna and Athens), including two—Prague and Budapest—from the former communist east, plus Frankfurt. The latter, another traditional financial

Table 12.2 Leading cities for network connectivity in accountancy, advertising, law and management consultancy

Rank	City	Accountancy	Rank	City	Advertising
1	London	1	1	London	0.75
2	Milan	0.67	2	Paris	0.75
3	Paris	0.66	3	Warsaw	0.63
4	Brussels	0.59	4	Brussels	0.62
5	Lisbon	0.57	5	Athens	0.6
6	Rome	0.57	6	Stockholm	0.6
7	Berlin	0.56	7	Madrid	0.6
8	Madrid	0.56	8	Milan	0.6
9	Oslo	0.55	9	Budapest	0.57
10	Barcelona	0.55	10	Vienna	0.56

Rank	City	Law	Rank	City	Management consultancy
1	London	1	1	London	0.67
2	Paris	0.7	2	Paris	0.65
3	Frankfurt	0.59	3	Zurich	0.55
4	Brussels	0.54	4	Madrid	0.55
5	Amsterdam	0.4	5	Rome	0.48
6	Munich	0.4	6	Amsterdam	0.47
7	Milan	0.39	7	Dublin	0.47
8	Madrid	0.37	8	Frankfurt	0.47
9	Warsaw	0.34	9	Milan	0.46
10	Düsseldorf	0.32	10	Stockholm	0.44

centre, is interesting as it is Germany's first ranked city in terms of global network connectivity. Germany has by far the largest economy in Europe but has no city in the top 10 in Table 12.1. This reflects the country's very "horizontal" urban system, relating to its federal political structure and to the fact that its capital city, Berlin, was a divided city during the Cold War and has yet to fully recover economically (Cochrane & Jonas, 1999; Krätke, 2001). Berlin appears in the next stratum along with (1) other political capitals of much smaller countries (Oslo, Helsinki, Copenhagen) including another from the east (Bucharest), and (2) another German city (Hamburg) and (3) another traditional financial centre (Geneva).

The other ranking in Table 12.1 shows the connectivity produced just by financial service firms. Notice the distribution of connectivity levels for these services is different from the general results. Although levels of connectivity are similar for the leading cities, city connectivities for finance fall away rapidly after the ninth rank indicating that financial service firms alone generate a much more concentrated pattern of connectivity in Europe. Realistically, therefore, there were just nine international financial centres in 2008 and since then, the ninth, Dublin, has reduced to below this standard. Looking at the specifics, the top two remain the same, London and Paris, but Madrid now ranks above Milan. The biggest riser is Frankfurt that, with Zurich, is now more connected than Brussels. Also Luxembourg, Munich and Düsseldorf join the top 25.

Table 12.3 Relative links to major world regions: top 25 cities

Rank	City	Europe	City	Northern America	City	Pacific Asia
1	Hamburg	3.52	Amsterdam	0.53	Frankfurt	2.46
2	Stockholm	2.86	London	0.1	London	1.86
3	Warsaw	2.67	Milan	0.07	Paris	1.41
4	Budapest	2.54	Zurich	0	Amsterdam	1.15
5	Copenhagen	2.45	Madrid	−0.02	Milan	0.76
6	Bucharest	2.41	Paris	−0.06	Madrid	0.6
7	Oslo	2.41	Dublin	−0.09	Prague	0.6
8	Lisbon	2.36	Frankfurt	−0.23	Geneva	0.34
9	Vienna	2.36	Rome	−0.3	Brussels	0.32
10	Zurich	2.27	Stockholm	−0.57	Dublin	0.07
11	Brussels	2.2	Brussels	−0.87	Warsaw	0.02
12	Rome	2.14	Berlin	−1.05	Budapest	−0.07
13	Barcelona	2.14	Oslo	−1.12	Zurich	−0.2
14	Berlin	2.09	Geneva	−1.14	Vienna	−0.24
15	Prague	1.84	Hamburg	−1.16	Lisbon	−0.57
16	Frankfurt	1.75	Barcelona	−1.2	Rome	−0.6
17	Madrid	1.62	Warsaw	−1.46	Athens	−0.67
18	Helsinki	1.33	Lisbon	−1.61	Stockholm	−0.76
19	Dublin	1.27	Prague	−1.63	Bucharest	−1.12
20	Paris	1.17	Athens	−1.71	Berlin	−1.14
21	Athens	1.02	Vienna	−1.85	Oslo	−1.23
22	Amsterdam	0.93	Helsinki	−2.24	Barcelona	−1.24
23	Milan	0.8	Budapest	−2.53	Helsinki	−1.3
24	Geneva	0.14	Copenhagen	−2.58	Copenhagen	−1.42
25	London	−0.01	Bucharest	−3.51	Hamburg	−1.48

Table 12.2 shows top ten city connectivity rankings for the other advanced producer services. They can be divided into two sets: in accountancy and legal services, London is the global leader and therefore completely dominates other European cities; advertising and management consultancy are archetypal American contributions to the professional services and therefore New York dominates globally. In these cases Paris joins with London as the top European stratum of cities.

The two London dominated services are, however, very different in all other respects. Accountancy is the most ubiquitous of the services treated in this analysis and legal services are the most concentrated. This creates contrasting city service structures: accountancy scores are much "flatter" than those for law. For instance, cities ranked tenth have connectivities of 0.55 and 0.32 respectively. The specifics are best represented by German cities: the capital Berlin appears in the accountancy list reflecting a general market attraction; Frankfurt, Munich and Düsseldorf are in the law list reflecting a finance market attraction.

For the two New York dominated services, Paris joins London in the top stratum due to the relatively low level of London's connectivities for these services: London drops to Paris's general level rather than vice versa. Below these two cities the scores for advertising have a flatter distribution that is dominated by capital cities, or more generally by cities with TV stations that are the main market for this service. Scores for management consultancy show a more concentrated pattern of connectivities and tend to mirror the financial services connectivities with its mixture of financial centres and capital cities.

Table 12.3 shows the geographical orientation results for the top 25 cities (as indicated by overall network connectivity in Table 12.1). Not surprisingly for the local region, in the first list for European concentration of connections all cities bar one have positive scores, which means that they are relatively more connected to other European cities than to cities outside Europe. The odd one out is London with a very small negative score and this reflects the fact that the more important cities in Table 12.1 are less focused on connections to their European neighbours. Paris, Milan and Madrid are ranked in the bottom ten but Brussels is predictably ranked higher as the "capital of Europe". It is not unusual for major cities of a world region to be more orientated away from their region; this is what makes them world or global cities, and London has been so categorised as "un-European" previously (Taylor & Derudder, 2004; Taylor & Hoyler, 2000). Conversely less connected cities from Table 12.1 have relatively high connections to other European cities: Hamburg, Copenhagen and Oslo fit into this category. Geneva is the big exception with relatively low European links (ranked 24th) and low general connectivity (23rd in Table 12.1); which is the converse of Warsaw with relatively high European links (ranked 3rd) and high general connectivity (ranked 6th in Table 12.1). Therefore the "localism" of European city hinterlands is only loosely related to general connectivity; specific city functions (Geneva in international finance) and even location (Athens bordering other regions) is sometimes important.

Orientations to Northern American cities are shown in the second list in Table 12.3 and indicate that the vast majority of leading European cities are relatively under-linked to Northern America. The pattern is fairly straightforward with the cities ranking high on financial connectivity in Table 12.1 also being relatively over-linked to Northern America; the respective top tens in these lists largely overlap. Those relatively under-linked to Northern America are capital cities of smaller countries, the bottom nine on the list match this description. The orientations to Pacific Asian cities shown in the third list in Table 12.3 have a similar financial bias to those on the previous list but with higher positive scores. This reflects a much more balanced pattern between under- and over-linkage. Zurich is the exception among international financial centres suggesting its long established embedding in old financial markets is not being transferred to new financial centres in Pacific Asia. At the under-linked end of the scale capital cities no longer feature as prominently as for Northern American connections. Previous research has shown that Pacific Asian cities are especially strong in financial services (Taylor, 2004) and this new finding shows how this global pattern is strongly reflected in the hinterworlds of leading European cities.

In conclusion, Europe, as the cradle of modernity and for more contemporary reasons, has many cities well integrated into the world city network. However, the degree of integration varies greatly with London, then Paris, the most integrated cities, especially in the core world regions of economic globalization. Patterns vary between different service sectors—Frankfurt and Zurich rise for financial services, Milan and Lisbon for accountancy, Warsaw and Athens for advertising, Frankfurt and Amsterdam for law, and Zurich and Rome for management consultancy. When it comes to geographical orientations, connectivities to Northern American and Pacific Asian cities largely reflect the pattern of financial centres in Europe: it is this finance/ Pacific Asia link in city connectivities that may well be the key discriminating factor in the future economic successes of European cities.

4 British Cities in the World City Network

The UK space economy has long been notorious for its primate pattern of cities centred on London and the South East. For much of the twentieth century, UK governments pursued regional policies specifically to counter "the drift to the South" resulting from the decline of the industrial cities and towns of northern Britain from their nineteenth century economic prime. But such policies proved to have limited impact on the economic forces creating London's primacy. With the rise of neoliberal globalization from the late 1970s, the prospects for the cities collectively known as "not-London" seemed to have been further reduced: the demise of regional policy was followed by government policy that precipitated the City of London's "Big Bang". This opened up the City to foreign banks and other financial services to ensure London would become a key locale for on-going economic globalization. In 1991 Saskia Sassen announced that London, with New York and Tokyo, was an archetypal "Global City".

With this global position added to London's national dominance, it seemed that London's UK primacy in the new world of globalization would be greater than ever. And this was confirmed by the first measurement of global network connectivities in 2000 (Taylor, 2004; Taylor, Catalano, & Walker, 2002): London was ranked first globally and with no other UK city in the top 100 (Beaverstock, Hoyler, Pain, & Taylor, 2001). Other studies have highlighted the economic underperformance of UK provincial cities compared to their European counterparts (e.g. Parkinson et al. 2004; 2006). However, at the beginning of the twenty-first century there appeared to be a revival of provincial UK cities. New measures of global network connectivity in 2004 showed that UK cities had experienced some of the most rapid increases in global network connectivities in the world: Edinburgh, Bristol, Cardiff and Leeds being particular noteworthy in this respect (Taylor & Aranya, 2006). Further work has indicated that the UK space economy, while not becoming fully "balanced", has been developing inter-city networks to complement the continuing London-headed urban hierarchy. In other words, since 2000, major global service providers have found it worth their while to make use of UK provincial cities as well as London (Taylor, Evans, Hoyler, Derudder, & Pain, 2009; Taylor, Hoyler, Evans, & Harrison, 2010). This is the context in which to consider UK cities in the 2008 world city network analysis.

Table 12.4 Connectivity of leading UK cities and their "localism"

Rank	City	GNC	With other UK cities
1	London	1	−1.31
2	Manchester	0.22	2.47
3	Edinburgh	0.21	3.28
4	Birmingham	0.21	3.32
5	Glasgow	0.19	3.04

Table 12.4 shows global network connectivities for the top 5 UK cities to illustrate how well the leading British cities are integrated into the world city network. The outstanding result is no surprise: the continuing dominance of London. While other UK cities are still not major players in the world city network, there are now some moderately important world cities that can be identified with about one fifth of London's connectivity. Manchester, Glasgow and Birmingham have been in competition to be the UK's "second city" for more than a century and they continue to be leading cities in globalization but are now joined by Edinburgh, Europe's newest financial centre. Manchester and Birmingham are the centres of the two major economic regions outside the South East, the North West and West Midlands respectively, and are reinventing themselves as new European and world cities. Edinburgh is the fast riser based upon being the capital city of Scotland, the UK's main political devolution (with its new service needs), as well as being home to successful banks (before the credit crisis when these data were collected; see Derudder, Hoyler, & Taylor, 2011). Glasgow has traditionally been the economic centre for Scotland but may now be being overtaken by its neighbour Edinburgh; however it is still of some importance within contemporary globalization. The overall message of this table is not that any UK city is seriously rivalling London but that leading British cities across the country are integrated into the world city network to a moderate degree.

Nevertheless, the second list in Table 12.4 indicates that we should not take this argument for worldwide integration of UK provincial cities too far. This shows measures of "localism" at the national level, the degree to which a city's connectivity is dominated by links to other cities within the country. The list emphasizes the separation of London from the rest; the city is strongly under-linked in its relations to other British cities. With a large negative score, London is shown to be very "un-local": the vast majority of its connections are beyond the UK. The other four cities have positive scores indicating the importance of domestic links relative to foreign connections. However, the results do show that it is Manchester that is the least local, confirming its position as the British provincial city most integrated into the world city network.

The first list in Table 12.5 measures the "traditional globalism" of UK cities by showing their combined connectivity to London and New York, NYLON, as the "main street dyad" of contemporary globalization (therefore there is no score for London in this list). The provincial cities here divide into two pairs with positive scores for Manchester and Edinburgh and negative scores for Birmingham and Glasgow. The second list in Table 12.5 shows city connections to what may be an

Table 12.5 Globalization orientations of leading UK cities

City	NYLON globalization	China globalization
London	–	0.95
Manchester	0.01	−0.08
Edinburgh	0.06	−0.06
Birmingham	−0.17	−0.22
Glasgow	−0.13	−0.05

emerging new globalism based upon Beijing, Shanghai and Hong Kong. Once again London stands out with its large positive connection to this Chinese tri-city centre; the other cities are all relatively under-linked for this China connection. There are no signs that leading provincial British cities are getting into place to benefit from the China globalization that may dominate the twenty-first century (Taylor, 2013).

In conclusion, the analysis of 2008 GaWC data has generally confirmed recent writings on UK cities in globalization: London continues to completely dominate the connections between the UK and the rest of the world economy as measured by business service links. Leading provincial cities are becoming moderately important service nodes in their own right but the primacy of London is as strong, or perhaps even stronger, with the coming of economic globalization.

5 German Cities in the World City Network

In contrast to the British case of extreme primacy in global network connectivity, German cities show a much more balanced integration into the locational networks of leading advanced producer service firms (Hoyler, 2011). This is in part due to the long history of territorial fragmentation and political decentralization of state power, which has led to the emergence of a polycentric urban system with complementary functional and sectoral specialization (Blotevogel, 2000). Metropolitan functions are distributed across a number of important cities and city-regions (Krätke, 2004; Blotevogel & Schulze, 2009), in particular those designated "European Metropolitan Regions" in recent spatial policy agendas (BBR, 2005).

Of the 14 German cities with over 500,000 inhabitants, there are five which show a global network connectivity of over 30 % of that of the worldwide leading city, London (Table 12.6). The highest ranked city in Germany is Frankfurt am Main, reflecting its role as a major international financial centre that attracts not only financial service firms but has become a national gateway for many other knowledge-intensive business services (Hoyler, Freytag, & Mager, 2008). While Frankfurt clearly stands out, Germany's three largest cities, Berlin, Hamburg and Munich, follow with only minor differences in their global network connectivity scores. For the capital Berlin this reflects an increased importance after the end of the Cold War, which had limited its potential for economic growth (Korcelli-Olejniczak, 2012). Hamburg, the major port and economic centre in northern

Table 12.6 Connectivity of leading German cities and their "localism"

Rank	City	GNC	With other German cities
1	Frankfurt	0.5	−0.05
2	Berlin	0.39	0.65
3	Hamburg	0.37	0.89
4	Munich	0.35	0.75
5	Düsseldorf	0.32	0.84

Germany, and Munich, the southern German manufacturing and high-technology hub (Lüthi et al., 2010) are followed closely by Düsseldorf, the principal advanced producer service location for the Rhine-Ruhr region, and by Stuttgart (ranked sixth with 0.27), the centre for corporate servicing of South-West Germany (Strambach, 2002). However, the remaining eight cities with over 500,000 inhabitants all achieve global network connectivity scores of over 0.05, with five cities showing over ten per cent of London's connectivity. Cologne leads this group of second-tier service centres, ahead of Leipzig, Dresden, Bremen and Hannover (see Hoyler, 2011).

The "localism" measure in Table 12.6 underlines the specific role of Frankfurt as Germany's "most international" city: as the only city with a (slight) negative score, its connectivity in advanced producer servicing is balanced between national and transnational links. This reflects its particular strengths in corporate law (rank 3 in Europe; Table 12.2), finance (rank 5; Table 12.1) and management consultancy (rank 8; Table 12.2). Other German cities make it into the European Top Ten only once: Berlin in accountancy (rank 7, Table 12.2), and Munich and Düsseldorf in corporate law (ranks 6 and 10 respectively; Table 12.2). The positive scores of these cities on the "localism" measure indicate the relative importance of domestic over foreign connections, with Berlin and Munich less domestically oriented than Hamburg and Düsseldorf (Table 12.6). Compared to British provincial cities (Table 12.4) however, German cities are significantly more "un-local", reflecting the export-orientation of the German economy and the associated need for cross-border service provision.

Table 12.7 focuses on selected geographical patterns of these non-domestic linkages to the traditional centres of global capital, New York and London (NYLON), and to the new emerging focus of contemporary globalization in China (Beijing–Hong Kong–Shanghai). The top five German cities show a remarkable degree of similarity in their aggregate advanced producer service connections to these two major poles of globalization: With the exception of Berlin, all cities are relatively over-linked to NYLON and to the Chinese city triad. Frankfurt once again is ahead of other German cities in terms of the intensity of these connections, followed by Munich, Düsseldorf and Hamburg. The relative strength of these linkages compared to provincial UK (Table 12.5) and Spanish (Table 12.9) cities may well signify a more favourable positioning of German cities in a changing global economic order.

In conclusion, although Frankfurt emerges as the leading German city in terms of global network connectivity, the analysis of the 2008 GaWC data has also confirmed

Table 12.7 Globalization orientations of leading German cities

City	NYLON globalization	China globalization
Frankfurt	1.29	1.09
Berlin	−0.12	−0.23
Hamburg	0.12	0.01
Munich	0.9	0.77
Düsseldorf	0.48	0.4

the relative strength of other German cities in their integration into worldwide advanced producer services networks. Berlin, Hamburg and Munich share similar overall levels of integration but are marked by sector-specific differences in the strength of their linkages. Düsseldorf and Stuttgart complete the leading group of six cities that act as prime strategic nodes in the organizational networks of major advanced producer service firms who operate parts of their business from/in Germany. The analysis confirms the enduring polycentric nature of the German urban system, even when viewed through the lens of globally operating advanced producer service firms.

6 Spanish Cities in the World City Network

Five big urban areas lead the Spanish city network; the two metropolises of international renown (Madrid and Barcelona) and three regional metropolises (Valencia, Seville and Bilbao). According to *Urban Audit* data, in 2009 Madrid had 6,271,638 inhabitants and Barcelona 4,440,629. This demographic feature has been interpreted as the doubled-headed character of the Spanish urban network, halfway between vertical and horizontal national city structures. Other urban areas of the country appear close to this leading group, of which only Bilbao falls just below the threshold of one million inhabitants.

The recent evolution of Spanish cities has been shaped by an intense economic-territorial restructuring process associated with globalization and technological change. Two key factors help explain the strengthening of big cities that has occurred over the last decades. First, the influence of the decentralized nature of the Spanish state, favouring the emergence of regional urban networks, more integrated locally and headed by regional centres that maintain intense relations with each other and with the two national metropolises. Second, the reinforcing view of Spanish cities as relevant global actors competing with other cities nationally and in Europe.

Although the last perspective has dominated recent urban policies and strategies in Spain, the empirical knowledge has not advanced in parallel. Certainly, Spanish cities have only marginally been considered within international research on world cities, as shown by a review of the main published works during the 1990s (Taylor, 2004). However, some general findings can be summarised from previous research conducted on global office networks of advanced business services. Madrid and

Table 12.8 Connectivity of leading Spanish cities and their "localism"

Rank	City	GNC	With other Iberian cities
1	Madrid	0.65	−0.63
2	Barcelona	0.42	−0.01
3	Valencia	0.12	0.53
4	Seville	0.11	0.61
5	Bilbao	0.09	0.67

Barcelona act as "classic gateway cities" in contemporary globalization, connecting the national to the world economy (Taylor, 2004). However, Madrid occupies a higher position in the world urban hierarchy and is considered a major global service centre for the key sectors of advertising, accountancy, banking/insurance and legal services (Beaverstock, Smith, Taylor, Walker, & Lorimer, 2000). Moreover, in a general context of an increasing concentration of advanced services in leading world cities, the net connectivity gains of Madrid in the early 2000s are confirmed against the losses of neighbouring cities, such as Lisbon or Barcelona (Taylor, Catalano, & Gane, 2003). The international roles also vary along the world city network configurations; Madrid has been classified within a "global route arena" (intercontinental linkages), highlighting its financial services connections with important Latin American cities (reinforcing historical linkages). Barcelona, partially in the same cities group, presents some characteristics similar to other European financial centres connected with the great banking cities of Asia Pacific. Other Spanish cities belong to "European urban arenas" of national scope that, unlike the main economies of the continent (Germany, UK and France), do not constitute a specific cluster (Taylor et al., 2002).

According to the 2008 GaWC data, only five of the bigger Spanish cities reach a global network connectivity above 0.05 (proportionally to the maximum connectivity of London; Table 12.8). Their position fits the population size distribution, Madrid leading the connectivity of Spanish cities with nearly two thirds of the highest global network connectivity. The leadership of Madrid as the centre of the Spanish economy is based upon its status as political capital and also reflects an increasing specialization in advanced tertiary activities (OECD, 2007). Other relevant factors include the strong performance of its real estate market and the success of cluster promoting policies, including the services of international fairs (Cuadrado-Roura & Rubalcaba-Bermejo, 1998; Sánchez-Moral, Calatrava, & Melero, 2008).

Barcelona, second in the ranking, is the other half of the traditional dual primacy pattern of the Spanish urban system, always striving to escape from the shadow of Madrid as the dominant local world city. Despite competition with Madrid to attract big companies, economic institutions or international organisations, which seems to favour the Spanish capital as the place to locate, the economic performance of Barcelona in globalization is clearly influenced by its history as capital of the main manufacturing region of the country. Nowadays, Barcelona has developed a successful urban development model that, without relinquishing the importance of high-tech manufacturing, reinforces the attractiveness for national and international advanced business services firms and especially for creative industries, being internationally recognized as a cultural-creative European hub (Boix, 2011).

Table 12.9 Globalization orientations of leading Spanish cities

City	NYLON globalization	China globalization
Madrid	0.67	0.71
Barcelona	0.15	0.04
Valencia	−0.55	−0.30
Seville	−0.42	−0.35
Bilbao	−0.66	−0.60

The best way to approach the rest of the Spanish cities is through the alternative analysis of "localism". Thus, the relative concentration of domestic connections within the country rises significantly in Bilbao, Seville and Valencia, second order metropolises displaying a certain international projection but mainly dealing with the articulation at the regional level of the territory and the economy. On the other hand, the "localism" data also provide new evidence about divergence at the top of the urban network. Madrid is the only city that could be considered "un-local", while in Barcelona, only slightly negative on this score, the weight of the domestic connections continues to be quite important.

The relative concentration of connections with New York and London, or "NYLON globalization", confirms the increase in distance between Madrid and the rest of the Spanish cities, including Barcelona (Table 12.9). Even more intense are the differences found in the connections with the new economic centre of gravity represented by the Chinese cities of Beijing, Hong Kong and Shanghai. The values of "China globalization" highlight an even greater preponderance of Madrid in articulating the flows in this area, while Barcelona is clearly penalised by this new criterion, being only very weakly positive on this measure. Other big Spanish cities show negative values.

In summary, despite the intermediate position of the urban structure of Spain, characterized by a doubled-headed city network and recent regional decentralization, there are strong differences in the integration of Spanish cities into the world city network. Madrid, a second order world city, acts as a "gateway city" connecting the national and the global economy. Barcelona shares this function to some extent but displays a lower degree of specialization in advanced producer services and fewer external connections of these activities. Although for regional metropolises local connections are even more important, due to their territorial articulation function, they also strive to achieve some international projection by means of attracting advanced business services, which are assumed to be strategic for their future urban development.

7 Conclusion: How Resilient Is Europe and Its Cities?

This snapshot of the network externalities of European cities in three countries in 2008 suggests a relatively healthy resilience in the face of economic crises. Europe as a whole has numerous cities comparatively well connected within the world city network due to its multiple states, and in the three states we focussed on, despite major differences in national urban structures, all the leading cities have reasonable overall connections and the major cities have strong connections to the competing centres of economic globalization (NYLON and China). But as we noted in the introduction we cannot yet know how the economic crisis will pan out as a global restructuring of the world-economy. We do have some new evidence from a later survey conducted in 2010, which we are just beginning to digest. For Europe this shows little change except for some specific, and expected, examples. Athens is the big loser with its world city network ranking dropping from 37th to 60th indicating a strong symptom of peripheralization. Also Edinburgh is the biggest loser within our three countries: Europe's "newest banking centre" was home to vulnerable banks resulting in a world city network ranking drop from 117th to 128th. Otherwise European cities still appear relatively resilient although Madrid dropping to 17th from 11th in world city network ranking may be cause for concern.

Acknowledgements This text is part of Spanish R&D&I projects CSO2009-10888 and CSO2011-2652

References

Baeten, G. (2001). The europeanization of Brussels and the urbanization of "Europe". *European Urban and Regional Studies, 8*, 117–130.

BBR (Bundesamt für Bauwesen und Raumordnung). (2005). *Raumordnungsbericht 2005*. Bonn: BBR.

Beaverstock, J. V., Hoyler, M., Pain, K., & Taylor, P. J. (2001). *Comparing London and Frankfurt as world cities: A relational study of contemporary urban change*. London: Anglo-German Foundation.

Beaverstock, J. V., Smith, R. G., Taylor, P. J., Walker, D. R. F., & Lorimer, H. (2000). Globalization and world cities: Some measurement methodologies. *Applied Geography, 20*, 43–63.

Blotevogel, H. H. (2000). Gibt es in Deutschland Metropolen? Die Entwicklung des deutschen Städtesystems und das Raumordnungskonzept der "Europäischen Metropolregionen". In D. Matejovski (Ed.), *Metropolen: Laboratorien der Moderne* (pp. 179–208). Frankfurt am Main: Campus.

Blotevogel, H. H., & Schulze, K. (2009). Zum Problem der Quantifizierung der Metropolfunktionen deutscher Metropolregionen. In J. Knieling (Ed.), *Metropolregionen: Innovation, Wettbewerb und Handlungsfähigkeit* (pp. 30–58). Hannover: Akademie für Raumforschung und Landesplanung (ARL).

Boix, R. (2011). Facing globalization and increased trade: Catalonia's evolution from industrial region to knowledge and creative economy. *Regional Science Policy and Practice*. doi:10.1111/j.1757-7802.2011.01057.x.

Cochrane, A., & Jonas, A. (1999). Reimagining Berlin: World city, national capital or ordinary place? *European Urban and Regional Studies, 6*, 145–164.

Cuadrado-Roura, J. R., & Rubalcaba-Bermejo, L. (1998). Specialization and competition amongst European cities: A new approach through fair and exhibition activities. *Regional Studies, 32*, 133–147.

de Groof, R. (Ed.). (2008). *Brussels and Europe*. Brussels: BRIO.

Derudder, B., Hoyler, M., & Taylor, P. J. (2011). Goodbye Reykjavik: International banking centres and the global financial crisis. *Area, 43*, 173–182.

Elmhorn, C. (2001). *Brussels: A reflexive world city*. Stockholm: Almqvist & Wiksell International.

Faludi, A. (Ed.). (2002). *European spatial planning*. Cambridge, MA: Lincoln Institute of Land Policy.
Fujita, M., & Thisse, J.-F. (2002). *Economics of agglomeration: Cities, industrial location and regional growth*. Cambridge: Cambridge University Press.
Hoyler, M. (2011). External relations of German cities through intra-firm networks—A global perspective. *Raumforschung und Raumordnung, 69*, 147–159.
Hoyler, M., Freytag, T., & Mager, C. (2008). Connecting Rhine-Main: The production of multi-scalar polycentricities through knowledge-intensive business services. *Regional Studies, 42*, 1095–1111.
Jacobs, J. (1969). *The economy of cities*. New York: Random House.
Jacobs, J. (1984). *Cities and the wealth of nations*. New York: Random House.
Jacobs, J. (2000). *The nature of economies*. New York: Modern Library.
Korcelli-Olejniczak, E. (2012). Marked by dynamics: Berlin and Warsaw in the process of functional change. In B. Derudder, M. Hoyler, P. J. Taylor, & F. Witlox (Eds.), *International handbook of globalization and world cities* (pp. 517–529). Cheltenham: Edward Elgar.
Krätke, S. (2001). Berlin: Towards a global city? *Urban Studies, 38*, 1777–1799.
Krätke, S. (2004). Urbane Ökonomien in Deutschland: Clusterpotenziale und globale Vernetzungen. *Zeitschrift für Wirtschaftsgeographie, 48*, 146–163.
Lüthi, S., Thierstein, A., & Goebel, V. (2010). Intra-firm and extra-firm linkages in the knowledge economy: The case of the emerging mega-city region of Munich. *Global Networks, 10*, 114–137.
OECD. (2007). *OECD territorial reviews: Madrid, Spain*. Paris: OECD.
Parkinson, M., Champion, T., Simmie, J., Turok, I., Crookston, M., Katz, B., & Park, A. (2006). *State of the English cities*. London: Office of the Deputy Prime Minister (ODPM).
Parkinson, M., Hutchins, M., Simmie, J., Clark, G., & Herdonk, H. (2004). *Competitive European cities: Where do the core cities stand?* London: Office of the Deputy Prime Minister (ODPM).
Sánchez-Moral, S., Calatrava, A., & Melero, A. (2008). Las funciones comando de Madrid en la economía global: una aproximación a través del proceso de atracción de capital extranjero. *EURE–Revista Latinoamericana de Estudios Urbano Regionales, 34*, 25–44.
Sassen, S. (1991). *The Global City: New York, London, Tokyo*. Princeton, NJ: Princeton University Press.
Sassen, S. (1994). *Cities in a world economy*. Thousand Oaks: Pine Forge Press.
Strambach, S. (2002). Change in the innovation process: New knowledge production and competitive cities—The case of Stuttgart. *European Planning Studies, 10*, 215–231.
Taylor, P. J. (2001). Specification of the world city network. *Geographical Analysis, 33*, 181–194.
Taylor, P. J. (2004). *World city network: A global urban analysis*. London: Routledge.
Taylor, P. J. (2013). *Extraordinary cities: Millennia of moral syndromes, world-systems and city/state relations*. Cheltenham: Edward Elgar.
Taylor, P. J., & Aranya, R. (2006). Connectivity and city revival. *Town and Country Planning, 75*, 309–314.
Taylor, P. J., Catalano, G., & Gane, N. (2003). A geography of global change: Cities and services, 2000–2001. *Urban Geography, 24*, 431–441.
Taylor, P. J., Catalano, G., & Walker, D. R. F. (2002). Measurement of the world city network. *Urban Studies, 39*, 2367–2376.
Taylor, P. J., & Derudder, B. (2004). Porous Europe: European cities in global urban arenas. *Tijdschrift voor Economische en Sociale Geografie, 95*, 527–538.
Taylor, P. J., Evans, D. M., Hoyler, M., Derudder, B., & Pain, K. (2009). The UK space economy as practised by advanced producer service firms: Identifying two distinctive polycentric city-regional processes in contemporary Britain. *International Journal of Urban and Regional Research, 33*, 700–718.

Taylor, P. J., & Hoyler, M. (2000). The spatial order of European cities under conditions of contemporary globalisation. *Tijdschrift voor Economische en Sociale Geografie, 91*, 176–189.
Taylor, P. J., Hoyler, M., Evans, D. M., & Harrison, J. (2010). Balancing London? A preliminary investigation of the 'Core Cities' and 'Northern Way' spatial policy initiatives using multi-city corporate and commercial law firms. *European Planning Studies, 18*, 1285–1299.
Taylor, P. J., Ni, P., Derudder, B., Hoyler, M., Huang, J., & Witlox, F. (Eds.). (2011). *Global urban analysis: A survey of cities in globalization*. London: Earthscan.
van den Berg, L., Braun, E., van der Meer, J., & Mingardo, M. (2007). The urban dimension in European policy: history, actors and programmes. In L. van den Berg, E. Braun, & J. van der Meer (Eds.), *National responses to urban challenges in Europe* (pp. 39–62). Aldershot: Ashgate.

Chapter 13
Geographical Distribution and Regional Specialization of Knowledge-Intensive Business Services: An Empirical Investigation Across European Regions

Blanca L. Delgado-Márquez and Marcos M. García-Velasco

1 Introduction

Tertiarization processes characterize structural changes at both national and regional levels. In this context, knowledge-intensive business services (KIBS) have made a profound contribution to the creation of employment, increases in production levels and the promotion of investment, especially in industrialised regions. During the expanding cycle of the European economy (1995–2007), the amount of net employment increased by 26 million in the EU-15 and more than a third were generated by KIBS (and the two other thirds by Other KIS and Less KIS, respectively). Thus, employment in KIBS has been one of the main drivers of employment over the last years. Moreover, during the period of economic crisis (2008–2010), whilst total employment decreased by more than five million, net employment in KIBS increased by 226,000 throughout the EU-27. As a consequence, the literature has paid an increasing attention to the study of KIBS over the last years (Alvesson, 2004; Cuadrado-Roura & Maroto-Sánchez, 2010; Fischer & Frolich, 2001; Gallego & Maroto, 2010; Koch & Stahlecker, 2006; Mudambi, 2008; Muller & Doloreux, 2009; Rubalcaba-Bermejo & Cuadrado-Roura, 2001; Vence & González, 2002).

Previous works have stressed the crucial role of such services in the dissemination of both tacit and explicit knowledge (Gertler, 2003). Nonetheless, while advances in Information and Communication Technologies (ICT) have fostered geographical distribution of knowledge-intensive services, this achievement mainly refers to explicit knowledge (i.e. knowledge that can be codified through data, formulae, books, papers, etc). Hence, diffusion of tacit knowledge (i.e. knowledge

B.L. Delgado-Márquez (✉) • M.M. García-Velasco
Department of International and Spanish Economics, Faculty of Economics and Business. Campus de Cartuja s/n. E-18071, Granada, (Spain)
e-mail: bdelgado@ugr.es; marcos@ugr.es

J.R. Cuadrado-Roura (ed.), *Service Industries and Regions*, Advances in Spatial Science, DOI 10.1007/978-3-642-35801-2_13,

that resides in the minds of individuals and cannot be codified) still faces strong geographical barriers (García-Velasco, 2005; Howells, 2002).

Interest in recognition of the increasingly central role of KIBS in Western economies began in the late 1970s and the early 1980s (Coffey, 2000). As a consequence, the location of knowledge-intensive business services has been extensively documented for the 1970s and 1980s when researchers turned their attention to the effects of tertiarization upon regional development (Beyers & Alvine, 1985; Coffey & Polése, 1987; Drennan, 1987; Illeris, 1996). A key focus of their attention was the potential for regional development centred on KIBS, and, in more general terms, the uneven spatial development that seemed to be reinforced by the spatial concentration of these services towards the top of the urban hierarchy. Whilst an increasing proportion of recent literature has adopted a non-geographic perspective to investigate the role of KIBS in innovation systems and, more generally, in the process of innovation (Antonelli, 1999; Attewell, 1992; den Hertog, 2000; Miles, 2005; Muller & Zenker, 2001; Strambach, 2001; Wood, 2006), simultaneously the attention paid to the spatial distribution of this type of service has not decreased. The profound and increasing interest in the geographical location of KIBS is justified by their widely-recognized central role as facilitators in local milieu and in flexible production as well as being vectors of information procurement and exchange (e.g., innovation and growth in manufacturing (and other) companies is linked to their access to, and use of, producer services (Cooke & Leydesdorff, 2006; MacPherson, 1997)).

Given the relevance of KIBS as a source of local and regional competitiveness (Gallego & Maroto, 2010, Martínez-Alcocer & Maroto-Sánchez, 2010) as well as the role played by location in their performance (e.g., Wood, 2006), this chapter serves to identify the main trends in the evolution of regional concentration and specialization in KIBS across European regions over the last years. Specifically, the territorial unit under analysis corresponds to NUTS 2 regions in the European Union (EU), and the period analysed ranges from 1995 to 2010—so that the study covers the effects of the enlargement of the EU (2004 and 2007), the effects of cohesion policy (1994–1999 and 2000–2006), economic expansion (1995–2007) and the beginning of the economic crisis (2008–2010), among others. This text seeks to answer the following questions: Where have KIBS located in Europe? In which regions has the share of employment in KIBS as compared to total employment been more important? What has been the nature of the geographical allocation of KIBS in EU regions? Which factors have contributed to explain the location of KIBS?

To deal with the aforementioned objectives, the empirical analyses are structured into different parts. The descriptive part of the analyses relies on the study of geographical concentration and regional specialization in European regions. This part considers employment as the main variable that can explain the spatial allocation of KIBS across European regions taking Eurostat as its source. In general, the territorial units taken in account have been NUTS 2 level for the 27 EU member countries. The period considered is 1995–2010, although data for earlier years have not been available for several countries, and there have been some methodological changes. Then several phases have been considered: 1995–2000, 2000–2004, 2004–2007 (enlargement), and economic crisis (2008–2010). One important caveat is that by considering employment and not GDP, this implies that regional differences or progress in productivity are not taken in account.

After the descriptive analyses, the chapter investigates the main factors that may help to explain the location of KIBS in European regions, such as the influence of certain variables related to economies of agglomeration and accessibility to clients and knowledge resources on the regional specialization patterns of KIBS. Before undertaking those empirical analyses, the following section aims to provide a theoretical review of the relevance of proximity in the location of KIBS. Finally, the last section summarizes the main conclusions.

2 Relevance of Proximity in Location of KIBS: Theoretical Background

The concept of geographical proximity has traditionally been linked to the spatial distance between actors. Previous research has stated that, *ceteris paribus*, the greater the distance between actors, the less intensive the positive externalities (Knox & Agnew, 1994; Krugman, 1991).

As described by previous works, KIBS seek to result in the creation, accumulation or dissemination of knowledge in general and tacit knowledge in particular. Given the role of KIBS as facilitators and disseminators of knowledge, geographical proximity is crucial for efficient knowledge transfer. Lo (2003) points out three distinctive features which are important for knowledge transfer: (1) in contrast to data and information, knowledge is linked to individuals and context, (2) the more implicit knowledge there is, the more difficult the transfer is without personal contact, (3) knowledge is limited to specific organisational and spatial territories; this may also be valid for codified knowledge.

Although transport facilities and information and communication technologies have experienced a rapid development that is reducing the barriers of distance in service production, many economists claim that spatial proximity is still crucial in economic processes because while *information diffuses rapidly across organisational and territorial borders, it is wrongly assumed that understanding does, too* (Morgan, 2004: 3).

The advantages derived from proximity in terms of the performance of KIBS can be classified into different groups. Firstly, geographical proximity is assumed to foster processes of innovation (e.g., Boschma, 2005). These processes rely heavily on the exchange of information and knowledge between different actors and are frequently based on learning-by-interacting (Howells, 2002). Specifically, KIBS are intended to result in the creation, accumulation or dissemination of knowledge in general, and tacit knowledge in particular (e.g., know-how, know-who). The transfer of tacit knowledge requires trust, common understanding, frequent communication, and face-to-face contact (Roberts & Andersen, 2000). These factors are favoured by short physical distances amongst the actors involved (e.g., Howells, 2002; Morgan, 2004). Thus, provided that the experience of common work and co-location is essential within the exchange process of implicit knowledge, geographical proximity can be conducive to efficient knowledge transfer. As innovation is a crucial element in the early development of firms in the KIBS sector, spatial proximity can be assumed to have a positive effect upon the post-entry growth of these firms.

Secondly, relations between firms providing KIBS and their clients are highly complex, i.e., both the provision and the development of new and innovative services take place mainly via close interaction between service providers and clients. Such interaction involves a high degree of customization, and the communication involved must be highly intensive; consequently, practices such as standardization, routinization, and supervision are difficult to apply (Alvesson, 2004; Miles et al., 1995). In this context, KIBS firms must keep in permanent contact with clients. Specifically, in the cases of certain KIBS firms offering tailor-made knowledge of a product for a particular customer, the relationship with the client may be crucial. Since the establishment and continuity of these relations is facilitated by spatial proximity, locally embedded firms may have better expectations for growth (Boschma & Weterings, 2005). Furthermore, intense interaction between service providers and clients, which is the norm in the provision of knowledge intensive services, is improved by short distances between agents (Illeris, 1994).

Thirdly, the socialisation and learning procedures necessary for the successful transfer of tacit knowledge require co-presence and co-location between transmitter and receiver (Roberts & Andersen, 2000). The provision of knowledge intensive services requires specialized knowledge and cumulative learning processes, which can mainly be achieved through intense interaction between service suppliers and clients (Johannisson, 1998; Strambach, 2002). A great deal of tacit knowledge is developed interactively and shared within networks, so that geography becomes relevant in the transfer of knowledge by KIBS firms. Moreover, the accumulative nature of knowledge generates spatial spillovers, as stated by Marshall (1890). Consequently geographical—as well as social, political, psychological—distance is an influential factor in the diffusion of tacit knowledge (Acs et al., 2002). The partly tacit character of knowledge is likely to be responsible for the importance of localised networks of personal contacts in innovation activities of firms in some metropolitan regions (Fischer & Frolich, 2001). From this point of view, the location of KIBS firms may be of great importance in major metropolitan regions (Wood, 2002).

Fourthly, the relevance of distance to clients in the locational decisions of KIBS firms obviously depends on the degree of tacitness in knowledge, and the greater the need for a customized service, the lesser the locational freedom. KIBS firms supply advanced services, which are usually tailored to the requirements of other companies or public agencies (Aslesen & Isaksen, 2007; Wood, 2002), and they may be considered to be critical in terms of achieving greater competitiveness for firms or increased efficiency in the public sector. In this case, a decisive factor in location could be the wish to be as close as possible to government institutions. Thus, KIBS firms are frequently located in capital cities or set up a branch close to them. Nonetheless, some authors have pointed out that a lower importance of the distance derived from an intensive use of the ICT in many activities has turned the output of these companies into a tradable good. This implies the reduction of the weight of demand location as an explanatory factor of its location in favour of other factors. Obviously, from these changes it is not deduced that the location pattern must be necessarily different, but rather that one of the factors explaining the location has partly lost its importance and, therefore the location pattern may be different.

Furthermore, location is important not only from the clients' perspective, but also in terms of accessibility to resources. In light of this question, KIBS firms try to locate near to information and knowledge sources (Doloreux et al., 2008). In this sense, purely neoclassical macroeconomic theory justifies business growth in terms of available factors so that the production of advanced services would be concentrated on those areas with plenty of productive factors (mainly human and technological capital) and would be less costly. Nonetheless, this approach needs to be complemented by new endogenous growth, thus reinforcing the role of knowledge and other intangible elements, i.e., factors that are hidden behind technological change and total factor productivity. Therefore, the second group of location factors includes a heterogeneous set of drivers such as access to highly-educated workers and accessibility to information and other resources.

It is important to note that tacit knowledge, because of its nature, is context-specific (Gertler, 2003). Proximity to related professional services (competitors) and experts (Keeble & Nachum, 2002; Sokol et al., 2008) and suppliers (Cook et al., 2007; Freel, 2006; Koch & Stahlecker, 2006) may also be important from this point of view. Proximity to government may sometimes be important due to the need for KIBS firms to know the modifications in public regulations as soon as possible.

Nonetheless, it is important to note that, in terms of greater accessibility to resources derived from geographical proximity, the most relevant resource is human capital. KIBS firms are attracted to locations with a great pool of educated people. The core activities of these companies are based on the intellectual skills of a very large proportion of the labour force and also are often based on the sale of products and on service work. A large number of employees typically have an academic education and relevant experience. Formal education is seen as very useful for facilitating theoretical and analytical abilities essential to such organizations (Alvesson, 2004). Thus, proximity to this factor becomes crucial when there is some degree of immobility in human capital, because of professional factors—their skills and abilities may be related to the environment—social or family reasons. Thus, KIBS firms are attracted to areas where there is a great availability of human capital (Cook et al., 2007; Keeble & Nachum, 2002; Koch & Stahlecker, 2006; Sokol et al., 2008).

Spatial proximity contributes to the foundation of new firms. Previous research stresses the fact that most new firms are small and have limited financial and personal resources in order to explain the fact that the vast majority of start-ups are established in the region that the company founders have been living and/or working in before (e.g., Brüderl et al., 1996; Cooper, 1985). Access to information, knowledge and other resources is facilitated by existing personal and social networks that are usually best developed within a short geographical distance (Illeris, 1994; Johannisson, 1998).

Proximity to clients, suppliers, competitors, human capital, and information sources is very important for an efficient provision of KIBS. Thus, agglomeration is crucial driver for location, and many advantages are inherent in this kind of location. Moreover, uncertainty and vulnerability derived from the process of globalization mean that KIBS firms are attracted towards agglomerations.

Therefore, from a theoretical viewpoint, the factors that influence the geographical location of KIBS can be classified into different groups, i.e., demand (clients) and supply factors (information and knowledge mainly). Nevertheless, accessibility seems to be the crucial element and therefore the concept of agglomeration economies is fundamental. As a result of a careful review of a significant amount of research on services, in general, and on KIBS, in particular (Beyers, 2005; Coffey & Shearmur, 2002; Wood, 2002), we are in a position to provide a classification of drivers that influence the spatial distribution of KIBS.

2.1 Agglomeration Economies

Agglomeration economies can be defined as the benefits associated with the co-location of firms and advantages created by the concentration and co-habitation of economic activities in terms of access to: markets, suppliers, a varied and qualified workforce, formal and informal networks, specialised services and industries, and technological infrastructure (Malmberg & Maskell, 2002; Maskell & Kébir, 2006).

Previous research has shown that firms benefit and become more productive from interaction with each other, due to interaction and communication externalities (Fujita & Thisse, 2002). Consequently, large agglomerations—especially urban ones—gain an advantage that stimulates further, cumulative, growth. Individual firms can benefit from upstream and downstream externalities that bring about co-location advantages.

2.2 Knowledge Spillovers

Knowledge spillovers are derived from agglomeration economies. This term refers to positive milieu-externalities created by: research and other investments in knowledge, inter-firm labour mobility, and local skill development processes (Feldman, 1994). Knowledge spillovers benefit co-located actors with investments made by neighbours, while spatially separated actors are not able to access these kinds of effects (Marshall, 1890).

2.3 Region and Urban Size

Beyond other drivers, region and urban size remain the principal organizing factors for KIBS across space. According to theories of urban agglomeration, advanced services would benefit from the scale economies offered by metropolitan centres. For example, the empirical evidence for the existence of agglomeration economies and localised knowledge spillovers is often found in metropolitan regions.

This geographical concentration or selective agglomeration is crucial for the success and competitiveness of firms offering knowledge services (Scott, 1988). The strong concentration of KIBS in larger cities and the capital region especially, and the fact that KIBS derive a considerable share of their business from local customers indicate a more intense state of competition in these urban areas. Furthermore, KIBS encourage the innovative development of cities by strengthening connections amongst strategic planning of firms (Wood, 2002).

Already two decades ago, Coffey and Polèse (1989) undertook an investigation motivated by the attention KIBS received as policy-levers for developing lagging regions, and came to the conclusion that the potential for KIBS to locate outside of major metropolitan areas was limited. Additionally, as expected, location in a large urban centre seems to be relevant in the use of decision functions, at least in certain specific branches.

Indeed, urban agglomerations consist of interlinked urban areas, which in turn can be decomposed into zones. For each of these levels, there are co-location advantages so that spatial concentrations can be observed at different levels of spatial resolution. This implies that a system of urban agglomerations has a radial result, with major centres, centres of urban areas as well as sub-centre concentrations (Anas et al., 1998).

2.4 Labor Market Effects

This factor refers to the quality of human resources and their level of skills and qualifications (Coffey & Shearmur, 2002). To have access to a labour force with better greater competences and skills as well as higher levels of experience has traditionally been considered as essential in order to develop advanced or highly technological activities (Illeris, 1996). In this sense, the relationship between the base of knowledge of human capital and the change and economic growth of an urban centre has a crucial relevance (Glaeser et al., 1995; Matthiessen et al., 2002).

2.5 Accessibility

A strong infrastructure for transport and communications are one of the criteria that drive a firm's decision about whether to concentrate on certain areas or regions. Nonetheless, while some studies assume that firms supplying KIBS make their location decisions as a response to the accessibility to customer demand (customer contacts) that each possible location offers, there is no commonly accepted measure for accessibility in the literature (e.g., Johansson & Klaesson, 2007; Mattsson, 1984; Weibull, 1976).

2.6 Other Local Qualitative Aspects Characteristics

Among these factors, we can cite governmental structures, cooperation among firms or the institutional context existing in the different geographical areas (Cooke et al., 2004). Nevertheless, due to the qualitative nature of these factors, they are difficult to measure and to approach from a statistical point of view (Doloreux et al., 2001).

The influence of these factors upon KIBS location depends on the characteristics of the activity (degree of customization, specialization, etc.), as well as on the mix of factors mentioned previously. Thus, it is difficult to determine a priori whether KIBS development may improve territorial cohesion or not. In this point, empirical analyses constitute crucial tools to ascertain the geographical distribution of KIBS and the effects upon regional development.

3 Spatial Allocation of KIBS Across European Regions: A Descriptive Approach

Geography of KIBS in EU regions is characterized by a high degree of spatial concentration and polarization of activity across relatively few regions. Some areas show a high degree of specialization and they are located in regions containing capital cities, areas around them (Brussels and London), and some other central cities (Amsterdam, Rotterdam, Frankfurt, Hamburg and Munich) overall. On the contrary, regions with low level of specialization locate in Objective 1 regions and regions of New Member Countries (except capital cities).

The evolution of spatial allocation of KIBS in EU regions has shown a reduction in the geographical concentration until 2007 at least, thus benefiting Objective 1 regions during the first stage (1995–2007), and regions of New Member Countries during the last decade. However, the current economic downturn seems to have decelerated regional convergence in KIBS employment.

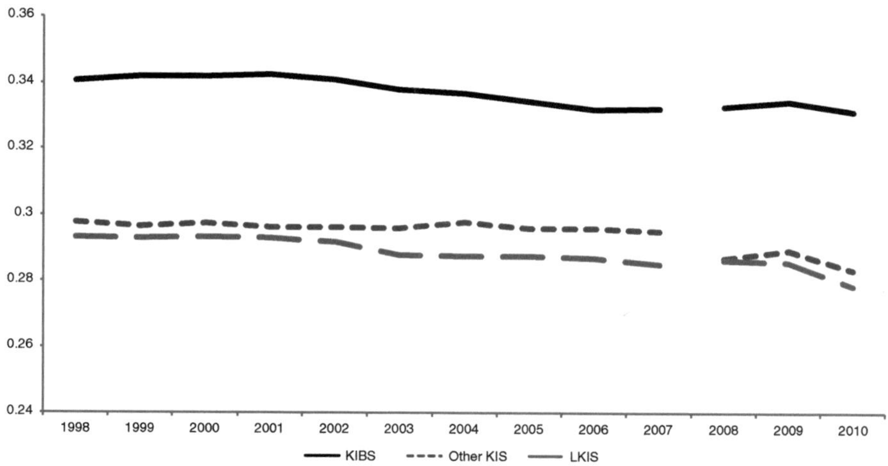

Fig. 13.1 Evolution of Gini index in KIS employment in EU regions

3.1 *Profiles of Regional Specialization in KIBS and Geographical Concentration*

Spatial allocation of KIBS in EU regions has been very concentrated. The Gini[1] index for KIBS achieves the highest scores among services (Fig. 13.1). Thus, 33 % of employment in KIBS (UE-27) was located in 34 regions—19 regions that hosted capital cities, and 6 regions in England (close to the London region), 5 in Benelux and 4 in Germany. This space contained 4.8 % of the area of EU, 18 % of the population and 26 % of the GDP in 2008. Several authors (e.g., Vence & González, 2002; Wood, 2002) have pointed out that KIBS have concentrated in major metropolitan regions, suggesting that this is where client-consultancy demand-supply interactions remain best developed.

It is evident that the greater the knowledge intensity in a service activity, the greater its geographical concentration. It can be observed in that values of the coefficient of variation[2] are reflected in the confidence intervals.[3] These values are greater in every category of KIBS than for Other KIS and Less KIS (L-KIS), and

---

[1] We have used data on employment/km^2 associated with KIBS as a proxy for describing how KIBS are agglomerated. Additionally, the Gini Index has also been estimated for each category of KIBS and Less KIS in order to compare and check. Therefore, if a sector was evenly spread into regions the value of the Gini index would be zero. However, on the contrary, the greater this value is, the greater the spatial concentration is.

[2] As we know, the coefficient of variation is the standard deviation for the mean divided by the mean. In our case, we can divide the mean plus standard deviation by the mean.

[3] The confidence interval of specialization quotient (SQ) is defined as $SQ_{i,q-KIBS,t} \in \left(\overline{SQ_{q-KIBS,t}} \pm s.d\right)$, where: $\overline{SQ_{q-KIBS,t}}$: average employment share for q-type of KIBS at period t, and $s.d.$: standard deviation for the mean.

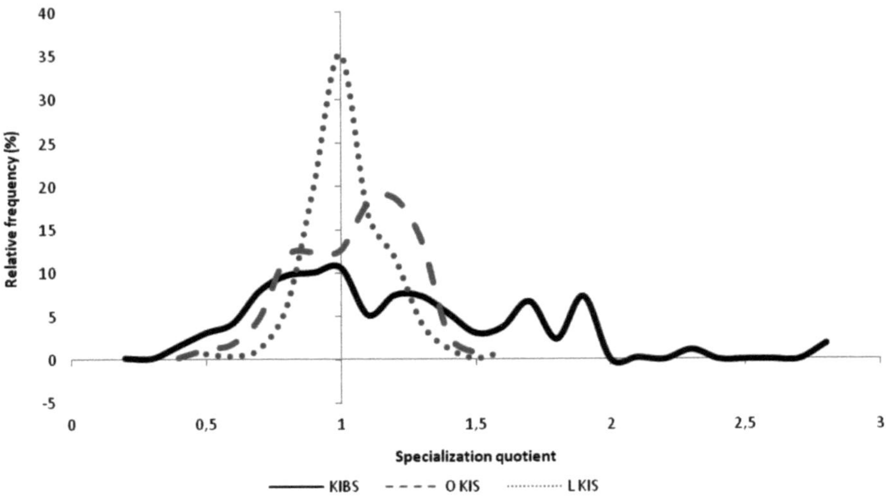

Fig. 13.2 Employment allocation in EU regions in 2010

Manufacturing (High Tech, Medium and Low Tech). Therefore, the confidence interval of specialization quotient[4] (SQ) ranges from 0.65 to 1.35 in KIBS roughly, wider than L-KIS (0.85–1.15) and O-KIS (see Tables 13.1 and 13.2 in the Annex of Tables).

Figure 13.2 represents the share of employment in every level of SQ in L-KIS, O-KIS and KIBS. This figure shows a shape for KIBS that is very different from a normal distribution as compared to O-KIS and L-KIS, which are more normal-shaped and sharpened.

[4] The specialization quotient in the q-type of KIBS can be defined as follows:

$$SQ_{i,q-KIBS,t} = \frac{\left(E_{i,q-KIBS,t} / \sum_{j=1}^{m} E_{i,j,t}\right)}{\left(\sum_{i=1}^{n} E_{i,q-KIBS,t} / \sum_{i=1}^{n}\sum_{j=1}^{m} E_{i,j,t}\right)} \quad i = 1, \ldots, n; j = 1, \ldots, m$$

where:

$\sum_{i=1}^{n} E_{i,q-KIBS,t} \neq 0$, that is, there will always be at least one employee working in the q-type of KIBS in the total of regions under study.

The remaining variables represent the following concepts:

n: total number of NUTS 2 regions; $n \in \mathbb{R}^+$

m: total economic activities at period t; $m \in \mathbb{N}^+$

$E_{i,q-KIBS,t}$: employment in q-type of KIBS in region i at period t.

$\sum_{j=1}^{m} E_{i,j,t}$: employment in all economic activities in region i at period t.

$\sum_{i=1}^{n} E_{i,q-KIBS,t}$: employment in q-type of KIBS in all NUTS 2 regions at period t.

$\sum_{i=1}^{n}\sum_{j=1}^{m} E_{i,j,t}$: employment in all economic activities in all NUTS 2 regions at period t.

The high spatial concentration in KIBS leads to great differences within EU regions. This may imply the existence of different groups according to the level of specialization.[5] The results of that distribution are collected in Tables 13.1 and 13.2 (see the Annex of Tables). The data reveal that a few regions with a high level of employment in KIBS have benefited from a large share of that employment. On the contrary, many regions have shown very low levels of employment in KIBS.

The geography of KIBS in EU regions has shown an irregular central-peripheral pattern, very different to the regular central-place model. Economic and political relevance seems to have influenced in the location of KIBS activity. KIBS employment has been found in capital city regions since in every country the importance of KIBS in the total employment structure in capital cities has been higher than in the other regions of the country. However, the level of economic development has also explained to some extent the regional allocation of KIBS in general.

In 1995, KIBS were very important in capital city regions and in some regions of developed countries. There, the share of employment in KIBS as compared to total employment was higher than in other regions. Consequently, these regions achieved high SQ values. On the contrary, the lowest levels of SQ in KIBS employment were located in Objective 1 regions.

In 2010, capital city regions—including those belonging to New Member Countries—and developed regions again showed the highest relative level in KIBS employment. On the contrary, those regions with the lowest SQ in KIBS belonged to 12 New Member Countries that joined the EU after the enlargement that took place in 2004 and 2007 with the exception of their capital cities (Map 13.1 registers SQ levels in KIBS in EU regions). Therefore, one out of two regions with the lowest value in SQ in KIBS belonged to a New Member State. In particular, many regions of Bulgaria, Hungary, Poland and Romania have shown the least values.

The Objective 1 regions of Cohesion Countries during the programming periods previous to 2007 (Portugal, Ireland, Greece and Spain) and Southern Italian regions have reached low levels of SQ, but they were higher than in the case of the regions of the 12 New Member States in general. Within this group, Ireland and some Objective 2 regions of Spain (Basque Country and Catalonia) have reached the medium index.

In the rest of the EU, only a few regions of developed countries located in central areas within France (6 regions out of 21) and peripheral regions of Germany (3 regions out of 33) and United Kingdom (5 regions out of 36) have obtained the lowest levels of SQ in KIBS employment.

---

[5] Due to the shape of SQ distribution in KIBS, we have classified regions into different groups according to confidence intervals and the shape of modes. The values within the confidence interval of KIBS have been split into three groups taking into account the values obtained by L-KIS. Interest in studying the top class and the various modes found have led to the consideration two additional higher groups -regions where the SQ ranges from 1.35 to 1.75 and regions whose SQ is greater than 1.75- and a lower group: regions whose SQ is less than 0.65.

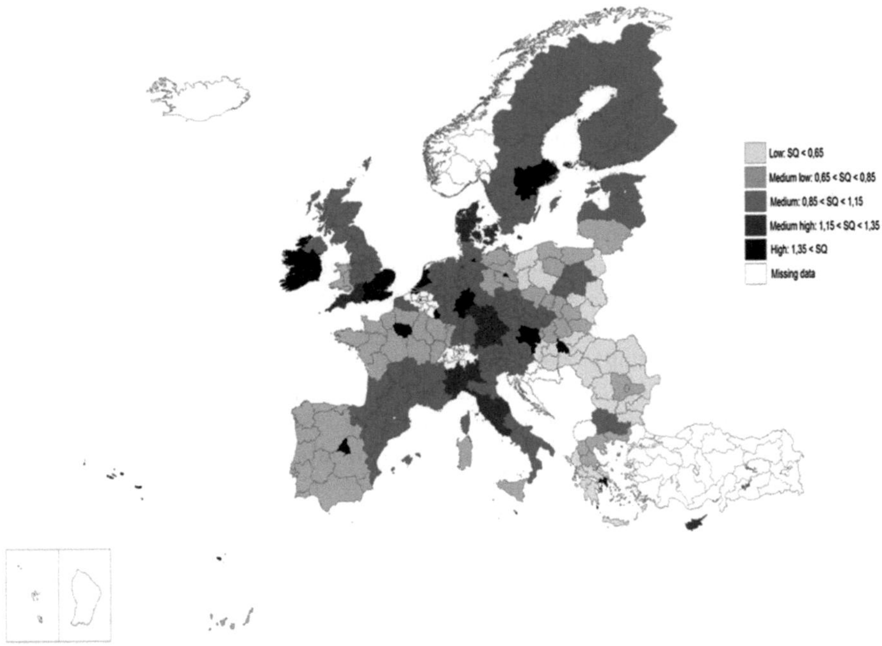

Map 13.1 Specialization quotient in total KIBS employment in 2010

3.2 Evolution in Regional Specialization in KIBS and Geographical Concentration

There have been several changes in the spatial allocation of KIBS employment in EU regions over the period 1995–2010. Throughout these years, the evolution of employment in KIBS in EU regions has been marked by several changes and periods of upheaval. Firstly, the EU expanded from 12 Member States to 27. Austria, Finland and Sweden joined the EU in 1995, 10 New Member Countries entered in 2004 and Bulgaria and Romania became members in 2007. During this period, there was a period of significant economic expansion until 2007 and a very profound crisis from 2008–2010. Moreover, there have been several programming periods of cohesion policy: 1994–1999, 2000–2006 and 2007–2013. During the first two periods, Cohesion Countries (Portugal, Ireland, Greece and Spain) benefited from the financial assistance of the EU cohesion policy; nonetheless, from 2007 onwards, those financial resources have lost relative importance in these countries, and have been reallocated towards the regions of New Member Countries. Thus, the analysis of the evolution of regional allocation of KIBS employment in EU-27 during this period should take in account these events and the methodological changes in data collection in 2000 and 2008, so that all of this period may be divided into several phases: 1995–2000, 2000–2004, 2004–2007, and 2008–2010.

In general, employment in KIBS has increased at a higher rate than total employment in the EU, so the weight of KIBS in total employment has also increased. The annual average rate of variation in KIBS employment was 3.2 % for EU-15 over the period 1995–2007, and 1.3 % for total employment. Nevertheless, growth rates in KIBS employment have been decreasing throughout the different phases. During the period 1995–2000, the average annual rate of variation was 3.7 for EU-15. Throughout the first decade of the twenty-first century, these rates were decreasing for EU-27, however they were greater in EU-15 during 2004–2007 than in the previous period (2000–2004). Nevertheless, throughout the current economic crisis, employment in KIBS has hardly increased over the period 2008–2010.

Behaviour has also been different according to the kind of region. Overall, Objective 1 regions belonging to Cohesion Countries (and Southern Italian regions) performed better with regard to growth in KIBS employment in the EU during the period 1995–2007 (within EU-15) than other regions. In those regions, the annual average rate of variation in this variable was 5.8 %, and the areas where the growth was more intense were located in Irish and Spanish regions. Thus, the SQ in KIBS employment—the share of employment in KIBS as compared to total employment in relation to the EU mean—rose in these regions on average whilst the relative values of regions in developed countries decreased (see Fig. 13.3). Therefore, the gap between these Objective 1 regions and the rest of EU regions with regard to SQ values in KIBS employment has narrowed over the 1995–2007 period. Nevertheless, this general evidence of convergence has masked significant differences in the evolution of SQ values amongst the Objective 1 regions themselves.

Amongst the causes for this different expansion one may identify several factors: the low level of SQ values in KIBS employment in these regions at the beginning; a favourable business cycle; and the effects of cohesion policy which may also have contributed to those results.

However, during the current economic crisis, employment in KIBS in these regions has decreased at a high rate between 2008 and 2010 (−3.7 %), affecting Mediterranean regions more strongly than others. In addition to the effects of a profound economic crisis, European financial resources towards these regions have reduced in relative terms for the programming period of EU cohesion policy 2007–2013.

Overall employment in KIBS in regions of New Member States has increased at a high rate. During the period 2004–2007, many regions of Poland, the Czech Republic, Estonia, Latvia, Lithuania, Cyprus, Malta and Slovakia—even some Romanian and Bulgarian regions—reached a high rate of growth in KIBS employment.

Throughout the current economic crisis, KIBS employment in many regions of New Member Countries managed to grow—mainly in regions of Poland, Romania and Slovenia—during 2008–2010 when the annual average rate of growth in KIBS employment was 3.2 %. Similar causes to the aforementioned case—growth in KIBS employment in Objective 1 regions during 1995–2007—may explain this expansion: a low initial level and effects of cohesion policy may have contributed to

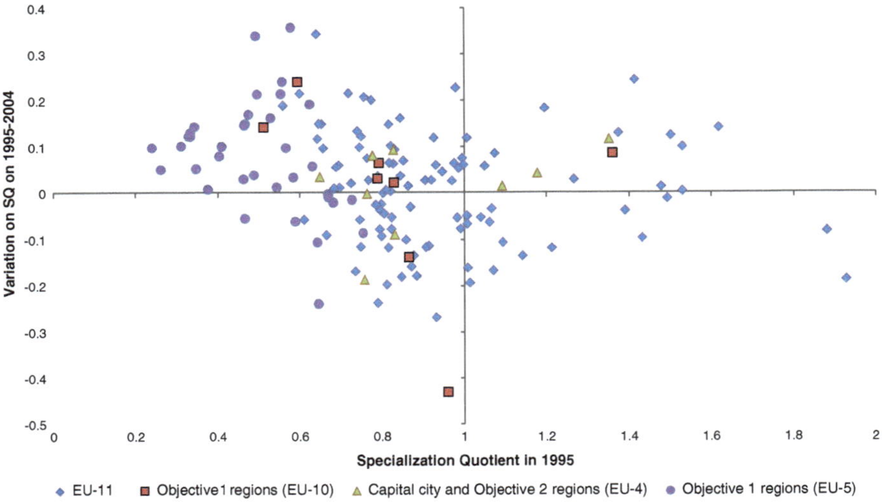

Fig. 13.3 Specialization quotient in KIBS and variation in SQ on 1995–2004 (EU-15)

these results. Since 2007, these regions have benefited most from the European financial resources devoted to cohesion policy. Overall SQ in KIBS employment has increased in these regions while the relative values of the other regions have decreased (see Fig. 13.4).

Thus, there has been a process of catching up and reduction of regional disparities in KIBS employment, but with different actors in each stage. At the same time, there has been a spatial reallocation of KIBS employment from their traditional sites towards Southern regions in the first stage and towards Eastern regions later. Among the regions that have most notably increased their position in the ranking of KIBS employment are the capital cities of Cohesion Countries (Athens, Dublin, Lisbon and Madrid) and New Member Countries (Bratislava, Budapest, Bucharest, Prague and Warsaw) and some other regions such as regions of Luxembourg or Germany (Bavaria, Bremen, Düsseldorf, Karlsruhe and Köln).

The evolution of the Gini index has been decreasing, in general terms, since 1998. Thus, there was a process of spatial dispersion in KIBS before 2007.[6] Traditional regions where KIBS were agglomerated lost presence in favour of peripheral regions in general, mainly in Spain, Portugal, Greece, Ireland, Italy, Latvia and Lithuania. Other areas that benefited from the geographical dispersion of KIBS were located in regions where KIBS were not very important in the past (e.g., in some regions of France, Germany, Sweden and the United Kingdom).

[6] Several changes in NACE in 2008 (from Rev. 1.1 to Rev. 2) have modified the definition of KIA (knowledge intensive activities). Specifically, the definition of some categories of KIBS has been altered.

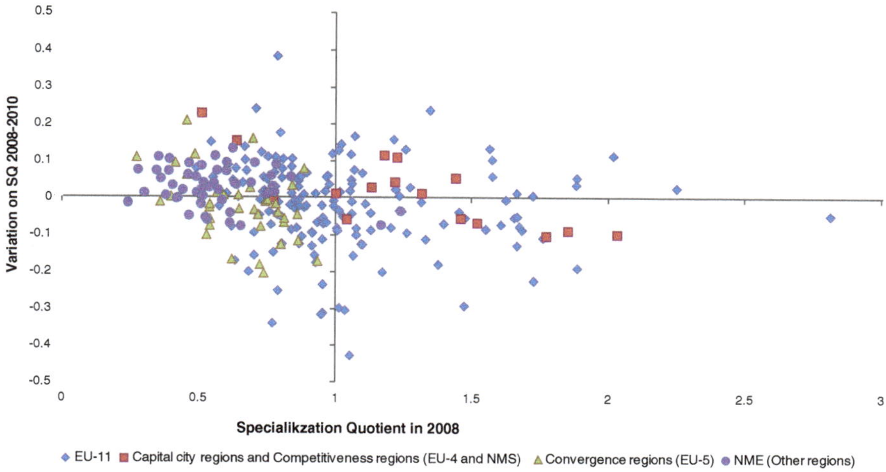

Fig. 13.4 Specialization quotient in KIBS and variation in SQ on 2008–2010 (EU-27)

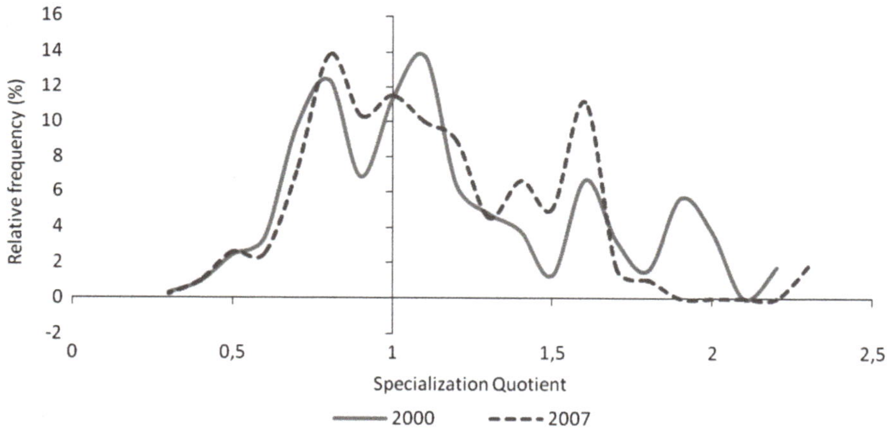

Fig. 13.5 Employment allocation in KIBS in EU regions

Regional differences in employment in KIBS across EU regions have narrowed. As a result, the shape of figures is progressively approaching a normal distribution (see Fig. 13.5) and there has been a progressive reduction in confidence intervals.

As a consequence, polarization diminished between 1995 and 2007. The superior group lost some members and reduced its level of SQ in KIBS employment throughout this period (see Fig. 13.3). Figure 13.3 shows that points with the highest SQ values in KIBS employment in 1995 have overall reduced their SQ value over this period. At the same time, regions with a low level of SQ were able to increase their level of specialization during the period of economic expansion, moving from a low level of specialization to a higher specialization profile. A reduction in regional differences has been the result of widening the share of

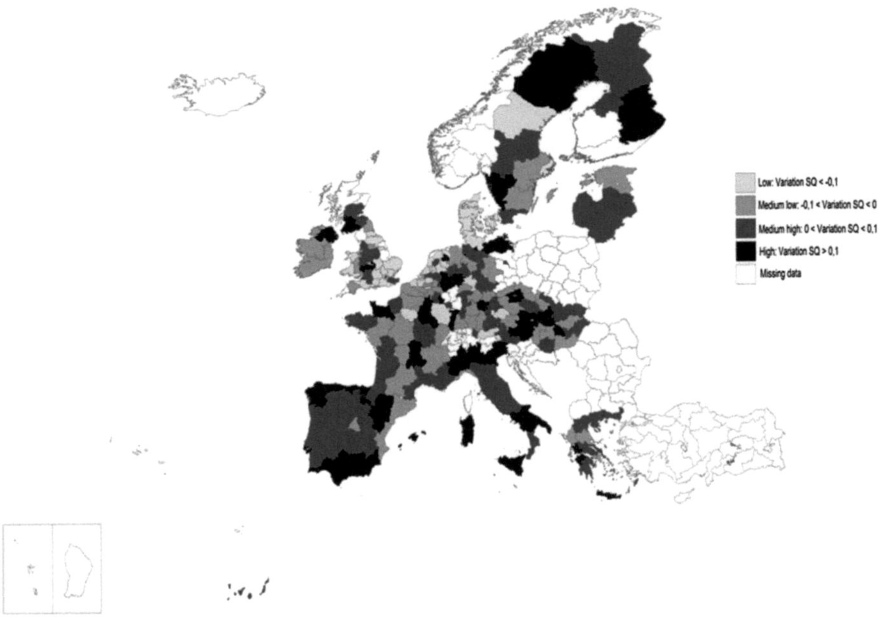

Map 13.2 Variation in specialization quotient in KIBS employment 2000–2007

employment in KIBS in relation to total employment in regions with a low SQ value at the beginning. The majority of regions in Cohesion Countries and other Objective 1 regions have increased their employment in KIBS at higher rates in general, thus elevating their SQ (see Map 13.2 registering variation in SQ in 2000–2007). These regions have benefited most from the process of expansion and this may be a result of cohesion policies during the period 1994–2006.

Enlargements in 2004 and 2007 to New Member States meant that regional disparities in KIBS employment increased for statistical reasons because the SQ values of regions in New Member States were lower.

Capital city regions of New Member Countries benefited from the enlargement, increasing their employment in KIBS at higher rates than other regions. During the period 2004–2007, many regions of New Member Countries also increased their SQ in KIBS.

A tendency towards a reduction in geographical polarization in KIBS employment seems to have been broken during the recent crisis. Taking into account the aforementioned need for caution, if one analyses the specialization profile, one can draw the conclusion that the economic cycle has changed the evolution of the specialization profile in KIBS. Regions of early Cohesion Countries have lost employment in KIBS and have reduced their SQ value in KIBS in general (see Fig. 13.4). Nevertheless, there has been a process of convergence within these regions (except for capital cities) leading to a convergence club during the period 2008–2010.

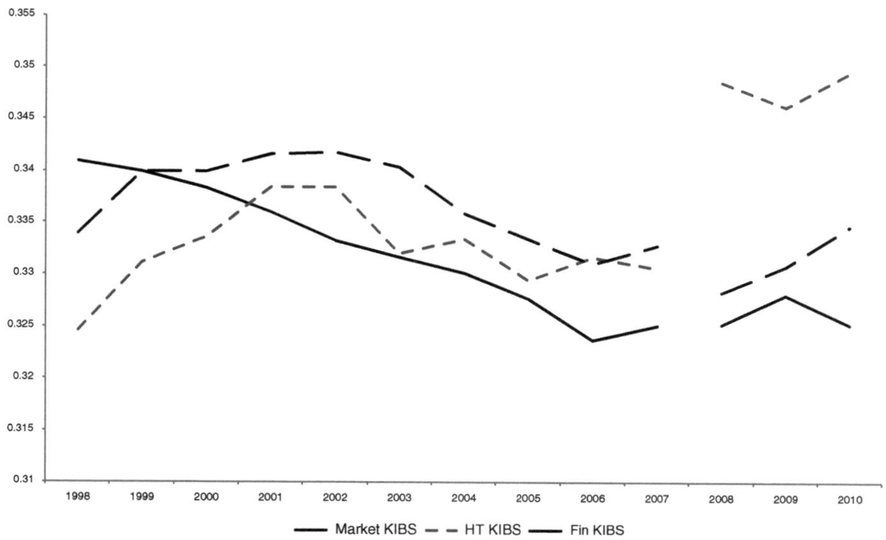

Fig. 13.6 Evolution of Gini index in KIBS employment in EU regions

However, regional differences in KIBS do not seem to have increased because on the other hand, regions of New Member Countries have increased employment in KIBS and SQ values in general. Polish, Romanian and Bulgarian regions benefited most during this period. Thus, the results of the convergence in KIBS employment are not clear in this current period of crisis.

3.3 *Regional Disparities Within KIBS*

If we now look separately at the categories of KIBS, one can observe important differences. Market KIBS employment has been the main category in KIBS employment (more than two out of three of jobs in KIBS have been created in this activity). Thus, the main regional differences are located in High-Tech KIBS and Financial KIBS. The greatest concentration is found in High-Tech KIBS and Financial KIBS (Fig. 13.6) as the Gini index shows.

Polarization is higher in High-Tech KIBS and Financial KIBS than in Market KIBS. This can be appreciated in the shape of their respective distributions (see Fig. 13.7) and the values of the confidence intervals.[7] The highest values of specialization are achieved by Financial KIS, where Luxemburg had over four

[7] The distribution of SQ in Market-KIBS is very similar to KIBS in general, with the same confidence intervals. Nonetheless, in the case of HT-KIBS and F-KIBS the range is roughly from 0.45 to 1.55, and their plots are the most different from those of a normal distribution.

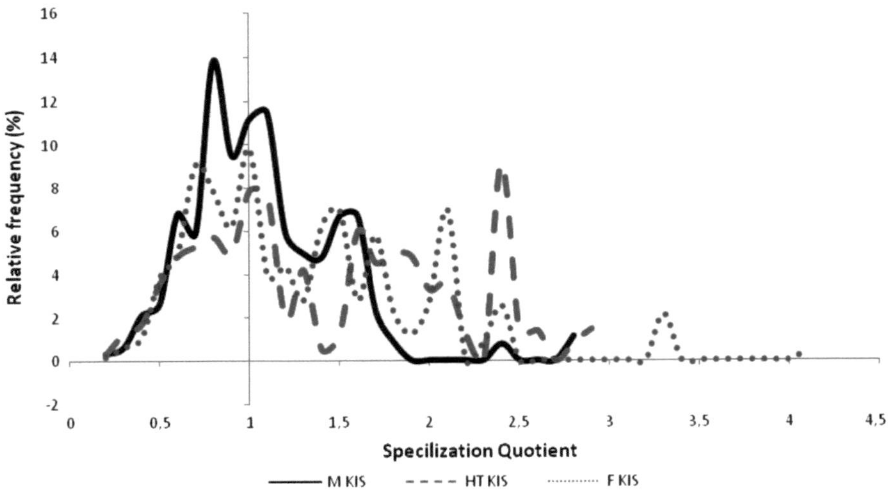

Fig. 13.7 Employment allocation in KIBS in EU regions in 2010

times the mean level of the EU-27 in 2010 and London three times that mean, thus proving to be the most specialized locations in Europe. Maps 13.3, 13.4, and 13.5 represent the levels of specialization in every category of KIBS in 2010.

There are few differences between Maps 13.1 and 13.3. Consequently, the overview for Market KIBS is very similar to KIBS in general.

The differences are most significant in the cases of High Tech KIBS and Financial KIBS. High Tech KIBS are concentrated in capital cities in general. Moreover, employment in this kind of KIBS is particularly relevant in Northern regions (Sweden, Denmark, Holland, Germany, Belgium, Luxembourg and the United Kingdom) in general. In these regions, the level of technology is very high, as some indicators reveal (I + D/PIB, patent applications, etc.). Furthermore, manufacturing firms in these areas may demand a great amount of this kind of KIBS due to the process of services externalization.

Financial KIBS are also located in capital cities. Other regions with a high share of employment in Financial KIS in 2010 are located in Belgium, Luxembourg, Germany and the United Kingdom. Map 13.5 registers the main financial sites of Europe.

Every category of KIBS shows different paths in terms of the evolution of the Gini index, but it is only evident that geographical concentration has decreased in Market KIBS. The results of considering different periods in these categories of KIBS do not show any important changes with respect to those mentioned for KIBS in general.

The development of employment in each category of KIBS shows some important differences. The most dynamic has been Market KIBS, whose annual average rate of growth was higher than 4 % during the period 1995–2007 (EU-15) and positive during the period 2008–2010. On the contrary, the growth of employment

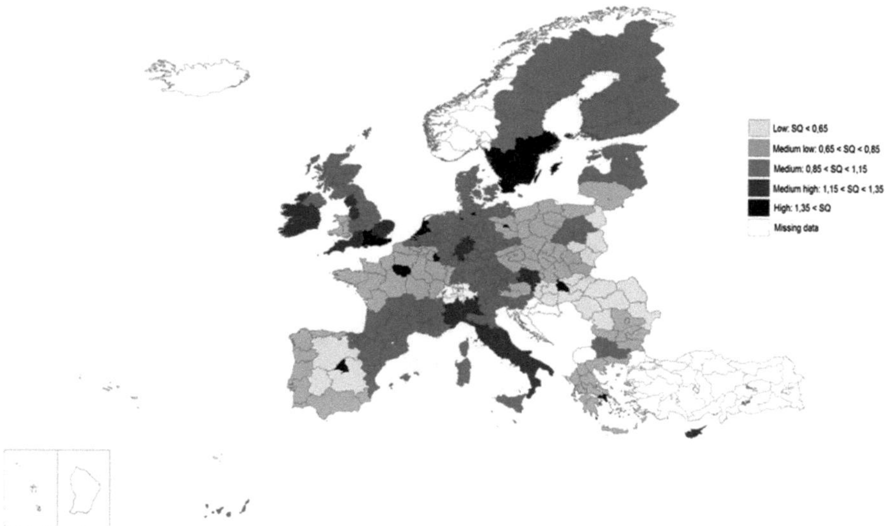

Map 13.3 Specialization quotient in market KIBS employment in 2010

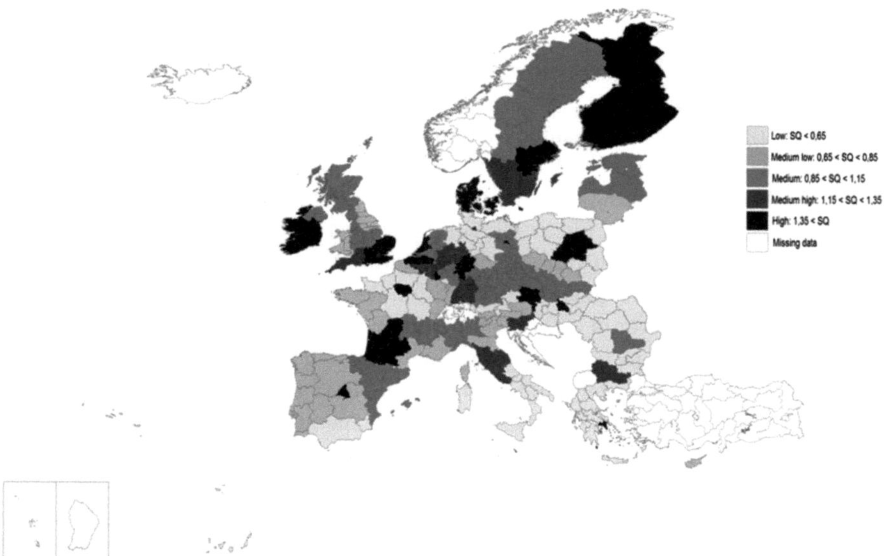

Map 13.4 Specialization quotient in high tech KIBS employment in 2010

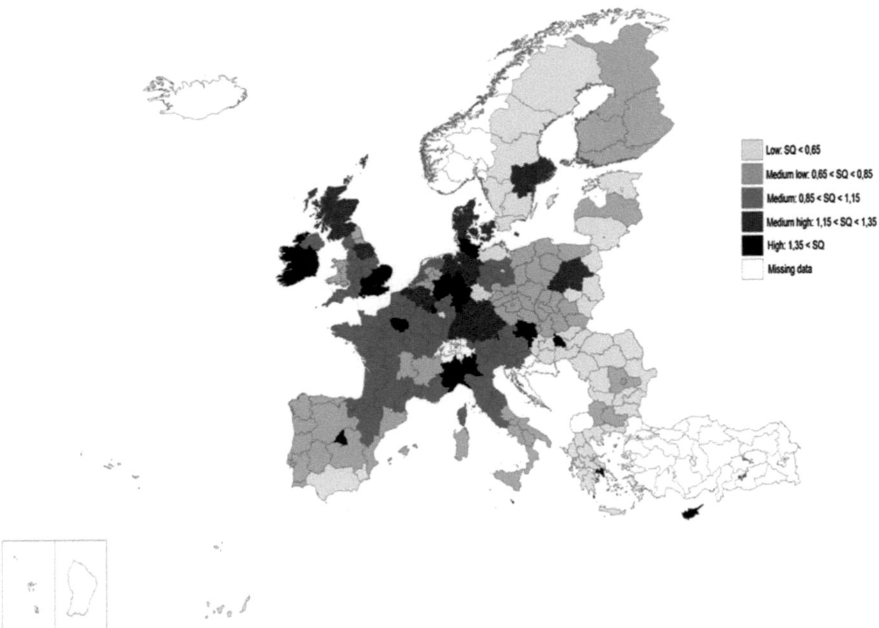

Map 13.5 Specialization quotient in financial KIBS employment in 2010

in Financial KIS has been very low during the expansion cycle and negative during the current economic crisis.

The spatial allocation of employment in each category of KIBS changed in EU regions during the period 1995–2010. SQ in the period of economic expansion (1995–2007) shows some significant differences among categories of KIBS, mainly between Market KIBS and Financial KIS. In the first case, regional differences tend to clearly reduce due to a process of catching up followed by regions with the lowest level that have augmented their SQ level in general. Therefore, its confidence interval was lower in 2007 than in previous years, and the number of regions included in high and in low categories has diminished as well as their share of employment in KIBS. The evolution in Market KIBS has been very similar to that of Total KIBS. Furthermore, the conclusions mentioned above could also be applied here. Regions in Cohesion Countries have benefited most from the growth in employment in Market KIBS (see Tables 13.1 and 13.2). The annual average rate of growth in Market KIBS employment has been higher than 4.4 % in this area during the period 1995–2007 (UE-15). Employment in Market KIBS increased notably in several regions of Italy (9 out of 21), Germany (8 out of 33), France (6 out of 22) and Spain (7 out of 17) mainly.

Figure 13.8 shows that points with the highest SQ values in Market KIBS employment in 1995 have generally reduced their SQ value over the period 1995–2004. At the same time, points with the lowest SQ values increased their

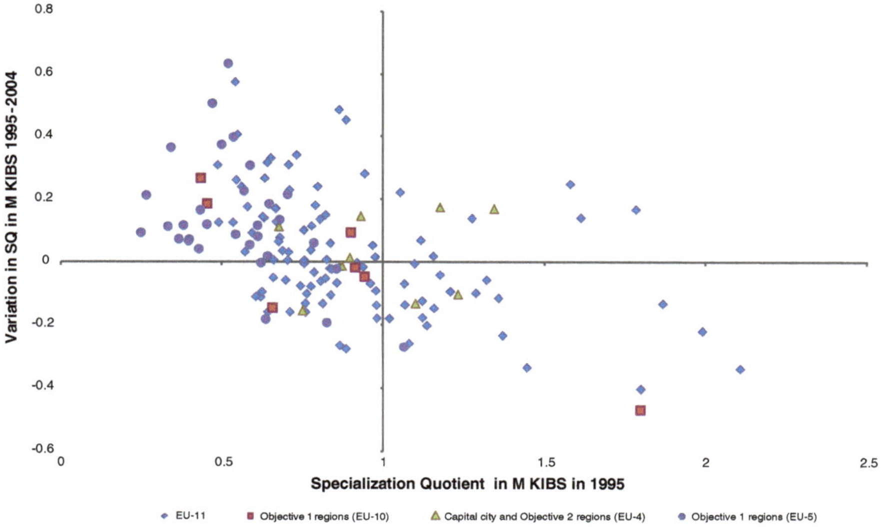

Fig. 13.8 Specialization quotient in market KIBS in 1995 and variation in SQ 1995–2004. EU-15 regions

SQ values during that period. Therefore, geographical polarization in Market KIBS employment has reduced during the expansion cycle.

However, regions in Cohesion Countries have lost a great deal of employment during the crisis (the annual average rate of variation: −4 %). Spanish regions and Southern Italian regions have been the main actors of this unstable behaviour.

Regions in New Member Countries have also benefited from expansion in Market KIBS employment in general, mainly before 2007 (enlargement to EU-25 and to UE-27), but also during the period of crisis. Nevertheless, the highest level of generation of net employment in Market KIBS during the crisis has been located in regions of the United Kingdom (14 out of 36), Germany (9 out of 33), France (3 out of 22) and Poland (3 out of 16).

Something similar to Market KIBS has happened to High-Tech KIS, but not so clearly, since the number of regions in high and low categories of SQ has increased as has its confidence interval. The main results of regional evolution in SQ in HT KIS have been that a lot of Southern regions (e.g., Italy, Spain and Portugal) have increased their SQ values between 1995 and 2007. Moreover, some regions of Germany and Holland have experienced the same phenomenon.

Employment in regions of Cohesion Countries has been the most dynamic during the 1995–2007 period, and has performed positively during the recent economic crisis. The main actors have been Spanish regions (located mainly on the Mediterranean Coast and in Madrid), Southern Italian regions and the capital cities of Cohesion Countries and New Member States. These regions started from a very low position. Therefore, it seems to have been a process of catching up stimulated by the assistance of cohesion policy.

Those regions with the highest SQ values in HT KIBS employment at the beginning of the period have generally reduced their SQ value during the period 1995–2007. At the same time, regions with the lowest SQ values have increased their SQ values in that period. Therefore, geographical polarization in HT KIBS employment has decreased in general.

Nevertheless, during the current crisis, the process of reduction in regional disparities seems to have ceased. Old Objective 1 regions from the Cohesion Countries appear to have lost positions in these years in favour of some central regions and New Member Countries regions (Polish regions mainly).

The behaviour of Financial KIS has differed from Market KIBS considerably, showing a polarisation tendency towards the extreme categories throughout 1995–2007, although it has mainly benefitted regions with the lowest levels. Nonetheless, it is important to highlight the fact that this situation may be generated by the use of a less labour-intensive technology in regions with medium profile, which would therefore lead to a lower amount of employment in the category of regional specialization. During this same period, employment in Financial KIS has risen at higher rates in many regions of New Member Countries mainly (e.g., Latvia, Lithuania, Poland and Slovakia), and also in Ireland, Spain and Luxemburg.

This tendency seems to have dramatically changed during the current economic crisis (2008–2010). Taking into account the need for caution mentioned above, regional differences in Financial KIS have clearly increased. The effects of the financial crisis have mainly affected regions at the lowest levels and particularly regions of the Cohesion Countries, which is reflected in their number, their share of employment, and their confidence intervals (see Tables 13.1 and 13.2). The reallocation of Financial employment has been very important for many regions in Belgium, Bulgaria, Estonia, Finland, Holland, Latvia, Portugal, Spain and the United Kingdom towards other regions of New Member Countries (the capital cities of the Czech Republic, Poland, Romania, Slovakia and Slovenia mainly), Luxembourg, some regions in the area around London and in some German regions with high SQ. Therefore, the latter regions have increased their employment in Financial KIS and have benefited most during the crisis.

Figure 13.9 shows that regions of EU-4 (and Southern Italian regions) have generally lost employment in Financial KIBS employment during the current crisis. On the contrary, regions of New Member States have increased employment in this activity in general.

4 Factors Influencing the Location of KIBS: Empirical Investigation

Specifically, this section investigates the potential drivers that may influence the location of KIBS in European regions during the last years. In order to carry out the empirical analysis, we adopted the approach proposed by Miles (2005), according

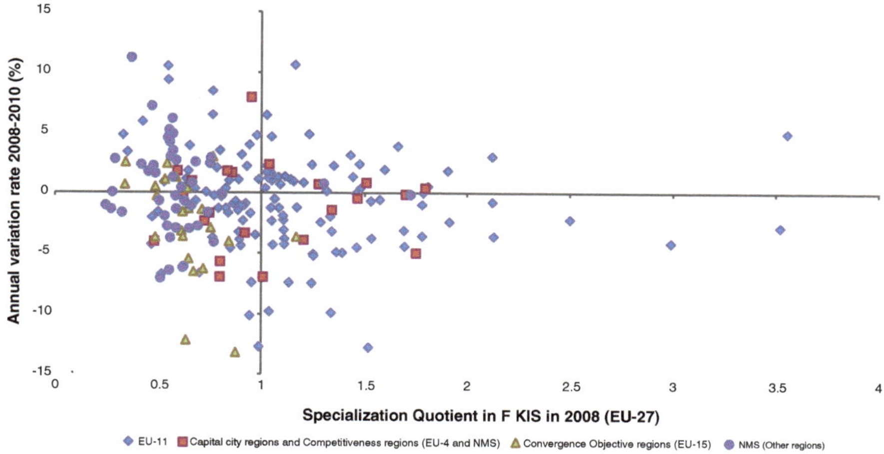

Fig. 13.9 SQ in financial KIS in 2008 and variation in F KIS employment on 2008–2010 (EU-27 regions)

to which the term *knowledge-intensive* can be interpreted in terms of labour qualification (Miles, 2005). Thus, a knowledge-intensive firm refers to a firm that undertakes complex operations of an intellectual nature where human capital is the dominant factor (Alvesson, 1995).

From an empirical viewpoint, due to the methodological break applied by Eurostat from 2008 onwards in the measurement of KIBS, two periods are analysed separately. On the one hand, we conducted an analysis for the period 2000–2007 (i.e. economic expansion) and, on the other hand, we explored the period 2008–2010 (i.e. beginning of the current economic crisis).

The previous literature dealing with the main variables influencing the location of KIBS points to several variables related to accessibility to clients and to knowledge and information and agglomeration economies. Therefore, this section serves to build a model that helps to explain the SQ in total KIBS (dependent variable) in relation to other variables, such as: capital city (CC), employment in total manufacturing (ETM), employment in high-tech manufacturing (EHTM), human resources in science and technology (HRST), GDP/km^2 (density of production, labelled DP), patent applications (PA), and accessibility (A), in each temporal moment t for each region I, according to the following specification:

$$SQ_{it} = \alpha_i + \beta_1 CC_{it} + \beta_2 ETM_{it} + \beta_3 EHTM_{it} + \beta_4 HRST_{it} + \beta_5 DP_{it} + \beta_6 PA_{it} + \beta_7 A_{it} + v_i + \varepsilon_{it}$$

where $i = 1, \ldots, N$ are the regions under study (N = 204), $t = 1, \ldots, L$ is the temporal length under analysis, v_i is the random component, and ε_{it} is the residual of the model.

In relation to the period 2000–2007, Table 13.3 (see Annex) reports the results obtained. We can observe that capital city status, employment in high-tech manufacturing, availability of human resources in science and technology, wealth concentration (measured by the variable GDP/km^2), patent applications and accessibility positively and significantly influence specialization in KIBS at a regional level in Europe. On the contrary, employment in total manufacturing exerts a negative influence on such specialization.

The second period under study covers the years 2008–2010. In this case, Table 13.4 (see Annex) includes the information. There are three main differences in comparison to the other period. Firstly, wealth concentration is not a relevant driver of regional specialization during the period 2008–2010, whilst in the previous years this variable played a relevant role in explaining the levels of regional specialization. Secondly, the variable "capital city" has a positive and stronger influence than in the previous years, which represents the polarisation of KIBS from peripheral areas towards more traditional regions (i.e. capital cities) from 2008–2010. Thirdly, the negative influence of the variable "employment" on total manufacturing is considerably stronger during this second period under analysis.

5 Conclusions

KIBS constitute a relevant source of employment, production, investment, and knowledge dissemination, especially for industrialised areas. Thus, they may have a relevant role in the spatial allocation of economic activities and regional development. Whilst previous studies have shown that KIBS were not equally distributed across European regions, is this description still valid nowadays? This chapter seeks to explore whether the patterns of regional concentration and specialization in KIBS have changed during the last 15 years. In order to achieve this aim, this study has investigated the evolution of both geographical distribution as well as regional specialization in KIBS (i.e., Total KIBS, High-Tech KIBS, Market KIBS, and Financial KIBS) in European regions during the last years. It is important to note that the period under study was not randomly chosen, as it involves both the enlargement of the EU undertaken in 2004 and in 2007 as well as the beginning of the current economic crisis.

The data analysed reveal that in 2010 KIBS was the most geographically concentrated activity. Within the context of KIBS, the greatest geographical concentration has been found in High-Tech KIS and Financial KIS.

The second topic addressed in this study is the regional specialization of European regions. By calculating the specialization quotient, European regions are classified into different profiles of regional specialization. There are significant differences among the groups of regions. Those regions with the highest values of SQ concentrated KIBS employment. Thus, KIBS were concentrated in capital city regions, Central and Northern regions of EU-27. The share of KIBS employment as compared to total employment in those areas was very high due to the agglomeration of economic activity and to a greater accessibility to markets and knowledge resources in these regions. On the contrary, Objective 1 regions during the programming period 2000–2006, and regions of New Member Countries (except capital cities regions) have obtained the lowest levels of SQ values in KIBS.

Nevertheless, our results concerning the evolution of spatial concentration, during the period 1995–2007, show a spatial diffusion process in KIBS location. Traditional regions where KIBS were agglomerated have lost participation in favour of peripheral regions as well as regions where KIBS were not very important in the past (e.g., in France, Germany, Sweden and the United Kingdom). Therefore, it seems that there has been a process of catching up and geographical diffusion during the economic expansion period, which reflects some of the effects of cohesion policy. Nonetheless, this process of spatial dispersion may have shifted during 2008–2010 due to the economic crisis, which has affected regions in Cohesion Countries in a more intense way.

Many regions in Cohesion Countries managed to increase employment in general and KIBS employment in particular during the period prior to the current crisis, thereby approaching the levels of regions in developed countries. However, through the period 2008–2010 many of them lost employment and have also lagged behind in the catching up process. On the contrary, regions in New Member Countries have generally made progress during the two periods.

Moreover, there has been a catching up process and a reduction in regional disparities in KIBS employment, but with different actors in each stage. At the same time, there has been a spatial reallocation of KIBS employment from the traditional sites towards Southern regions in the first stage and later towards Eastern regions. Within the regions that have most increased their position in the ranking of level in KIBS employment one can find the capital cities of Cohesion Countries (EU-4) and New Member Countries and a few other regions.

Reduction of regional disparities in KIBS has not been observed for Financial KIBS, which seem to have experienced a process of geographical polarisation. However, employment in Financial KIBS has risen at higher rates in many regions in New Member States, Ireland and Portugal.

Furthermore, our exploration of factors influencing location of KIBS confirms the main conclusions of previous relevant literature. The results show that the variables representing accessibility to clients and resources of knowledge and economies of agglomeration, such as patent applications, human resources, transport accessibility and density of GDP serve to explain the regional distribution in KIBS of European regions.

Finally, we conducted an empirical analysis in order to address the factors that may help explain the location of KIBS across European regions for the periods 2000–2007 and 2008–2010 separately. The data show that capital city status, employment in high-tech manufacturing, availability of human resources in science and technology, patent applications and accessibility positively and significantly influence the specialization in KIBS at a regional level in Europe. On the contrary, employment in total manufacturing exerts a negative influence on such specialization. The main difference between both periods is that, whilst the density of production (measured by the variable GDP/km^2) exerts a positive and statistically significant influence during the period of economic expansion, this influence does not hold for the period of current economic crisis (i.e. 2008–2010).

In summary, employment in KIBS is concentrated geographically in EU regions in capital cities, Central and Northern regions due to the agglomeration of economic activity and to the greater accessibility to markets and knowledge resources in these regions. However, a process of spatial dispersion has occurred during the economic expansion period (1995–2007), coinciding with the process of enlargement (2004) and with cohesion policy (1994–1999 and 2000–2006). Areas where employment in KIBS has increased at higher rates have been regions in Cohesion Countries (EU-4), and Objective 1 in 1995–2007 and in New Member Countries in 2004–2007. This process seems to have been decelerated during the economic crisis, because KIBS employment has reduced in regions of Cohesion Countries and have increased in Mew Member Countries regions. Thus, while the application of cohesion policy for the period 2007–2013 is expected to benefit the latter regions, the data for 2008–2010 point to such a conclusion.

Respect to policy implications, in a context characterized by the global competition in the KIBS sectors, policy makers are seeking to build competitive advantage for their nations or regions. In this sense, nations that participate in the production and delivery of knowledge intensive business services should recognize certain strategic issues that may give them a sustainable competitive advantage. These issues involve policy implications.

First, investment in infrastructures and ICTs. KIBS count on suppliers who can deliver services with seamless efficiency. Physical infrastructures may be less critical for certain types of knowledge services (e.g., medical transcriptions), but they are vital for the delivery of other services that require regular face-to-face interactions (e.g., medical services; educational programs). Furthermore, cross-border connectivity is clearly facilitated and mediated by ICT through teleconferencing, e-mail, video-conferencing and virtual networks (Jones, 2005). ICTs have led to often rapid and unexpected changes in competitive positions of firms and countries (Kautonen et al., 2009). Therefore, focused investment on infrastructures and ICTs must be warranted so that business can be conducted (Daniels, 2004).

Second, the strategic role of small and medium-sized enterprises (SME) (Bryson & Rusten, 2005). SMEs, constituting as much as 90 % of enterprises in many countries around the globe, have been identified as the driving force behind innovations and entrepreneurial investments, job creation, international trade, and new product and service developments. However, SMEs are usually confronted with enormous challenges. SMEs are usually more vulnerable to managerial, financial and technological challenges as compared to bigger companies. Thus, it is imperative that nations pay special attention to supporting SMEs.

Third, research and development (R&D). This variable represents knowledge services at their most intense level. While there have been huge improvements in R&D investments, enhancing greater university-industry collaborations should be promoted.

Acknowledgments Authors are grateful for the useful comments received during the 51st Congress of the European Regional Science Association held in Barcelona (Spain) in 2011 from several participants, as well as for the graphical assistance with the maps provided by J. M. Trujillo.

Annex of Tables

Table 13.1 Share in employment by level of specialization quotient (%)

		UE-15			UE-27		
		1995	2000	2004	2004	2007	2010
HT-KIS	Superior (SQ > 1.75)	8.04	16.24	11.37	18.49	11.63	31.97
	High (1.75 < SQ < 1.35)	8.96	19.45	21.83	15.75	27.43	12.26
	High medium (1.35 < SQ < 1.15)	10.25	12.01	9.99	11.77	8.80	6.58
	Medium (1.15 < SQ < 0.85)	21.79	21.04	29.63	26.90	25.01	19.15
	Low medium (0.85 < SQ < 0.65)	17.14	16.41	15.33	12.90	13.41	11.00
	Low (SQ < 0.65)	5.45	9.73	8.57	11.19	10.23	11.41
	ND	28.38	5.12	3.27	3.00	3.51	7.63
	Total	100	100	100	100	100	100
M-KIS	Superior (SQ > 1.75)	9.93	12.09	10.17	11.34	9.12	4.18
	High (1.75 < SQ < 1.35)	5.42	13.91	14.55	18.44	19.28	24.95
	High medium (1.35 < SQ < 1.15)	8.80	11.90	10.77	13.23	15.03	13.10
	Medium (1.15 < SQ < 0.85)	18.56	28.50	37.54	33.31	36.56	32.08
	Low medium (0.85 < SQ < 0.65)	18.86	23.32	19.46	14.96	12.18	16.36
	Low (SQ < 0.65)	8.65	6.98	5.93	7.32	7.68	8.63
	ND	29.77	3.31	1.57	1.39	0.15	0.71
	Total	100	100	100	100	100	100
Fin-KIS	Superior (SQ > 1.75)	3.26	10.82	10.87	16.28	13.00	19.2
	High (1.75 < SQ < 1.35)	9.81	21.03	19.79	21.19	25.11	21.72
	High medium (1.35 < SQ < 1.15)	17.17	15.06	20.25	15.42	10.73	7.14
	Medium (1.15 < SQ < 0.85)	18.28	23.43	21.75	19.70	22.79	20.27
	Low medium (0.85 < SQ < 0.65)	16.28	16.72	11.98	12.40	12.63	16.47
	Low (SQ < 0.65)	6.36	7.54	12.51	4.60	11.44	10.32
	ND	28.84	5.39	2.85	10.41	4.30	4.88
	Total	100	100	100	100	100	100
T-KIBS	Superior (SQ > 1.75)	7.09	11.63	3.25	13.33	10.66	12.78
	High (1.75 < SQ < 1.35)	8.88	14.83	21.78	13.87	17.23	18.82
	High medium (1.35 < SQ < 1.15)	5.48	10.35	12.17	17.65	15.79	14.75
	Medium (1.15 < SQ < 0.85)	23.00	31.65	34.18	29.25	32.07	25.51
	Low medium (0.85 < SQ < 0.65)	20.75	22.09	22.81	16.31	16.72	19.23
	Low (SQ < 0.65)	6.52	6.19	4.22	8.24	7.53	8.90
	ND	28.27	3.25	1.58	1.35	0.02	0.01
	Total	100	100	100	100	100	100
O-KIS	Superior (SQ > 1.75)	2.58	0.72	0.27	0.24	0.36	
	High (1.75 < SQ < 1.35)	3.69	8.41	9.97	13.34	16.79	3.48
	High medium (1.35 < SQ < 1.15)	13.23	27.07	23.98	24.68	24.75	33.16

(continued)

Table 13.1 (continued)

		UE-15			UE-27		
		1995	2000	2004	2004	2007	2010
	Medium (1.15 < SQ < 0.85)	30.08	38.79	41.72	34.80	31.44	43.66
	Low medium (0.85 < SQ < 0.65)	20.67	17.78	18.31	21.14	24.06	16.41
	Low (SQ < 0.65)	1.88	3.01	3.34	3.76	2.57	3.25
	ND	27.88	4.22	2.41	2.04	0.03	0.04
	Total	100	100	100	100	100	100
L-KIS	Superior (SQ > 1.75)		0.07	0.06			
	High (1.75 < SQ < 1.35)	1.00	1.24	1.45	1.42	1.27	2.41
	High medium (1.35 < SQ < 1.15)	12.64	11.38	11.66	15.33	13.74	16.37
	Medium (1.15 < SQ < 0.85)	55.61	77.07	77.04	71.43	75.34	72.71
	Low medium (0.85 < SQ < 0.65)	8.11	6.81	8.26	9.47	8.50	7.45
	Low (SQ < 0.65)	0.17	0.14		1.03	1.12	1.03
	ND	22.48	3.36	1.52	1.26	0.04	0.03
	Total	100	100	100	100	100	100

Source: Eurostat. Self elaboration
ND no data

Table 13.2 Number of regions by level of specialization quotient

		UE-15			UE-27		
		1995	2000	2004	2004	2007	2010
HT-KIS	Superior (SQ > 1.75)	2	8	5	12	11	25
	High (1.75 < SQ < 1.35)	7	20	21	22	31	13
	High medium (1.35 < SQ < 1.15)	17	18	13	19	18	13
	Medium (1.15 < SQ < 0.85)	39	37	59	68	61	42
	Low medium (0.85 < SQ < 0.65)	39	47	42	50	49	38
	Low (SQ < 0.65)	24	42	39	61	57	61
	ND	83	39	32	35	40	75
	Total	211	211	211	267	267	267
M-KIS	Superior (SQ > 1.75)	6	5	6	8	7	4
	High (1.75 < SQ < 1.35)	7	13	11	21	20	28
	High medium (1.35 < SQ < 1.15)	9	23	16	22	28	22
	Medium (1.15 < SQ < 0.85)	35	47	69	88	97	84
	Low medium (0.85 < SQ < 0.65)	48	64	61	61	54	67
	Low (SQ < 0.65)	45	42	37	56	56	52
	ND	61	17	11	11	5	10
	Total	211	211	211	267	267	267
Fin-KIS	Superior (SQ > 1.75)	4	8	8	17	13	15
	High (1.75 < SQ < 1.35)	6	19	21	25	30	22
	High medium (1.35 < SQ < 1.15)	15	18	26	27	17	18
	Medium (1.15 < SQ < 0.85)	37	47	44	52	56	49
	Low medium (0.85 < SQ < 0.65)	39	40	31	43	40	57
	Low (SQ < 0.65)	28	38	49	64	68	59
	ND	82	41	32	39	43	47
	Total	211	211	211	267	267	267

(continued)

Table 13.2 (continued)

		UE-15			UE-27		
		1995	2000	2004	2004	2007	2010
T-KIBS	Superior (SQ > 1.75)	2	5	3	9	6	7
	High (1.75 < SQ < 1.35)	12	16	19	19	25	25
	High medium (1.35 < SQ < 1.15)	4	12	11	24	23	22
	Medium (1.15 < SQ < 0.85)	41	58	63	83	87	67
	Low medium (0.85 < SQ < 0.65)	58	74	77	63	65	78
	Low (SQ < 0.65)	40	34	31	62	59	66
	ND	54	12	7	7	2	2
	Total	211	211	211	267	267	267
O-KIS	Superior (SQ > 1.75)	5	2	1	2	2	
	High (1.75 < SQ < 1.35)	9	16	21	30	44	12
	High medium (1.35 < SQ < 1.15)	24	48	38	52	53	79
	Medium (1.15 < SQ < 0.85)	54	73	84	90	79	105
	Low medium (0.85 < SQ < 0.65)	48	44	47	68	72	59
	Low (SQ < 0.65)	17	16	13	18	15	10
	ND	54	12	7	7	2	2
	Total	211	211	211	267	267	267
L-KIS	Superior (SQ > 1.75)			1	1		
	High (1.75 < SQ < 1.35)	4	4	5	6	6	6
	High medium (1.35 < SQ < 1.15)	23	18	20	36	32	31
	Medium (1.15 < SQ < 0.85)	109	154	158	185	194	193
	Low medium (0.85 < SQ < 0.65)	20	22	20	29	30	32
	Low (SQ < 0.65)	1	1		3	3	3
	ND	54	12	7	7	2	2
	Total	211	211	211	267	267	267

Source: Eurostat. Self elaboration
ND no data

Table 13.3 Estimates[a,b] for the drivers of regional specialization (2000–2007)

Variable	Coefficient
Capital city	0.052* (0.653)
Employment in total manufacturing	−0.417** (−3.317)
Employment in high-tech manufacturing	0.315*** (3.231)
Human resources in science and technology	0.396* (2.571)
GDP/km^2	0.289* (2.114)
Patent applications	0.377** (3.239)
Accessibility	0.233** (3.032)
R^2	0.887
Adjusted R^2	0.851

[a]Dependent variable: specialization quotient for Total KIBS
[b]These coefficients correspond to standardized coefficients
$^{*}p < 0.05$; $^{**}p < 0.01$; $^{***}p < 0.001$

Table 13.4 Estimates[a,b] for the drivers of regional specialization (2008–2010)

Variable	Coefficient
Capital city	0.064* (0.895)
Employment in total manufacturing	−0.539*** (−4.792)
Employment in high-tech manufacturing	0.339** (3.889)
Human resources in science and technology	0.268* (2.518)
GDP/km^2	0.164 (1.346)
Patent applications	0.372** (3.577)
Accessibility	0.249** (3.617)
R^2	0.910
Adjusted R^2	0.881

[a]Dependent variable: specialization quotient for Total KIBS
[b]These coefficients correspond to standardized coefficients
$^{*}p < 0.05$; $^{**}p < 0.01$; $^{***}p < 0.001$

References

Acs, Z. J., de Groot, H. L. F., & Nijkamp, P. (2002). *The emergence of the knowledge economy: A regional perspective*. Berlin: Springer.

Alvesson, M. (1995). *Management of knowledge-intensive companies*. Berlin: de Gruyter.

Alvesson, M. (2004). *Knowledge work and knowledge-intensive firms*. New York: Oxford University Press.

Anas, A., Arnott, R., & Small, K. (1998). Urban spatial structure. *Journal of Economic Literature, 36*, 1426–1464.

Antonelli, C. (1999). The evolution of the industrial organization of the production of knowledge. *Cambridge Journal of Economics, 23*, 243–60.

Aslesen, H. W., & Isaksen, A. (2007). Knowledge-intensive business services and urban industrial development. *Service Industrial Journal, 27*(3), 321–338.

Attewell, P. (1992). Technology diffusion and organizational learning: The case of business computing. *Organization Science, 3*, 1–19.

Beyers, W. (2005). Services and the changing economic base of regions in the United States. *Service Industries Journal, 25*(4), 461–476.

Beyers, W., & Alvine, M. (1985). Export services in postindustrial society. *Papers in Regional Science, 57*, 33–45.

Boschma, R. (2005). Proximity and innovation: A critical assessment. *Regional Studies, 39*(1), 61–74.

Boschma, R., & Weterings, A. (2005). The effect of regional differences on the performance of software firms in the Netherlands. *Journal of Economic Geography, 5*(5), 567–588.

Brüderl, J., Preisendörfer, P., & Ziegler, R. (1996). *The success of newly founded firms: An empirical study analysing the chances and the risks of firm start-ups*. Berlin: Duncker & Humblot.

Bryson, J. R., & Rusten, G. (2005). Spatial divisions of expertise: Knowledge intensive business service firms and regional development in Norway. *The Service Industries Journal, 25*(8), 959–977.

Coffey, W. (2000). The geographies of producer services. *Urban Geography, 21*, 170–83.

Coffey, W., & Polése, M. (1987). Trade and location of producer services: A Canadian perspective. *Environment & Planning, 19*, 597–611.

Coffey, W., & Polèse, M. (1989). Producer services and regional development: A policy-oriented perspective. *Papers of Regional Science Association, 67*, 13–67.
Coffey, W., & Shearmur, R. (2002). Agglomeration and dispersion of high-order service employment in the Montreal Metropolitan Region, 1981–1996. *Journal of Planning Literature, 17*(1), 85–168.
Cooke, P., Heidenreich, M., & Braczyk, H. J. (Eds.). (2004). *Regional innovation systems* (2nd ed.). London: Routledge.
Cooke, P., & Leydesdorff, L. (2006). Regional development in the knowledge-based economy: The construction of advantage. *Journal of Technology Transfer, 31*, 5–15.
Cook, G. A., Pandit, N. R., Beaverstock, J. V., Taylor, P. J., & Pain, K. (2007). The role of location in knowledge creation and diffusion: Evidence of centripetal and centrifugal forces in the city of London financial services agglomeration. *Environment and Planning, 39*, 1325–1345.
Cooper, A. C. (1985). The role of incubator organizations in the founding of growth-oriented firms. *Journal of Business Venturing, 1*(1), 75–86.
Cuadrado-Roura, J. R., & Maroto-Sánchez, A. (2010). *Servicios y regiones en España*. Madrid: FUNCAS.
den Hertog, P. (2000). Knowledge-intensive business services as co-producers of innovation. *International Journal of Innovation Management, 4*, 491–528.
Daniels, P. (2004). Book review: Internationalization, technology and services (Editors: M. Miozzo & I. Miles). *Journal of Economic Geography, 4*, 475–476.
Doloreux, D., Amara, N., & Landry, R. (2008). Mapping regional and sectorial characteristics of knowledge-intensive business services: Evidence from the province of Quebec (Canada). *Growth and Change, 39*(3), 464–496.
Doloreux, D., Shearmur, R., & Filion, P. (2001). Learning and innovation: Implications for regional policy—An introduction. *Canadian Journal of Regional Science, 24*(135), 41.
Drennan, M. (1987). New York in the world economy. *Survey of Regional Literature, 4*, 7–12.
Feldman, M. P. (1994). Knowledge complementarity and innovation. *Small Business Economics, 6*, 363–372.
Fischer, H. M., & Frolich, J. (Eds.). (2001). *Knowledge, complexity and innovation systems*. Berlin: Springer.
Freel, M. (2006). Patterns of technological innovation in knowledge-intensive business services. *Industry and Innovation, 13*(3), 335–358.
Fujita, M., & Thisse, J. F. (2002). *Economics of agglomeration: Cities, industrial location, and regional growth*. Cambridge: Cambridge University Press.
Gallego, J., & Maroto, A. (2010). Análisis de la localización regional de los servicios intensivos en conocimiento a nivel europeo. *Investigaciones Regionales, 17*, 71–92.
García-Velasco, M. (2004). *Regional specialization and territorial concentration in knowledge intensive services in Europe* (Working paper).
García-Velasco, M. (2005). Regional specialization and territorial concentration in knowledge intensive services in Europe. Paper presented at the XVth International RESER Conference held between the 22nd and the 24th of September, 2005, in Granada (Spain).
Gertler, M. S. (2003). Tacit knowledge and the economic geography of context, or the undefinable tacitness of being (there). *Journal of Economic Geography, 3*, 75–99.
Glaeser, E. L., Scheinkman, J. A., & Shleifer, A. (1995). Economic growth in a cross-section of cities. *Journal of Monetary Economics, 36*, 117–143.
Howells, J. R. L. (2002). Tacit knowledge, innovation and economic geography. *Urban Studies, 39*, 871–888.
Illeris, S. (1994). Proximity between service producers and service users. *Tijdschrift voor Economische en Sociale Geografie, 85*(4), 294–302.
Illeris, S. (1996). *The service economy: A geographical approach*. Chichester: Wiley.
Johannisson, B. (1998). Personal networks in emerging knowledge-based firms: Spatial and functional patterns. *Entrepreneurship & Regional Development, 10*(4), 297–312.

Johansson, B., & Klaesson, J. (2007). Infrastructure, labour market accessibility and economic development. In C. Karlsson, W. P. Andersson, B. Johansson, & K. Kobayashi (Eds.), *The management and measurement of infrastructure—performance, efficiency and innovation* (pp. 69–98). Cheltenham: Edward Elgar.

Jones, A. (2005). Truly global corporations? Theorizing 'organizational globalization' in advanced business-services. *Journal of Economic Geography, 5*, 177–200.

Kautonen, M., Daniels, P., & Monnoyer, M. C. (2009). Service competitiveness and cohesion: An introduction to the special issue. *The Service Industries Journal, 29*(4), 401–405.

Keeble, D., & Nachum, L. (2002). Why do business services cluster? Small consultancies, clustering and decentralization in London and Southern England. *Transactions of the Institute of British Geographers, 27*(1), 67–90.

Knox, P., & Agnew, J. (1994). *The geography of the world-economy* (2nd ed.). London: Arnold.

Koch, A., & Stahlecker, T. (2006). Regional innovation systems and the foundation of knowledge-intensive business services. A comparative study in Bremen, Munich and Stuttgart (Germany). *European Planning Studies, 14*(2), 123–146.

Krugman, P. (1991). *Geography and trade*. Cambridge: MIT.

Lo, V. (2003). *Wissensbasierte netzwerke im finanzsektor. Das beispiel des mergers & acquisitions-geschäfts*. Wiesbaden: Deutscher Universitatsverlag.

MacPherson, A. (1997). The role of producer service outsourcing in the innovation performance of New York state manufacturing firms. *Annals of the Association of American Geographers, 87*, 52–71.

Malmberg, A., & Maskell, P. (2002). The elusive concept of localization economies: Towards a knowledge-based theory of spatial clustering. *Environment and Planning, 34*(3), 429–449.

Marshall, A. F. (1890). *Principles of economics*. London: Macmillan.

Martínez-Alcocer, J. G., & Maroto Sánchez, A. (2010). Análisis de la localización regional de los servicios intensivos en conocimiento a nivel europeo. *Investigaciones Regionales, 17*, 71–91.

Maskell, P., & Kébir, L. (2006). What qualifies as a cluster theory? In B. T. Asheim, P. Cooke, & R. Martin (Eds.), *Clusters and regional development* (pp. 131–152). London: Routledge.

Matthiessen, C., Schwarz, A., & Find, S. (2002). The top-level global research system 1997-99: Centers, networks and nodality. An analysis based on bibliometric indicators. *Urban Studies, 39*(5–6), 903–927.

Mattsson, L. G. (1984). Some applications of welfare maximization approaches to residential location. *Papers of Regional Science, 55*(1), 103–120.

Miles, I. (2005). Knowledge intensive business services: Prospects and policies. *Foresight, 7*, 39–43.

Miles, I., Kastrinos, N., Bilderbeek, R., den Hertog, G., Flanagan, K., & Huntink, W. (1995). *Knowledge-intensive business services: Their role as users, carriers, and sources of innovation*. Lusembourg: Report to the EC DG XIII Spring EIMS Programme.

Morgan, K. (2004). The exaggerated death of geography: Learning, proximity and territorial innovation systems. *Journal of Economic Geography, 4*, 3–21.

Mudambi, R. (2008). Location, control and innovation in knowledge-intensive industries. *Journal of Economic Geography, 8*, 699–725.

Muller, E., & Zenker, A. (2001). Business services as actors of knowledge transformation: The role of KIBS in regional and national innovation systems. *Research Policy, 30*, 1501–1516.

Muller, E., & Doloreux, D. (2009). What we should know about knowledge-intensive business services. *Technology in Society, 31*(1), 64–72.

Roberts, J., & Andersen, B. (2000). Knowledge and innovation in the new service economy. In B. Andersen et al. (Eds.), *Knowledge and innovation in the new service economy*. Cheltenham: Elgar.

Rubalcaba-Bermejo, L., & Cuadrado-Roura, J. R. (2001). Spain: Knowledge-intensive services. A paradigm of economic change. In P. Wood (Ed.), *Consultancy and innovation. The business service revolution in Europe*. Londres: Routledge.

Scott, A. J. (1988). Flexible production systems and regional development: The rise of new industrial spaces in North America and Western Europe. *International Journal of Urban and Regional Research, 12*, 171–186.

Sokol, M., Van Egeraat, C., & Williams, B. (2008). Revisiting the informational city: Space of flows, policentrity and the geography of knowledge-intensive business services in the emerging global city-region of Dublin. *Regional Studies, 42*(8), 1133–1146.

Strambach, S. (2001). Innovation processes and the role of knowledge-intensive business services (KIBS). In K. Koschatzky, M. Zulicke, & A. Zenker (Eds.), *Innovation networks: Concepts and challenges in the European perspectives* (pp. 53–68). Heidelberg: Physica.

Strambach, S. (2002). Change in the innovation process: New knowledge production and competitive cities – The case of Stuttgart. *European Planning Studies, 10*(2), 214–231.

Vence, X., & González, M. (2002). Los servicios y la innovación. La nueva frontera regional en Europa. *Economía Industrial, 347*, 41–66.

Weibull, J. W. (1976). An axiomatic approach to the measurement of accessibility. *Regional Science Urban Economics, 6*(4), 357–379.

Wood, P. (2002). Knowledge-intensive services and urban innovativeness. *Urban Studies, 39* (5–6), 993–1002.

Wood, P. (2006). Urban development and knowledge-intensive business services: Too many unanswered questions? *Growth and Change, 37*(3), 335–361.

Methodological Notes

- Source of data: Eurostat and Espon.
- Territorial units: We have considered regions at NUTS 2 level. It is important to note that there are lacks of data for these regions at some years. Due to this situation, we have eliminated in the analyses those regions with missing data. However, the final sample provides more than 90% of representativity of the total population.
- Time series: There is a methodological break since 2008. However, the descriptive analyses have considered four periods from a transversal perspective, and we have taken into account establishing comparisons among categories of regional specialization according to the specialization quotient rather than comparing among absolute numbers, which could have led to problems of comparability among data.
- Sectors: We have taken knowledge-intensive sectors, and mainly with private. Other public knowledge-intensive sectors have not been included into the analyses explicitly.
- Classification of KIBS: Despite the intense controversy about the classifications of KIBS, we have used the traditional classification to avoid the presence of breaks in the available data series.

Chapter 14
A New Service Typology: Geographical Diversity and Dynamics of the German Service Economy

Johannes Glückler and Ingmar Hammer

1 Introduction

Although there is a wide consensus about the increasing predominance of the service economy in the developed economies, and although the structural shift from manufacturing to services has been broadly discussed, the inherent heterogeneity within the service economy still seems to be little understood. Services comprise a large set of seemingly different activities that vary with respect to qualification, productivity, innovativeness, growth, distinctive locational structures and geographical dynamics. This diversity is usually studied in one of two different ways. Micro approaches usually rely on case studies within selected service branches in order to reveal the particularities of these activities (e.g. Daniels & Bryson, 2005; Glückler 2004; Léo & Philippe, 2007; Rusten, Bryson, & Gammelsæter, 2005) whereas macro approaches attempt to capture the diversity of services by classifying services into homogenous statistical categories of similar attributes (e.g. OECD, 2000).

After a decade or more of research on service taxonomies, scholars and policy makers still lack pragmatic tools to capture and monitor the development of regional service economies. Service typologies are affluent yet incomplete and often partial in their focus. Many regional monitors are based on idiosyncratic definitions and sometimes incomparable classifications of service sectors (Chadwick, Glasson, & Lawton Smith, 2008; Jung, 2006; Laafia, 2002; Wood, 2006). This can be observed, to take a few examples, in studies on media, consulting, logistics etc. (Cook & Pandit, 2007; IHK München & Oberbayern, 2003). As a consequence, interregional comparisons and the development of evidence-based regional are rather difficult. Apart from the use of partial and sometimes inconsistent taxonomies, there is an additional debate about the boundary between service and non-service sectors (e.g. European Commission, 2005). In this context of taxonomic confusion and in an interest to help ease comparative regional

J. Glückler (✉) • I. Hammer
Institute of Geography, University of Heidelberg, Heidelberg, Germany

J.R. Cuadrado-Roura (ed.), *Service Industries and Regions*, Advances in Spatial Science, DOI 10.1007/978-3-642-35801-2_14,

monitoring and regional policies, we develop an integrative typology of the service economy. This chapter is a more detailed and refined elaboration of a previous article in which we design a three-dimensional service typology (Glückler & Hammer, 2011). Here, we develop a more refined and accurate methodology to obtain a more concise and valid classification of service types that capture different sectoral and spatial dynamics relevant for regional policy. The following sections pursue three concrete objectives:

First, we develop a classification which integrates several partial service typologies into a multidimensional taxonomic framework which is meant to deliver a useful differentiation of service types in the economy. Different terminologies and conceptual distinctions of service activities have been developed (Cook, Goh, & Chung, 1999; Lovelock, 1983), but only a minority of service classifications has made the effort to match its theoretical concepts with statistical data and to translate terminologies into standard industry codes (Browning & Singelmann, 1975; Illeris, 1996; Miles, 1993; Miles & Watt, 1995).

Second, we develop an own classification to distinguish between different, yet important forms of service types which have been comprehensively discussed in the scientific literature. Particularly, our typology enables us to distinguish service industries by their knowledge-intensity and demand orientation. In the absence of an own classification scheme for technology-intensive business-orientated services, we draw on existing typologies to distinguish non-technological service (e.g. management consulting) from technological services (e.g. computer software). Despite the theoretical stimulation of this classification and the usage of an own classification scheme, we seek to develop full value by matching important dimensions of services with a concise statistical classification of service types at the three-digit level of the European standard industry classification NACE.

Third, we demonstrate empirically how this typology captures significant differences in the spatial and sectoral dynamics of the service economy in Germany. By translating conceptual service types into the sectoral classification of NACE categories we offer the opportunity to empirically assess the capacity of the new typology to capture a significant part of the heterogeneity of sectoral and regional service development. The paper uses employment statistics for Germany to analyse services development in terms of growth, locational structure and geographical dynamics of regional employment. In particular, the paper analyses how reliably the typology manages to separate employment decline from growth, geographical agglomeration from decentralized distribution and regional expansion from local clustering. The final section will discuss the limitations and opportunities of this approach.

2 Toward a Pragmatic Classification of Service Types

Service research has a long tradition and reaches back at least to the well-known contributions on the structural transformation from agriculture over manufacturing to services (Clark, 1940; Fisher, 1939). However, many conceptions of characteristics and types of services have been developed without translations into statistical industry standards such that empirical analysis is often limited to case examples rather than broader assessments of the service economy. The current NACE standard[1] was introduced in 1970 and adapted only slowly to the new realities of an ever more service-based European economy. Although 70 % of European value-added is generated in the service sector, the industry classification distinguishes many sub-classes in manufacturing but only broad classes of service activities. Many European countries began only toward the end of the 1990s to implement systematic observations of services development and published annual statistical reports. In Germany, for instance, the federal service statistics were launched only in 2000 and the first report was based on figures from 2001 (Gans-Raschke, 2006). Fortunately, the major revision of the European industry classification NACE Rev. 2 will provide more classificatory detail to service activities from 2008 onwards (Eurostat, 2008).

The first step toward a service typology is to define the boundary between service activities and non-service activities. Instead of pursuing a functional approach to services (e.g. Dicken, 1992; Illeris, 1996) which would be difficult to break down to the level of employment data, this approach follows a sectoral perspective. Based on the NACE standard of industry classification, business sectors are qualified as primary, industrial or service activities at the three-digit level. Within this framework, the service sector is defined as the sum of NACE groups G (wholesale and retail etc.) to O (other community, social and personal services).[2] In contrast to mainstream definitions, however, recent work by the European Commission (European Commission, 2005) includes utilities as network services into the service sector. In line with this extension, the service economy is defined here to comprise the NACE groups E *and* G to O.

The second step in conceptualizing a typology is to define the relevant criteria and to join similar services into more or less homogenous service types. Service typologies are as affluent as the range of criteria that are considered

[1] The *Nomenclature générale des activités économiques dans les Communautés Européennes* (General Industrial Classification of Economic Activities within the European Communities) was first developed in 1970. Since this first taxonomy was not compatible with other international industry standards, a joint United Nations Statistical Office/Eurostat working group got involved in the third revision of the International Standard Industrial Classification of All Economic Activities (ISIC Rev. 3), which was adopted by the United Nations Statistical Commission in February 1989. Subsequently, a working group promoted by Eurostat with representatives of Member States developed a revised version of NACE, called NACE Rev. 1., which was established in 1990. In 2002, the minor update NACE Rev. 1.1 was established. NACE Rev. 1.1 introduced a few additional items and changes to some titles. The aim of the update was to reflect new activities which did not exist before (e.g. call centres) and activities which had manifestly grown in importance. In 2002, a new revision was initiated until 2007.

[2] The NACE groups are defined as follows: E. Electricity, gas and water supply, G. Wholesale and retail trade; repair of motor vehicles, motorcycles and personal and household goods, H. Hotels and restaurants, I. Transport, storage and communication, J. Financial intermediation, K. Real estate, renting and business activities, L. Public administration and defence; compulsory social security, M. Education, N. Health and social work, O. Other community, social and personal service activities.

(Buckley, Pass, & Prescott, 1992; Cook et al., 1999; Illeris, 1996; Lovelock, 1983): services have been distinguished by the type of delivery, the nature of service activity, the type of client relationship, the user sectors, the method of finance, the degree of standardization and many other characteristics. There have been various macro approaches to service classifications in which individual dimensions have been combined to more complex typologies (e.g. Miles, 2007; Viitamo, 2007; cf. Illeris, 1996, ch. 3, for an extended discussion of earlier approaches). These approaches are limited to the extent that they define rather broad types often only at the level of one or—at most—two digits of the standard statistical classification of economic activities. Our attempt is to build on earlier work, combine partial perspectives into an integrated typology and achieve a higher resolution by matching service types to three-digit industry codes of economic sectors.

Within the debate about the multiple distinctive aspects of services, three dimensions appear especially critical and receive full attention by researchers as well as by regional analysts and policy makers: demand orientation, knowledge-intensity, and technology-intensity. Each of these dimensions is expected to imply different economic effects. In the past, business services have experienced higher growth in employment and revenues than consumer services. Knowledge-intensive services seem to yield higher value-added and offer more opportunities for innovation than operational services both for themselves and for their client industries. Finally, technology-based services are hypothesized to yield higher productivity gains than non-technological or technology-using services. In consequence, the general aim of a new typology is to conceive meaningful service types which capture significant differences in growth, innovation and productivity gains. This paper focuses on the analysis of growth differentials in sectoral employment and the geographical dynamics of sectoral growth in Germany.

2.1 Demand Orientation

One of the first criteria to classify services is the demand orientation and the position within the supply chain (Illeris, 1996). While some services are delivered directly to final consumers, other services are used by intermediary clients such as firms and other organizations (Ronning, 2003). Research on business services has proliferated over the last decade because the demand for intermediary services had grown tremendously. Without going into the details of the different explanations of business service growth, three major trends may be observed.

First, corporations have been focusing increasingly on the core competences and have begun to externalize (or outsource) formerly internal support and administrative services (Fink, Köhler, & Scholtissek, 2004; Prahalad & Hamel, 1990). In their study of the headquarters of large corporations located in London, Aksoy and Marshall (1992) found downsizing and externalization of employment up to 90 % during the 1980s and 1990s. If all growth in business services was merely the result of the "hollowing out" of head offices in the manufacturing sector—as Williams,

Williams and Haslam (1990) and Marshall (1994) call this process—service growth would only be a statistical effect of reassigning activities from the industrial to the service sector. Second, there has also been an emergence of new services in the market. Given the ever more global competition, accelerating innovation cycles and an ongoing specialization in product markets, management labour has experienced an increasing division of labour (Wood, 2002a). As a consequence, corporate management has produced new demand for expert services, e.g. financial services, and has stimulated an ongoing growth and diversity of business services. A final element driving external service demand for organizations is the development of information and communication technologies with two major effects: new technologies have led to the innovation of new additional service offerings for business, as for instance, the delivery tracking of commodities along the supply chain or real-time inventory assessment. Moreover, the proliferation of computer-support for many business processes has leveraged external demand for business services comprising hardware implementation and maintenance, software, data management and technological solutions to business processes.

More difficult than identifying new sources of intermediary demand is it to clearly define statistical delineations between consumer and business services. Any taxonomic approach faces a mixed-market problem because empirically, many service sectors are "both-and" (Illeris, 1996) in that they sell to final consumers *and* intermediate firms (Miles, 1993). Financial services, transport and logistics as well as different retail businesses are sectors that broadly sell to both, intermediary and final consumers. Extant classifications (e.g. Ganz 2005; Haas & Lindemann, 2000; Hertog et al. 2006) are problematic because of their lack of transparency about the empirical boundaries between business and final demand. So far, there is no comprehensive typology of business and consumer services that would provide a clear rationale to qualify e.g. inland water transports (61.2) as business or consumer services? One feasible solution to this problem is to collect data on real demand for every service industry although this task has not been done for many years and only for highly aggregated sectors (e.g. Fontaine, 1987).[3] We use input–output statistics from the national accounting system in Germany in order to measure the relative proportions of intermediary (business) and final (consumer) demand for each of the 93 three-digit service branches (Destatis, 2005). Within the input–output system, we define business demand as purchases by all economic sectors, private organizations and fixed investments. In contrast, final demand is represented by the consumption through households for each of the 93 service sectors. Based on this transformation, we are able to assign business and consumer demand for each service industry. Database activities, for instance, sell 98 % of

[3] In the mid 1980s, Fontaine (1987) assessed the French demand structure for rather broad service categories empirically (quoted in Illeris, 1996, ch. 3). Another way to assess the demand orientation of an industry would be to use input-out-tables. To our knowledge, however, the demand structure has so far not been analyzed for services at the NACE three-digit level.

combined output[4] to business (98 %) and only 2 % to final consumers. By using a threshold, we classify all services as business services if they sell more than 50 % of combined output to other business sectors. In turn, services are defined as consumer services if their output to households is higher than that to business sectors. This dichotomization implies some weaknesses: on the one hand, the composition of demand may vary over years and produce different percentages e.g. for final consumption; on the other hand, the composition of demand is certainly specific for each regional and national economy and therefore only valid for the German context. Other countries may be characterized by different demand structures and the classification of certain service sectors may therefore differ. The resulting classification complies with the level of the three-digit industries and it incorporates the whole range of knowledge-intensive and also operational services such as cleaning, security, facility management etc. (Fig. 14.1).

2.2 Knowledge Intensity

The qualification and expertise necessary for the provision of services has been a second important dimension in services research. Many researchers have looked at the particular conditions of knowledge-intensive services and their development over time and space (e.g. Balaz, 2004; Bryson & Rusten, 2005; Hauknes & Antonelli, 1997; Rubalcaba, 2007). The term knowledge-intensive generally alludes to high levels of individual expertise of service professionals as well as the general complexity and specificity of the service offerings (Tether & Hipp, 2002; Wood, 2002b). One approach to capture knowledge-intensity is to investigate the quality of human capital in service products by measuring the share of employees with a tertiary education in overall employment (Haas & Lindemann, 2000). Generally, the higher the knowledge-intensity of a service the more difficult it is to standardize these services. Problems become more specific and solutions grow more complex. The specificity and sometimes uniqueness of expertise services require and enable more innovativeness and competitiveness because standardization and cost strategies are hard to realize. Knowledge-intensity stimulates innovation in order to generate rents from temporary knowledge monopolies. This is true not only for service providers but often also for service users in agriculture and manufacturing. One typology available that has provided a classification of sectors by knowledge-intensity has been developed by the *Niedersächsisches Institut für Wirtschaftsforschung*. Although it has been used by several research organizations in Germany (e.g. ZEW, DIW, Fraunhofer ISI etc.), for the framework studies on the German innovation system carried out by the

[4] Combined output refers to the sum of intermediary and final demand, excluding sales to the public sector and exports.

Federal Ministry of Education and Research (Legler & Frietsch, 2006) and by ourselves in an earlier approach (Glückler & Hammer, 2011), we found the typology intransparent with respect to the raw employment information entering the classification.[5] In addition, we began to be interested in adopting a more widely accepted approach to capture the knowledge-intensity of business sectors.

Therefore, we make use of Richard Florida's conception of the creative class as encompassing those people who perform non-routine tasks, who identify with their professions and who use expertise to develop customized solutions to complex problems (Florida, 2002). Based on the crucial assumption that "*it is what people actually do, rather than their industry affiliations or their educational attainment, that makes them economically productive*" Boschma & Fritsch, 2009, 393), we consider those professionals as knowledge workers who require individual creativity, problem-solving capacities and discretionary measures to accomplish with specific tasks (Glückler, Ries, & Schmid, 2010). In its empirical assessment, Florida first distinguishes three groups of creative people: the creative core, the creative professionals and the bohemians. Second, he matches concrete professions with these three levels of creative people. The group of bohemians comprises writers, photographers or sportspeople etc., the group of creative professionals includes employees such as consultants or professionals for technical problem solving, and the creative core consists of engineers, mathematicians, teachers or academics. To determine the knowledge-intensity of the 93 service branches, we use a matrix of employment data that cross tabulates the number of employees for each of the 93 sectors in each of 334 statistical professions (as defined by the Federal Employment Agency (KldB88).[6] We calculate the ratio of creative employment, i.e. the number of jobs in occupations that belong to the creative class, for every service branch and use again a threshold of 50 % to distinguish operational from knowledge-intensive services. Services are classified as knowledge intensive if the share of employment in creative occupations surpasses 50 %, in case of less than 50 % services are classified as operational. In contrast to existing classifications, such as the classifications developed by the *Niedersächsische Institut für Wirtschaftsforschung* (NIW/ISI-List), which focus on the qualification of employees in every branch (share of academic graduates as well as employees with special skills like planning, design or construction), we focus on the percentage of the creative class in every service branch.

Given the knowledge intensity and demand-orientation for every service industry, we are able to assign every service sector in a matrix. Figure 14.1 displays the distribution of service sectors by their relative composition of demand and knowledge-intensity. No single sector operates without any creative employment. Obviously, even in operative services like transportation or facility management

[5] In the work of Legler and Frietsch (2006), the procedure of classifying knowledge-intensity has only been accessible for us in an incomplete way. We could not retrieve a documentation of the criteria and values for classification.

[6] For a detailed discussion and documentation of this methodology see Glückler et al. (2010).

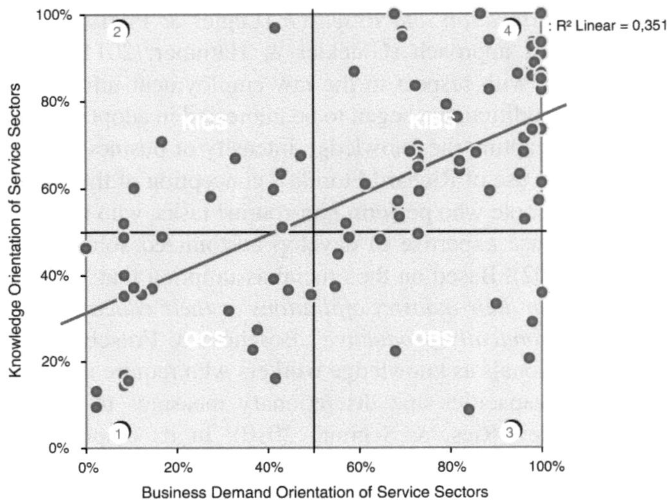

Fig. 14.1 Knowledge intensity and demand orientation for the complete service industry

there are highly qualified people who are responsible for calculations, management or governance. Figure 14.1 also demonstrates that the level of demand orientation and the share of creative employment are statistically associated. With an increasing business orientation, the knowledge-intensity also rises ($R^2 = 0.34$). Typical examples of operational and consumer orientated services are hotels, bars, restaurants or cafes as well as retail businesses. These services predominantly employ skilled personnel without university grades. Notably, even knowledge-intensive consumer oriented services never reach the high ratios of knowledge-intensity that business oriented services achieve. As a consequence, consumer services tend to be rather operational whereas business services tend to be more knowledge-intensive. On average, while 41 % of the workforce in consumer services is knowledge-intensive, 69 % of total employment in business services is knowledge-intensive. With a few exceptions such as insurance and pension funding, this finding suggests that business services pose more complex and unique tasks that require higher competences and problem-solving capacities than consumer-oriented services. Nonetheless, some operational business services show very little need for creative professionals. These findings apply to transport services, industrial cleaning, security activities and labor recruitment and provision of professionals.

2.3 Technology Intensity

The third dimension of our analysis is technology-intensity. While many services simply use technologies such as information and communication technologies to transmit part or all of their service, other services are focused on developing or improving these technologies (e.g. software firms and databank development). The technology dimension is particularly critical for the economics of services because services are often unable to experience productivity gains (Baumol, 1996). While work is an input in the manufacturing sector, which may be replaced through the application of technology at the same or higher output, work is both, input *and* output in most services. Generally, increases in productivity over time may be achieved through increased capital per worker, improvements in technology, labour skill or management, and economies of scale as output rises (Heilbrun, 2003). In services, however, these sources of productivity gains are often not available. Despite these restrictions, several OECD studies provide evidence for sometimes remarkable productivity increases in technological services (Wölfl, 2003). Services that focus strongly on the development or application of technology seem to be sources of productivity growth. An economic perspective which is sensitive to productivity gains requires classifying services according to their technology-intensity (Freel, 2006; Tether & Hipp, 2002; Viitamo, 2007). So far, traditional service classifications have largely disregarded the technology-intensity. Only recently, new research has proposed a promising attempt to classify the service sector into distinct productivity types (Viitamo, 2007).

Similar to our approach, Viitamo recombines three dimensions—capital intensity, degree of standardization, degree of tangibility—into eight service types with distinct productivities. Despite its conceptual appeal, however, one of the critical problems of these approaches is that they are difficult to match with NACE industry codes and therefore hard to assess empirically. In contrast, the OECD (Hauknes & Antonelli, 1997) have suggested a typology based on NACE codes that distinguishes non-technological or technology-using from technology-based services. In the absence of an own classification scheme, the approach taken in this paper has been adapted from the OECD's technology classification of manufacturing industries: they have used the research and development (R&D) intensity as the defining criterion to distinguish technologically intensive from less intensive services (Hauknes & Antonelli, 1997). This definition, however, is not fully independent of knowledge-intensity because R&D intensity inherently requires highly qualified human capital. This latent effect between the intensities of knowledge and technology in services causes the third dimension of our typology to be empirically incomplete.[7]

[7] Operational services are defined by a lower share of employees with a tertiary education. Conceptually, it is therefore difficult to qualify as technology-intensive as long as R&D intensity is the defining criterion. To our knowledge, no alternative research has been published on the technology-intensity of operational services. While operational services are subject to standardization and the use of technology, they are usually not found to be producers of technology. In some cases, however, services such as logistics and retailing have been found to develop new technologies internally. Hence, it would be possible to empirically find technology-intensive firms within these service sectors. It is certainly an area for future research to conceive alternative measures of technology-intensity.

2.4 A New Service Typology

A combination of the three dimensions demand, knowledge-intensity and technology-intensity produces a cube with eight theoretical service types of which five are empirically defined (Fig. 14.2): These types are operational (OCS) and knowledge-intensive consumer services (KICS) on the one hand, and operational (OBS), knowledge-intensive (KIBS) and technological knowledge-intensive business services (TKIBS) on the other. In contrast to our prior classification where we recombined existing classifications (Glückler & Hammer, 2011), this revised typology is rooted in original employment statistics for Germany and therefore represents a fully coherent classification of services sectors. As a consequence of our approach to reassess the composition of demand (based on input–output statistics), and the composition of knowledge-intensity (based on the share of jobs considered as part of the creative class), several service sectors have been reclassified.

To ensure transparency, some major changes are discussed. Apparently, the most important reallocations occur in OCS and KIBS (Table 14.1). On the one side, operational consumer services lose more than a half of the sectors to other types and on the other hand, knowledge-intensive service sectors quadrupled in number of sectors. This accounts for a remarkable shift of service branches from consumer services to knowledge intensive business sectors. While in the case of KICS and OBS the number of sectors in each type remains constant, some sectors change from OBS to KIBS and KICS, and other industries are reclassified from OCS to KICS. We checked for all reallocations and found a remarkably improved fit of the overall classification. Many prior misclassifications now appear reassigned correctly to the corresponding service type: while cargo handling was classified as consumer service in the old typology, it is now correctly reclassified as a business service. In line with this example, many previously misclassified sectors have been corrected. We read this result a substantial improvement of the typology[8] although some problematic assignments remain primarily because of missing data in the national accounts statistics (e.g. space transport).[9] In contrast to OCS, KICS, OBS,

[8] Other examples are scheduled air transport or veterinary activities or wholesale that have changed correctly from operational consumer services (OCS) to knowledge-intensive business services (KIBS).

[9] Our service typology is based on the three-digit level of sector classification; input–output tables were used to distinguish between consumer and business services. But the national account system does not break all branches down to the three digit-level. For some branches, demand orientation is only available at the two-digit level. Space transport for example relates to non-scheduled flights, data is only available for both services together (62.2-3). As the national account system joins non-scheduled flights and space transport (62.2-3) both are classified as consumer services, although space transport should be classified as a business-oriented service sector. Given the fact that the national account system does not provide data on a three-digit level for every service branch, we use two-digit data for the following ten sectoral groups to estimate value at the three-digit level: wholesale and commission trade, retail trade, financial intermediation, insurance and pension funding, activities auxiliary to financial intermediation, computer and related activities, public administration and defense, sewage and refuse disposal, activities of membership organization as well as other service activities (NACE-Code: 51, 52, 65, 66, 67, 72, 75, 90, 91 and 93). For the stated economic activities we assigned the ratio of demand orientation to the three-digit level.

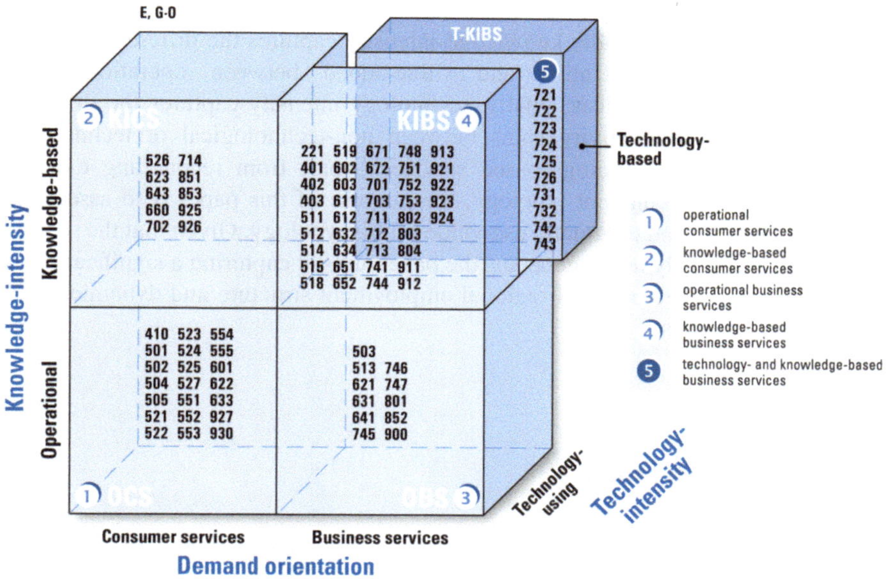

Fig. 14.2 Five service types in the service economy. [In the standard European industry classification, NACE 22.1 (publishing) forms part of the manufacturing sectors (22 publishing, printing and reproduction). Within this complex, publishing is an activity that essentially mediates the generation of content between creators and the final producers of physical or digital media (printing). In compliance with existing empirical studies on the media services sector (München & Oberbayern, 2003), we identify publishing (22.1) as a service and have therefore included it into the service classification. This corresponds also with the 2008 Revision of NACE Rev.2: publishing activities are now explicitly separated from printing and are classified into section J - Information and communication activities.]

Table 14.1 Change of service sectors between the old and the new classification

	# Sectors in the old classification	# Sectors changing with OCS	# Sectors changing with KICS	# Sectors changing with OBS	# Sectors changing with KIBS	# Sectors in the new classification
OCS	46	–	3	6	18	21
KICS	15	2	–	1	8	10
OBS	12	0	2	–	4	11
KIBS	10	0	1	0	–	41
T-KIBS	10	0	0	0	0	10

the sectoral composition of TKIBS remains identical. Since we lack original information on the R&D-intensity for each service sector, we rely on the Haukness and Antonelli's (1997) classification of technology-intensive services.

The typology provides a complete classification of all service branches and thus produces a complete picture of the service economy. The typology has been developed from three partial classifications which are widely accepted in services research and which respond to critical aspects of economic analysis: The demand orientation responds to growth differentials between consumer and business services over the last decade; knowledge-intensity captures the different levels of innovativeness, standardization and value-added between operational and knowledge-intensive services; finally, technology-intensity captures the different opportunities for productivity gains between non-technological or technology-using services and technology-based services. Apart from reconciling existing typologies into one coherent typology, the purpose of this paper is to assess the empirical viability and pragmatic use-value of this typology. Given that the service economy is extremely heterogeneous, the paper aims at capturing a significant part of the diversity in sectoral and regional employment structure and dynamics with the five service types.

3 Data and Methods

The empirical analysis uses the employment statistics of the German *Bundesagentur für Arbeit*. The statistics include the full population of employees subject to social insurance contribution, a group that corresponds with around 65 % of the total active workforce in Germany. Other forms of work for which social security is not mandatory and which are therefore not included in these data, are state officers, self-employed, freelancers, unpaid mutual help and people employed in policy programs of so-called mini-jobs (Destatis, 2007). Employment data were collected for the years 1999, 2002 and 2005[10] and were clustered by three-digit NACE codes into the five service types deduced in the previous section. The data are subject to some limitations. Data for the city-states Bremen, Hamburg and Berlin are incomplete. In addition, some sectors such as space technologies had to be excluded because of a lack of data for most regions. For reasons of anonymity, data are kept blind whenever a sector is represented by less than three local plants in a spatial unit (here NUTS-3). These missing data may produce errors in aggregate counts which are mostly small but may lead to 50 % deviation in very rare cases. The analysis covers service employment in Germany at the NUTS-3 level[11] of 412 regions (*Kreise*). In later analyses, these regions are categorized into nine settlement types as defined by the BBR classification of settlement structures.[12] The settlement types are defined by the size and density of the regional population as well as by the functional centrality of the settlements (Table 14.2). Service

[10] The month of reference was 31 June for each year.

[11] The Nomenclature of Territorial Units for Statistics, (NUTS) is a geocode standard for referencing the administrative divisions of countries for statistical purposes. The standard was developed by the European Union, and thus covers the member states of the EU in detail. NUTS level 3 corresponds with the administrative level of *Kreise* in Germany.

[12] The BBR (*Bundesministerium für Bauwesen und Raumordnung*) is the German Federal Office for Building and Regional Planning.

Table 14.2 Service employment by service types, in different years (own calculations)

	Employees 1999	Employees 2002	Employees 2005	Share in services	Abs. change (1999–2005)	Rel. change (1999–2005)
Total economy	27,402,367	27,505,182	26,112,916		−1,289,451	−4.71 %
Non-services	9,955,345	9,337,564	8,471,595		−1,483,750	−14.90 %
Services	17,447,022	18,167,618	17,641,321	100.0 %	194,299	1.11 %
OCS	3,932,027	3,991,360	3,749,951	21.3 %	−182,076	−4.6 %
KICS	3,447,037	3,655,674	3,664,114	20.8 %	217,077	6.3 %
OBS	1,737,185	1,873,914	1,888,420	10.7 %	151,235	8.7 %
KIBS	7,531,262	7,732,854	7,451,143	42.2 %	−80,119	−1.1 %
TKIBS	799,511	913,816	887,693	5.0 %	88,182	11.0 %

employment varies considerably in its regional distribution and growth dynamics at the level of the *Länder* (NUTS 1).

There have been strong rates of decline in the Eastern regions, a pronounced growth in the South and a more heterogeneous development in central and northern Germany. Eastern regions have lost jobs in the order of 10 % and higher since 1999. Even the high-growth types of KIBS and TKIBS have been under decline in the East. Only KICS experienced an increase over the period. In contrast, the southern regions of Bavaria and Baden-Wuerttemberg as well as North Rhine-Westphalia have experienced a far above average job growth in services and the smallest rates of decline in non-service jobs[13] in Germany. The more balanced regions of central and northern Germany grew moderately in the growth sectors of the service economy (OBS, KIBS and TKIBS) and lost a considerable amount of jobs in the non-service sectors. On balance, these regions suffered from net job losses. In consideration of this regional development, we focus our empirical analysis on the moderate-growth regions in Central Germany, namely North Rhine-Westphalia, Hesse, Rhineland-Palatinate, Lower Saxony as well as on the Eastern regions under economic pressure, namely Saxony, Saxony-Anhalt, Brandenburg and Thuringia. Together these regions represent more than half of Germany's population, 56 % of total service employment and almost 10 of the 18 million service workers.

The aim of the regional analysis is to compare differentials in employment growth and in the geographical dynamics of this growth. Employment growth is measured as the growth rate of the number of employees per sector j between 1999 and 2005 in percent. Two other measures are used to assess the geographical dimension of job growth. The *coefficient of localization* (Isard, 1960) measures the extent to which an industry is localized compared with the spatial (multi-regional) distribution of all economic activities: $CL_j = \frac{1}{2}\sum_{i=1}^{n}\left|\frac{E_{ij}}{E_j} - \frac{E_i}{E}\right|$, where E indicates the number of employees in region i and industry j. The coefficient CL_j for

[13] Here, non-service jobs represent the residual sectors of agriculture, fishery and forestry, industrial manufacturing, mining and construction (NACE Rev. 1.1 groups A, B, C, D and F).

any industry j is calculated by subtracting the employment share of region i in total employment ($\frac{E_i}{E}$) from the employment share of this region in the respective industry j ($\frac{E_{ij}}{E_j}$). The sum of the positive (or the negative, not both) deviations represents the coefficient which varies between 0 and 1. If most jobs in a sector are highly clustered in only a few regions, the coefficient converges to one. If jobs spread equally across the regions, the coefficient approaches zero. In order to analyse the geographical dynamics, the change rates of the coefficient of localization between 1999 and 2005 are computed and then compared for each service type. Increases in the coefficient of localization indicate processes of spatial concentration and clustering whereas value decreases reflect processes of geographical expansion and diffusion.

4 The New Service Typology in Practice

In this section we aim at assessing the empirical value of the suggested typology to capture part of the heterogeneity of services development in time and space. Concretely, we expect the five service types to systematically follow differential sectoral growth, different structures of geographical localization and different geographical dynamics of locational patterns.

4.1 Sectoral Employment Dynamics in Germany

Research on growth sectors in the service economy has either focused on specific industries such as advertising (e.g. Daniels, 1995; Faulconbridge, 2006), accountancy (e.g. Chaston, Megicks, & Williams, 2005; Daniels, Thrift, & Leyshon, 1989), management consulting (e.g. Armbrüster, 2006; Keeble, Bryson, & Wood, 1992), law firms (e.g. Warf & Wije, 1991), among many others. Or it has taken a rather broad view of services: while some studies put emphasis on the knowledge-intensity (Tether & Hipp, 2002; Windrum & Tomlinson, 1999) others focus on the business orientation of services (Daniels & Bryson, 2005; Miles, 2007; Rubalcaba & Kox, 2007; Wood, 2002b). Partly, business services and knowledge-intensive services overlap in what is called knowledge intensive business services (e.g. Bryson & Rusten, 2005; Chadwick et al., 2008; Strambach, 1994). However, many business services are not knowledge-intensive and many knowledge-intensive services are not directed to firms. Our typology offers a more fine-grained analysis of employment dynamics because the two dimensions of demand-orientation and knowledge-intensity are now integrated and recombined into one typology. We first report descriptive statistics for the distribution of employment dynamics across the five service types in Germany for the period 1999–2005. Afterwards, we seek a more structural account of the diverse growth rates between service activities.

The employment statistics illustrate the overall significance of service work in Germany. In 2005, more than two thirds of the German workforce (67.6 %) was employed in sectors of the service economy (Table 14.2): The majority of service jobs is found in knowledge-intensive business services such as publishing, financial activities, real estate activities, banking, health and educational services (KIBS). Operational consumer services are the second large type (OCS). Here, many jobs are related with retail sale, food services or transport activities. In comparison with operational services, knowledge intensive services (KIBS, KICS, and TKIBS) employ still a substantial share of service workers (68 %). Otherwise operational business services (OBS) and technology intensive business services (TKIBS) have experienced strong growth since 1999. TKIBS grew strongest since 1999 closely followed by operational business services. However, differences in employment growth are less obvious between operational and knowledge-intensive services. In absolute terms, employment grew in all service types between 1999 and 2002, a period characterized by strong economic growth during the "new economy" wave. Between 2002 and 2005, however, only KICS and OBS continued to grow, but at a much lower growth rate than between 1999 and 2002. At the same time, all other types suffered from employment decline. In OCS and KIBS the employment loss outweighs prior growth leading to a net loss over the whole period. Notably, job losses were the highest in operational consumer services. A big share of overall employment growth rests on a few service sectors in each service type. The largest employment group, KIBS, illustrates the heterogeneity of these dynamics: apart from continuously growing (e.g. renting of automobiles or financial services) and shrinking sectors (e.g. real estate activities or monetary intermediation), there is another set of sectors experiencing oscillations of growth and decline during the period 1999–2005 (e.g. publishing, wholesale on a fee or contract basis, sea and coastal water transport). The few growth sectors accounted for more than 92 % of all new jobs in this type. Across the entire German service economy, only 13 out of 93 sectors accounted for 86 % of net job growth.

Employment growth in the knowledge-intensive consumer services and operational business services (KICS and OBS) have been higher in absolute numbers than in the other services (OCS, KIBS and TKIBS), mainly because service work in these sectors is very labour-intensive (human health activities, social work activities, transportation, retail sale or library, archives, museums and other cultural activities). Together, OBS and KICS represent nearly one third of the entire workforce in the service economy. In relative terms, however, business services like OBS and TKIBS have grown remarkably stronger than consumer services. The increasing demand for business services is most visible in the steep increase of jobs in operational business services (OBS) and technological knowledge-intensive business services (TKIBS). Within this sectors, labour recruitment and provision of personnel account for a major part of job growth. A few service sectors accounted for the lion share of job growth in the entire service economy: social services (KICS), health services (KICS), consulting (KIBS) and software development (TKIBS). Conversely, also the declining sectors spread across the types, such as specialized retailers (OCS), credit institutes (KIBS), or publishers (KIBS).

What is the underlying association to account for these growth differences? Employment growth proves to be statistically independent from the employment size of a sector. This is important to note since it refutes simple life cycle effects of young and small sectors growing strongly and big and older sectors undergoing decline. Instead, job growth in services varies significantly across the service types. Statistical tests for mean differences demonstrate that the average growth rates of the sectors differ significantly such that each type seems to entail specific employment effects (Table 14.3). Operational consumer services lost employment in all NUTS-1 regions (*Länder*) under research. Operational business services (OBS) and technology-intensive business services (TKIBS) experienced the strongest job growth in the northern and central regions of Germany as well as in the new Eastern *Länder*; especially TKIBS overtook all other service types with growth rates between 24 % and 132 %. However, the dynamics of job growth differ between regions in the East of Germany and the other regions under research. Especially in Saxony and Thuringia, TKIBS such as hardware and software consultancy or data processing doubled in employment between 1999 and 2005. Similar to TKIBS, operational business services grew strongly although this development varies between the Western and Eastern regions. In the former, OBS grew more than 10 % while in the latter operational business services saw only slight employment growth. The slow growth of OBS in the new Eastern *Länder* of Germany has also been found by other studies (Geppert, 1999), although with another classification of operational business services. In our previous work we found high growth rates in KIBS as well as in all business services (Glückler & Hammer, 2011). However, the improved service typology leads to higher heterogeneity within KIBS mainly because the number of service sectors has quadrupled. Different growth dynamics of single service sectors such as financial intermediation or real estate activities weakens the results concerning the growth dynamics.

4.2 Locational Structure of Services

The former section has demonstrated differential sectoral growth for the defined service types. The next two sections will examine the differential locational structure and geographical dynamics of service employment in the German *Länder*. Service sectors are not distributed equally across the regions of a territory because they differ in the spatial reach of delivery. The economic base-model offers a regional income perspective and distinguishes two types of services: they are basic if they increase regional income by means of exportation beyond the region (e.g. call centres, tourism). In contrast, they are non-basic if demand is only local and can therefore not support interregional trade (e.g. restaurants, retail). Given the farther reach of basic services, they are expected to display higher levels of spatial concentration than non-basic sectors. This simple model has been modified because the qualification of basic or non-basic depends on the

Table 14.3 Test for mean differences of employment growth rates (in percent) by service types

Region	Type	No. sectors	Mean	S.E.	*F*	d.f.
Lower Saxony	OCS	21	−3.94	10.12		
	KICS	9	9.77	9.63		
	OBS	11	20.89	22.76		
	KIBS	41	0.48	23.44		
	TKIBS	10	24.39	62.41	3.093*	91
North Rhine-Westphalia	OCS	21	−0.35	18.28		
	KICS	9	4.59	12.81		
	OBS	11	11.78	15.90		
	KIBS	41	0.36	23.35		
	TKIBS	10	46.03	81.63	4.356**	91
Hesse	OCS	21	−1.03	18.68		
	KICS	9	4.24	13.06		
	OBS	11	11.33	24.78		
	KIBS	41	5.70	20.77		
	TKIBS	10	31.41	52.54	2.935*	91
Rhineland	OCS	21	1.68	21.91		
Palatinate	KICS	9	10.93	12.15		
	OBS	11	5.51	32.23		
	KIBS	40	9.95	50.46		
	TKIBS	9	64.55	103.81	2.841*	89
Brandenburg	OCS	21	−2.97	25.28		
	KICS	9	−6.88	15.29		
	OBS	11	−0.74	34.67		
	KIBS	41	−4.32	44.10		
	TKIBS	9	71.02	140.67	3.730**	90
Saxony	OCS	21	−6.23	31.77		
	KICS	9	−10.60	14.05		
	OBS	11	4.71	25.75		
	KIBS	41	−5.89	37.84		
	TKIBS	10	132.01	360.69	2.923*	91
Saxony-Anhalt	OCS	20	−16.61	20.74		
	KICS	9	−2.70	18.50		
	OBS	10	2.25	25.48		
	KIBS	40	−1.44	40.77		
	TKIBS	9	34.16	48.21	3.343*	87
Thuringia	OCS	21	−14.91	20.47		
	KICS	9	−10.32	18.64		
	OBS	11	3.30	38.16		
	KIBS	39	−11.51	23.57		
	TKIBS	9	126.85	366.17	2.899*	88

Note: *d.f.* degrees of freedom, *S.E.* standard error
$^{*}p < 0.10$; $^{**}p < 0.01$

geographical scale of a region. It is easier for a service to be basic if a region is defined at the municipal level rather than at the national level. Moreover, many services which are not directly exported may still leverage exports in the basic sectors indirectly (Illeris, 2005). Many producer services, for instance, enhance the competitiveness of manufacturing sectors and thus contribute to industrial exports indirectly. Our typology provides an alternative perspective by focusing on the knowledge-intensity of services. High qualifications are less frequent and less equally distributed in society and space. Our typology expects knowledge-intensive services (KICS, KIBS and TKIBS) to display higher levels of geographical concentration than operational services (OCS and OBS).

The degree of locational concentration of a sector is measured by the coefficients of localization (c_L) at the level of NUTS-3 regions (*Kreise*). This coefficient is subject to a scale effect. The more people employed in a service sector, the smaller is its geographical concentration. Accordingly, many large sectors (e.g. retailing, gastronomy or logistics) follow more even distributions across the regions in Germany. Despite this scale effect, the locational concentration is also associated with the type of service as defined in our typology. Operational services (OCS and OBS) are distributed more evenly across the regions whereas knowledge services are significantly more concentrated in space (KICS, KIBS, TKIBS). This difference is obvious for North Rhine-Westphalia, Hesse and Saxony-Anhalt. In the other countries, the spatial distribution of services differs. In Saxony, Rhineland-Palatinate, Brandenburg, Saxony and Thuringia, consumer services are more distributed than business services, a result similar to prior investigation (Glückler & Hammer, 2011). A test for mean differences of the coefficients of localization demonstrates that operational business services (OCS) are most evenly distributed across space with a coefficient ranging from $c_L = 0.23$ in Saxony to $c_L = 0.31$ in Rhineland-Palatinate (Table 14.4). These findings correspond with the economic-base model as well as with results presented by Alecke und Untiedt (2008) who found that consumer services are widely distributed across space. Given the fact, that consumer services primary serve households, the spatial distribution of this service type can be explained by the spatial distribution of demand. The distribution of operational services differs in Germany. Employment in operational business services spreads across space with a $c_L = 0.29$ in Saxony-Anhalt and a $c_L = 0.37$ in Rhineland-Palatinate and Thuringia. This includes services such as wholesale trade, post and courier activities or cargo handling and storage. Knowledge-intensive services are spatially more concentrated than operational services. For KIBS the coefficient of localization ranges from $c_L = 0.35$ to $c_L = 0.43$ and even higher for TKIBS. However, the increased spatial concentration of knowledge-intensive service sectors does not apply to KICS. Depending on the federal state, KICS are sometimes more concentrated than OBS, for example in North Rhine-Westphalia ($c_L = 0.48$) or Saxony-Anhalt ($c_L = 0.48$), sometimes they are less concentrated such as in Brandenburg ($c_L = 0.48$) or Lower Saxony ($c_L = 0.48$). This finding does not fully correspond with the spatial implications of the economic base model. While the economic base model does also find operational services to be basic, e.g. call centres and tourism, our typology illustrates that knowledge-intensity increases

Table 14.4 Test for mean differences in coefficients of localization between the service types and by region

Region	Service type	No. sectors	Mean	S.E.	*F*	d.f.
Lower Saxony	OCS	21	0.27	0.16		
	KICS	9	0.30	0.15		
	OBS	11	0.31	0.18		
	KIBS	41	0.38	0.22		
	TKIBS	10	0.50	0.17	2.932**	91
North Rhine-Westphalia	OCS	21	0.26	0.18		
	KICS	10	0.37	0.27		
	OBS	11	0.30	0.19		
	KIBS	41	0.38	0.21		
	TKIBS	10	0.45	0.13	2.074*	92
Hesse	OCS	21	0.28	0.18		
	KICS	9	0.32	0.14		
	OBS	11	0.31	0.16		
	KIBS	41	0.40	0.18		
	TKIBS	10	0.41	0.15	2.177*	91
Rhinland-Palatinate	OCS	21	0.31	0.19		
	KICS	9	0.36	0.18		
	OBS	11	0.37	0.23		
	KIBS	41	0.43	0.23		
	TKIBS	10	0.53	0.18	2.260*	91
Brandenburg	OCS	21	0.26	0.17		
	KICS	9	0.29	0.18		
	OBS	11	0.32	0.23		
	KIBS	41	0.40	0.21		
	TKIBS	10	0.47	0.22	2.845**	91
Saxony	OCS	21	0.23	0.16		
	KICS	9	0.31	0.19		
	OBS	11	0.32	0.22		
	KIBS	39	0.35	0.18		
	TKIBS	10	0.46	0.14	2.915**	89
Saxony-Anhalt	OCS	20	0.24	0.15		
	KICS	9	0.36	0.22		
	OBS	10	0.29	0.14		
	KIBS	40	0.39	0.19		
	TKIBS	10	0.53	0.21	4.792***	88
Thuringia	OCS	21	0.24	0.14		
	KICS	9	0.36	0.21		
	OBS	11	0.37	0.23		
	KIBS	39	0.38	0.18		
	TKIBS	10	0.57	0.21	5.381***	89

Note: *d.f.* degrees of freedom, *S.E.* standard error
$^*p < 0.10$; $^{**}p < 0.05$; $^{***}p < 0.01$

the mean level of geographical concentration significantly. In this respect, the base model complements with our perspective on locational service concentration by capturing also the clustering of some operational service sectors.

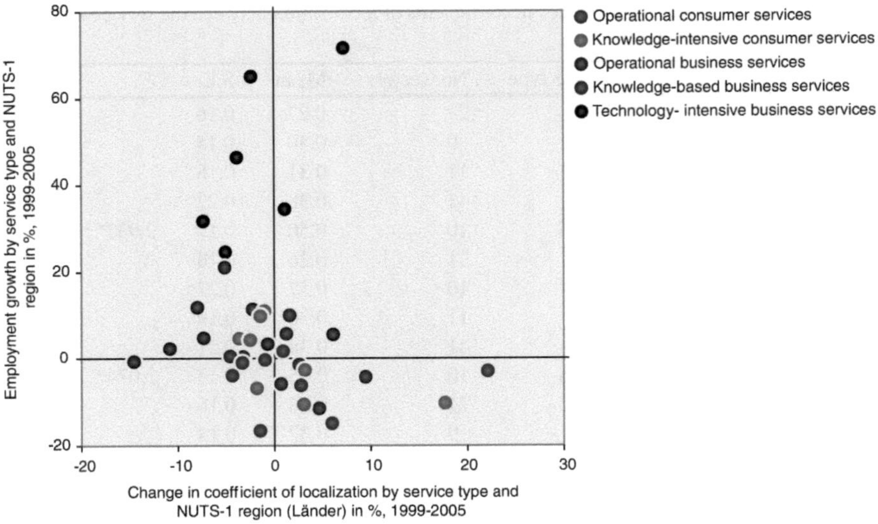

Fig. 14.3 Changes in employment and the coefficient of localization (in percent) between 1999 and 2005 for the five service types in eight German regions (Länder)

4.3 Spatial Dynamics of Locational Structure

How does the locational pattern of concentration change over time? One first assumption would be to expect growing service sectors to expand over space and to experience dynamics of relative decentralization. Do regions benefit equally from the high growth in operational, knowledge-intensive and technological services? In contrast to this expectation, however, the employment growth of a service sector is statistically independent from its geographical dynamics. Growth sectors do not necessarily expand spatially. Instead, the analysis of the geographical dynamics suggests different effects of concentration and decentralization for all five service types. While employment in OCS, KICS and KIBS had experienced a continuous spatial concentration, employment in OBS and TKIBS underwent a process of geographical diffusion. The job growth of OBS and TKIBS leads to a spatial distribution of new service jobs (Fig. 14.3). "Labour recruitment and provision of personnel," for instance, has evolved mainly in urban areas because of agglomeration effects. With the increasing suburbanization those services have been leaving the inner cities toward the urban fringes. Low market entry barriers promote this process (Buch, Groll, & Niebuhr, 2008; Jahn & Wolf, 2005). These results are significant for all eight *Länder* we analyzed. Notwithstanding, knowledge-intensive business services (KIBS) have continued to grow in urban centers and the coefficient of localization has slightly, yet further increased between 1999 and 2005 (1.5 %). This finding is in line with earlier research on the relation between strong sectoral growth at increasing rates of geographical concentration, for instance, in the management consulting business (Glückler, 2004). Much in contrast, OBS had also grown strongly but instead of concentrating in space, new jobs emerged in all

Table 14.5 Employment growth in the German NUTS-3 regions by service and settlement types, 1999–2005

	Area types	OCS	KICS	OBS	KIBS	TKIBS
Agglomeration	Central cities	−9.3 %	−4.4 %	−0.1 %	−4.1 %	3.4 %
	Highly agglomerated counties	0.3 %	10.9 %	13.3 %	1.9 %	22.4 %
	Agglomerated counties	−3.4 %	9.0 %	14.7 %	1.2 %	5.1 %
	Rural counties	−7.0 %	9.3 %	−5.7 %	−6.6 %	−9.6 %
Urbanized area	Central cities	−9.2 %	6.3 %	11.7 %	−3.7 %	3.0 %
	Agglomerated counties	−3.3 %	8.8 %	14.0 %	−0.3 %	9.9 %
	Rural counties	−6.6 %	6.7 %	2.1 %	−6.4 %	−0.4 %
Rural area	Rural counties with higher density	−6.6 %	6.7 %	13.8 %	−5.4 %	2.1 %
	Rural counties with lower density	−8.7 %	3.9 %	−0.4 %	−12.7 %	−15.2 %

kinds of regions following a process of spatial decentralization. The contrary spatial development of KIBS and OBS as described above, is also found in other countries, for instance in the French metropolitan area (Léo & Philippe, 2007).

The differential geographical dynamics of employment produce different effects for different settlement types and regions in Germany. OCS declined in the central cities of the agglomerated areas and those of their urban hinterland, while they stagnated in the suburban hinterland of the central cities (Table 14.5). KICS grew in all settlement types except for the central cities where they declined remarkably by −4.4 %. OBS grew strongest in the more peripheral areas, i.e. urbanized areas and more sparsely populated regions thus producing a strong effect of geographical diffusion of employment. In contrast, KIBS and TKIBS grew strongest in the densely populated and agglomerated regions.

High-growth sectors such as OBS and TKIBS do not necessarily follow the same pattern of geographical diffusion. Much in contrast, while the former does expand over space, the latter type seems to yield job growth predominantly in agglomerated urban contexts. This finding is important for labour market analysis and regional policy. Regional governments and policy agencies should obtain a more detailed understanding of the geographical effects of sectoral growth in order to detect opportunities and avoid misleading policies to develop activities that run counter their geographical trends.

5 Discussion and Implications for Future Research

In this paper we have further elaborated on an integrative service typology developed in prior research (Glückler & Hammer, 2011) and which responds to important debates about differential economic effects on service activities on the one hand, and to the demand for pragmatic use and analytical power in regional analysis, on the other. By combining three crucial dimensions—demand orientation, knowledge-intensity and technology-intensity—we have defined five different service types: operational and knowledge-intensive consumer services (OCS and

KICS), operational business services (OBS) and knowledge-intensive as well as technological knowledge-intensive business services (KIBS and TKIBS). Based on the European standard of industry classification NACE, we categorized 93 three-digit sectors into these five service types in order to test for significant differences in employment growth, locational structure and geographical dynamics in these sectors. In the context of eight regional economies of Germany, we analyzed service employment between the period 1999–2005 and found considerable support for the typology. The rates of employment growth, locational structures and the geographical processes of concentration and diffusion were shown to be significantly different for each of the service types. The empirical analysis revealed that business services, both OBS and TKIBS proved to be the high-growth sectors in the German service economy—although their growth rates differed significantly. Growth, however, was not associated with uniform geographical development. While OBS expanded across the regions and led to a decentralization of jobs TKIBS, KICS and KIBS developed nearly at the same levels of spatial concentration. In contrast to OBS operational consumer services (OCS) are increasingly concentrated spatially due to consolidation effects. In sum, a service typology of only five basic types manages to capture a considerable proportion of diversity in the heterogeneous service economy. Regional analysts and policy makers may benefit from a pragmatic service typology which translates coherently into the industrial classification standard of the European Union (NACE). It will help identify growth sectors and understand the different geographical dynamics of clustered vs. expansive growth.

In the future, the service typology will have to be adjusted to the European Commission's revision of the industry classification NACE Rev. 2 which was used from 1 January 2008 onwards (Eurostat, 2008). The detail of the revised classification has substantially increased (from 514 to 615 classes). Especially for service activities, this increase is visible at all levels, including the highest one, while for other activities, such as agriculture, the increase in detail affected mostly the lower level of the classification. In future research, the typology may benefit from being adapted to this more fine-grained industry classification of NACE Rev. 2. Another opportunity to improve the significance of the service typology is to disaggregate at the four or five-digit level of the NACE code. This, however, would render its use more costly and complicated. Further application of this typology to other regions in Europe would prove helpful in order to better assess its usefulness for regional analysis.

The typology could not absorb the entire heterogeneity of the 93 service sectors within each of the five service types and it did not work equally well for the observed regions. Part of these variances may be related to differences in the state of regional development and settlement structures in the eight federal regions of Germany. Future research should address more intensively the diversity and heterogeneity inherent in the service economy (Tether & Hipp, 2002). Some of the weaknesses of the typology are based on ambivalent service sectors. To take two examples: renting of transport equipment belongs to KIBS although much of their business is with final demand than other business. Similarly, real estate activities are assigned to KIBS but many of the activities are for consumers. These and other sectors that serve both, final users as well as other businesses are difficult to classify when the result should be a dichotomized classification. Furthermore, many activities like wholesale (KIBS) employ highly qualified as well as less qualified people, a problem which we do not solve completely with our classification and thus produce noise in the typology. This research has focused on the stylized data about regional and employment statistics. It has to be acknowledged that the locational structure and the geographical dynamics of service sectors are subject to influences not considered here. Institutional, legal and policy frameworks, regulation and fiscal incentives, the structures of competition, demand and labour markets as well as power relations and governance regimes along the services value chains may impact locational choice and processes of firm foundation and relocation. Future research is needed, therefore, to compare service development internationally in order to achieve a more profound understanding of the geography of the service economy. The basic typology seems to be a useful taxonomic tool to inform ongoing research in this direction.

References

Aksoy, A., & Marshall, N. (1992). The changing corporate head office and its spatial implications. *Regional Studies, 26*(2), 149–162.

Alecke, B., & Untiedt, G. (2008). Die räumliche Konzentration von Industrie und Dienstleistungen in Deutschland. *Jahrbuch für Regionalwissenschaft, 28*(1), 61–92.

Armbrüster, T. (2006). *The economics and sociology of management consulting*. Cambridge: Cambridge University Press.

Balaz, V. (2004). Knowledge-intensive business services in transition economies. *Service Industries Journal, 24*(4), 83–100.

Baumol, W. J. (1996). Children of performing arts, the economic dilemma. The climbing costs of health care and education. *Journal of Cultural Economics, 20*, 183–206.

Boschma, R. A., & Fritsch, M. (2009). Creative class and regional growth: Empirical evidence from seven European countries. *Economic Geography, 85*(4), 391–423.

Browning, H. L., & Singelmann, J. (1975). *The emergence of a service society: Demographic and sociological aspects of the sectoral transformation of the labour force in the U.S.A.* Austin: Texas University.

Bryson, J., & Rusten, G. (2005). Spatial divisions of expertise: Knowledge intensive business service firms and regional development in Norway. *Service Industries Journal, 25*(8), 959–977.

Buch, T., Groll, F., & Niebuhr, A. (2008). *Zeitarbeit regional—Überraschend dynamisch in Hamburg und Ulm*. Nürnberg: IAB.

Buckley, P. J., Pass, C. L., & Prescott, K. (1992). Internationalization of service firms: A comparison with the manufacturing sector. *Scandinavian International Business Review, 1*, 39–56.

Chadwick, A., Glasson, J., & Lawton Smith, H. (2008). Employment growth in knowledge-intensive business services in Great Britain during the 1990s—Variations at the regional and sub-regional level. *Local Economy, 23*(1), 6–18.

Chaston, I., Megicks, P., & Williams, J. (2005). The exploitation of market knowledge within relationship orientated small accountancy practices. *Service Industries Journal, 25*(7), 845–860.

Clark, C. (1940). *The conditions of economic progress*. London: MacMillan & Co.

Cook, D., Goh, C.-H., & Chung, C. (1999). Service typologies: A state of the art survey. *Production and Operations Management, 8*(3), 318–338.

Cook, G. A. S., & Pandit, N. R. (2007). Service industry clustering: A comparison of broadcasting in three city-regions. *Service Industries Journal, 27*(4), 453–469.

Daniels, P. W. (1995). Internationalisation of advertising services in a changing regulatory environment. *The Service Industries Journal, 16*, 276–294.

Daniels, P. W., & Bryson, J. (2005). Sustaining business and professional services in a second city region. *Services Industries Journal, 25*(4), 505–524.

Daniels, P. W., Thrift, N. J., & Leyshon, A. (1989). Internationalisation of professional producer services: Accountancy conglomerates. In P. Enderwick (Ed.), *Multinational service firms* (pp. 79–106). London: Routledge.

Destatis. (2005). *Input–output-Rechnung der Bundesrepublik Deutschland.* Fachserie 18 Reihe 2, Sonderauswertung. Wiesbaden: Statistisches Bundesamt.

Destatis. (2007). *Bevölkerung und Erwerbstätigkeit -Struktur der sozialversicherungspflichtig Beschäftigten.* Wiesbaden: Statistisches Bundesamt.

Dicken, P. (1992). *Global shift—The internationalization of economic activity.* London: Sage Publications Ltd.

European Commission. (2005). *European forum on business related services.* Brüssel: European Commission.

Eurostat. (2008). *NACE Rev. 2. Statistical classification of economic activities in the European Community.* Brussels: European Commission.

Faulconbridge, J. (2006). Stretching tacit knowledge beyond a local fix? Global spaces of learning in advertising professional service firms. *Journal of Economic Geography, 6*(4), 517–540.

Fink, D., Köhler, T., & Scholtissek, S. (2004). *Die dritte revolution der wertschöpfung.* München: Econ.

Fisher, A. (1939). Production: Primary, secondary, tertiary. *The Economic Record, 15*, 24–38.

Florida, R. (2002). *The rise of the creative class: And how it's transforming work, leisure, community and everyday life.* New York: Basic Books.

Fontaine, C. (1987). *L'Expansion des services: Un quart de siècle en France et dans le monde dé veloppé, I-III.* Paris: Rexervices.

Freel, M. (2006). Patterns of technological innovation in knowledge-intensive business services. *Industry and Innovation, 13*(3), 335–358.

Gans-Raschke, H.-J. (2006). *Die bedeutung der dienstleistungsstatistik für die berechnung der wertschöpfung in den volkswirtschaftlichen gesamtrechnungen.* Wiesbaden: Statistisches Bundesamt.

Ganz, W. (2005). Research in the Services Sector. Stuttgart: Fraunhofer Institut für Arbeitswirtschaft und Organisation.

Geppert, K. (1999). *Unternehmensbezogene Dienstleistungen im Land Brandenburg - Strukturen, Defizite und Entwicklungsmöglichkeiten.* Beiträge zur Strukturforschung. Berlin: Duncker & Humblot.

Glückler, J. (2004). *Reputationsnetze. Zur Internationalisierung von Unternehmensberatern. Eine relationale Theorie.* Bielefeld: Transcript.

Glückler, J., & Hammer, I. (2011). A pragmatic service typology: Capturing the distinctive dynamics of services in time and space. *Service Industries Journal, 31*(6), 941–957.

Glückler, J., Ries, M., & Schmid, H. (2010). *Kreative ökonomie. Perspektiven schöpferischer arbeit in der stadt Heidelberg* (Heidelberger Geographische Arbeiten, Bd. 131). Heidelberg: Geographisches Institut.

Haas, D., & Lindemann, S. (2000). *Dienstleistungsstandort Bayern.* München: Institut für Wirtschaftsgeographie der LMU München im Auftrag des Bayerischen Staatsministeriums für Wirtschaft, Verkehr und Technologie.

Hauknes, J., & Antonelli, C. (1997). *Knowledge intensive services—What is their role?* Paper for the OECD Business and Industry Policy Forum on realising the potential of the service economy: facilitating growth, innovation and competition, STEP Group, Oslo.

Heilbrun, J. (2003). Baumol's cost disease. In R. Towse (Ed.), *A handbook of cultural economics* (pp. 91–101). Cheltenham: Edward Elgar.

Hertog, P. d. (2006). Research and Development Needs of Business Related Service Firms. Utrecht: dialogic.

Illeris, S. (1996). *The service economy. A geographical approach*. Chichester: Wiley.

Illeris, S. (2005). The role of services in regional and urban development: A reappraisal of our understanding. *Service Industries Journal, 25*(4), 447–460.

Isard, W. (1960). *Methods of regional analysis. An introduction to regional science*. Cambridge: Published jointly by the Technology Press of the Massachusetts Institute of Technology and Wiley.

Jahn, E. J., & Wolf, K. (2005). *Entwicklung der leiharbeit und regionale disparitäten* (IAB Kurzbericht Nr. 14/2005). Nürnberg.

Jung, H.-U. (2006). *Der Dienstleistungsstandort Osnabrück im überregionalen Vergleich. Kurzstudie im Auftrag der WFO Wirtschaftsförderungsgesellschaft Osnabrück mbH*. Osnabrück: Niedersächsisches Institut für Wirtschaftsforschung.

Keeble, D., Bryson, J., & Wood, P. (1992). Entrepreneurship and flexibility in business services: The rise of small management consultancy and market research firms in the United Kingdom. In K. Caley, F. Chittenden, E. Chell, & C. Mason (Eds.), *Small enterprise development: Policy and practice in action* (pp. 43–58). London: Chapman.

Laafia, I. (2002). *Nationale und regionale Beschäftigung in Hightech- und wissensintensiven Sektoren in der EU. 1995–2000*. Brüssel: Eurostat.

Legler, H., & Frietsch, R. (2006). *Neuabgrenzung der Wissenswirtschaft. Forschungsintensive Industrien und wissensintensive Dienstleistungen* (Studie No. 22-2007). Berlin: BMBF.

Léo, P.-Y., & Philippe, J. (2007). Executives and business services: Key factors of French metropolitan growth. *Service Industries Journal, 27*(3), 215–232.

Lovelock, C. H. (1983). Classifying services to gain strategic marketing insights. *Journal of Marketing, 47*, 9–20.

Marshall, J. N. (1994). Business reorganization and the development of corporate services in metropolitan areas. *The Geographical Journal, 160*(1), 41–49.

Miles, I. (1993). Services in the new industrial economy. *Futures, 25*, 653–672.

Miles, I. (2007). Business services and their users: A literature review. In L. Rubalcaba, & H. Kox (Eds.), *The new service economy. Challenges and policy implications for Europe* (pp. 39–61). Cheltenham: Edward Elgar.

Miles, I., & Watt, S. (1995). *Technological change in services and international trade competitiveness*. Genf: United Nations Conference on Trade and Development

IHK München und Oberbayern. (2003). *Der Medienstandort München*. München: IHK München.

OECD. (2000). *The service economy*. Paris: OECD Publications.

Prahalad, C., & Hamel, G. (1990). The core competence of the corporation. *Harvard Business Review, 68*, 79–91.

Ronning, G. (2003). *Unternehmensnahe dienstleistungen im bundesländervergleich*. Tübingen: Landesgewerbeamt Baden-Württemberg.

Rubalcaba, L. (2007). *The new service economy. Challenges and policy implications for Europe*. Cheltenham: Edward Elgar.

Rubalcaba, L., & Kox, H. (2007). *Business service in European economic growth*. Houndmills: Palgrave.

Rusten, G., Bryson, J. R., & Gammelsæter, H. (2005). Dislocated versus local business service expertise and knowledge: The acquisition of external management consultancy expertise by small and medium-sized enterprises in Norway. *Geoforum, 36*, 15.

Strambach, S. (1994). Knowledge intensive business services in the Rhine-Neckar area. *Tijdschrift voor Economische en Sociale Geografie, 85*, 354–365.

Tether, B. S., & Hipp, C. (2002). Knowledge intensive, technical and other services: Patterns of competitiveness and innovation compared. *Technology Analysis and Strategic Management, 14*(2), 163–182.

Viitamo, E. (2007). *Productivity of business services—Towards a new taxonomy*. Lappeenranta Lappeenranta University of Technology Finland, Department of Industrial Management Research Report.

Warf, B., & Wije, C. (1991). The spatial structure of large U.S. law firms. *Growth and Change, 22*(4), 158–174.

Williams, K., Williams, J., & Haslam, C. (1990). The hollowing out of British manufacturing and its implications for policy. *Economy and Society, 19*(4), 456–490.

Windrum, P., & Tomlinson, M. (1999). Knowledge-intensive services and international competitiveness: A four country comparison. *Technology Analysis and Strategic Management, 11*(3), 391–408.

Wölfl, A. (2003). *Productivity growth in service industries. An assessment of recent patterns and the role of measurement*. Paris: OECD.

Wood, P. (2002a). Knowledge-intensive services and urban innovativeness. *Urban Studies, 39*, 993–1002.

Wood, P. (2002b). *Consultancy and innovation: The business service revolution in Europe*. London: Routledge.

Wood, P. (2006). Urban development and knowledge-intensive business services: Too many unanswered questions? *Growth and Change, 37*(3), 335–361.

Chapter 15
On the Difficulty of Comparing the Spatial Distribution of Service Industries Across Nations: Contrasting Spain and Canada

Mario Polèse and Fernando Rubiera-Morollón

1 Introduction: Why Might Services Locate Differently in Different Nations?

This paper compares the spatial distribution of employment in service industries in Spain and in Canada. There is no particular reason for the choice of these two nations beyond the fact that the authors are familiar with them, a necessary pre-condition if the authors are to usefully analyze empirical results. An abundant literature has developed over the last few decades on the location of service industries (Coffey & Shearmur, 1997; Daniels, 1985; Illeris, 1996). Much of this literature has focused on so-called high-order services, sometimes also applying other labels such as producer services, business services, knowledge-intensive services, and so on (Hanell & Neubauer, 2006; Rubiera, 2005; Shearmur & Doloreux, 2008; Wernerheim & Sharpe, 2003). This is understandable for this is where most of the growth in service employment has occurred since the 1960s, although in more recent times employment growth in the health sector has accelerated in industrialized nations, a reflection of aging populations.

As the label "high order" implies, such services tend to concentrate at the top of the urban hierarchy Indeed, high-order, knowledge-rich, services almost define the modern metropolis. Studies for Spain, Canada, and other industrialized have confirmed, time and again, the positive relationship between city-size and employment in high-order, knowledge-rich, services (see studies cited above). Typical examples are banking, securities traders, consulting in various fields (management, engineering, etc.), advertising, and software development, as well as an almost infinite (and

M. Polèse (✉)
INRS, University of Quebec, Montreal, Canada

Centre Urbanisation Culture Société, Montreal, Canada
e-mail: mario.polese@ucs.inrs.ca

F. Rubiera-Morollón
University of Oviedo, Oviedo, Spain

J.R. Cuadrado-Roura (ed.), *Service Industries and Regions*,
Advances in Spatial Science, DOI 10.1007/978-3-642-35801-2_15,

growing) range of scientific and technical services. The reason they are also alternatively referred to as producer or businesses services is that they act as intermediary inputs into production, a reflection of the growing complexity and knowledge-content of modern production processes.

The reasons for the concentration of such services in large metropolitan areas are not difficult to explain. On the input side, the diversity and rapidly changing nature of talents and know-how mean that only the largest cities will provide the necessary labor pool. A large management consulting firm, for example, is constantly combining and recombining talent, depending on the nature of the contract. The same holds for a large advertising firm which, depending on the ad campaign, may need an opera singer 1 day, a cartoonist the next, and an animal trainer the day thereafter, and so on. Such industries are, in other words, dependant on a constant stream of face-to-face meetings with a wide (and changing) range of individuals. No wonder that the largest management consultancies and ad agencies are headquartered in a few "world" cities (Taylor, 2004). On the output side, the low cost of service delivery (at least in many cases) further facilities concentration, reinforced in recent times by the introduction of IT (information technology) which allows the almost costless delivery of reports, blueprints, video clips, and other information-rich products to distant places. A securities broker in New York or London can take an order over the phone or via the Internet, with little consideration the client's location. The effect of IT has, on balance, been to consolidate the concentration of high-order services in large metropolitan areas (Gaspar & Glaeser, 1998; Hall, 1999)

On the other hand, for services for which the consumer must be present at the place of production (or sale), city-size matters less. Classical central place theory (Christaller, 1935) remains entirely adequate to explain the spatial distribution of most such consumer services, with retailing the prime example. Most services, health and education included, remain constrained by distance and market access, although IT has opened new doors (i.e. distance learning). Other services again are subject to specific size and distance constraints, with accommodation and other tourist-related services an obvious example. Here, the consumer is willing to travel over often considerable distances to partake the service. The amenity attributes of place (climate, natural beauty, architecture, etc.) are the principal considerations rather than city-size, although the latter can be an attractive force in some cases. Summarizing, we can identify four types of services in terms of spatial distribution:

- Services sensitive to agglomeration economies, exhibiting hierarchal distributions. As noted earlier, the prime examples are business, financial services, and other information-rich services. Certain transport (i.e. air travel) and communication services (i.e. TV and radio broadcasting) equally fall in this class, due to the weight of scale economies in the case of former (air transport hubs) and the weight of diversified information inputs for the latter. Wholesaling and distribution services are also subject to important scale economies, with a growing trend to concentration. Agglomeration economies (size) are also a factor for higher order education and health services. However, for these two cases location may also be influenced by public sector choices, a first indication that national institutional frameworks can affect distributions.

- Services that need to be close to the consumer, and thus exhibit fairly even spatial distributions. Retailing, household, and other personal services are typical examples, as well as primary health and education services.
- Services sensitive to geography and natural attributes, with accommodation and tourist-related services the prime examples. Geography also matters for transport services, with country-size and population density important considerations.
- Services sensitive to public sector choices and institutional (and historical) context. The distribution of public service (public administration) employment is the prime example. One would not expect such employment to be distributed in the same manner in a highly centralized state such as France as in a highly decentralized nation such as the United States. Indeed, the very definition of what constitutes public administration employment may vary from one nation to another, a point to which we shall return.

For the two latter classes, the effects of national differences are fairly easy to conceptualize. Thus, in nations such as France and the United States with significant climatic differences (cold North, warm South), climate will have a visible impact on location decisions of firms and households. The attraction of "sunny" locations is today well documented, both in Europe and in America (Cheshire & Magrini, 2006; Rappaport, 2007; Rappaport & Sachs, 2003). On other hand, in a nation such as Canada with few "sunny" locations, climate will matter less, and city-size more. By the same token, the effects of country-size on the distribution of transport services are fairly easy to predict. In a vast nation such as Canada with numerous isolated communities (each with its own airport), city-size should matter less than in the more densely settled nations of Europe.

For high-order information-rich services, reasons why the basic rules of location (notably, city-size) might apply differently are more difficult to conceptualize. A priori, there is no reason why agglomeration economies should apply differently to, say, management consulting activities in nation x and nation y. It is difficult to imagine why the positive relationship with city-size should not hold. However, the relationship might be stronger or weaker across nations, depending on development levels and institutional context. Let us consider each factor in turn, starting with information-rich producer services (consulting, advertising, computer services, etc.). We may reasonably assume that the (intermediate) consumption of such services is a positive function of GDP levels, more developed economies consuming proportionality more producer services than less developed economies. Thus, in nations characterized by major internal development disparities (Italy is a prime example in Europe), we would expect the effects of such disparities to weaken the hierarchical relationship between city-size and location for producer services.

Let us now considered the impact of institutions, a notion that encompasses not only political and administrative structures, but also the legal and legislative framework that governs the functioning of the economy and of society as a whole.

Different nations have different institutions. This, more than anything else, makes comparisons across nations difficult. Here, we encounter a major difference between the study of location of manufacturing activity and of services. What constitutes a particular manufacturing activity and *mutatis mutandis* where it locates is little affected by institutions. A textile mill is a textile mill everywhere; an automobile assembly plant is an automobile assembly plant. Not so for many service industries, where definitions are more often fuzzy and porous, subject to nation quirks. Consider the financial sector. A priori, "banks" seems like a clearly defined activity with the same meaning across nations. However, a closer look at the banking sector reveals a broad variety of legislative frameworks across states with differing definitions of what banks can and cannot do. For example, Canadian banks are not allowed, at least in principal, to engage in investment banking, an activity left to portfolio mangers and similar institutions; while investment banking is a major factor in German banks. US banks are State or locally chartered, constraining their market areas, unlike Spanish or Canadian banks which function nationally. Locally chartered credit unions (*Cajas* in Spain), quasi-banks, are a major presence in some nations, but not in others. The financial sector is highly regulated in all nations, affecting both the organization and the spatial distribution of establishments. In short, rigorous comparisons are impossible.

Finance is not the only regulated sector. Other examples are health, education, telecommunications, and air travel. Should one expect the same location patterns for health-related employment in nations in which the health sector is largely public as in nations where it is largely private? Probably not. In other cases, dividing lines between sectors are highly porous, allowing firms to move between industry classes depending on local context and self-interest. A prime example is the distinction between retail and wholesale trade. If, for example, in country x it is more advantageous to declare oneself a retailer (for fiscal or other reasons) than a wholesaler, the latter sector will be proportionally smaller in nation y, which makes no such fiscal distinction.

2 Data and Methodology

As for any cross-country study, the primary challenge is the comparability of data and of observations. Comparing industry classes in the service sector creates particular challenges, as noted in the foregoing discussion. The comparability of observations is no less problematic. No international rule exists for defining urban areas, cities, or metropolitan areas. No two nations apply exactly the same rules. No rigorously comparable dataset exists, even among the nations of the EU, although Eurostat has made considerable progress. Ideally, from an economic analytic point of view, urban areas should conform to integrated labor markets (commuting sheds). Statistics Canada is one of the few national statistical agencies to systematically apply a labor

Table 15.1 Basic information, Canada and Spain (2001)

	Canada		Spain	
Total population (2001)	30,007,894		40,847,371	
Area (km^2)	9,984,670		504,782	
Density (inhabitants per km^2)	0.33		80.92	
Service sector in total employment (%)	72.59		57.44	
Five largest metropolitan areas	Toronto	4,675,481	Madrid	4,802,868
	Montreal	3,327,089	Barcelona	4,334,202
	Vancouver	1,986,913	Valencia	1,405,220
	Ottawa	1,051,242	Seville	1,124,859
	Calgary	951,395	Bilbao	984,144
Population in urban areas over 500,000	14,935,856 (49.77 %)		18,459,762 (45.19 %)	
Number of cases (n)	9		13	
Population in urban areas between 100,000 and 500,000 inhabitants	5,013,036 (16.70 %)		7,350,027 (18.01 %)	
Number of cases (n)	25		35	
Total urban population: urban areas, 10,000 inhabitants and over	23,658,741 (78.84 %)		30,807,451 (75.42 %)	
Number of cases (n)	147		226	

market rule to the definition of urban areas, where all municipalities and spatial units falling within a common commuting shed are classified as belonging to the same urban area. According the 2001 Canadian census, Canada had 147 urban areas, classified as either CMAs (Census Metropolitan Areas: population over 100,000) or CAs (Census Agglomerations: population above 10,000, but below 100,000).

Spain has no equivalent system of urban (agglomeration) classified areas. Data is available by municipalities only, which may or may not fall within a common labor shed. The challenge, thus, is building a comparable system for Spain, grouping municipalities which are part of the same urban area. The initial dataset comprised 453 urban observations for Spain. Taking into account geographical proximity and information on labor market connections, a new database was built comprising 226 aggregated metropolitan areas with populations of 10,000 or above. Although not rigorously comparable with the Canadian system, the number of urban observations in Spain is comparable, given the population size of the two countries, with similar city-size distributions (Table 15.1). The five largest urban areas have comparable populations; we may thus reasonably assume that associated agglomeration economies are not dramatically different and that the city size variable measures comparable realities. In addition to population data, urban areas in both nations with less than 500,000 inhabitants are classified according to distance from a larger metropolis, using a scheme employed in previous studies (Polèse & Shearmur, 2004, 2006; Polèse, Rubiera-Morollon, & Shearmur, 2007). For Spain, observations are also classified as coastal/ interior and as administrative capitals (or not) of provinces or autonomous communities.

The construction of a comparable employment dataset by service industry class was largely guided by the nature of available data for Spain. In both Spain and

Canada, data was drawn from the 2001 census. The nation with the more aggregated industry classes essentially sets the rules for the others. In Canada, highly disaggregated data by industry is available at the urban level (several hundred classes), but for only sixteen (16) broad industry classes in Span. Thus, the Spanish classes become the benchmark. Canadian industry employment data was reclassified to fit into nine (9) broad (Spanish) service industry classes listed below:

1. Retail Trade, Wholesaling, and Distribution
2. Accommodation and Tourist-related Services
3. Transportation, Storage, and Communication
4. Finance and Insurance
5. Business and Scientific Services and Real Estate
6. Public Administration
7. Education
8. Health Services
9. Household and other Personal Services

A more detailed listing of industries contained in each class is given in Appendix. We are fairly confident that the nine (9) classes capture the same industries in each nation. This, however, does not do away with the issues raised in introduction on the inherent difficulty of comparing service industries. The highly aggregated nature of the classes above does serve to diminish some problems (i.e. all financial institutions are grouped together), but in turn creates others. The grouping of "real estate" with business and scientific services is particularly unfortunate. Real estate-related services tend to exhibit a very different spatial logic from that of business services. In North American classifications, real estate management is generally classified with the financial sector, unlike the Spanish system. In short, the perfect classification system does not exist.

The data are analyzed in using various tools, descriptive statistics as well as correlation and regression analysis. The point of departure is the relationship between service employment location (relative employment concentrations) and city-size. The relationship is visually illustrated using location quotients[1] showing relative service employment concentrations by city-size class for Spain and for Canada. For both the Canadian (n = 147) and the Spanish urban systems (n = 226), location quotient results are regressed against city population size for each industry group. This completed by a correlation analysis of location quotients by paired industry groups.

The relationship between city-size and location quotients is plotted for selected industry groups in both nations. The latter part of the analysis focuses on Spain and on high-order services. Regression analysis residuals are mapped and analyzed. Finally, a more complete regression model for higher-order service groups is attempted for Spain.

[1] The location quotient (LQ) is defined as: $LQij = (Eij/Ej)/(Ei/E)$; where E is employment, i the industry and j the urban region.

Table 15.2 Relationship between population size and location quotients: Spain and Canada, for eight industry groups $LQ_{xa} = \alpha + \beta LOG(POPSIZE)_a + u_x$

	B (coefficient)		R^2	
Industry group	Canada	Spain	Canada	Spain
(1) **Retail Trade, Wholesaling, and Distribution**	−0.019*	0.008	0.290*	0.003
(2) **Accommodation and Tourist-related Services**	−0.045**	0.013	0.070**	0.001
(3) **Transportation, Storage and Communication**	0.030	0.061**	0.018	0.107**
(4) **Financial and Insurance**	0.092**	0.104**	0.303**	0.252**
(5) **Business & Scientific Services and Real Estate**	0.137**	0.121**	0.578**	0.469**
(6) **Public Administration,**	0.047	0.078**	0.009	0.039**
(7) **Education**	0.017	0.073**	0.009	0.107**
(8) **Health Services**	−0.043*	0.113**	0.022**	0.137**

Significant at **0.01; *at 0.05

3 Results I: The Relationship with City-Size

Table 15.2 shows regression results for the relationship between city-size (log) and location quotients (log) for employment by industry group[2] for Spain (n = 226) and Canada (n = 147). Correlation coefficients (between paired industry groups) are given in Table 15.3 for both nations. Figures 15.1, 15.2, 15.3, 15.4, and 15.5 show location quotients by city-size class in Spain and Canada for five industry groups, focusing on those of most interest. Finally, scatter gram results for the relationship between city-size and location quotients are shown in Fig. 15.6 for three industry groups.

The regression results (Table 15.2) illustrate both the similarities and the differences between the two nations. As expected, the two highest-order industry groups (4 and 5), both exhibit a positive and significant relationship between relative employment concentration and city, but slightly stronger in both cases in Canada than in Spain. In each country, the relationship is stronger for business and scientific services, the most knowledge-intensive group, which again is consistent with expectations. The relationship is visually illustrated in Fig. 15.1, which reveals an almost perfectly hierarchical relationship in both nations. By the same token, the plotted scatter grams (Fig. 15.6) for group 5 show largely similar patterns for both nations. In short, whether in Canada or in Spain, knowledge-intensive services are, as a rule, sensitive to agglomeration economies as measured by city-size.

For financial services (group 4), the relationship, although positive, is less symmetrical, especially for Spain. Figure 15.2 reveals a relative concentration of financial service employment in mid-sized urban areas (populations between 100,000 and 250,000) as well in several smaller places (populations above 25,000), with location quotients well-above those for Canada. However, it is interesting to note that no city below one million exhibits a quotient above unity (the national average) suggesting a partially bi-modal distribution in Spain (unlike Canada) with both a strong

[2] Household and Domestic Services are excluded because of low numbers in Canada.

Table 15.3 Pearson correlation matrix: location quotients by industry group, Spain (2001) (226 observations)

	Retail & wholesale	Accommodation	Transport. and com.	Financial & insurance	Business & scientific	Public admin.	Education	Health
Retail & wholesale	1	0.042	0.260**	0.227**	0.106	0.002	0.186**	0.133*
Accommo-dation	0.042	1	0.167*	0.068	−0.105	−0.029	−0.257**	−0.151*
Transport and Com.	0.260**	0.167*	1	0.511**	0.241**	0.137*	0.077	0.205**
Financial & insurance	0.227**	0.068	0.511**	1	0.703**	0.318**	0.444**	0.543**
Business & Scientific Services	0.106	−0.105	0.241**	0.703**	1	0.334**	0.573**	0.636**
Public admin.	0.002	−0.029	0.137*	0.318**	0.334**	1	0.577**	0.581**
Education	0.186**	−0.257**	0.077	0.444**	0.573**	0.577**	1	0.762**
Health services	0.133*	−0.151*	0.205**	0.543**	0.636**	0.581**	0.762**	1

*Significant at 0.05 (bilateral); **significant at 0.01 (bilateral)

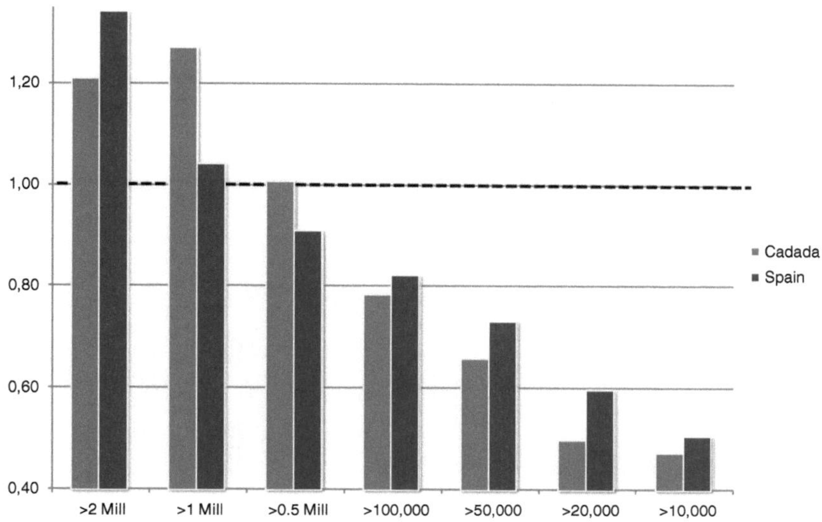

Fig. 15.1 Business & Scientific Services and Real Estate, Canada and Spain (2001). Location quotients

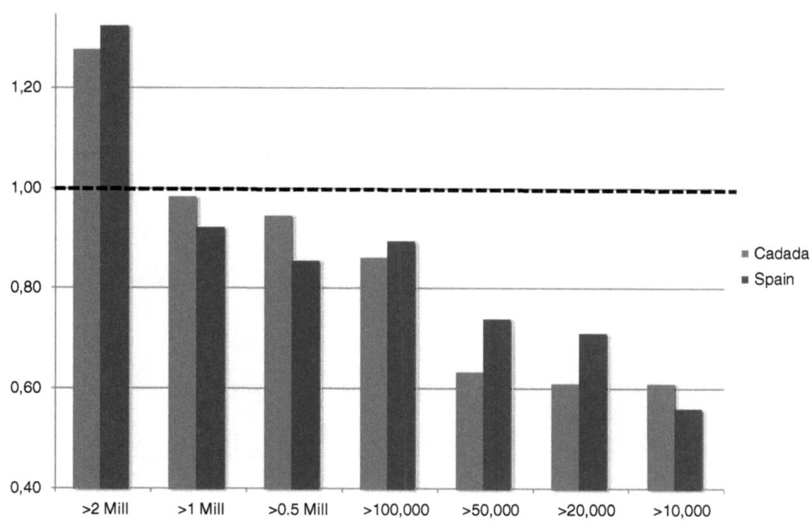

Fig. 15.2 Finance and insurance, Canada and Spain (2001). Location quotients

concentration in the very largest cities and then in a group of mid-sized cities. A possible explanation (to which we shall return) is the presence of *Cajas* (local credit unions), which are traditionally located in provincial capitals of which there are fifty-two (52) in Spain. Credit unions are less prevalent in Canada. Also, Canada's administrative geography is different with only ten provinces (Table 15.4).

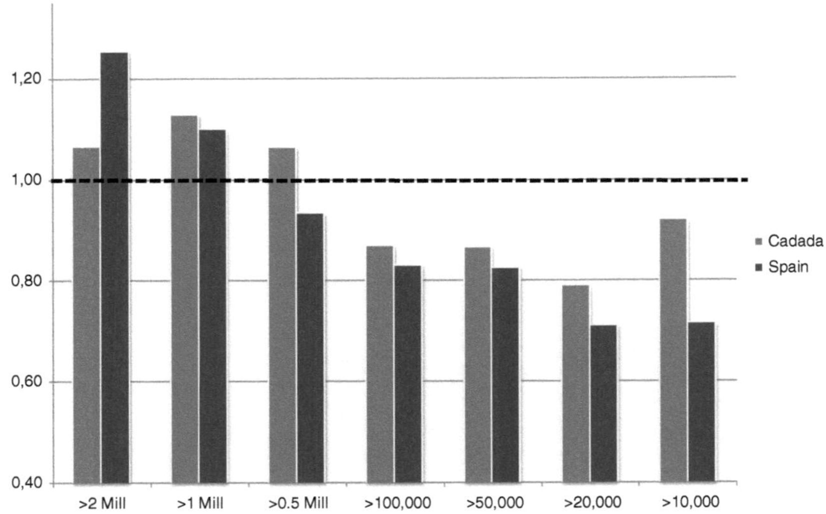

Fig. 15.3 Transport, storage and communication, Canada and Spain (2001). Location quotients

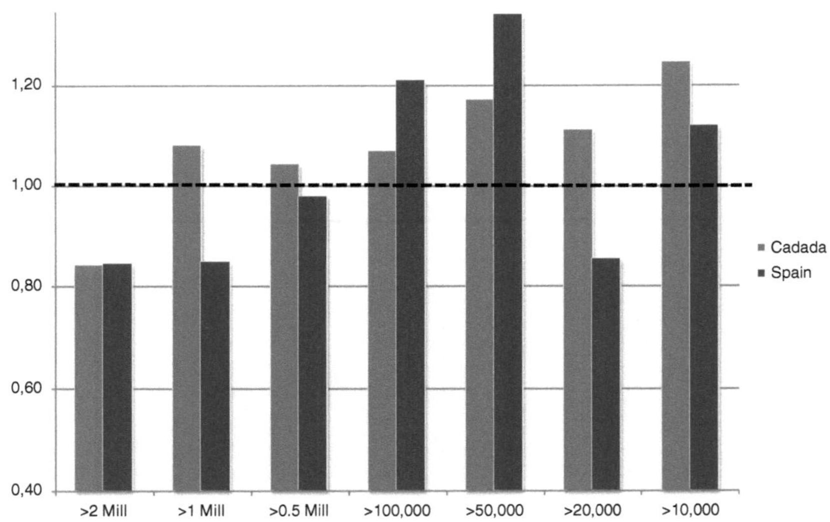

Fig. 15.4 Accommodation and tourism services, Canada and Spain (2001). Location quotients

Staying with these two high-order groups, the correlation matrix (Table 15.3) reveals that their respective location quotients are positively related in both nations, but more strongly in Spain than in Canada (with coefficients, respectively, of 0.70 and 0.50). Thus, in both nations, high-order services, whether business or financial services will often cluster in the same large cities. This, again, is as expected. But, this is where the similarities end. Taking group 5, the most knowledge-intensive

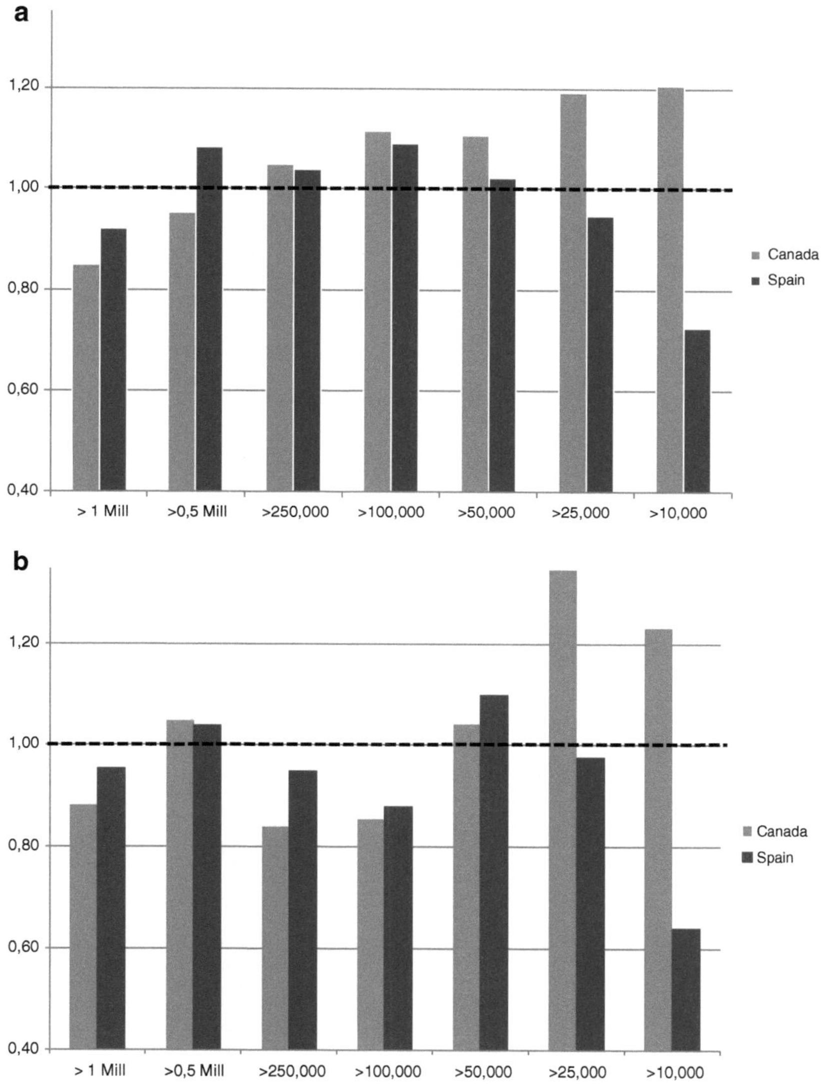

Fig. 15.5 Health services, Canada and Spain (2001). Location quotients. (**a**) All places; (**b**) Central places (less than 1 h driving from a metropolis over 500,000 inhabitants)

group, as our benchmark, the next highest correlation for Canada is with group 3 (transportation and communications), which is not unexpected since this group includes knowledge-intensive activities such as broadcasting and also because favored cities will often be local transport hubs. However, in Spain although group 5 also shows a positive relationship with group 3, as expected, correlations are even stronger (unlike Canada) with group 6 (public administration) and

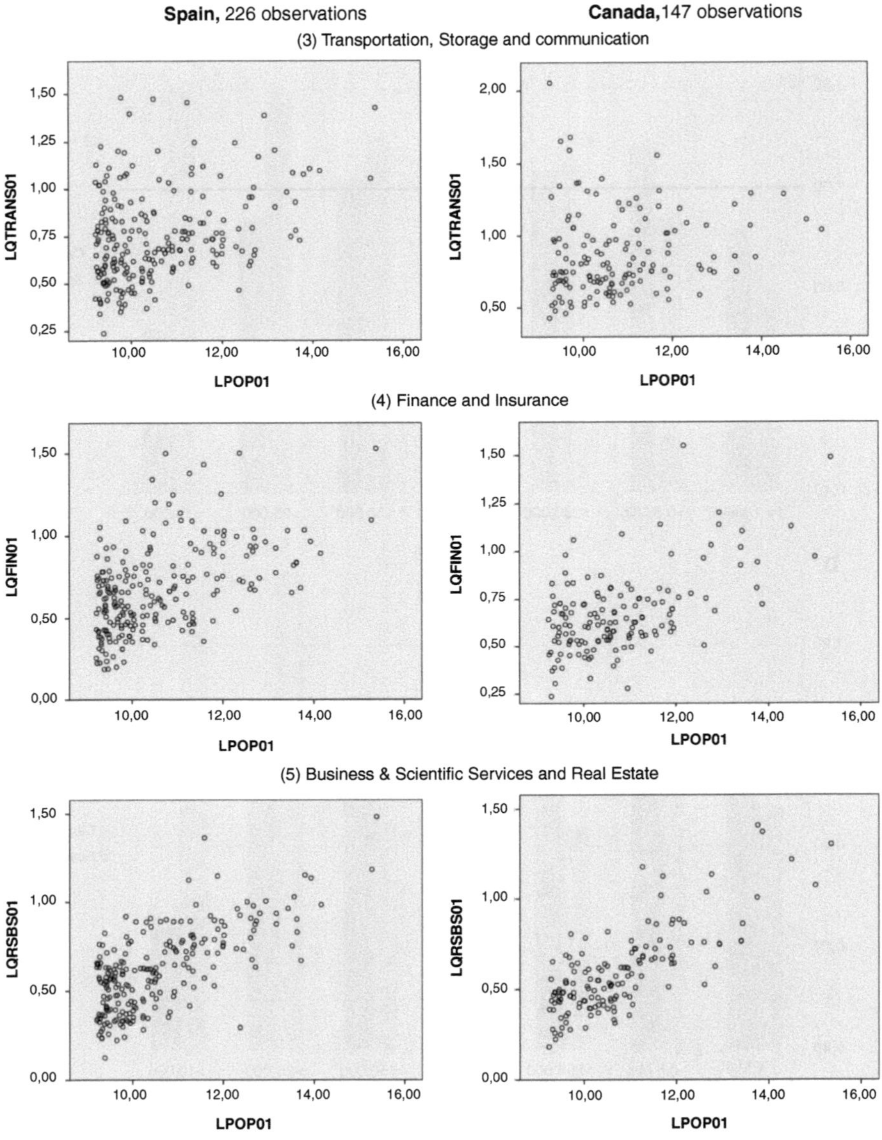

Fig. 15.6 Relationships between location quotient and population size. Plotted results (LOG transformations) for three industry groups, Canada and Spain

especially with groups 7 and 8, respectively education and health services. This would seem to suggest that the geography of services (retailing and tourist-related activities aside) is much more "deliberate" in Spain, so to speak, with health, education, and knowledge-intensive services often co-located in the same urban

Table 15.4 Pearson correlation matrix: location quotients by industry group, Canada (2001) (147 observations)

	Retail & wholesale	Accommo-dation	Transport. and com.	Financial & insurance	Business & scientific	Public admin.	Education	Health
Retail & wholesale	1	0.111	0.062	0.101	−0.064	−0.200*	0.157	0.318*
Accommodation	0.111	1	0.171*	−0.099	−0.086	0.073	0.067	0.055
Transport and Com.	0.062	0.171*	1	0.143	0.280**	0.299**	0.077	−0.051
Financial & insurance	0.101	−0.099	0.143	1	0.496**	−0.034	0.101	−0.021
Business & Scientific Services	−0.064	−0.086	0.280**	0.496**	1	0.237**	0.138	−0.212**
Public admin.	−0.200	0.073	0.299**	−0.034	0.237**	1	0.269**	0.161
Education	0.157	0.067	0.077	0.101	0.138	0.269**	1	0.350**
Health services	0.318**	0.055	−0.051	−0.021	−0.212**	0.161	0.350**	1

*Significant at 0.05 (bilateral); **significant at 0.01 (bilateral)

areas.[3] It is difficult not to interpret this as the reflection of a different institutional context.

Geography also plays a role. Its impact is, in part, illustrated in Fig. 15.3, which shows a largely hierarchical distribution for transportation and communications services in Spain, in turn confirmed by a positive and significant regression coefficient (Table 15.2), but not for Canada. The fairly high quotients for smaller Canadian cities (thus, the absence of a significant relationship with city-size) reflects the relative concentrations of transport services, but also some communications services, in small outlying cities far from a major urban center. Many such cities house not only a local airport, but also local radio and televisions stations. The influence of geography is also apparent for group 2 (accommodation), showing a significant but negative regression coefficient for Canada, where we would again expect accommodation services to be proportionately more present in small outlining cities (note the positive correlation with transportation, group 3). In Spain, by comparison, the accommodation sector appears to locate within its own universe, almost in opposition to others, with no significant relationship to city-size and with negative correlations with the education and health sectors. This result is not difficult to interpret. The accommodation sector in Spain is in large part oriented toward tourism, concentrated in coastal areas. However, as in Canada, it is positively correlated with industry group 3 (transportation and communications) which, again, is not surprising.

The indirect impact of tourism also helps to explain the difference between Canada and Spain for industry group 1, largely dominated by employment in retailing. The retailing demand engendered by tourists (and other semi-permanent residents) largely wipes out any city-size effect; thus, the absence of a significant relationship (Table 15.2). For Canada, the relationship is significant and negative, a reflection of the often disadvantaged status of small cities. In numerous small towns, retailing and personal services account for a disproportionate share of local employment, simply because there is little else in terms of employment opportunities, with transfer payments (pensions, unemployment insurance...) an important source of local income. This impression is strengthened by the positive correlation with the health sector. Smaller communities will, as a rule, have higher shares of older persons.

The most intriguing result is for health, but also the most difficult to explain. The spatial distribution of employment in health services seems to evolve in opposing directions in Spain and in Canada. The relationship with city-size is significant in both cases, but in opposite directions, positive in Spain and (mildly) negative in Canada (Table 15.2). The impression of spatial "opposites" is reinforced by the positive correlation with business and scientific services in Spain, but negative in

[3] It is possible that Spain's generally lower level of tertiarization introduces a bias. The nine services accounted for 57 % of Spain's total employment in 2001, compared to 73 % in Canada. However, it is difficult to see what the bias might be since the denominator in the location quotient equation is a constant in both nations.

Canada. In Spain, the health sector exhibits significant positive correlations with public administration (group 6) and education (group 7), while in Canada the relationship only holds up (but less strongly) for education. On the whole, public service sector employment (groups 6–8) appears fairly evenly distributed over space in Canada, with little relation to city-size, and with even a slight negative relation, as noted, for health services. The negative relation with size is clearly visible in Fig. 15.5a for health services, with the smallest class showing the highest location quotient, the opposite of the result for Spain.

How might this difference be explained? The most obvious explanation is geography. In Canada, because of distance, many small outlying communities will have hospitals or clinics since large urban centers are too far to service the local population. To test for this hypothesis only "central" cities were considered; that is, the largest cities (500,000+) and those within 1 h travel time (Fig. 15.5b). Clearly, the geography explanation must be rejected. The difference between smaller cities (populations under 50,000) in Spain and in Canada is now even greater. In other words, small towns in Canada, including those near large urban centers, have proportionately much higher shares of local employment in health services than in Spain. Location quotients, it is useful to recall, are calculated in relation to the national average, eliminating any bias attributable to national differences in health employment. In Canada, employment in health services in small cities is, as a rule, above the national average (a quotient above unity), but well blow the national average in Spain.

No simple explanation comes to mind. Visibly, Spain has a different philosophy than Canada with respect to the location of services provided (mainly) by the public sector, with health services the most notable example. The difference might, for example, reflect a difference in the nature of the service or in the coverage provided. A totally universal health care system, as in Canada, should in principle not be affected by spatial differences in income, while where the service is billed and/or provided by the market (at least in part) would be. The difference might equally be due to differences in the organization and management of the service, perhaps due to resource constraints; with a more centralized structure in country x compared to country y. Whatever the explanation might be in the Canadian-Spanish case, the spatial distribution of health care services provides an example of the probable impact of institutional factors and of national context.

4 Results II: Factors Other Than City-Size

In this section, we focus on Spain and the three industry groups representative of high-order services (groups 3–5). All three, we saw, show a positive relation with city-size (Table 15.2), exhibiting generally hierarchical distributions (Figs. 15.1, 15.2, and 15.3), consistent with theory and with the literature. The absence of a significant relationship in Canada for group 3 (transportation and communications) was fairly easy to explain, a reflection of its vast geography. For the other two groups, the relationship was systematically hierarchical in both nations, but somewhat less so in

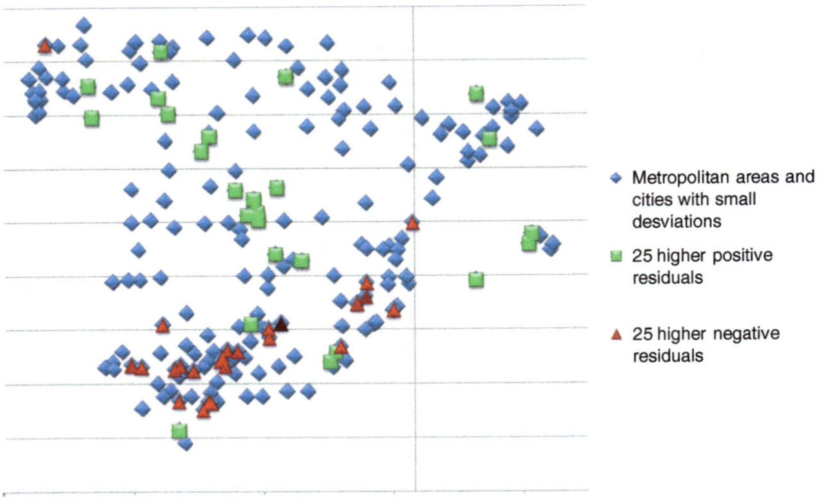

Fig. 15.7 Transportation and communications services. Mapped Regression Residuals, Spain (2001)

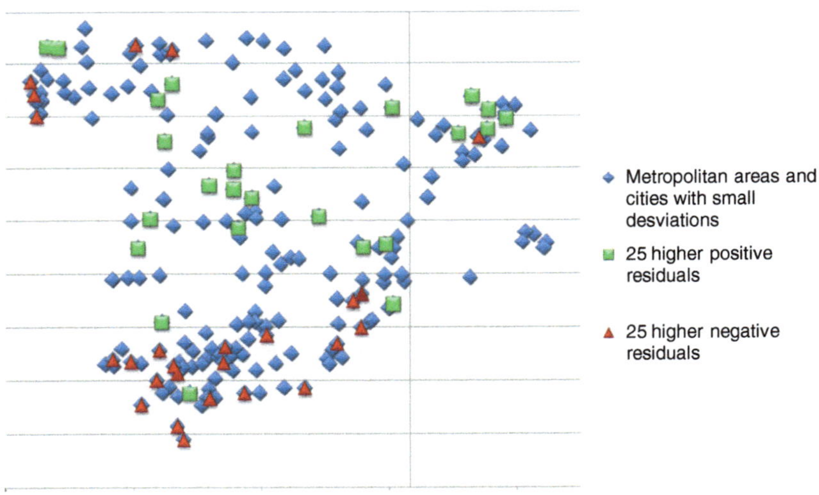

Fig. 15.8 Finance and insurance. Mapped Regression Residuals, Spain (2001)

Spain. Why might such high-order services deviate from the norm? The answer may lie in an examination of the residuals. The residuals for the three relevant regressions (formulation in Table 15.2) were mapped for Spain (Figs. 15.7, 15.8, and 15.9). Following an examination of the geography of residuals,[4] three new regressions were carried out, introducing variables suggested by the maps (Table 15.5).

[4] Note that the residuals exhibited normal distributions in all three cases, as would be expected.

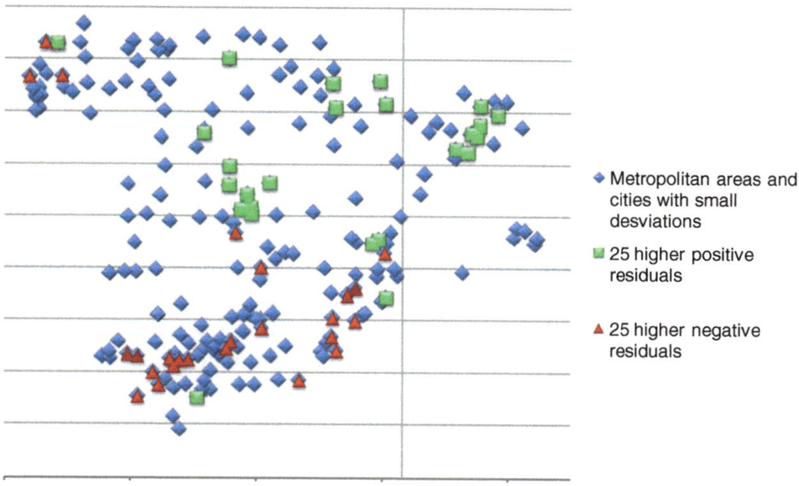

Fig. 15.9 Business & Scientific Services. Mapped Regression Residuals, Spain (2001)

Table 15.5 Relationship between residuals and four variables: three industry groups, Spain (226 observations) (2001); $\text{RESIDUALS}_i = \alpha + \beta_1\text{CAPITAL} + \beta_2\text{COASTAL} + \beta_3\text{CENTRAL} + \beta_4\text{EASTWEST} + \beta_5\text{NORTHSOUTH} + u$

	(3) Transport and communications	(4) Finance and insurance	(5) Business & Scientific Services and Real Estate
CAPITAL: Capital of province (1 if so, 0 otherwise). 52 capitals	−0.024	0.092**	0.016
COASTAL: Coastal location (1 if coast, 0 otherwise)	0.030	−0.115**	−0.031
Central: (1 if 1 h drive from a metropolis of 0.5 million +, 0 otherwise)	−0.038	−0.036	0.019
EASTWEST: East (values from 0 to 1, west–east coordinates)	0.230*	0.141	0.064*
NORTHSOUTH: North (values from 0 to 1, south–north coordinates)	0.151*	0.260**	0.108*
R^2	0.044*	0.229**	0.061*

*Significant at 0.05; **significant at 0.01

Before looking at Spanish results, a word is in order on the spatial distribution of residuals in Canada for industry groups 4 and 5 (the reader may which to consult the scatter grams again: Fig. 15.6). For business and scientific services, the most knowledge-intensive and also the most hierarchical, an examination of the highest positive residuals revealed an above-average presence of Western Canadian urban areas, often small and often specialized in mining or petrol exploration. Such extractive activities are large employers of engineers, geologists, and other related scientific and technical personnel, driving up the location quotients for industry group 5. This is not a variable which would make much sense on Spain. Capital cities are also overrepresented, which is not surprising, a variable which should equally play in Spain. For financial services (group 4), no particular logic is discernable in Canada for observations with predicted results above (or below) the regression line. Residuals are a mixed bag, both in terms of region or type of city, an indication that they are essentially random; there is little more to say about the location of most financial services beyond the attraction of size and the weight of agglomeration economies. The Spanish case is rather different, as we shall see,

Starting with transportation and communications services (Fig. 15.7), the mapped residuals reveal quite clear regional trends. These services are manifestly underdeveloped (given city-size) in the south, which is somewhat surprising given the weight of tourism in much of this part of Spain. On the other hand, it is useful to recall the aggregate nature of industry group 3. What this result suggests is that tourism has a limited impact (i.e. on only some sub-sectors such as air travel). It is also a reminder of the generally less-developed nature of the Spanish south. The highest values (residuals) seem to follow the arteries linking Madrid with Mediterranean ports in the southeast, with Asturias and Cantabria in the North (national highway IV), and with Barcelona to the northeast. In short, as in Canada, deviations form the norm (as predicted by city-size) can generally be traced to geography. Given, the nature of the service, this is not surprising.

There is less reason to believe why geography should intervene for knowledge-intensive services represented by groups 4 and 5 (putting aside exceptions such as oil exploration). There is no a priori reason to think why geography should matter for financial services. Yet Fig. 15.8 reveals definite regional patterns. Again, southern Spanish cities are underperformers (given their size). Another cluster of negative residuals is observable in the extreme northwest (Galicia, notably), traditionally one of Spain's poorest regions. Here, what we are probably seeing is not so much the impact of geography as of development differences. On the basis of Fig. 15.8, it would appear that financial services are sensitive to income levels, an entirely reasonable assumption. Investment banking and portfolio management are not generally services in high demand among poor populations. In Spain regional income disparities act, it appears, as a counterweight to agglomeration economies based on pure population size, reducing the predictive power of the latter for financial services.

The interference of regional development disparities shows up even more clearly for business and scientific services (Fig. 15.9). With only minor exceptions, all the high positive residuals lie within an approximate rectangle bounded by Madrid, Valencia, Barcelona, and Bilbao: the Spanish economic heartland. This result highlights the "center-periphery" dimension of certain relationships and, more specifically, the role of the spatial distribution of demand in the location of high-order services. In nations such as Spain, characterized by significant internal development differences with often deep historical roots, we should not expect to observe neatly hierarchical distributions for high-order services.

Table 15.6 All variables considered and location quotients regressions: three industry groups, Spain (226 observations) (2001); $LQ_{xa} = \alpha + \beta_0 LOG(POPSIZE)_a + \beta_1 CAPITAL + \beta_2 COASTAL + \beta_3 CENTRAL + \beta_4 EASTWEST + \beta_5 NORTHSOUTH + u$

	(3) Transport and communications	(4) Finance and insurance	(5) Business & Scientific Services and Real Estate
LOG(POPSIZE): Population size in logarithm	0.076**	0.088**	0.119**
CAPITAL: Capital of province (1 if so, 0 otherwise). 52 capitals	0.060	0.130*	0.021
COASTAL: Coastal location (1 if coast, 0 otherwise)	0.019	−0.103**	−0.030
Central: (1 if 1 h drive from a metropolis of 0.5 million +, 0 otherwise)	0.040	−0.034	0.190
EASTWEST: East (values from 0 to 1, west–east coordinates)	0.232*	0.143	0.064*
NORTHSOUTH: North (values from 0 to 1, south–north coordinates)	0.154*	0.257**	0.108*
R^2	0.149**	0.428**	0.501**

*Significant at 0.05; **significant at 0.01

In Table 15.5, the residuals for the three industry groups are regressed against four new variables, one institutional and three geographic. Unsurprisingly, the two geographic variables referring to (west–east and south–north) coordinates are significant for the transport and communications sector, although with a low R^2. For financial services, both the role of provincial capitals (undoubtedly via the presence of local *Cajas*) and of geography comes out, the latter via the repulsive effect of coastal locations and the attraction of the north. For business and scientific services, the development effect comes out via the simultaneous pull of the east and the north, although with, again, a small R^2, undoubtedly the reflection of a rather crude variable specification on our part. An income or GDP variable would perhaps have shown better results. Table 15.6 summarizes the results in a single regression.

5 Conclusions

Taking the Spanish and Canadian urban systems, the spatial distribution of service employment for eight major industry groups was examined, using location quotients, correlation and regression analysis, as well as mapping techniques. On the whole, the findings are consistent with the literature and with exceptions. In both nations, high order-services, specifically financial and knowledge-intensive business services, exhibit hierarchical distributions, concentrated at the upper end of the urban hierarchy. By the same token, city-size matters less for consumer-oriented services such as retailing and accommodation services, with even an observed negative relation for Canada.

Once one goes beyond these fairly predicable results, a number of differences emerge between Spain and Canada. In most cases, differences can be traced to differences in geography and in institutional context. Regulatory differences, as well as differences in administrative or political models, can influence location choices for services sectors such as finance, transportation, communications, education, and health.

The impact of geography is evident for accommodation and other tourist-related services. In Spain, with a well-developed tourist sector concentrated mostly in coastal areas (generally along the Mediterranean), spatial distributions are largely independent of other factors, with only weak links to other service sectors. Geography also matters for transportation and communications services, with higher concentrations in smaller cities in Canada, a reflection of greater distances.

The role of institutional and development differences is more difficult to interpret. Services provided (at least) in part by the public sector (education, health...) exhibit different spatial patterns in the two nations. On the whole, such services are more spatially clustered in Spain, positively associated with city-size. No such pattern is discernable in Canada, where publically-funded services are often proportionally more present in smaller cities. The most visible difference is for health services, which are positively related with city-size in Spain, but negatively in Canada. Geography is not the explanation for the difference equally holds for cities falling within a short distance from larger cities. Visibly, health services in Spain are not organized along the same lines as in Canada.

The spatial distribution of financial, businesses, and scientific services in Spain in part mirrors internal development differences, with higher relative concentrations (given city-size) in cities located in the richer northern and eastern parts of the nation. No equivalent pattern is discernable in Canada. The location of financial services in Spain is also sensitive to the administrative status of cities (provincial capital or not), most probably a reflection of the presence of *Cajas* (credit unions) in capitals. Again, no similar pattern is discernable in Canada. However, the capital-city factor is positively associated with businesses and scientific services in both nations.

The general impression for Spain (as compared to Canada) is: (1) of a more "structured" (or deliberate) pattern of urban service location, focused on a limited number of large and mid-sized cities and (2) of a more "dualistic" pattern, with marked regional differences, which in turn impact local demand levels for different services. The former is in part, we suggest, a reflection of institutional choices and differences, while the second mirrors internal differences in development.

Description of Industry Classes

1. *Retailing, Wholesaling & Distribution*
Food, beverage, and other wholesalers
Grocery and general merchandise stores
Gasoline stations, automobile dealers, automotive parts, repair and maintenance
Shoe and clothing stores
Furniture and home appliances stores
Pharmacies, other stores and retailers
2. *Accommodation*
Hotels and other traveler accommodation
Rooming and boarding houses
RV and camper parks and recreational camps
Restaurants, other food services, and drinking places
3. *Transport, Storage, and Communication*
Air, rail, truck, and water transportation
Urban transport
Warehousing and storage
Radio and television broadcasting
Telecommunications
Postal service & couriers
4. *Finance & Insurance*
Banks & other credit institutions
Securities trading & Portfolio Management
Insurance carriers and related activities
5. *Business & Scientific Services and Real Estate*
Software, IT services & data processing
Accounting & Management consulting
Advertising and related services
Architects & engineering, scientific and technical services
Legal services
Managers and lessors of real estate
6. *Public Administration*
National/federal government administration
Provincial and regional public administration
Municipal and local public administration
7. *Education*
Elementary and secondary schools
Community colleges and other post-secondary
Universities
8. *Health*
Hospitals & out-patient care centers and homes
Offices of physicians and other health practitioners
Clinics, medical and diagnostic laboratories
9. *Personal*
Personal, domestic, and laundry services

References

Cheshire, P. C., & Magrini, S. (2006). Population growth in European cities: Weather matters—but only nationally. *Regional Studies, 40*(1), 23–37.
Christaller, W. (1935). *Die Zentralen Orte in Süddeutschland*. Jena: Fischer Verlag.
Coffey, W., & Shearmur, R. (1997). The growth and location of high order services in the Canadian urban system, 1971-1991. *The Professional Geographer, 49*(4), 404–418.
Daniels, P. (1985). *Service industries: A geographical perspective*. New York: Methuen.
Gaspar, J., & Glaeser, E. (1998). Information technology and the future of cities. *Journal of Urban Economics, 43*, 136–156.
Hall, P. (1999). *Cities in civilization. Culture, innovation, and urban order*. London: Phoenix Giant.
Hanell, T, & Neubauer, J. (2006). *Geographies of knowledge production in Europe* (Nordregio Working Paper # 2006:3). Nordregio, Stockholm. www.nordregio.org.
Illeris, S. (1996). *The service economy: A geographical approach*. U.K., Wiley: Chichester.
Polèse, M., Rubiera-Morollon, F., & Shearmur, R. (2007). Observing regularities in location patterns: An analysis of the spatial distribution of economic activity in Spain. *European Urban and Regional Studies, 14*(2), 157–180.
Polèse, M., & Shearmur, R. (2004). Is distance really dead? Comparing industrial location patterns over time in Canada. *International Regional Science Review, 27*(4), 1–27.
Polèse, M., & Shearmur, R. (2006). Growth and location of economic activity: The spatial dynamics of industries in Canada 1971-2001. *Growth and Change, 37*(3), 362–395.
Rappaport, J. (2007). Moving to nice weather. *Regional Science and Urban Economics, 37*(3), 375–398.
Rappaport, J., & Sachs, S. (2003). The United States as a coastal nation. *Journal of Economic Growth, 8*, 5–46.
Rubiera, F. (2005). *Los servicios avanzados a empresas, dinámicas de localización, patrones de especialización y efectos sobre el desarrollo regional*. Madrid: Thomson-Civitas.
Shearmur, R., & Doloreux, D. (2008). Urban hierarchy or local milieu? High-order producer service and (or) knowledge-intensive business service location in Canada, 1991-2001. *The Professional Geographer, 60*(3), 333–355.
Taylor, P. J. (2004). *World city network: A global urban analysis*. London: Routledge.
Wernerheim, M., & Sharpe, C. (2003). High-order producer services in metropolitan Canada: How footloose are they? *Regional Studies, 37*, 469–490.

Chapter 16
The Importance of Creative Services Firms in Explaining the Wealth of European Regions

Rafael Boix, Blanca De-Miguel-Molina, and José Luis Hervas-Oliver

1 Introduction

This paper shows the important impact that a specific set of services, those belonging to the creative industries, have on regional economic development and wealth generation. Creative industries are a set of knowledge-based activities focused on the generation of meaning, contents and aesthetic attributes through the use of creativity, skill and talent, and have the potential to create wealth from trade and intellectual property rights. A key hypothesis in this paper is that creative services firms are a "growth driver" that promotes wealth in the regions where they are located. This is due to the fact that firms in creative industries introduce new ideas that are subsequently transferred to other firms of the economy, increasing the output of the whole economy. The objective of the research is to provide causal evidence of the impact of creative services on regional wealth.

The two points of departure of the article are the theoretical framework provided by Potts and Cunningham (2008) and the empirical experiment by De Miguel, Hervás, Boix, and De Miguel (2012). Potts and Cunningham (2008) propose four alternative models to better understand how creative industries may be linked with the whole economy. In two of these models, the creative industries are thought of having normal or negative impacts on the economy; in another, they act as high-order systems impacting on the generation of innovation and facilitating technical change; and in a fourth (known as the "growth model"), on which we focus, the creative industries are conceived as having a positive impact on the output of the economy.

De Miguel et al. (2012) compared the growth effect in European regions. Using Eurostat's Structural Business Statistics data, the authors found that an increase of

R. Boix (✉)
Department of Economic Structure, University of Valencia, Valencia, Spain
e-mail: rafael.boix@uv.es

B. De-Miguel-Molina • J.L. Hervas-Oliver
Department of Business Organisation, Politecnic University of Valencia, Valencia, Spain

J.R. Cuadrado-Roura (ed.), *Service Industries and Regions*,
Advances in Spatial Science, DOI 10.1007/978-3-642-35801-2_16,

1 % in the employment share of creative industries caused a differential increase of about 1,400 euros in GDP per capita. However, they neglected a crucial aspect: the creative industries are composed of both manufacturing and services enterprises. If we differentiate between those firms engaged in creative services and those in creative manufacturing, the correlation with GDP growth is negative for creative manufacturing (−0.34), but positive for creative services (0.64) (Fig. 16.1). Since the share of creative services in a region's productive structure is much larger than the share of creative manufacturing, this causes the positive behaviour of the aggregated indicator. Consequently, in order to understand the specific influence of the creative services industry we must treat it separately from that of creative manufacturing.[1]

By focussing on the effect of creative services, the paper will redress the fact that the role of services as contributors to regional development is still generally undervalued, with services traditionally being seen as consumers rather than as generators of economic wealth (Alexander & Akehurst, 2005). A failure to recognise their role would undermine an economic planning process. Therefore, this paper contributes additional evidence to this important discussion.

Hitherto, the literature on the broader category of knowledge intensive services (KIS) and their effect on regions (e.g. Miles, 2001; Müller & Zenker, 2001; Wood, 2002) has neglected the specific role of creative services within regions. However, the importance of creativity and creative services was highlighted by UNCTAD (2010, p. 3), which said "a new development paradigm is emerging that links the economy and culture, embracing economic, cultural, technological and social aspects of development at both the macro and micro levels. Central to the new paradigm is the fact that creativity, knowledge and access to information are increasingly recognized as powerful engines driving economic growth and promoting development in a globalizing world. The emerging creative economy has become a leading component of economic growth, employment, trade and innovation, and social cohesion in most advanced economies." Our paper contributes to addressing a neglected area which UNCTAD says is of great importance, by focusing on the relationship between the creative services and the wealth of regions.

The structure of this paper is as follows: Sects. 2 and 3 introduce the theoretical framework, explaining the relationship between creative industries and knowledge-intensive services and the relationship between creative industries and wealth. Section 4 explains the methodology for comparing at the regional level the causal link between creative services and regional wealth. Section 5 presents the results. The paper ends with the conclusions and discussion in Sect. 6.

---

[1] In the latest classifications of activities (ISIC Rev.3.1 and NACE Rev.2) the only creative industries classified as manufacturing belong to the "fashion" sector, which is commonly assimilated to clothing and footwear industries. Within these industries it is difficult to separate those firms that focus on the intangible part (fashion design) from those more addressed to bulk production and, in many cases, a firm performs both activities.

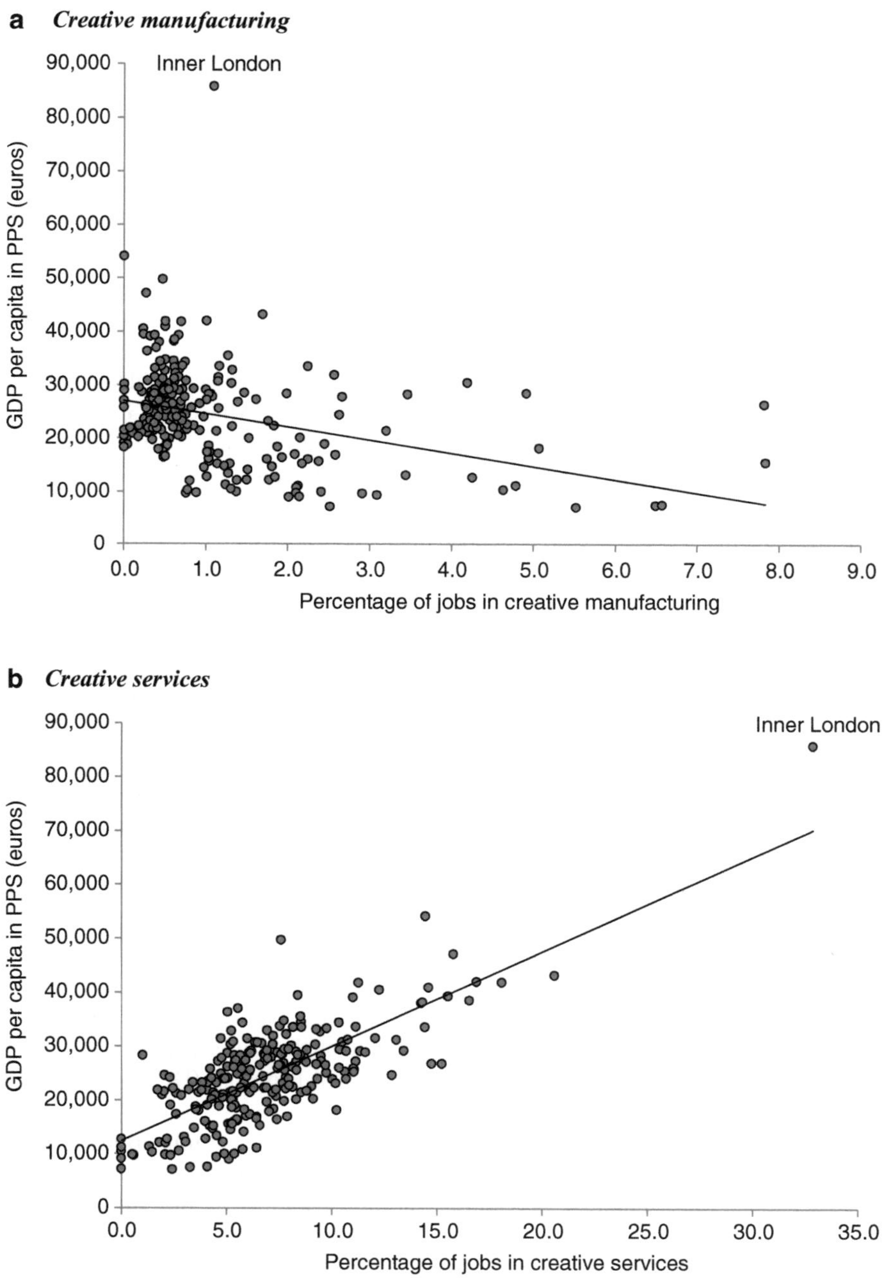

Fig. 16.1 Correlation between GDP per capita in power purchasing standard (PPS) and the percentage of creative manufacturing and creative services jobs in European regions in 2008 (250 regions with data available). *Source*: Elaboration from Eurostat

2 Creative and Knowledge-Intensive Services: Towards a Compatible Definition and Taxonomy

The British Department for Culture, Media and Sport (DCMS, 2009) defines **creative industries**, as "those industries that are based on individual creativity, skill and talent, and which have the potential to create wealth and jobs through developing intellectual property." The most comprehensive taxonomy of creative industries, which is also particularly appropriate to cross-country comparisons, has been proposed by UNCTAD (2010). This classification includes both manufacturing and service industries, although the majority of the sectors included in creative industries are services, especially knowledge-intensive services (KIS).[2]

There is a lack of literature that theoretically links creative services and knowledge-intensive services (KIS), exceptions being short discussions in Miles and Green (2008), Sunely, Pinch, Reimer, and Macmillen (2008) and Müller et al. (2009). In practice, creative services are typically identified as a sub-group of knowledge-intensive services, and their importance is often related to the ever-increasing dependence of manufacturing industries on the service sector (Redondo-Cano & Canet-Giner, 2010). Using the nomenclature of the OECD and Eurostat (2009), some knowledge intensive services can be categorised as "high-tech knowledge-intensive services" (such as audiovisual, broadcasting or computer programming services) and the remainder are "rest of knowledge-intensive services" (e.g. publishing, architecture or advertising).[3] Empirical research that relates creative services and KIS includes that of Chapain, Cooke, De Propris, McNeil, and Mateos (2009) which looked at their spatial co-location and Stoneman (2009) which focused on their linkages with soft innovation.

What differentiates creative services from the "rest of knowledge intensive services (KIS)?" First, the "knowledge economy" literature focuses on the use, production and management of knowledge and information as the main force for economic development (Drucker, 1969; Machlup, 1962; OECD, 1996) and has found strong theoretical applications in new growth theory (Lucas, 1988; Romer, 1986). On the contrary, the "creative industries" literature focuses on creativity, skills and talent and their potential for wealth and job creation (a good example is the recently emerged industry of videogames). The "creative industries" term originated in Australia with the report "Creative Nation: Commonwealth Cultural Policy" (DCA, 1994), although it was popularised by the Department of Culture, Media and Sports in the United Kingdom (DCMS, 1998) and elaborated upon by UNCTAD (2010).

A second differentiating factor derives from the practical implications of each approach. A good way to illustrate the difference is to relate the "knowledge economy" and the "creative economy" through the theory of "differentiated

[2] A detailed review of the literature on creative industries exceeds the scope of this paper. Good critical surveys can be found in UNCTAD (2010), O'Connor (2007) and Flew and Cunningham (2010).

[3] Strictly speaking, some concrete creative services, such as for example cultural retail, are in practice assigned by Eurostat to the "less-knowledge-intensive group" (LKIS).

knowledge bases" (Asheim & Hansen, 2009). The underlying idea behind this theory is the characterization of a specific (or critical) knowledge input on which an innovation activity is based. Asheim, Boschma, and Cooke (2011) distinguish three types of knowledge bases: (1) an "analytical base" derived from the production and use of explicit (codified) knowledge that originates from science and technology; (2) a "synthetic base," where knowledge is created through a more inductive process of testing, experimentation and practical work; and (3) a "symbolic base," where knowledge is related to the creation of contents, desires and aesthetic attributes of products.

The "knowledge economy" is considered to mainly refer to activities based on analytical and synthetic knowledge bases due to the fact that the economic activities are classified according to the intensity of use of research and development (as measured by the units of added value they contributed), and also according to intensity of use of human capital (as measured by the relative use of people with university degrees and by the use of human resources in science and technology). This analysis results in the differentiation of high, medium and low technology manufacturing (e.g. pharmaceutical industry, automotive industry and furniture respectively), and knowledge-intensive and knowledge-non-intensive services (e.g. telecommunications and wholesale trade respectively).

The "creative economy" is thought mainly to have a symbolic base involving the creation of new realities and artistic or cultural expressions in the form of contents, desires and aesthetic attributes. Since most of the workers in creative firms are talented and skilled, these activities tend to be classified as knowledge-intensive. This point is noted by UNCTAD (2010, p. 3), defining creative industries as "a set of knowledge-based activities, focused but not limited to the arts, potentially generating revenues from trade and intellectual property rights," and by the European Commission (2010, p. 13) and Power and Nielsén (2010, p. 7), where "creative and cultural activities are knowledge-driven industries that drawn to specialized labour markets and to clusters."

Table 16.1 contains NACE Rev.2 codes of the creative services activities in the UNCTAD (2010) list, showing the relationship with knowledge-intensive services. KIS that are defined as creative service industries are: "Audiovisual," "Broadcasting," "Computer Programming," "R&D," "Publishing," "Architecture and Engineering," "Advertising," "Design and Photography," "Arts, Entertainment and Recreation."[4]

[4] We intentionally avoid discussing how the notion of creative industries conceptually relates to regarding other concepts such as cultural industries or arts, as well as to different taxonomies. This discussion is addressed in Pratt (2007) and UNCTAD (2010). The justifications for our use of the UNCTAD taxonomy are that it is derived from a broad and rigorous debate about an appropriate taxonomy, and that it is more comprehensive than single country based taxonomies such as that of DCMS (2009). Two thirds of the creative industries are shared among the various taxonomies so that, in this sense, the empirical differences are moderate.

Table 16.1 Classification of services in terms of creativity and knowledge intensity, based on NACE Rev.2

Knowledge-intensive services	Creative	Non-creative
High-tech knowledge-intensive services (HTKIS)	59 audiovisual 60 programming and broadcasting 62 computer programming 72 R&D	61 telecommunications 63 information service activities
Rest of knowledge-intensive services (RKIS)	58 publishing 71 architecture and engineering 73 advertising 74 design, photography 90–93 arts, entertainment and recreation (section R)	50–51 water and air transport 64–66 financial and insurance 69–70 legal and accounting; head offices; management consultancy 75 veterinary activities 78 employment 80 security and investigation 84–88 public administration and defence, compulsory social security, education, human health and social work
Less-knowledge-intensive services (LKIS)	4779 retail sale of second-hand goods in stores	45–47 (except 4779) wholesale and retail trade 49 land and pipelines transport 52–53 warehousing, postal and courier 55–56 accommodation and food service 68 real estate 77 rental and leasing 79 travel agency 81 services to buildings and landscape 94–96 membership organisations, repair of computers and personal and household goods, other personal service 97–99 domestic personnel; undifferentiated goods; extraterritorial organisations

Source: Elaborated from UNCTAD (2010) and Eurostat (2009)
Note: 58—publishing includes: 581 publishing of books, periodicals and other publishing activities; and 582 software publishing
59—audiovisual includes: 591 motion picture, video and television programme activities; and 592 sound recording and music publishing activities
60—programming and broadcasting includes: 601 radio broadcasting; and 602 television programming and broadcasting activities
62—computer programming includes: 6201 computer programming activities; 6202 computer consultancy activities; 6203 computer facilities management activities; and 6209 other information technology and computer service activities
71—architecture and engineering includes: 711 architectural and engineering activities and related technical consultancy; and 712 technical testing
72—R&D includes: 721 research and experimental development on natural sciences and engineering; and 722 research and experimental development on social sciences and humanities
73—advertising includes: 731 advertising; and 732 market research
74—design, photography includes: 741 specialised design activities; 742 photographic activities; 743 translation and interpretation; and 749 other professional, scientific and technical activities
R—arts, entertainment and recreation includes: 90 creative, arts and entertainment activities; 91 libraries, archives, museums and other cultural activities; 92 gambling and betting activities; and 93 sports activities and amusement and recreation activities

3 Linking Creative Industries to Regional Wealth

Potts and Cunningham (2008) propose four models of how the creative industries might relate to the broader economy: "the welfare model," "the competitive model," "the growth model," and "the innovation model."

In the "welfare model," creative industries are conceived as being affected by "Baumol's disease" (Baumol & Bowen, 1966) and their rate of productivity growth is less than in the rest of the economy. They have a negative impact on an economy's production, such that they consume more resources than they produce ($dY/dCI < 0$, where Y is production and CI creative industries) and their growth comes at the cost of aggregate economic growth. However, the commodities produced are welfare enhancing ($dU/dCI > 0$, where U is the utility). In this model, policy prescriptions would focus on subsidies and price maintenance in order to protect the creative industries.

In the "competitive model," the creative industries are just another industry, and a change in their size or value has a proportionate effect on the rest of the economy. They are neutral (i.e. do not have more effect than do other activities) in respect of technological change, innovation or productivity growth: $dY/dCI = 0$ and $dU/dCI = 0$. This implies that the marginal benefit of a redirection of resources towards these industries is zero, and requires the same policy treatment as the rest of industries.

In the "growth model" the creative industries are a "growth driver" and their impact on the economy is more than proportional ($dY/dCI > 0$). This could be due to supply-side effects, such as the fact that their productivity is higher than other industries, or because creative industries introduce new ideas that are then transferred to other sectors of the economy, or because such industries facilitate the adoption and retention of new ideas in other sectors. Or/and there could be demand-side effects, such as where a growth in the GDP (Y) causes a proportionate increase in demand for creative industries services. Policy implications depend on the magnitude of each effect (supply and demand) but may include an awareness of a need to deal with creative industries as a "special sector" due to their effects on the whole economy.

The "innovation model" is based on the Schumpeterian tradition found in business and strategic literature. It reconceptualises the creative industries as a higher-order system that operates on the economic system, similar to science, education and technology in the national systems of innovation approach. Therefore, the main effects of creative industries are not their direct effects on production or wealth, but, rather, their contribution to the technical change.

4 Empirical Design

4.1 The Model

The preliminary evidence provided in the introduction suggests that creative services play a role indicated by the Potts and Cunningham's "growth model." Unfortunately, the authors only provide a general framework and not the concrete mechanisms through which creative services affect the wealth of regions. In De Miguel et al. (2012), this was achieved by means of an empirical model consisting of a linear equation which compared the effects on the wealth of regions of agglomeration and productive structure, including the share of employment in creative industries in a region. To analyse the effects of agglomeration and structure, use was made of Eurostat and OECD classifications of activities by knowledge intensity, and differentiating creative and non-creative activities. Despite its simplicity, the model exhibited a high performance, explaining 60 % of the variance. However, further insights that draw on our conceptual discussion in Sect. 2 indicate that the equation is more complex than it first appears since differences in regional wealth are due to four forces: creativity (symbolic knowledge), other forms of knowledge (analytic and synthetic), the productive structure, and the effects of agglomeration economies.

We depart from the same framework, but we introduce two improvements. First, in the light of previous evidence (Fig. 16.1), our variable of interest is the more focused "creative services" (as measured by share of employment in creative services in respect of each region's total employment), and "creative manufacturing" is assumed to be a standard industry included in low-tech manufacturing activities.[5] Second, it is possible that the original equation was miss-specified since the only type of agglomeration economies included were localization economies derived from the existence of "average" regional clusters. Therefore, we will include as control variables the other two types of agglomeration economies: internal scale economies and urbanization economies.

4.2 Data and Variables

Our sample comprises 250 European regions at NUT2 from Eurostat's Structural Business Statistics (SBS), Science and Technology Statistics (STS) and Economic Accounts (ESA) databases, and corresponds to 2008.[6] SBS, in combination with the new NACE, provides a good source of data for this research, as the information is disaggregated from two to four digits. This detail is not usually required as the new NACE is particularly designed to deal with the requirements of the knowledge economy, so that creative industries are properly captured at the two digits level in most cases (Table 16.1). The activities of the NACE code R (Arts, Entertainment and Recreation) are not available in the SBS database and have been obtained from the STS database, which means that it also includes employees from the public sector.

[5] If creative manufacturing is estimated separately it exhibits an average negative impact on regional wealth. However, that does not change the evidence and implications of the general results.

[6] The countries for which data was not available, such as Greece, Luxembourg and Malta, were not included. Data for the year 2001 have also been used for the design of the exogenous instrumental variables.

The differences among databases have been taken into account to compute the total number of employees.

In the framework suggested by Potts and Cunningham (2008) the effect on output is captured using Gross Domestic Product (GDP) data. As we will compare differences in output across regions, GDP must be divided by the number of inhabitants, or employment, in each region. Thus, following De Miguel et al. (2012) we focused on GDP per capita. This variable mixes productive efficiency and income per capita, and is an indicator traditionally used as a proxy for the regional wealth in cross-country and cross-region studies (Barro & Sala-i-Martin, 1991; Quah, 1996).

To calculate the regional productive structure, data on employment was extracted from the previously mentioned codes, and the percentage of employment in each service sector with respect to total regional employment was calculated. The statistical calculations use the services groupings that appear in Table 16.1.

The term *agglomeration economies* denotes "all economic advantages accruing to firms from concentrated location close to other firms: reduced production costs due to large plant size; the presence of advanced and specialized services; the availability of fixed social capital (e.g. infrastructures); the presence of skilled labour and of managerial expertise, and of a broad and specialized intermediate goods market" (Capello, 2006, p. 18). The seminal works by Ohlin (1933) and Hoover (1937), most of the other classical texts, and the recent book of Capello (2006) differentiate three families of agglomeration economies: "internal to the firm" (scale economies), external "localization economies" (external to the firm but internal to the industry), and "urbanization economies" (external to the firm and external to the industry). As in De Miguel et al. (2012), we use as a proxy for "localization economies" the sum of regional clustered activities at the two digit level. It is considered that an activity is clustered when its Location Quotient for firms in the industry is above 1:

$$LQ_{ij} = \frac{\text{Firms in the NACE code } j \text{ in region } i/\text{Firms in the NACE code } j \text{ in the EU27}}{\text{Firms in the region } i/\text{Firms in the EU27}} \tag{16.1}$$

Although there is some correlation between those indicators used to identify regional employment structures and those used to denote localization economies, they employ different concepts and are measured using different data (employment in the first case and number of firms in the second one). Results in Fig. 16.2 present for 250 European regions the share of employment in creative services in each region's total employment, and the relative specialisation of each region (LQ) in creative services. The figure demonstrates an unequal distribution of the importance of creative services across the European regions. The regions that specialise more in creative services in terms of employment structure and LQ tend to be those that have a large metropolitan area, such as London, Paris, Amsterdam, Brussels or Madrid.

Following Lazzeretti, Boix, and Capone (2009), proxies used for "urbanization economies" include: the total population in the area (market potential); population

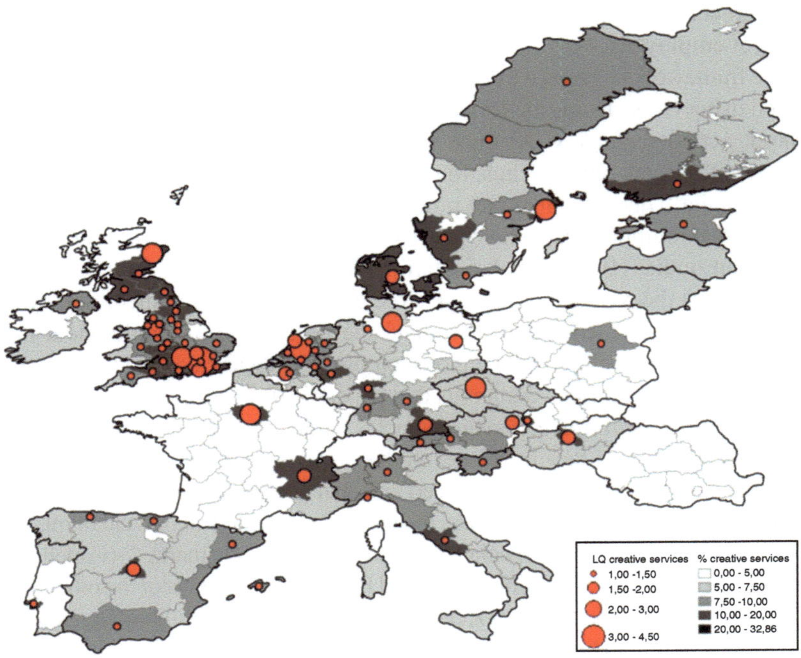

Fig. 16.2 Shares of creative services in regional employment, and relative specialisation of each region (location quotient) in creative services, for 250 EU regions. *Source*: Elaboration from Eurostat

density (population per km^2) which favours knowledge spillovers; and diversity of the productive structure at two digits which fosters cross-fertilization across sectors (inverse of the Hirschman–Herfindahl index calculated for employment in 60 sub-sectors in the economy in 2001, $IHHI_j = 1/\left(\sum_i L_{i,j}/L_{i,j}\right)^2$. Following again Lazzeretti et al. (2009), the proxy used for "internal economies" is the average firm size in the region (average number of employees by firm in the region). This captures scale economies and the organization of the production. Descriptive statistics are provided in Table 16.2.

4.3 *Disentangling the Supply-Side Effects: Simultaneity and Causality*

The equation used as the basis of the regression model was:

$$\frac{GDP_i}{\text{Population}_i} = \beta_0 + \beta_1 \text{Creative Services}_i + \beta_2 \text{Knowledge Structure}_i + \beta_3 \text{Agglomeration}_i + \varepsilon_i, \quad (16.2)$$

where *creative services* is the percentage of a region's i employment found in creative services. *Knowledge structure* includes four variables that are the share of employment in the region in: "high-tech services," "rest of knowledge-intensive services," "less-knowledge-intensive services," and "manufacturing" (and in every case creative industries belonging to these groups have been removed to avoid double counting). *Agglomeration* includes nine variables. These include five variables related to localization economies in the region (the number of clusters of respectively creative services, high-tech services, other KIS, less-KIS, and manufacturing); three variables related to urbanization economies (population, population density, and productive diversity); and one variable related to internal economies (average firm size).

In the Potts and Cunningham (2008) growth model, an increase in the share of creative industries caused an increase in the output (through a supply-effect), but an increase in regional production or wealth also translated into an increase of demand for creative industries services. In Eq. (16.2) both effects are simultaneous, and it is not possible to know the direction of the causality. In the absence of a robust theoretical model, three basic solutions are suggested in the econometric literature: the use of time-lagged variables, Granger tests in time-dynamic models, and instrumental variables.

As the equation is time-static, the use of instrumental variables is the most suitable methodology. This involves a system of two equations in which the second one is an auxiliary equation where the share of employment in creative services is the dependent variable. Justification for the content of the instrumental equation is provided by Lazzeretti et al. (2012) who in their article explain the reasons for the clustering of employment in creative industries. The authors introduce three main determining forces, namely: "culture and heritage," "the influence of agglomeration economies," and "the presence of a creative class."

$$\begin{aligned}\text{Creative services}_i = \beta_0 + \beta_1 \text{ Heritage}_i + \beta_2 \text{Agglomeration}_i \\ + \beta_3 \text{ Creative Class}_i + \varepsilon_i\end{aligned} \quad (16.3)$$

Heritage is measured using cultural endowments (the presence of UNESCO goods per million inhabitants). *Agglomeration* economies variables include the average firm size in the region, the average firm size in creative services in the region, the productive diversity in the creative services string and the population. Florida's *Creative class* is measured by patents per million inhabitants, R&D expenditures in relation to GDP and the percentage of the population that is creative class. Instruments for the percentage of employment in creative services are calculated for the year 2001 in order to assure their exogeneity. The fit (R^2) of the instrumental regression is 0.85.[7]

[7] We refer to Lazzeretti et al. (2012) for the detail in the elaboration of the variables used as instruments.

5 The Distinct Role of Creative Services in European Regions: Main Results

5.1 Results of the Regression Analysis

OLS and instrumental variable regressions were estimated, verifying the statistical significance of the model in Eqs. (16.2) and (16.3) (Table 16.3). The first column of Table 16.3 shows the estimates of the model only using the structure variables, the second only using the variables of agglomeration (including "localization," "urbanization" and "internal scale economies"), and the other two columns show the parsimonious estimation of the integrated model, combining all the variables and removing the statistically non-significant and collinear variables.

The results show that our initial hypothesis is confirmed: an increase in 1 % in the percentage of employment in creative industries in the region translates to an increase of 0.39 % in GDP per capita (that is, an increase of 1,479 euros per capita) (Table 16.3, column 3, $p < 0.01$). Therefore, creative services are a "growth driver" that promotes wealth in the regions where they are located. Furthermore, when compared with the other indicators, the share of employment in creative industries has the higher causal impact on the differences in GDP per capita of the regions.

This impact is slightly higher than the 1,424 euros provided by De Miguel et al. (2012) due to the fact that the latter also includes creative manufacturing which, having a negative impact, reduces the size of the coefficient. The instrumental variables estimates of the model show quite similar results (Table 16.3, column 4). As the endogeneity test (Durbin–Wu–Hausman test) does not reject the exogeneity of the creative services, it is preferable to use the results of column 3 because the OLS estimates are more efficient.[8] Therefore, our initial hypothesis on the existence of a positive supply-side effect of creative industries services on output per capita is proved.

There are two other relevant results. First, the effects of the employment structure seem to be more important than the effects of agglomeration, even if both are correlated. In fact, most of the variables of agglomeration become statistically and economically non-significant when are included in the same equation the variables of the employment structure (Table 16.3, column 3), exceptions being the number of clusters of less-knowledge-intensive services ($\beta = 526, p < 0.01$) and the population density ($\beta = 1.62, p < 0.05$). The good performance of the instrumental equation also suggests that the effects of agglomeration could be translating to the rest of the economy through the economic structure.

Second, whereas the share of persons employed in "high-tech non-creative services" does not have a statistically significant effect on the GDP per capita the

[8] This does not mean actually that the variable is exogenous, only that their effects on the consistency of the estimates are not relevant. In this case, OLS produce the best linear unbiased estimator.

Table 16.2 Descriptive statistics

Variable	Mean	Std. Dev.	Min	Max
GDP per capita in PPS	24,465	9,005	7,100	85,800
%Employment in creative services	6.88	3.83	0.01	32.86
%Employment in high-tech services[a]	0.88	0.78	0.01	4.43
%Employment in rest of knowledge-intensive services[b]	28.25	6.45	13.98	42.71
%Employment in less-knowledge-intensive services[c]	27.77	4.17	14.55	45.42
%Employment in manufacturing	16.40	7.40	0.01	35.99
Number of clusters of creative services	2.70	2.23	0.00	8.00
Number of clusters of high-tech services[a]	0.85	0.64	0.00	2.80
Number of clusters of rest of knowledge-intensive services[b]	2.85	1.88	0.00	7.00
Number of clusters of less-knowledge-intensive services[c]	5.27	2.10	1.00	9.00
Number of clusters of manufacturing activities	9.39	2.96	3.00	16.00
Population	1,934,258	1,531,182	27,153	11,700,000
Population density (population/km^2)	363.14	890.89	3.30	9,405.70
Productive diversity	16.73	5.62	3.43	26.23
Average firm size in the region	8.21	7.02	1.00	44.22

Notes:
[a]Includes only telecommunications and information service activities as the rest (motion picture, video and television, sound recording and music, broadcasting, computer programming, and scientific research and development) are included in "creative services"
[b]Excluding publishing, architectural and engineering activities, advertising, and arts, entertainment and recreation, included in "creative services"
[c]Excluding retail sale of other goods in specialized stores, included in "creative services"

share of persons employed in "rest of knowledge-intensive services" does have a positive effect (166 euros, $p < 0.01$). This is due to the fact that in the first case the shares of these services are very similar across the EU regions whereas in respect of the non-creative RKIS there is more heterogeneity. Therefore, and shading the results by Leydesdorff and Fritsch (2006) and Leydesdorff, Dolfsma, and Van der Panne (2006), when the creative services are isolated from both categories, RKIS (Rest of knowledge-intensive services) seem to be more important than KIHTS (knowledge-intensive high-tech services) in explaining differences in wealth.

5.2 *A Further Insight into the Effects of Creative Services by Sub-sector, and the Incidence of Co-Location*

A further question revolves around whether the relation between creative services and GDP per capita holds for every kind of creative service or only for some of them. Table 16.4 shows the correlation coefficients between the shares of creative

Table 16.3 Final estimates. P-values in brackets

	(1) OLS Robust[d]	(2) OLS Robust[d]	(3) OLS Robust[d]		(4) IV Robust[d,e]
Dependent variable: GDP per capita in PPS	Coefficient	Coefficient	Coefficient	Elasticity (dY/dX)	Coefficient
Constant	−2,242.33	9,442.75	6,178.05		5,751.38
	(0.617)	(0.000)	(0.000)		(0.000)
% Employment in creative services	**1,707.73**		**1,479.48**	**0.3909**	**1,604.91**
	(0.000)		(0.000)		(0.000)
% Employment in high-tech services[a]	−503.72				
	(0.451)				
% Employment in rest of knowledge-intensive services[b]	**287.86**		**166.61**	**0.2006**	**156.32**
	(0.000)		(0.005)		(0.007)
% Employment in less-knowledge-intensive services[c]	203.21				
	(0.072)				
% Employment in manufacturing	99.48				
	(0.148)				
Number of clusters of creative services		**1,323.03**			
		(0.000)			
Number of clusters of high-tech services[a]		**−2,482.05**			
		(0.001)			
Number of clusters of rest of knowledge-intensive services[b]		91.09			
		(0.789)			
Number of clusters of less-knowledge-intensive services[c]		**526.24**	**526.26**	**0.1168**	**519.85**
		(0.025)	(0.003)		(0.003)
Number of clusters of manufacturing activities		−182.65			
		(0.337)			
Population		0.0004			
		(0.111)			
Population density (population/km^2)		**4.283**	**1.62**	**0.0185**	1.307
		(0.000)	(0.039)		(0.095)
Productive diversity		**515.78**			
		(0.000)			

(continued)

Table 16.3 (continued)

	(1) OLS Robust[d]	(2) OLS Robust[d]	(3) OLS Robust[d]		(4) IV Robust[d,e]
Dependent variable: GDP per capita in PPS	Coefficient	Coefficient	Coefficient	Elasticity (dY/dX)	Coefficient
Average firm size in the region		132.02			
		(0.167)			
R^2	0.5906	0.5258	0.6179		0.6162
R^2-adj	0.5822	0.5079	0.6116		–
Mean VIF	1.35	1.78	1.36		–
Durbin–Wu–Hausman endogeneity test (*p*-value)	–	–	–		0.13
Obs	250	250	250		250

Notes:
[a]Includes only telecommunications and information service activities as the rest (motion picture, video and television, sound recording and music, broadcasting, computer programming, and scientific research and development) are included in "creative services"
[b]Excluding publishing, architectural and engineering activities, advertising, and arts, entertainment and recreation, included in "creative services"
[c]Excluding retail sale of other goods in specialized stores, included in "creative services"
[d]Huber–White robust estimators used to prevent the problems of normality and heteroskedasticity
[e]Instruments for the percentage of employment in creative services are calculated for the year 2001 in order to reinforce exogeneity. They include cultural endowments (UNESCO goods by million inhabitants), average firm size in the region, average firm size in the creative services in the region, productive diversity in the creative services string, population, patents per million inhabitants, R&D expenditures on GDP and percentage of creative class

services in regional employment and GDP per capita. The relevant results are obtained by relating sectors to GDP per inhabitant. First, every creative service is significantly correlated with the GDP per capita, and the correlations range from 0.33 to 0.67.

Second, by taking into account only correlations higher than 0.5, the results show that some creative services are more important than others in explaining differences in the wealth of regions. These sectors are computer programming (HTKIS), advertising (RKIS), publishing (RKIS), audiovisual (HTKIS), architecture & engineering (RKIS), R&D (HTKIS) and creative retail (LKIS). These results lead us to believe that the wealth of a region depends, to a great extent, on a wide range of knowledge-intensive creative services.

Third, from the results in Table 16.4, we also observe that there is a positive and statistically significant correlation between the different creative services (from 0.2 to 0.8). Thus, we can conclude that there is a tendency for creative service to co-locate with one another, which is in line with the findings of Wernerheim (2010) for the services industries in Canada, and with De Propris, Chapain, Cooke, MacNeill, and Mateos-García (2009) for the creative industries in the UK. Taking correlations of more than 0.5 as strong correlations, the results show that:

Table 16.4 Co-location of different creative services sub-sectors

	GDP in pps	Retail (creative)	Publishing	Audiovisual	Broadcasting	Computer programming	Architecture and engineering	R&D	Advertising	Design, photography	Arts, entertainment and recreation
GDP in pps	1										
Retail (creative)	0.5091[a]	1									
Publishing	0.6600[a]	0.2791[a]	1								
Audiovisual	0.6169[a]	0.2770[a]	0.7512[a]	1							
Broadcasting	0.3847[a]	0.1833[a]	0.5993[a]	0.6581[a]	1						
Computer programming	0.6873[a]	0.3092[a]	0.7248[a]	0.6041[a]	0.4315[a]	1					
Architecture and engineering	0.5300[a]	0.3641[a]	0.4408[a]	0.3729[a]	0.3068[a]	0.5011[a]	1				
R&D	0.5256[a]	0.3262[a]	0.4801[a]	0.4121[a]	0.2420[a]	0.6763[a]	0.4260[a]	1			
Advertising	0.6733[a]	0.3022[a]	0.7966[a]	0.7399[a]	0.5889[a]	0.7563[a]	0.4168[a]	0.5013[a]	1		
Design, photography	0.3716[a]	0.2044[a]	0.4924[a]	0.5716[a]	0.4061[a]	0.5337[a]	0.4625[a]	0.3270[a]	0.4727[a]	1	
Arts, entertainment and recreation	0.3354[a]	0.2250[a]	0.4429[a]	0.4655[a]	0.3298[a]	0.4831[a]	0.2677[a]	0.3791[a]	0.4069[a]	0.4367[a]	1

Correlation coefficients between the shares of creative services sub-sectors in aggregate employment of 250 regions
[a]Statistically significant at 5 %

- Publishing strongly co-locates with audiovisual (HTKIS), broadcasting (HTKIS), computer programming (HTKIS), and advertising (RKIS);
- Audiovisual (HTKIS) strongly co-locates with publishing (RKIS), broadcasting (HTKIS), computer programming (HTKIS), advertising (RKIS), and design and photography (RKIS);
- Broadcasting (HTKIS) strongly co-locates with publishing (RKIS), audiovisual (HTKIS) and advertising (RKIS);
- Computer programming (HTKIS) strongly co-locates with publishing (RKIS), audiovisual (HTKIS), architecture and engineering (RKIS), R&D (HTKIS), advertising (RKIS), and design and photography (RKIS);
- Architecture and engineering (RKIS) strongly co-locates with computer programming (HTKIS);
- R&D (HTKIS) strongly co-locates with computer programming (HTKIS) and advertising (RKIS);
- Advertising (RKIS) strongly correlates with publishing (RKIS), audiovisual (HTKIS), broadcasting (HTKIS), computer programming (HTKIS) and R&D (HTKIS);
- Design and photography (RKIS) strongly co-locates with audiovisual (HTKIS) and computer programming (HTKIS);
- Finally, cultural and creative retail, as well as arts, entertainment and recreation, co-locate with the remaining sectors, although the coefficient is in every case lower than 0.5.

6 Conclusions

The main objective of this research has been to investigate whether creative services firms are a "growth driver" that promotes regional wealth. The paper disentangles the differences between "creative services" and "rest of knowledge-intensive services," and establishes a robust framework to understand to what extent different types of services contribute to the wealth of European regions.

An initial hypothesis of a supply-side effect of creative services on output per capita was confirmed using a robust procedure. An increase in 1 % in the percentage of employment in creative industries in the region translated into an increase of 0.39 % in GDP per capita, that is to say 1,479 euros in GDP per capita. This was a higher effect than was found for the presence of "rest of knowledge-intensive services," "manufacturing" or "agglomeration economies." All the sub-sectors in creative services proved to be positively and significantly correlated with GDP per capita, and several patterns of co-location between these sub-sectors were detected, which generated diverse profiles across the regions.

The contribution and results are relevant, not only because there is a lack of research concerning the role of creative services firms within regions, but also because our findings provide additional evidence to support the idea that services, particularly those related to the creative processes, are not consumers but rather net generators of economic wealth.

At this point, further conclusions are constrained by the limitations of the research. First, we focused on the impact of creative services on the wealth of regions (growth model), but not on their systemic effect on innovation in regional economies (innovation model). Second, it is important to know whether the wealth and innovation generating effects of creative services are confined within a region or do they spill over to other regions. Third, additional research on the combinations of creative services in regions could suggest ways to reinforce patterns of complementarity amongst creative services, and between them and other sectors, while taking into account regional diversities. Fourth, while the analysis has focused on the meso-level of the region, micro-level investigation at the level of the firm could provide additional evidence regarding profits, location decisions, heterogeneous behaviours, and/or evolutionary patterns.

Thus, if creative services impact basically on wealth and have highly local effects, they could be a significant objective for regional-driven policy. If, however, the geographical effects are supra-regional, then national policy or coordination between regions could play an important role. If the effects are focused on concrete segments of firms, the scope of the policy changes radically. On the other hand, if the wealth impacts of creative services derive basically from the supply-side, public policies should aim to provide the appropriate conditions for enterprise development and interaction, rather than provide subsidies and price policies to protect industries. Finally, if their effects on innovation spill over to the rest of the local economic system, other strategies such as financial support to creative services firms could be effective.

In any case, the field of study on creative services requires further research, looking at more types of effects, going deeper into industry and firm detail, and studying comparable long term series in order to capture time-dynamic effects.

Acknowledgements The authors would like to thank Ministry of Science and Innovation and Universitat Politècnica de València (Spain) for financially supporting this research (ECO2010-17318 MICINN Project and Research Project n. 2677-UPV). We also would like to thank an anonymous referee and Frank Pyke by their valuable comments.

References

Alexander, N., & Akehurst, G. (2005). Identifying the role of services in the regional economy and servicing the identity of the region. *The Service Industries Journal, 25*(4), 421–428.

Asheim, B. T., Boschma, R., & Cooke, P. (2011). Constructing regional advantage: platform policies based on related variety and differentiated knowledge bases. *Regional Studies, 45*(7), 893–904.

Asheim, B. T., & Hansen, H. K. (2009). Knowledge bases, talents, and contexts: on the usefulness of the creative class approach in Sweden. *Economic Geography, 85*(5), 425–442.

Barro, R. J., & Sala-i-Martin, X. (1991). Convergence across states and regions. *Brookings Papers on Economic Activity, 1*, 137–158.

Baumol, W. J., & Bowen, W. G. (1966). *Performing arts, the economic dilemma*. Cambridge: MIT.

Capello, R. (2006). *Regional economics*. Abingdon: Routledge.

Chapain, C., Cooke, P., De Propris, L., McNeil, S., & Mateos, J. (2009). *Creative clusters and innovation: Putting creativity on the map*. London: Nesta.
DCA (Department of Communications and the Arts, Australia). (1994). *Creative nation: Commonwealth cultural policy*. Available in an electronic version at http://www.nla.gov.au/creative.nation/contents.html.
DCMS. (1998). *The creative industries mapping document*. London: DCMS.
DCMS. (2009). *Creative industries economic estimates statistical bulletin*. London: DCMS.
De Miguel, B., Hervás, J. L., Boix, R., & De Miguel, M. (2012). *The importance of creative industry agglomerations in explaining the wealth of European regions. European planning studies 20*(8), pp. 1243–1262.
De Propris, L., Chapain, C., Cooke, P., MacNeill, S., & Mateos-García, J. (2009, August). *The geography of creativity* (NESTA Interim Report).
Drucker, P. (1969). *The age of discontinuity: Guidelines to our changing society*. New York: Harper and Row.
European Commission. (2010). *Green paper on cultural and creative industries: Unlocking the potential of cultural and creative industries*. Brussels: DG Education and Culture.
Eurostat. (2009). *'High-technology' and 'knowledge based services' aggregations based on NACE Rev. 2*. Luxembourg: Eurostat.
Flew, T., & Cunningham, S. (2010). Creative industries after the first decade of debate. *The Information Society, 26*(2), 113–123.
Hoover, E. M. (1937). *Location theory and the shoe and leather industries*. Cambridge: Harvard University Press.
Lazzeretti, L., Boix, R., & Capone, F. (2012). Why do creative industries cluster? An analysis of the determinants of clustering of creative industries. In L. Lazzeretti (Ed.), *Creative industries and innovation in Europe: Concepts, measures and comparative case studies*. London: Routledge, 45–64.
Leydesdorff, L., Dolfsma, W., & Van der Panne, G. (2006). Measuring the knowledge base of an economy in terms of triple-helix relations among "technology, organization and territory". *Research Policy, 35*(2), 181–199.
Leydesdorff, L., & Fritsch, M. (2006). Measuring the knowledge base of regional innovation system in Germany in terms of a triple helix dynamics. *Research Policy, 35*(10), 1538–1553.
Lucas, R. E. (1988). On the mechanics of economic development. *Journal of Monetary Economics, 22*, 3–42.
Machlup, F. (1962). *The production and distribution of knowledge in the United States*. Princeton: Princeton University Press.
Miles, I. (2001). *Services innovation: A reconfiguration of innovation studies* (PREST discussion paper series, 01-05). Manchester.
Miles, I., & Green, L. (2008). *Hidden innovation in the creative industries*. London: NESTA.
Müller, K., Rammer, C., & Trüby, J. (2009). The role of creative industries in industrial innovation. *Innovation: Management, Policy & Practice, 11*, 148–168.
Müller, E., & Zenker, A. (2001). Business services as actors of knowledge transformation: The role of KIBS in regional and national innovation systems. *Research Policy, 30*(9), 1501–16 (Special issue).
O'Connor, J. (2007). *The cultural and creative industries: A review of the literature*. London: Arts Council England.
OECD. (1996). *The knowledge based economy*. Paris: OECD.
Ohlin, B. (1933). *Interregional and international trade*. Cambridge: Harvard University Press.
Potts, J., & Cunningham, S. (2008). Four models of the creative industries. *International Journal of Cultural Policy, 14*(3), 233–247.
Power, D., & Nielsén, T. (2010). *Priority sector report: Creative and cultural industries*. European Commission, Europe Innova-European Cluster Observatory.
Pratt, A. (2007). An economic geography of the cultural industries. In A. Leyson, L. McDowell, & R. Lee (Eds.), *An economic geography reader*. London: Sage.
Quah, D. (1996). Regional convergence clusters across Europe. *European Economic Review, 40*, 951–958.

Redondo-Cano, A., & Canet-Giner, M. T. (2010). Outsourcing agrochemical services: economic or strategic logic? *Service Business, 4*(3–4), 237–252.
Romer, P. M. (1986). Increasing returns and long-run growth. *Journal of Political Economy, 94*, 1002–1037.
Stoneman, P. (2009). *Soft innovation: Towards a more complete picture of innovative change.* London: NESTA.
Sunely, P., Pinch, S., Reimer, S., & Macmillen, J. (2008). Innovation in a creative production system: The case of design. *Journal of Economic Geography, 8*, 675–698.
UNCTAD. (2010). *Creative economy* (Report 2010). Geneva/New York: UNDP-UNCTAD.
Wernerheim, C. M. (2010). The tendency of advanced services to co-locate and the implications for regional government policy. *The Services Industries Journal, 30*(5), 731–748.
Wood, P. (Ed.). (2002). *Consultancy and innovation.* London: Routledge.

Chapter 17
Managing the Japanese Service Industry: Location, Competitiveness and Internationalization

Patrik Ström

1 Introduction

Japan has been a leading industrial power during the last 50 years. The manufacturing industry has developed world leading competitiveness and productivity. The service industry on the other hand has been less competitive and internationalized in comparison with other OECD countries (Ström, 2005). Increased regional and global competition changes the role of Japan in the knowledge intensive economy. The competitive advantage that have sustained the establishment of East Asian global and regional production networks are not particularly well suited for high-end services, rather the focus has been on manufacturing and process orientation (Yeung, 2009). Services are becoming a more important part of the future competitiveness of the economy. Japan is facing a number of challenges to foster a competitive knowledge economy with a high degree of services. The chapter seeks to explore the advanced service industry and its connections to the manufacturing sector through two perspectives. First, structural and cultural reasons in the business environment could help to explain the limited success of the service industry. Service internalization and bundling of services have made it difficult for independent firms to compete. Japanese firms need to rethink their global strategy and build networks that are in line with knowledge creation and globalization. Service firms must clearly define their strategy for localization and internationalization. Manufacturing firms need to develop their offering of bundled services in order to move from free to fee. Companies that are well placed to produce value-added for clients through the product life cycle utilizing the broad base of intra- and inter-firm competence would enhance their competitiveness. Second, uneven regional economic development has proven to be a hurdle for the development of a knowledge intensive service economy. Knowledge flows between companies in regional economic clusters has proven difficult.

P. Ström (✉)
Centre for International Business Studies, University of Gothenburg, Gothenburg, Sweden
e-mail: patrik.strom@geography.gu.se

J.R. Cuadrado-Roura (ed.), *Service Industries and Regions*,
Advances in Spatial Science, DOI 10.1007/978-3-642-35801-2_17,

2 Theoretical Frame

The locational dimension for producer services has been discussed both from a narrow regional perspective (e.g. Beyers & Lindahl, 1996; Ishimaru, 1994; Keeble & Nachum, 2002; Sjøholt, 1993) and a more outward perspective (Bryson, 2000; O'Farrell, Wood, & Zheng, 1998; Ström & Wahlqvist, 2010), where the regional perspective is connected with firms' internationalization processes. Keeble and Nachum (2002) show that professional business service firms are highly dependent on integrated clusters, with high levels of collaboration, knowledge transfer and labour mobility. Their study shows that service firms in the vicinity of good transportation infrastructure can often extend their client area. Bryson (2000) calls the increased need for knowledge diffusion, the availability of expertise and information, *relationship capitalism.* In combination with reputation and image, this creates much of the foundation for localization for business services (Amin, 1999). However, it is important to note that the factors influencing proximity might differ between different sub-sectors. Beyers and Lindahl (1996) show for example that firms in more rural areas can create solid economic positions. Cost-driven externalization does not explain the rapid growth in producer services. Specialized expertise in combination with many other factors of cost-and non-cost drives the development. In relation to the cross-cultural dimension of development of producer or business services, studies of countries like Japan are needed to understand the underlying development. A less individual and entrepreneurial setting might impact the development of service firms in a different way. The use us external advise is not viewed in the same way as in the western economic context. This also connects to the issue of competitive advantage in business services. Beyers and Lindahl (1999) show that the idea of cost-oriented strategies does not help to explain the development of successful service firms. Rather, many service firms show that it is possible to mix strategies and still not be inferior. This might be due to the continuously changing business environment for services. However, the study clearly shows that differentiation in services is a common way to achieve a favorable niche position. Beyers and Lindahl (1999:18) conclude: *They do this via their creativity, their abilities to undertake research and development, their abilities to respond quickly to client needs (i.e., to be flexible in the process of identifying how the service firm will respond to client needs/demands), and their personal attention to client needs.* Alvstam and Jönsson (2000) demonstrates the increased possibilities for consultants to work from pleasant distant locations with favourable infrastructure connections. Bryson and Rusten (2008) discuss the virtual organization forms that new information technology creates. Networks of expertise constantly form and reform into project teams. These teams can be localized or dispersed, often with a focus on prosperous business regions.

Services can be produced and consumed over longer distances due to digital communications and other forms of technology and other forms of value-chain functions. This has opened new structures where firms might selectively use different locational factors to facilitate their strategies (see Bryson & Rusten, 2008).

New technologies have also created what has been called a “second global shift” (Bryson, 2007). The underlying reasons for this geographical structure can be both based on considerations of pure cost, but also the fact that firms want to tap into specific agglomerations of knowledge found in urban concentrations (Ström, 2006; Wood, 2006). This strategic positioning can be described in the following way:

> “... many services are a result of changing strategies concerning holding versus buying knowledge, and different types of services... In fact, due to the scarcity of available labour resources, firms are now using mergers and acquisitions as a way of obtaining the necessary staff for their projects. Furthermore, the boundaries for sourcing services are not limited to national borders” (Hermelin and Rusten, 2007:6). This shows how firms are changing and constantly evaluating options of organization and location to better enhance their competitive advantage.

Beyers (2002) states that there is a need for increased regional knowledge on how and where key industries and location will play a role in the service economy. This has connection to the development of new classification structures to gain an understanding of the economic transformation, but also impacts the training of a new kind of labour force. Additionally, the interrelatedness of regions through networks and trade are important aspects for the future to better understand how they help to shape the economic geography of an economy where services and connections to other sectors are becoming increasingly important. Within this transformation, the consumption side of services needs better attention. In the (2012) paper that was given as a key-note lecture at the European Association for Research on Services (RESER) 20th anniversary conference many of these research areas were revisited. According to Beyers (2012) the service economy has not only showed resilience, but also been transformative in a number of ways. Research imperatives are services as engines of regional economic growth, trade and international business of services and the role of cultural and creative industries. Additionally, it follows earlier research in the 1990s, where the service economy was analyzed through production and consumption patterns (e.g. Beyers & Lindahl, 1996). Further in-depth case studies to enhance knowledge of firm level development and geographical impact is vital. This can be done both through qualitative and quantitative research on the regional level. This relates to trade and international business where classifications are in need of development to better understand the interconnectedness and transformations that exist in a dynamic service economy. The international connectedness of regions is still an area of future studies, where regional case-studies could provide new insights. The liberalization of service trade and investment together with an increased will of service providers to move abroad would need additional research. Finally, the methodology of using new ways of cluster analysis seems to be a promising tool for the advancement of service research. Not least in catching the dynamics of the rapidly changing new sub-sectors of the cultural and creative industry.

The theory of service research comprises many areas in relation to geographical levels and industry sectors. The overview points to the need for regional and international comparisons to develop understanding of how the interconnected global service economy is developing (Bryson, Rubalcaba, & Ström, 2012; Daniels, 2012).

This becomes even more apparent when seeing new emerging markets like China and India challenging traditional service economies in the OECD. Additionally, the cultural impact on service development might be different in non-western based business environments. Therefore, theory development based on the Asian business context is sought for the future (e.g. Yeung, 2009). Thus far, despite the growth of the service economy in Asia, it has rendered rather limited attention in the literature.

3 The Japanese Service Industry

During the high-growth post war era, it was stated that the future lay in the service economy. The Ministry of Trade and Industry (MITI, later renamed METI) wanted to promote the change into a society where services such as finance, trade, business services etc. should be valuable parts. Even though effort was put into transforming the economy, it is obvious that Japan did not manage to catch up with the most developed countries. Saxonhouse (1985) puts stress on the fact that the service sector in Japan has continuously lagged behind the American and Western European service sectors, in terms on productivity. For example, even though the manufacturing productivity in 1979 had caught up with the Western manufacturing industry, the service sector productivity was only about two thirds of the Western counterparts. Only in 1991 the service sector, excluding the wholesale industry, overtook manufacturing and the Japanese economy has had problems in increasing service productivity (Wainai, 2001; Wölfl, 2003). The exclusion of the wholesale industry is conducted in order to generate a picture of the intermediary role of services in the economy. Andresso-O'Callaghan and Bassino (2003) have confirmed the incremental move towards services in the Japanese economy. A large part of the growth has been generated through technological development and liberalization of world trade, along with an increase of intermediary services in production. Today, Japan still lags other OECD economies in terms of service industry development (see Table 17.1). In terms of productivity growth within the service industry between 1995 and 2003, it achieved only 34 % of the productivity growth rate of the U.S. and only 62 % of the productivity growth rate of the U.K (JETRO, 2007). Increased competition and deregulation are measures that might increase the service productivity (Ström, 2005). However, it is important to remember the difficulties of comparing productivity measurements across different nations and industries. Different forms of organizational structures and modes of service delivery together with cultural aspects makes this kind of comparisons highly complex (Gadrey, 2002; Maroto-Sánches & Cuadrado-Roura, 2009). Despite these difficulties the results points towards the complexity of the Japanese service industry.

A study by METI shows that the Japanese investments in R&D in the non-manufacturing sectors significantly lag the corresponding investments in the USA. Only 12 % of the R&D funding for industrial expenses went into the non-manufacturing side in Japan. In the US 43 % of the total R&D expenses ended up in the non-manufacturing sectors (METI, 2007).

Table 17.1 Services as share of GDP and employment in selected OECD countries

	Contribution to GDP (%)		Employment share (%)	
Country	1997	2007	1997	2007
Canada	66.6	66.7	74.1	75.9
Denmark	71.2	72.4	69.5	73.6
France	73.3	77.4	70.3	74.2
Germany	67.7	68.7	62.3	67.7
Italy	67.3	70.4	60.5	65.5
Japan	65.5	70.1	61.6	67.9
Sweden	68.4	70.3	71.3	76.1
UK	68.7	76.3	71.4	76.3
USA	72.8	76.9	73.4	78.8

Source: OECD in figures (2009)

However, despite domestic economic problems during 1990s that was called the lost decade, the service sectors have increased their contribution to GDP and employment share. Sluggish economic performance has pushed firms such as the banking and finance industry to reposition their organizations through mergers and acquisitions. These restructuring efforts have also had repercussions on the foreign operations of Japanese companies (Ström, 2006). The Japanese economy has also tried to open up and promote service oriented FDI (FT, 2010a). A recent example of this growing interest comes from China. Chinese investors are moving into the Japanese economy to acquire expertise, technology, distribution channels, quality and brand names (FT, 2010b). Succession problems in family run business are also part of the opportunities that Chinese firms see in Japan. Over the last 5 years this trend has taken off and investments are found within both services and manufacturing. The establishment of financial subsidiaries of Chinese wealth funds such as Citic is one example. These firms are service providers themselves, but also help to transform the domestic Japanese business environment. Additionally, a more flexible labor market has also favored external advice and a demand for changing recruitment on shorter basis (Ishimaru & Daniels, 2007; JETRO, 2005a). The development has led to a significant increase in new firms within this recruitment service area (JETRO, 2005b). One explanation behind this lagged development is found within service internalization and bundling where firms have service operations within larger organizational structures, and that services are supplied to clients for free in a bundled package (Ishimaru & Daniels, 2007; Ström, 2004). Industrial companies might include services along with their products, often through the product life cycle (Bramklev & Ström, 2011). This creates value for customers, but without necessarily achieving a monetary gain for the supplier. Banks can also supply complementary services when lending money to clients. This is something that would have rendered income through special fees for Western banks.

The Japanese economy show the shift towards services even when the domestic economy is struggling and service sector has delivered a net increase of employment almost every year (Ström, 2004). This indicates that the strong dominance of manufacturing in the general economy and in government policies (Friedman, 1988;

Johnson, 1982) needs to be reconsidered (Ono, 2001). Another important aspect to remember is that it is not only low-skilled and low-paid service jobs that have increased during the economic slowdown in the 1990s, but there is also a positive development for knowledge intensive services, such as management consultancy, market research and other forms of professional services (Ishimaru & Daniels, 2007). The support for advanced services is also strong from the policy side of government. METI launched a new agenda labeled "Strategy for Creating New Services" in 2004, where business services were specifically targeted. METI see a strong demand of these services through focusing on transactions between organizations. The positive effects on manufacturing and non-manufacturing activities are also important and finally, that these services are a driving force in service trade liberalization (JETRO, 2005b)

3.1 Financial and Business Services in Japan

This section of the chapter uses financial and business services to show the development of these important knowledge intensive services within the Japanese economy. They are vital parts for both direct and indirect economic growth. Due to the very high savings rate, of which the lion's share has been put into ordinary bank accounts, the Japanese banks rapidly became the dominant players in the world (Dicken, 2011). This enabled them to move abroad and set up branch offices, mainly to support Japanese companies, and later to create local business opportunities. Haga (1997, 1999) has studied the spread of Japanese banks in Asia and the network that tie banks together in Asia. He concludes that the spread of Japanese banks follows a hierarchical pattern, where New York, London and Hong Kong are the primary locations. The locational choices were associated with the keiretsu connections in order to support Japanese customers abroad (Haga, 1997:124). Strategic considerations in relation to Japanese production networks in Asia and the possibility to be located at important knowledge-hubs have been vital for Japanese financial service providers (Ström & Yoshino, 2009). The complex web of production networks has also created services to be organized in-house and through related external providers. Expansion of these services to non-Japanese clients has proven to be difficult (Ström & Yoshino, 2009). The change that has occurred is that Japanese banks are closing operations in the West, but changing representative offices into branch offices in the lucrative Asian market (Haga, 1997:126).

The domestic restructuring of the Japanese financial industry has left Japanese banks in a stronger position after clearing out non-performing loans. After a sustained period of reduced international activity, the new mega-banks and investment banks such as Daiwa and Nomura are expanding the international operations again (Financial Times, 2010a). The recent financial crisis has become an opportunity for Japanese banks and investment firms to strengthen their global presence.

Nomura has bought the European and Asian parts of Lehman Brothers. It has both expanded the market reach and also brought in new competence to the firms that have been difficult to attract before (Ström & Yoshino, 2009).

The sogo shosha has a clear role when discussing Japanese business services. Their operations stretch both in geographical and sectoral terms. The foundation for their existence has been the trading operations where they played a crucial role in bringing in resources and intermediate products to Japanese companies (Dicken, 2011; Sjöberg & Söderberg, 2001). However, within these service conglomerates, a great variety of services such as, financing, consultancy, retail, just to mention a few exist. Porter, Takeuchi, and Sakakibara (2000) label this a strategy of trying to be "everything-to-everyone," which can leave the company without a focused strategy.

The presence of keiretsu-affiliated companies such as the sogo shoshas has also been stated as being one factor of explanation when Japanese firms in general choose to invest abroad (e.g. Mardas, Pournarakis, & Varsakelis, 1998; Ström, 2004). Mardas et al. (1998) show that the networks are vital parts of the internationalization of Japanese business. A keiretsu network seems to be of equal importance for both manufacturing and service operations, and they often follow each other closely abroad. This is supported by the findings of Mardas et al. (1998), by stating:

> Since the early 1990s the Japanese foreign direct investment takes an intensive form in services and especially in trading and financial activities. This evolution in the pattern of the Japanese foreign direct investment is the outcome of the keiretsu network. The trading companies and the financial firms, members of the groups, follow their partners in Europe, in a process of upward integration of the group presence in Europe. The trading and financial firms use their group partners as market potential as well as sources of information and risk minimization (Mardas et al., 1998:12).

Dicken and Miyamachi (1998) give an account of the functions and geography of the sogo shoshas. They conclude that the trading companies often control or own shares in manufacturing companies. This resembles more of a holding company, making the trade intermediation function only one part of the structure. Additionally, the sogo shoshas are highly diverse in terms of both the goods they deal in, but also in the kind of services and functions they offer to customers. The fact that they are tightly involved in financing both trading operations and larger projects, positions them across what can be called merchant and money-dealing capital. The sogo shosha is said to be "involved at each stage of the production/commodity chain, from the purchase of raw materials through to marketing the final product and including the specialized financial and non-financial inputs (Dicken & Miyamachi, 1998:75)." Regarding the embeddedness of the sogo shosha within the Japanese keiretsu structure, the empirical data suggest that this is still present, but it is becoming more important for firms to extend their client base. Furthermore it seems that the keiretsu connections are stronger within the domestic market. The very character of the operations of the sogo shosha has created a geographically dispersed office network, often under the control of Tokyo headquarters or more regional oriented head offices. Compared with Japanese manufacturing firms trying to take in local managers, the sogo shoshas have relied more on Japanese staff. One explanation might be the complex business structure that demands special cultural awareness.

The Research Institute Companies in Japan represent yet another specific form of business services (Ström, 2006). They are both think tanks related to a larger keiretsu group or specific company, and they are also consultancy companies performing various forms of research and consulting covering a vast competence area. Some firms might be more related to larger manufacturing groups and others are more financially based. In this case the firms are often part of securities operations or investment banks. These firms are different from the traditional big four in accounting and consultancy, and they are not as specialized as most western management consultancies. Their specific company relationships or associations can be a hurdle for getting non-Japanese clients.

An issue that is vital for the understanding the Japanese business services is the interconnectedness that exists with the manufacturing sector. A larger share of service internalization can be one explanation why the service sector seems to be comparatively less developed and internationalized. Japanese firms have traditionally focused on the product and services have been looked upon as add on. The important issue for future competitiveness is to make use of the service encapsulation and find a suitable pricing strategy (Bramklev & Ström, 2011). This has created a situation where few firms within the service sectors are seen as stand-alone operations. The strong interconnection between services and manufacturing is articulated through the large part of service investments made by manufacturing firms. The lion's share is made within various forms of services supporting the distribution chain. Some of these service investments pre-date the investments in the production of goods.

Kuwatsuka (2000) shows that manufacturing firms do much of the investments in Singapore, in order to support the network of subsidiaries in the region. Services such as purchase and various forms of business consulting can be supplied from regional headquarters in Singapore. These kinds of services have been moved to Singapore in order to expand the sales functions in the region. This newly formed structure of "branch economy" with a high content of services is a new phenomenon for Japanese firms. According to Kuwatsuka, the tendency has been more oriented towards keeping these services at the headquarter location. He concludes with that research exploring the relationship between agglomeration and corporate networks are essential in order to find out more of the service structure within Japanese manufacturing firms. These findings of an increased importance of regional headquarter design was also supported by a later study by Ström (2006). Firms interviewed expressed the ongoing strategy of moving more towards regional independence.

3.2 Internationalization of Services

Japanese firms within the knowledge intensive sector have traditionally been internationalized to a lesser extent than similar firms in other OECD economies. The Japanese firms have usually followed their Japanese clients into new markets. This process is similar to what has been the case among Western service providers. However, the Japanese firms have not been as successful in developing their client base in foreign markets. The main reason is that the business environment for services has favored internalization in larger companies or keiretsu groups (Ström & Mattsson, 2005). In changing market conditions, the networks and potential of leveraging resources among firms have proven to be valuable platforms for the internationalization process.

Comparative studies of Japanese firms in the UK and Singapore show a similar pattern of internationalization (Ström, 2006; Ström & Mattsson, 2006). Only after a long presence in the host markets has the development of increasing the client base started. This has often happened when traditional Japanese clients have come under pressure to restructure their operations. The other alternative has been to establish representative offices with a limited possibility to act in the host market. According to the firms, there are several reasons for entering foreign markets in which the long-term potential is seen to be great. The issue of competence and knowledge absorption was put forward by all firms irrespective of their service activities. This shows the importance of the so-called untraded interdependencies that are discussed in economic geography theory. These locational advantages are difficult to quantify, and Japanese firms have found it difficult to tap into, and connect with, local clusters of knowledge production (Morgan, Sharpe, Kelly, & Whitley, 2002). It seems the traditional business structure of these knowledge-producing firms create hurdles in their internationalization.

Within the knowledge-driven and service-based economy, the importance of international contacts has been a driving force behind rapid internationalization. These relations are more important than geographical distance and cultural similarities. Larger corporations such as the sogo shoshas have tried to develop new service areas to take advantage of new possibilities in the market. Examples are the growth of Japan's venture industry and increasing interest in the biotechnology industry. New companies have developed in the market of venture capital, and some firms such as *Globis* are trying to push entrepreneurship through education programs. The MBA program uses both traditional business cases from established business school, but more unconventional, cases from their own investments (FT, 2010). Recent studies show that Japanese firms are trying to utilize international relationships to develop their market potential in Asia and globally (Alvstam, Ström, & Yoshino, 2009). This becomes especially important for the Japanese firms since Chinese firms are increasingly competing in several sectors. With strong governmental support there is a move towards high end knowledge production of goods and services where the aim is to move away from "made in China, to created in China" (Ernkvist & Ström, 2008; Keane, 2007). The combination of relations to larger Japanese business groups and contact points at international knowledge hubs is vital in building new business relationships. The complexity of location decision making and how knowledge-producing firms need to find a strategic fit for their long-term development in Asia is important for evaluating the location strategy among knowledge-intensive firms (Ström & Yoshino, 2009). Since these production networks have predominantly been manufacturing based in terms of global success, this is a challenging task.

4 Regional Economic Development

The regional economic development in Japan has traditionally focused on promoting economic growth along the pacific belt. Within this belt the main industrial centers are found. Apart from the capital region around Tokyo (Kanto) comprising of larger cities such as Yokohama and Kawasaki, the large economic centers around Nagoya and Osaka-Kyoto (Kansai) are the most prominent areas. Other vital hubs of industrial activity are to be found around Hiroshima and Fukuoka. The economic growth in these areas has traditionally been used to support more peripheral regions. As the transformation of the Japanese economy has taken place after the high-growth postwar era, regional policies have seen new challenges. Areas such as Kyushu, northern Tohoku and Hokkaido, where resource and agricultural industry have been prominent have seen their economies come under pressure. Larger cities like Sapporo seem to be faring relatively well, but more peripheral areas have problems, and there are also examples of municipalities that have gone bankrupt. The economic transformation has also been problematic for larger cities within the Pacific belt. Increasing FDI and outsourcing of production among large Japanese manufacturing firms primarily to China, has left areas earlier depended to the electronics industry to struggle for sustained competitiveness. Other areas, such as Nagoya, where the automotive sectors have developed further and Toyota becoming the world's largest producer of cars, have seen increased regional growth. This has pushed initiatives to upgrade regional competitiveness through focusing on the knowledge economy and entrepreneurship.

The debate on the knowledge economy and the rise of the creative class (Florida, 2002) as an engine of economic growth engaged Japan as a tool for leaving the problems of the lost decade in the 1990s. A study of the situation in Japan reveals that there are differences compared with other OECD countries. Table 17.2 shows the different employment sectors that constitute the creative class in Japan. 25.1 % of Japan's total workforce qualifies as part of the Creative Class. Approximately 8.4 % of the 25.1 % are to be found in what Florida (2002) labels the Super-Creative Core (see Fig. 17.1).

The share of the creative class in Japan is lower than in the USA and Sweden to give a comparison within OECD (Ström & Wahlqvist, 2010; Tinagli, Florida, Ström, & Wahlqvist, 2007). This is an indication of the lower degree of service industry development. Service internalization and bundling might be one explanation,

Table 17.2 Employment structure of the working population in Japan

	Category A	Category B	Category C	Category D	Category E
Type of employment within the statistical account	Science, health, education, culture and technology artistic oriented occupations such as authors, musicians and artists	Contains managers, both in the private and the public sector, public sector, naturally, governmental legislators such as politicians can be found	Wholesale & Retail and Bank & Insurance and personnel engaged in Health and social security insurance	Food industry and hotel	Others

Source: Undergraduate Field Project (2006) and Japan Statistics Bureau

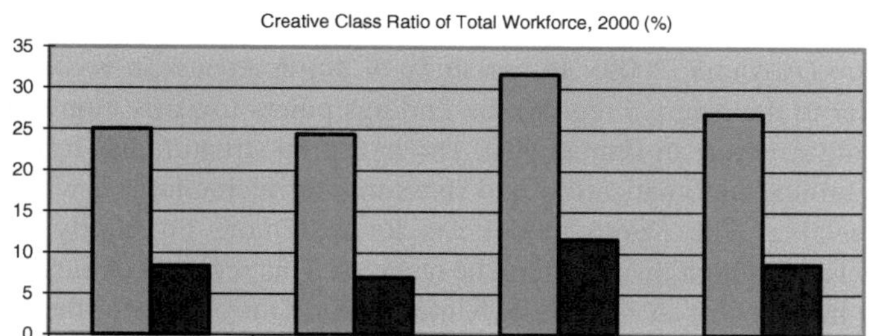

	JAPAN	HOKKAIDO	TOKYO	KYOTO
Creative Class	25.1	24.4	31.7	27
Super Creative Core	8.4	7	11.7	8.7

Fig. 17.1 Share of creative class employment in Japan. *Source*: Undergraduate Field Project (2006) and Japan Statistics Bureau

but it cannot explain the entire difference. It points toward the limited development of independent service providers in different service sectors. For comparative purposes Fig. 17.1 shows Tokyo and Kyoto in relation to Hokkaido and the national average. Other prefectures with high share of the creative class were Fukuoka, Osaka and Kanagawa. They all had Creative Class ratios above average. Kanagawa has a Super-Creative Core rate of 12.8 %, a figure that is higher than the one of Tokyo, due to a very high number of scientists and artistic related occupations.

The study also shows regional differences that resembles the difference in economic power and importance. Tokyo is dominating the creative economy. This is similar to what other studies have fund in countries where the capital region is centralizing a vast amount of administrative functions, compared to areas of more federal political and economic structure (Tinagli et al., 2007). Tokyo is the main hub of the knowledge or creative economy in Japan and can compete with other larger global cities. The potential and the capabilities for the development of the service economy are there. Apart from finance, and other business services, creative sectors as manga, computer games and other related creative industries are strong in Tokyo (Izushi & Aoyama, 2006) and where specific areas like Akihabara drive the development (Nobuoka, 2010). Hokkaido is one example of the difficulty of peripheral regions. On the one hand there are intensive work to develop cluster initiatives and meeting places within industry and between the public sector. Large universities are also part of these initiatives. On the other hand it is difficult to sustain these efforts and see long term competitiveness. Interviews with representatives of industry and public sector show that despite high quality of life, solid infrastructure and a committed work to attract new firms in the knowledge economy, results are meagre.

Kyoto is an example of an area that is both part of traditional Japan, but also at the center of activities in the knowledge economy through firms in computer games and bio-tech and world class academic institutions. Kyoto also displays a high rate of artistic occupations such as photographers, painters and designers and occupations related to classical industries such as kimono production. This connection to historical industries is also creating social norms that form the regional culture (Aoyama, 2009). In her study of entrepreneurs in Kyoto and Hamamatsu (closer to the Nagoya region) the findings points towards more open and outward looking structure in Hamamatsu. The industrial structure has formed entrepreneurship around automatization and mass-market technologies, with a strong global perspective. The Kyoto model has focused more on family oriented business models, with high quality on niche markets. This regional culture seem to influence even larger companies such as Nintendo and Omron, where the focus has been to create value in a different way than traditional production oriented firms in Japan. Nintendo has successfully outsourced hardware production and focused on the game experience and game development (Interview, Nintendo, 2005). This shows the contextual impact on companies and entrepreneurs. In relation to services, it is possible that a niche oriented approach is vital in the long run, through creating a high degree of specialization (Beyers & Lindahl, 1996). Furthermore, it strengthens the perception of regional characteristics and that knowledge intensive production of goods and services are difficult to develop through regional governmental initiatives.

5 Innovation Systems Approach

Over the years, Japan has tried to develop national innovation systems for supporting the formation of new firms and industries (Holroyd, 2008). These efforts have been implemented to strengthen the institutional setting needed for knowledge-producing firms and their internationalization. The program has been initiated by the national government, and various regional activities have been launched. Initiatives in the area of information technology, R&D, and bio-tech have seen strong support on regional levels. Cluster initiatives to support platforms for information sharing and knowledge creation have been established. Universities and other research and educational institutions have also been involved in upgrading the innovation systems (Holroyd, 2008).

Large corporations have been prioritized through the innovation system and the long term guaranteed employment, where researchers could easily move between R&D and manufacturing (Edgington, 2008). It has helped to sustain a business environment where service functions to a lesser extent been externalized and where innovation within services often have been tightly connected with manufactured goods (Ström, 2005). An example of this kind of innovation is provided by Bramklev and Ström (2011), where the case of packaging is used to show how service innovation is developed through the interaction of manufacturing firms and service providers. The study shows that service innovation often is highly technology or process driven with Japanese industry. This has also been facilitated through long-term business relations, product development and the *keiretsu* structure forming complex inter-company relationships (e.g. Gerlach, 1992; Kensy, 2001). Initiatives to develop SMEs in regional peripheral areas are not unproblematic. Studies of Japanese attempts to strengthen entrepreneurship indicate a mixed result (Sakakibara, 2003). Large sums of money have been spent but it has been difficult to achieve sustained results over the long run. Sapporo is one example of a city region that has worked hard to develop its attractiveness for high-tech and service-producing industries, but thus far has only had limited national and international success. The difficulty is often associated with creating a sustained economic development with firms moving in and new local firms within knowledge intensive industries appearing. Strong regional economic growth engines such as Tokyo and Kanto area; Nagoya and Hamamatsu and Kyoto and Kansai are vital.

Regional initiatives in the Kyoto area have showed sustained success in the field of life and bio-sciences (Ibata-Arens, 2008). Well-functioning social networks among people and firms, together with long-term strategies have helped a large number of firms gain strong momentum in innovation and international R&D. In a detailed case study, Ibata-Arens (2008) analyzes the key success factors behind one of these successful firms, Peptide. The local institutional setting, together with international contacts the founder established in Germany, China, and elsewhere proved to be vital. By interacting with researchers at conferences and by joining universities abroad, a network for internationalization evolved. In the global knowledge economy speed of internationalization has increased. Firms might move on to the global scene shortly after establishment, something referred to as *born globals*. This phenomenon in combination with competition from emerging market multinational has been studied by Mathews (2006). Competition in the global economy is today more dependent on linkages through networks to build competitive advantage. Utilization of these links can then create leverage on the market and through constant learning processes firms from less competitive environments can grow stronger.

In general, two categories of Japanese companies are present with various degree of success on the global market. The first category includes firms that are already on the international market, but need to develop their connection to the local relational or institutional business environment. Secondly, there are rapidly growing firms that take advantage of regional or national growth initiatives. These firms rely more on personal contacts around the world, and the speed and geography of their internationalization process is very different.

The data on Japanese firms show that the different conceptual perspectives are all present. Firms need to handle more neo-classical aspects of internationalization along with issues related to space and management (Ström & Schweizer, 2011). It seems questions of locational advantages and embeddedness are vital along with the managerial capability of the firms. It is the combination of these perspectives that can help us understand the internationalization process and location strategy. Connection related to persons and face-to-face interactions are vital in service operations (Gertler; 2003; Ström & Wahlqvist, 2010). The actor perspective on the development of service and knowledge intensive production should help to understand why certain areas and regions have a continues flow of entrepreneurship and knowledge creation. An issue of great importance for the Japanese government and regional authorities is the fact that Japan seems to be losing as a preferred location for business activities in Asia. A study by METI 2011 shows that Japan has lost its top position for R&D center and Asian headquarters between 2007 and 2009. This is a problem in the long-run, since it limits the position of Japan as an important center for international business center. Since research points to the importance of knowledge-hubs for the development of a competitive service industry, this has potential long term implications. It is primarily the rapid development in China that has changed the playing field of Asian business hubs.

6 Conclusion

This chapter looks at the fact that Japan, being the third largest economy in the world seems to have difficulties of fully utilizing the potential its service industry, both in terms of regional development and internationalization. There are potentially different reasons behind this development and the chapter presented some areas of explanatory value, such as the business environment and the limited usage of external advice that has been one of the more important drivers of advanced service growth in the west. Additionally the chapter has elaborated on the economic transformation from the regional economic development perspective and the innovation systems approach. Japan has seen a strong transformation of its economy during the last two decades and more of the economic growth has been concentrated around larger urban areas. This has created new challenges to upgrade and support regional initiatives in the service dominated economy. There are different outcomes of these attempts and it shows that it is difficult to reach a sustained growth in more peripheral areas. Large amounts of government support money have been directed towards the formation of innovation initiatives, but it seems that the possibility to steer the development is limited. Instead it is necessary to find regions with settings well in line with service industry development and build on these. This can be through the connection of advanced manufacturing and services, but also in more R&D intensive sectors such as bio-tech and creative industries. Regional dynamics are also influenced by the social and cultural norms that exist, which in turn can be highly important when generating venture capital bases. Regional policy must also be better positioned to see the potential in the service sectors and find policy tools to develop this further. Bringing actors together in regional business systems or connecting value adding global production networks marks the challenge for the development of the Japanese service industry.

The chapter also shows that a more stringent strategy of companies to unbundle services and charge for the concrete value, will increase competitiveness and productivity. Many of the leading manufacturing firms in Japan are already using a product-life-cycle perspective, but the potential gains for customers have not been factored into the pricing mechanism. A clearer strategy focus might also be a solution for the development and internationalization of the Research Institute Companies. In the future they might compete with western service providers in Asia, where they already have solid market knowledge. With a strategy focus from both government and private industry it is possible that the advanced Japanese service industry would see a stronger internationalization capability. The latest developments within the financial industry where Japanese firms like Nomura have been able to gain international presence show the good foundation in some sub-sectors of the industry. This should be seen as an attempt to combine the extensive research conducted on the service industry in the western business context with what is characterizing the development in Japan and Asia. The suggested research areas discussed by Beyers (2012) are vital in understanding the future resilience and development of the global service economy and its repercussions on the regional and national economic geographies.

References

Alvstam, C. G., & Jönsson, A. (2000). On promoting regional economic growth in industrial peripheries. Hyper-footloose business services—The case of Swedish teleworkers in Spain. *NETCOM – Networks and Communication Studies, 14*(1–2), 25–52.

Alvstam, C., Ström, P., & Yoshino, N. (2009). On the economic interdependence between China and Japan—Challenges and possibilities. *Asia Pacific Viewpoint, 50*(2), 198–214.

Amin, A. (1999). 'Placing globalization' theory, culture and society (1997). In J. Bryson, N. Henry, D. Keeble, & R. Martin (Eds.), *The economic geography reader*. Chichester: Wiley.

Andresso-O'Callaghan, B., & Bassino, J.-P. (2003). Measuring the contribution of services to Japanese growth (1975–1995). *Asian Business & Management, 2*, 223–237.

Aoyama, Y. (2009). Entrepreneurship and regional culture: The case of Hamamatsu and Kyoto, Japan. *Region Studies, 43*(3), 495–512.

Beyers, W. B. (2002). Services and the new economy: Elements of a research agenda. *Journal of Economic Geography, 2*, 1–29.

Beyers, W. B. (2012). The service industry research imperative. *The Service Industries Journal, 32*(4), 657–682.

Beyers, W. B., & Lindahl, D. P. (1996). Explaining the demand for producer services: Is cost-driven externalization the major factor? *Papers in Regional Science, 75*(3), 351–374.
Beyers, W. B., & Lindahl, D. P. (1999). The creation of competitive advantage by producer service establishments. *Economic Geography, 75*(1), 1–20.
Bramklev, C., & Ström, P. (2011). A conceptualization of the product/service interface: Case of the packaging industry in Japan. *Journal of Service Science Research, 3*(1), 21–48.
Bryson, J. R. (2000). Spreading the message. Management consultants and the shaping of economic geographies in time and space. In J. Bryson, P. W. Daniels, N. Henry, & J. Pollard (Eds.), *Knowledge space economy*. London: Routledge.
Bryson, J. R. (2007). The 'second' global shift: The offshoring or global sourcing of corporate services and the rise of distanciated emotional labour. *Geografiska Annaler, 89B*, 31–43.
Bryson, J. R., Rubalcaba, L., & Ström, P. (2012). Services, innovation, employment and organisation: research gaps and challenges for the next decade. *The Service Industries Journal, 32*(4), 641–655.
Bryson, J. R., & Rusten, J. (2008). Transnational corporations and spatial divisions of 'service' expertise as a competitive strategy: The example of 3 M and Boeing. *The Service Industries Journal, 28*, 307–323.
Daniels, P. (2012). Service industries at a crossroads: Some fragile assumptions and future challenges. *The Service Industries Journal, 32*(4), 619–639.
Dicken, P. (2011). *Mapping the changing contours of the world economy* (6th ed.). London: Sage.
Dicken, P., & Miyamachi, Y. (1998). 'From noodles to satellites': The changing geography of the Japanese Sogo Shosha. *Transactions of the Institute of British Geographers, 23*.
Edgington, D. (2008). The Japanese innovation system: University-industry linkages, small firms and regional technology clusters. *Prometheus, 26*(1), 1–19.
Ernkvist, M., & Ström, P. (2008). Enmeshed in games with the government: Governmental policies and the development of the Chinese online games industry. *Games and Culture, 2008*(3), 98–126.
Financial Times. (2010a, September 8). *Government set out to lure foreign investors.*
Financial Times. (2010b, September 8). *Big neighbour seeks expertise and technology.*
Financial Times. (2010c, December 9). *Years of stagnation leave market bereft.*
Florida, R. (2002). *The rise of the creative class.* New York: Basic Books.
Friedman, D. (1988). *The misunderstood miracle.* London: Cornell University Press.
Gadrey, J. (2002). The misuse of productivity concepts in services: Lessons from a comparison between France and the USA. In J. Gadrey & F. Gallouj (Eds.), *Productivity, innovation and knowledge in services; new economic and socio-economic approaches.* Cheltenham: Edward Elgar.
Gerlach, M. L. (1992). *Alliance capitalism: The social organization of Japanese business.* Berkeley: University of California Press.
Gertler, M. (2003). Tacit knowledge and the economic geography of context, or the undefinable tacitness of being (there). *Journal of Economic Geography, 3*, 75–99.
Haga, H. (1997). Banking centres and the network of international banks in Pacific Asia. *Asian Geographer, 16*(1–2), 1–20.
Haga, H. (1999). Changing networks of Japanese banks among mega-cities in the world. In A. G. Aguilar & I. Escamilla (Eds.), *Problems of megacities: Social Inequalities, environmental risks and urban governance.* Mexico: Universidad Nacional Autonoma de Mexico.
Hermelin, B. & G. Rusten (2007), The Organizational and Territorial Changes of Services in a Globalized World. *Geografiska Annaler* 89B, pp.5–11
Holroyd, C. (2008). Reinventing Japan Inc: Twenty-first century innovation strategies in Japan. *Prometheus, 26*(1), 21–38.
Ibata-Arens, K. (2008). The Kyoto model of innovation and entrepreneurship: Regional innovation systems and cluster culture. *Prometheus, 26*(1), 90–109.
Ishimaru, T. (1994). The spatial characteristics of manufacturing cities and service cities in Japan. *Jimbun-Chiri/Human Geography (Kyoto), 44*(2), 284–298.

Ishimaru, T., & Daniels, P. (2007). Understanding the goods-producer services gap: The case of Japan. *Bulletin of Fukuoka University of Education, 56*(1), 1–12.
Izushi, H., & Aoyama, Y. (2006). Industry evolution and cross-sectoral skill transfers: a comparative analysis of the video game industry in Japan, the United States, and the United Kingdom. *Environment and Planning A, 38*(10), 1843–1861.
JETRO. (2005a, July). Special report. *JETRO Japan Economic Monthly*. www.jetro.go.jp. Accessed 23 Feb 2011.
JETRO. (2005b, December). Business topics. *JETRO Japan Economic Monthly*. www.jetro.go.jp. Accessed 23 Feb 2011.
JETRO. (2007). *Focus Newsletter, Japan moves to improve efficiency of service sector*. www.jetro.go.jp. Accessed 23 Feb 2011.
Johnson, C. (1982). *MITI and the Japanese miracle*. Stanford: Stanford University Press.
Keane, M. A. (2007). *Created in China: The great new leap forward*. London: Routledge.
Keeble, D., & Nachum, L. (2002). Why do business service firms cluster? Small consultancies, clustering and decentralization in London and Southern England. *Transactions of the Institute of British Geographers, 27*(1), 67–90.
Kensy, R. (2001). *Keiretsu economy—New economy?* New York: Palgrave.
Kuwatsuka, K. (2000, December 5–9). *Geographical agglomeration of Japanese electronics firms in Singapore: Installed non-manufacturing functions and corporate organization*. Paper presented at global conference on economic geography, National University of Singapore, Singapore.
Mardas, D., Pournarakis, M., & Varsakelis, N. C. (1998). Japanese foreign direct investment in the European union: An extension of the Keiretsu network*? The Mid-Atlantic Journal of Business, 34*(3), 189–203.
Maroto-Sánches, A., & Cuadrado-Roura, J. R. (2009). Is growth of services an obstacle to productivity growth? A comparative analysis. *Structural Change and Economic Dynamics, 20*, 254–265.
Mathews, J. A. (2006). Dragon multinationals: New players in 21st century globalization. *Asia Pacific Journal of Management, 23*(1), 5–28.
METI. (2007, October). *Towards innovation and productivity improvement in service industries*. Service Unit. www.jetro.go.jp. Accessed 23 Feb 2011.
Morgan, G., Sharpe, D. R., Kelly, W., & Whitley, R. (2002). The future of Japanese manufacturing in the UK. *Journal of Management Studies, 39*(8), 1023–44.
Nobuoka, J. (2010). User innovation and creative consumption in Japanese culture industries: The case of Akihabara, Tokyo. *Geografiska Annaler: Series B, 92*(3), 205–218.
Nintendo. (2005, November). Major video game company Manager, Corporate Communication Department November 2005 Kyoto
O'Farrell, P., Wood, P. A., & Zheng, J. (1998). Regional influences on foreign market development by business service companies: Elements of a strategic context explanation. *Regional Studies, 32*(1), 31–48.
OECD. (2009). *OECD in figures 2009*. Paris: OECD.
Ono, H. (2001). Restructuring strategy of Japan's service sector in the twenty-first century. In S. Masuyama, D. Van den Brink, & C. S. Yue (Eds.), *Industrial restructuring in East Asia, towards the 21st century*. ISASS: Singapore.
Porter, M. E., Takeuchi, H., & Sakakibara, M. (2000). *Can Japan compete?* London: Macmillan.
Sakakibara, K. (2003). Japanese entrepreneurs and their firms: Survey results. *International Journal of Technology Management, 25*, 531–538.
Saxonhouse, G. R. (1985). Services in the Japanese economy. In R. P. Inman (Ed.), *Managing the service economy—Prospects and problems*. Cambridge: Cambridge University Press.
Sjöberg, Ö., & Söderberg, M. (2001). The Sogo Shosha: Finding a new role? In M. Blomström, B. Gangnes, & S. La Croix (Eds.), *Japan's new economy, continuity and change in the twenty-first century*. Oxford: Oxford University Press.

Sjøholt, P. (1993). The dynamics of services as an agent of regional change and development. *The Service Industries Journal, 12*(2), 36–59.
Statistics Bureau: Director-General for Policy Planning & Statistical Research and Training Institute. (2006, June 26). http://www.stat.go.jp/data/kokusei/2000/shosai/index.htm.
Ström, P. (2004). *The 'lagged' internationalization of Japanese professional business service firms: Experiences from the UK and Singapore* (Series B, no. 107). Department of Human and Economic Geography, School of Business, Economics and Law, Göteborg University.
Ström, P. (2005). The Japanese service industry: An international comparison. *Social Science Japan Journal, 8*(2), 253–266.
Ström, P. (2006). Internationalization of Japanese professional business service firms: Dynamics of competitiveness through urban localization in Southeast Asia. In J. W. Harrington & P. W. Daniels (Eds.), *Knowledge-based services, internationalization and regional development*. Aldershot: Ashgate.
Ström, P., & Mattsson, J. (2005). Japanese professional business services: A proposed analytical typology. *Asia Pacific Business Review, 11*(1), 49–68.
Ström, P., & Mattsson, J. (2006). Internationalisation of Japanese professional business service firms: Client relations and business performance in the UK. *The Service Industries Journal, 26*(3), 249–265.
Ström, P., & Schweizer, R. (2011). Space oddity—On managerial decision making and space. In R. Schlunze, N. Agola, & W. Baber (Eds.), *Spaces of international economy and management launching new perspectives on management and geography*. London: Palgrave Macmillan.
Ström, P., & Wahlqvist, E. (2010). Regional and firm competitiveness in the service based economy—Combining economic geography and international business theory. *Tijdschrift voor economische en sociale geografie, 101*(3), 287–304.
Ström, P., & Yoshino, N. (2009). Japanese financial service firms in East and Southeast Asia: Location pattern and strategic response in changing economic conditions. *Asian Business & Management, 8*(1), 33–58.
Tinagli, I., Florida, R., Ström, P., & Wahlqvist, E. (2007). *Sweden in the creative age* (Occasional Papers 2). Göteborg: Department of Human and Economic Geography. School of Business, Economics and Law, Göteborg University.
Undergraduate Field Project. (2006, June 27). *The creative class in Japan, 3 T; talent, technology and tolerance*. Sapporo: Tokai University.
Wainai, K. (2001). Productivity in the service sector in Japan. In: Asian Productivity Organization (Ed.), *Productivity measurement in the service sector* (pp. 63–76). Tokyo: Asian Productivity Organization.
Wölfl, A. (2003). *Productivity growth in service industries: An assessment of recent patterns and the role of measurement* (STI working paper 2003/07). Paris: OECD.
Wood, P. (2006). Urban revival and knowledge-intensive services: The case of the English 'core cities'. In J. W. Harrington & P. W. Daniels (Eds.), *Knowledge-based services, internationalization and regional development*. Aldershot: Ashgate.
Yeung, H. W.-C. (2009). Regional development and the competitive dynamics of global production networks: An East Asian perspective. *Regional Studies, 43*(3), 325–351.

Chapter 18
Services in Regional Development Policies: The French Case

Camal Gallouj and Faïz Gallouj

Regional economic policy has traditionally been targeted at manufacturing industry. In most European countries, and at least until the 1980s, services, and in particular business services, were largely ignored in regional policies, which gave priority to models based on development of (and by) large manufacturing firms. It was not until the 1980s that services, whose economic role was finally beginning to be recognised, were gradually taken into account by regional and national economic actors. Thus various forms of support programmes for services, initiated by a range of different actors (central government, regional authorities, *départements*, municipalities, chambers of commerce and industry, etc.), proliferated in France in the 1990s. Nevertheless, this increased interest in support services, as well as the number of actors involved, certainly give rise to questions as to the possible contradictory effects on local economies.

This chapter is divided into three sections. In the first, we locate policies on (business) services in the general (historical) context of local or regional development policies in France. In Sect. 2, the various policies and incentives put in place in support of services are summarised and an interpretative framework developed. The third section, finally, is given over to consideration of the contradictory or possibly cumulative effects of the incentive measures under investigation here.

C. Gallouj (✉)
University of Paris 13 (CEPN-CNRS), Paris, France
e-mail: Camal@Gallouj.com

F. Gallouj
Lille 1 University (CLERSÉ-CNRS), Lille, France

J.R. Cuadrado-Roura (ed.), *Service Industries and Regions*,
Advances in Spatial Science, DOI 10.1007/978-3-642-35801-2_18,

1 Local (and Regional) Policies in Favour of Services: An Historical Perspective

In both France and most OECD countries, services (and particularly business services) were for a long time largely absent from the regional and urban development policy agenda (Gallouj, Leloup, Merenne-Schoumaker, & Moyart, 2006; Marshall & Bachtler, 1987). Such policies focused primarily on the manufacturing sector and, to some extent, still do (see Levet, 2006; Portnoff 2010). As far as France is concerned, it is now customary to divide the history of regional development policy into three major periods (see Lacour, Delamarre, & Thoin, 2005): the period of national modernisation (from the 1950s to the early 1970s), the period of local and regional development (early 1970s to end of 1980s) and, finally, the period of renewal and refocusing (1990 to the present). In what follows, we describe these various periods, taking care to identify any possible signs of willingness to take account of (business) services.

1.1 The Period of National Modernisation (1950s to Early 1970s)

During this period (which covers the post-war decades of economic growth and prosperity known in France as the "Trente Glorieuses[1]"), the aim of development policies was to channel growth. French economic development was to be accelerated by giving top priority to *the expansion of manufacturing*. The principal measures introduced during this period were, to quote Noin (2009), intended to encourage "the decentralisation of manufacturing industry in order to obtain a better geographical balance by bringing pressure to bear on the location of manufacturing activities." The measures included both sticks (regulation of the establishment and expansion of manufacturing plants in the Ile-de-France) and carrots (financial subsidies to establish plants outside the Paris/Ile-de-France region).

The initial policies introduced during this period were small-scale initiatives with a strong sector focus (manufacturing primarily). While business services were not yet on the agenda, it is nonetheless true that, virtually concurrently with the measures targeted at manufacturing, measures aimed at certain specific services, such as tourism[2] for example, were introduced. Furthermore, the various measures initially targeted at industry were gradually extended beyond the initial focus on manufacturing industry. More generally, it can be argued that some of the policies of the time that targeted the business environment can be regarded as beneficial to

[1] "Glorious Thirty".

[2] Central government is, after all, heavily involved in exploiting the potential for tourism, both directly (coastal infrastructure: Languedoc-Roussillon in 1963, then Aquitaine in 1966) and indirectly (planning/development in mountain areas).

services, and particularly to business services. This applies, for example, to the decentralisation of the service sector and measures aimed at improving the transport and telecommunications infrastructure (development of motorway system and domestic flights).

The process of service-sector decentralisation, which began relatively late, drew its inspiration from the similar measures that had earlier been applied to the manufacturing sector. Thus it was based on regulation restricting office development in the Paris region (except la Défense) and the payment of subsidies to service providers setting up businesses in assisted areas.

Other measures introduced during this period can be regarded as indirect support for (business) services, even though this was not their declared aim. We are referring here in particular to all urban development policies, including the establishment of the so-called *métropoles d'équilibre*, which were intended to be counterweights (balances) to the dominant capital region of Paris, new towns and, to a lesser extent, medium-sized towns. After all, to quote Monod and Castelbajac (2010), the *métropoles d'équilibre* were supposed "to provide firms and the population of their region with *the services* that they had hitherto sought in the capital." This observation is particularly true of business services, even if only through the familiar phenomenon of the "short circuit" (see Gallouj, 1996; Gallouj et al., 2006). However, it should be noted that attempts to establish business and administration centres in these metropolitan areas, where business services were supposed to be developed, met with variable success (Devisme, 2000). The main purpose of the new towns was to contribute to the restructuring of the Paris conurbation and the establishment of new urban hubs (Cergy, Evry…), again with important service functions (Lacour et al., 2005).

1.2 The Period of Local and Territorial Development (Early 1970s–End of 1990s)

This second period is regarded by some authors as a period of retreat from spatial planning (Madiot & le Mestre, 2001). And yet it also saw the emergence and recognition of regional institutions, in advance of decentralisation and the focus on new regional development issues (Lacour et al., 2005). It was also during this period that the EU burst on to the scene and began to intervene much more heavily in regional development policies. The policies put in place were still very much oriented towards manufacturing industry. This was the case, for example, with the industrial redevelopment areas (*pôles de conversion industrielle*) (1984) or industrial estates/parks (*zones industrielles*) (1986). After all, even if service activities were not, in theory, excluded, most of the areas and sites in question were largely unsuited to the specific characteristics and needs of business services (particularly when they required high skill levels) and in particular to their methods of operating (to which we shall return).

The new issues and challenges facing spatial planning related essentially to new technologies, small and medium-sized enterprises and industries (SMEs and SMIs) and local and territorial development. In this sense, they were concerned to some extent with indirect support for business services. After all, the policies in question, even though they were once again not aimed directly at service activities, were significantly more favourable towards them. Rediscovery of the advantages of small and medium-sized enterprises and industries, combined with new thinking on local development, put the spotlight on services, and particularly business services. In the first half of the 1980s, a number of measures targeted specifically at business services were introduced: incentives to develop firms' demand for business services (*Fond Régional d'Aide au Conseil*/Regional Support Fund for Business Consultancy) or strengthening of firms' internal service functions (*Aide au Recrutement des Cadres*/management recruitment grants) etc.

1.3 The Period of Renewal and Refocusing (1990 to the Present Day)

The 1990s were characterised by strong tensions between the local and the global. To quote Lacour et al. (2005: p. 92), the decade saw "a volume of legislation that was unprecedented in the history of French development." Once again, business services did not receive any specific attention, but the various programmes, measures and initiatives put in place during this period (boosting of local productive systems (or clusters), digital infrastructure programmes, Universities 2000 plan, regional competitiveness clusters, reassertion of the role of large cities etc.) may to some extent be analysed as forms of indirect support for business services.

However, more detailed analysis shows that the dominant model for local productive systems (the dominant cluster model) in France still focused on manufacturing activities (textiles, mechanical subcontracting, agro-food etc.). The same applies to the competitiveness clusters which, with just a few exceptions, were concerned primarily with the development of manufacturing industry. They were intended "to make it possible to strengthen France's manufacturing potential, to create conditions favourable to the emergence of new activities with high international visibility and hence to improve the attractiveness of the regions." However, in France (as well as in many European countries) there is renewed interest from institutions in the role of business services in regional development. According to some authors, this interest is largely related to recognition of the contribution made by business services to competitiveness and to the capacity for innovation of their customers and, more generally, to the performance of local economy (Viljamaa, Kolehmainen, & Kuusisto, 2010; Cuadrado Roura & Maroto-Sanchez, 2010). The interest for business services now appears to be developing in two main directions: innovation and sustainable development (Gallouj and Djellal, 2010). The inclusion of services, in particular business services, in national and European surveys on innovation (CIS: Community Innovation Survey) and the inclusion of service innovations in the framework of OSEO-ANVAR funding (cf. Darnige, 2010) contribute to a better recognition of the role of business services within local and regional innovation systems (Muller & Zenker, 2001).

Over and above the strict question of innovation, the period of renewal and refocusing has also seen the re-emergence of the environment and sustainable development as objects of discourse. This being so, it might be thought that new prospects are opening up for the promotion and development of services (and business services), if only to the extent that they are considered to have a smaller environmental footprint than most manufacturing industries (Djellal & Gallouj, 2010).[3]

It can be concluded from this brief historical survey that services and more particularly business services, although not totally excluded, have not received any particular attention from the traditional actors in regional development policies. This does not mean, however, that these activities have been totally absent from debates and practices. Thus, besides the standard, highly regulated financial support measures, many campaigns, programmes and experiments, initiated by a very diverse range of actors (including large firms in some cases), have sought to invigorate the business services infrastructure. It is these experiments and practices that are the subject of the next section.

2 Regional Policies in Favour of Services: A Synthetic Interpretative Framework

In recent years, there has been a proliferation of theoretical and, particularly, empirical studies on services and policies to develop them (Mas Verdu, Ribeiro, & Roig Dobon, 2010; OCDE, 1995; Viljamaa et al., 2010). These studies essentially reflect practices that can be said to have improved very considerably since the 1990s. The actors in economic development now have at their disposal a relatively wide range of possibilities, which are summarised in Table 18.1.

In Table 18.1, we divide actions designed to promote services into four main groups: support measures aimed principally at the supply side (Sect. 2.1); measures aimed principally at demand (Sect. 2.2); support measures aimed specifically at linking supply and demand (Sect. 2.3); and infrastructure development and regional and urban planning programmes (Sect. 2.4).

[3] However, this idea is discussed and qualified by Jean Gadrey (2010). See also Fourcroy, Gallouj, and Decellas (2012).

Table 18.1 An interpretative framework for measures introduced by local institutions and authorities to promote services

Type of support	Example of practices
A. Support aimed mainly at "supply" Main characteristics: Support for the development of networks/groups of service providers and professions Assistance provided by these networks and groups (consolidation, monitoring) Contribution to the supply of resources and knowledge of common interest Two sub-categories can be identified: A1: actions and support measures for individual sectors or occupations A2: cross-cutting actions and support measures	– Role in the establishment of facilities, infrastructure, offices, incubators for service firms, business centres adapted to needs of business service providers – Facilitating the personal connections that lead to the establishment of networks and groups of service providers (clubs for professions, interdisciplinary associations, "communities," etc.) – Support for projects of common interest, joint promotion, shared investments – Studies of sectors, their markets, their problems and the perception current or potential users have of them – Specific information aimed at certain service providers (or all sectors). Service monitoring centres – Continuing training specific to certain service providers – Action to ensure that initial training (e.g. in business schools) includes some of the specificities of services – Action with service providers to promote quality, shared thinking on this topic and on ISO standards and certification – Assistance in setting up service firms – Idea of regional innovation and technology transfer centres (CRITS) targeting services – Encouraging service firms to set up locally – Organising supply directly (services centres) – Encouraging exploitation of niche markets and the export of services – Encouraging diversification and setting up of new service firms – Encouraging the development of customised, innovative solutions
B1. Support aimed mainly at "demand" (users) Main characteristics: Various financial incentives Moving from faulty or imperfect information to satisfactory information. Establishing a culture and "routines" based on the sensible use of services.	– Establishing the habit of using services by providing them directly, whether paid for, free or at reduced cost, or by offering incentives in the form of state subsidies (eg. FRACs—*fonds régionaux d'aide au conseil*) – Advice on use of services, advice on consultancy, information and advice on public measures – Promoting use of services by service sector (incl. local and regional authorities, public and quasi-public sector) – Strengthening of firms' managerial capacities (particularly in SMEs) through training and external recruitment

(continued)

Table 18.1 (continued)

Type of support	Example of practices
	– Developing a policy of attracting outside investment as well as promoting the use of (local) services by these same firms
B2. Linking supply with demand Main characteristics: Mutual knowledge, mutual adaptation. Creating exchange networks with a shared language	– Salons, forums and other initiatives designed to bring together service providers and users – Clubs including users – Directories and other data bases of use to customers – Various forms of personal intermediation – Development of intermediation functions ("middlemen")
C. Actions targeted at these service markets' external environment Infrastructure policies Relational policy and actions Main characteristics: Extension of networks listed above to other private and public actors General service culture in society, creation and general diffusion of knowledge on services	– General infrastructure (not specific to business services), improvement of communication networks, urban renewal, quality of life, etc. – Extension of links to other actors, both public (higher education, research, local authorities, Direction Régionale de l'Industrie, de la Recherche et de l'Environnement/DRIRE or Regional Office for Industry, Research and Environment) and private (banks, insurance companies, various opinion leaders) – "External" communications aimed at general public (image of services) and public authorities, information and awareness campaigns – Intercessions with public authorities on behalf of services, arguing that they should not be disadvantaged relative to manufacturing industry.

Sources: Various internal documents (institutions and local and regional authorities, chambers of commerce and other self-managing public bodies etc.)

2.1 Supply-Side Incentives and Policies

Incentives and policies designed to support the supply side (i.e. service providers) may be targeted, on the one hand, at firms and service providers from outside the region and, on the other hand, at firms already operating in the region (the aim here being to encourage the foundation and development of local service providers). In the first case, it should be noted that policies offering incentives to providers to set up in certain regions were introduced at a relatively early stage (from the 1970s onwards). Their objective was to combat unequal development by encouraging the expansion of local service provision, principally in areas that were lagging behind in this regard. In the second case, the opportunities offered to local (and regional) actors in development are significantly more extensive. More specifically, the measures put in place are generally of three main types:

- Improvement and strengthening of the supply side (by encouraging the foundation of new service firms, offering incentives to diversify and specialise, raising service providers' levels of skill and expertise, implementing quality approaches, putting in place qualification and certification programmes[4] and incentives to innovate and introduce new technologies, etc.).
- Incentives to develop exports of services (thereby recognising the very real role services—or at least some of them—play in the regional economic base).
- Organisation of supply, establishing interactions between and groupings of actors (in order to encourage exchanges, to achieve a critical mass and obtain economies of scale, etc.).

In other words, the overall thrust of these measures is to facilitate the emergence of networks and groupings of services providers and professions,[5] to support the actions taken by these networks and groupings and to contribute to the supply of resources and knowledge of common interest. There are many experiments following this trend in most European countries. For example, Toivonen (2007) presents a particularly successful experiment in Finland supporting the provision of services in the Tampere region. At the beginning of 2000, a public–private partnership (urban agglomeration of Tampere, university and private companies) was set up which led to the launch of a structure for the development of business services (Professia Ltd). This structure, which had the immediate role of incubator, had four main missions: promoting and encouraging the creation of business service companies (in particular in the information and communication technology sectors), improving the quality of existing business services, producing and disseminating information and research on business services and providing services to businesses (with the aim of "opening up new markets").

2.2 *Demand-Side Incentives and Support Measures*

Policies intended to support demand aim to generate and maintain a sufficient volume of outlets for service firms. Consequently, they seek to promote demand for services or to encourage firms to make use of outside resources and expertise. In other words, the aim is to establish a culture, if necessary by offering financial incentives, in which the judicious use of services becomes a matter of routine.

These policies providing direct support for demand take three main forms:

- The first consists of a direct subsidy for demand, that is for firms making use of services. This is one of the most familiar and widely used measures in most

[4] These actions may also, to some extent, fall within the scope of measures to link supply and demand (see Sect. 2.3).

[5] Most of the policies introduced to support the supply side may be aimed at particular sectors or occupations or, conversely, be more cross-cutting in nature.

European countries and in the OECD more generally. In France, the main tool deployed in this type of policy has been the *Impulsion Conseil* (Boosting Consultancy) programme, which replaced the *Fond Régional d'Aide au Conseil* (Regional Support Fund for Business Consultancy or FRAC), which was launched in 1982 and made available throughout the French regions from 1984 onwards.

- The second aims to develop competences within firms. The policies in question here seek to develop firms' own managerial competences, particularly in SMEs and SMIs, in such a way as to strengthen their ability to adapt and innovate. The most widely used form of support in this type of policy involves encouraging the recruitment of skilled personnel by giving small and medium-sized businesses subsidies to enable them to recruit highly qualified managerial staff (which may, for example, ensure effective interaction with service providers). This type of support is of course much more wide-ranging than this, and may include the provision of training for managers already in post or even the creation of internal training departments. The main tool of this type in use in France is the *Aide au Recrutement de Cadres hautement spécialisés* (Grant for the Recruitment of Highly Qualified Managers, or ARC), which was launched on a trial basis in Nord-Pas-de-Calais in 1986 and then introduced throughout the country in 1987.
- The third and final form focuses on supporting the formulation and emergence of new demand, whether from local authorities and other institutions themselves or from firms outside the locality. After all, the manufacturing sector has often been at the heart of studies on business services. The role that the service sector itself has played in the emergence and development of a local supply of business services has frequently been neglected, both by researchers and by practitioners in local and regional development. However, the sector is increasingly emerging as a major consumer of services and, more generally, the share of these activities in the turnover of service firms has been increasing sharply in recent years (Gallouj & Vidal, 2010).

Thus over and above the subsidies analysed in the preceding sections, a measure of this kind may involve increasing the purchase of services by local authorities and institutions in the broad sense (and intermediate organisations more generally). In other words, these public bodies are in a position to stimulate local and regional demand by giving priority to local providers. The purchasing practices of this type of organisation can, after all, encourage the development of the skills and expertise of local service providers. In some cases, the presence of these institutions and the demand they generate may actually be a *sine qua non* for the survival of a number of local service providers (the value of decentralising and relocating government bodies can be better understood from this point of view).

Measures in support of firms from outside the locality, intended to encourage them to increase their inputs in terms of locally provided services, are often cited in most studies, both empirical and theoretical. They refer to the Anglo-Saxon notion of the "branch plant economy." After all, many studies focused on the manufacturing sector have described the negative impact of large national and international firms on regional development and the emergence and development of a local market in business services. Thus most of these studies analyse the effects on local economies of branch plants. They show that these plants have little decision-making autonomy, establish few links with the local economy and therefore have a relatively weak multiplier effect. Thus the aim of policies aimed at this type of firms is to offer them incentives to develop an approach of the import substitution type, whereby they would reduce their imports of services from outside the locality and increase their share of locally produced services so as to avoid the phenomenon of the branch plant economy.

2.3 *Bringing It Together: Linking Supply with Demand*

The development of markets for services is often limited by the lack of information and transparency that is characteristic of this type of market (De Bandt & Gadrey, 1994; Djellal & Gallouj, 2007). Consequently, any local and regional policy must include measures aimed at improving market functioning. The aim should be to establish trust or—and this sometimes amounts to the same thing—to activate local personal networks in such a way as to facilitate contractual relations and exchanges between service providers and clients.

This being so, it would appear necessary to establish the preconditions for the development of services and to create more opportunities for contact and interaction between supply and demand. Thus according to Illeris (1989), local authorities can provide information, set up data banks on the supply of services and establish local organisations (or encourage initiators of such projects) whose role it is to facilitate exchanges of ideas and experiences, thereby producing synergies. These same local authorities should also concentrate primarily on establishing themselves as intermediaries or local transmitters of information in the areas in question. Illeris highlights what he calls their function as "middlemen", whose purpose is to assist SMEs–SMIs in their relations with service providers. "In areas where they know firms and their difficulties, their function as middlemen consists of giving information and advice on possible solutions to the problems firms face, which frequently involves initiating contacts between firms and service providers." This liaison function is performed in Belgium, for example, by the WaBAN network, whose members are "business angels" from the predominantly French-speaking Walloon region of Southern Belgium. The aim is to facilitate the linking of demand (project initiators) with supply (providers of finance) with a view to accelerating the process of launching new economic activities in the Walloon region (Moyart, 2003).

In practice, what can be observed, particularly in the Nord-Pas-de-Calais region, are attempts to clean up, upgrade and attach a "seal of approval" to service occupations. Such attempts may take the form of certification schemes, the production and diffusion of information on the local and regional supply of services (directories, groupings of service providers, etc.) or the publication of regional sectoral charts setting out the responsibilities and activities of each of the regional players (in order to avoid overlapping competences and to raise the visibility of each player). Moreover, as Romano (1995, p. 172) rightly points out, it is clear that those managing support programmes such as the FRAC (recently replaced by Impulsion Conseil), are above all "essential intermediaries whose job it is to reassure CEOs, assist in the choice of experts, keep a record of expertise and skills, in short make a market based on trust function properly..."

It is increasingly the case that this type of support is being accompanied by information and advertising programmes promoting the support measures, as well as by programmes of general information on the regional supply of business services. What appears to be emerging is a veritable supply of "advice about advice." This is a practice that is found in a number of other countries, incidentally: "assistance in formulating demand by providing information and advice to firms on the choices they face, which more and more countries are offering, is an important corollary of a policy based on subsidising demand." (OCDE, 1995). The publication by the French Ministry of Industry as early as 1995 of a first "practical guide on consultancy for SMIs" fell within the scope of this type of policy. The aim, after all, was to foster mutual knowledge and adaptation and establish exchange networks with a shared language.

2.4 Infrastructure Development and Planning Policies

The reference here is to actions (they can be described as "accompanying measures") that indirectly benefit both the supply and the demand sides by improving the general environment in which the "market for services" operates. The aim is to make regions and individual localities more attractive through infrastructure development and urban planning policies (upgrading telecommunications networks, improving the housing supply, developing cultural services, leisure activities and personal and domestic services, etc.) as well as by developing a service culture within the local economic system, and indeed more widely.

Infrastructure development and (inter and intra) urban planning policies should be considered in conjunction with studies of the factors determining the location of services (Aguilera, 2003; Lasch, 2005; Merenne-Schoumaker, 2002; Mouhoud, 2009). Implementation of such policies requires prior knowledge of these determining factors both within and between urban areas, identification of the mobility patterns of service provision at these same levels and an understanding of these services' "market areas" (Pouyanne et al., 2008). Nevertheless, given the circumstances that prevailed in the first decade of the twenty-first century, the measures put in place focused essentially on the following three areas:

(a) The development and improvement of telecommunications networks (traditional, advanced and mobile) and transport systems. Such facilities do, after all, play an important role in the development of services, since they enhance the opportunities for personal contacts and exchanges of information between economic agents (service providers, clients, government departments etc.). They make services more accessible to firms. Moreover, they enable business services providers to extend their area of operations, particularly when they are strongly export-oriented.[6]

(b) The urban and interurban planning, which is an essential factor in the location and growth of services (Philippe, 1991). After all, "urban development and planning have a genuinely operational function, since services are subject to production constraints, which the site must satisfy" (Bailly, Maillat, & Rey, 1984, p. 195). Housing provision and the diversity of that provision, in terms of both locations and functionality (adaptability, modularity), play an important role in the development and competitiveness of services.[7] The principles driving the location and development of services are not uniform and each type of service requires different facilities.[8]

(c) The improvement of the physical urban space (urban renewal and rehabilitation, redevelopment of traditional urban centres, improvement in accessibility, development of the provision of so-called support services, establishment of high-level business and commercial centres), of the socio-cultural environment and, more generally, of the amenities and facilities required for a good quality of life (health, culture, education, leisure etc.). These measures are intended primarily to make the locality in question (town, *département*, region, etc.) more attractive to skilled employees. They help to attract (skilled) workers from outside the locality and to stabilise the local supply of (skilled) labour.

Infrastructure development and planning policies, with which we are concerned here, can be envisaged for services as well as for firms in general (including the demand side). However, it is clear that such policies cannot be considered in isolation. They constitute the necessary but not sufficient conditions for the development of services. Consequently, they have to be combined with other forms of incentive such as those mentioned above.

---

[6] Thus with regard to the Rhône-Alpes region, the studies by the Laboratoire d'Economie des Transports (LET) have shown that high-speed trains (TGV), by reducing travel times and costs, have created new opportunities for the region's dynamic entrepreneurs to access the Paris market. Consequently, journeys linked to service operations have increased considerably which, according to the authors, constitutes "an attack on the Paris market by provincials".

[7] In what might be called the intercity or inter-regional competition paradigm, the housing supply also serves to make the local environment more attractive, thereby encouraging firms from outside the locality, particularly service firms, to move into the area.

[8] Advertising agencies and management consultants require a central location, while market research companies and IT services (which take up more space and sometimes need storage space as well) prefer to be located in business parks or enterprise centres on the outskirts of towns and cities. Some services have a preference for being located near university and research centres, while others (lawyers and bailiffs, for example) prefer locations near the courts. Services such as recruitment consultants prefer to be located close to transport facilities (underground lines etc.), particularly in order to be accessible to potential job applicants.

Over and above infrastructure development, it would seem necessary to put in place exchange networks with a language shared by public and private actors alike. The aim here is to encourage the production and general diffusion of information about services and to develop a general service culture within society. After all, most local economic organisations and institutions are still characterised by a very strong bias towards manufacturing, which is supposedly the engine at the heart of regional economies, with services relegated to the role of passive follower (and beneficiary) of the development and growth thus generated.

Thus policies seeking to reduce this manufacturing bias may take three main forms, which in essence have been put forward on many occasions, most notably by the chambers of commerce and other self-managing public bodies:

- Extension of relational strategies (see Sect. 2.3) to other actors, both public (higher education, research, local authorities, DRIRE, etc.) and private (banks, insurance companies, various opinion leaders etc.).
- "External" communications targeting the general public (with the aim of improving the image of services) and the public authorities.
- Intercessions with public authorities on behalf of services, arguing that they should not be disadvantaged relative to manufacturing industry (which is the case today, particularly with regard to support for innovation and R&D, to give but one example).

Taken as a whole, these measures seek to extend the networks referred to in the preceding sections (supply-side and demand-side networks, etc.) to other public and private actors. The aim here is to encourage the production and dissemination of information about services and to develop a general service culture in society at large. The recent creation (2009) in France of the Commission Permanente de Concertation pour les Services (CPCS—Permanent Negotiation Commission for Services) is clearly a result of this policy. This new type of commission (which brings together some ten ministers and experts from the academic world and the Institut National de Statistiques (INSEE—National Institute of Statistics and Economic Studies) as well as representatives from about ten leading service sectors) aims to provide information to the general public and government bodies on the situation and the problems faced by services and to put forward proposals (see Box 1).

Box 1. Missions and Aims of the Commission Permanente de Concertation pour les Services (CPCS)

The missions of the CPCS are to:

- Provide information to the government on the situation in the services sector in France and on the contribution of services to the development of the economy and employment
- Examine all issues of interest to this sector that could improve competitiveness

(continued)

- Propose innovation policy actions required for the services

In this respect, the CPCS will

- Draw up an economic analysis of the situation regarding service activities based on the work carried out by the Commission des Comptes des Services
- Formulate proposals to encourage the development of services and employment in the service sector
- Organise with professionals the negotiations required for the development and evaluation of public policies for services, in particular regarding competitiveness and innovation
- Co-ordinate and carry out prospective surveys in all fields of interest to services
- Co-ordinate and carry out impact studies of the competitiveness of companies, laws or draft laws and national or EU regulations that concern services directly or indirectly

Source: Decree of 27 August 2009

3 Public Policies for the Development of Business Services: Preliminary Questions and Contradictory Effects

The policies providing support for services outlined in the preceding sections have many positive effects on local and regional economies. However, their implementation raises a number of questions. At all events, certain "precautions" are required, since these policies can also have significant contradictory effects. Among the important questions that might be asked, it seems to us that the following four are the ones that are usually highlighted by the actors involved: which service activities should be supported (Sect. 3.1)? which actors have what legitimacy (Sect. 3.2)? what type of support should be given priority: direct or indirect (Sect. 3.3), tangible or intangible (Sect. 3.4)?

3.1 Which Service Activities Should be Supported?

The question of the type of services to support is often a central concern for political decision-makers, whether at national or regional level. In practice, policies in support of services have very often been focused on so-called mobile services (i.e. those that can choose their location or are not—or not to any great extent—subject to any location constraints) or on "exporting" services. Incidentally, this latter category is not without its problems, firstly because there is not necessarily any consensus as to what constitutes "exporting" services and, secondly, because this category does not take account of services that contribute indirectly to exports by supporting export activities. More recently, with the recognition of the close connection between innovation and knowledge-intensive service activities (Gallouj and Djellal, 2010; Rubalcaba, 2006), priority has been given to actions aimed at these services (OCDE, 1995).

However, policies of this type do entail some significant risks. In the long term, they may lead to the most dynamic regional providers being "creamed off" to the benefit of the capital region. After all, an increasing share of the Parisian market in their turnover may encourage regional service providers initially to open branches in Paris and then to relocate altogether.[9]

Moreover, and in the same vein, public policy has two apparently contradictory aspects. Within the local development paradigm, the aim of public policy is to develop the regional SME–SMI fabric and, as regards business services, to develop the demand for this type of firm and, consequently, the local supply. Conversely, within what might be called the inter-city or inter-regional competition paradigm, it also seeks to make the local environment more attractive in order to encourage firms from outside the region, particular service firms, to move into the locality. This then raises the question of local autonomy or local control of development processes.

3.2 What Actors and What Legitimacy?

If we look at regional development in France from an historical perspective, then it becomes clear that a very wide range of actors have been involved. Many different actors play a role, or can play a role, at various levels of the policies supporting business services, and indeed more generally (Knoepfel, Larrue, & Varone, 2006). Despite a number of recent changes, it is clear that the state, by virtue of its central and territorial administrative bodies and the financial and cognitive resources (knowledge and expertise) that it controls, is the main actor in regional development. Since the 1990s, however, it has conceded some of its power and freedom of manoeuvre to European and regional actors.

(a) At European level, even though it may be considered that the "Services Directive" should lead to major changes in the coming years, it still seems too early to draw any conclusions about the local impact, including in cross border areas (see Box 2).

[9] Policies that seek to make regional services more specialised may have similar outcomes, since specialisation forces providers to extend their operations into more distant markets.

Box 2. Local Effects of the Services Directive
In France, insofar as the drafting and adoption of regulations under the services directive are centralised at national level, no requirements are imposed at local level (cf. executive summary of 20 January 2010). However, although it is taking (too?) long to transpose this directive (in particular in France), some regional effects can already be seen which affect certain specific services sectors. Below are just three significant examples:

- The reform of the urban planning rules relating to commercial establishments has had the effect of encouraging chain stores to set up in new catchment areas. This development should lead to a diversification of retailing, in particular in areas with least competition.
- There should also be a certain number of (regional) changes in home care services, in particular those subject to "quality approval" once the barriers (banned by the directive) regarding exclusivity, type of legal forms for business entities or the requirement for a head office in the country have been removed.
- As for tourism, intervention in the classification procedures (awarding stars) by departmental tourism commissions (consultative commissions where competitors may be represented) has been removed, which should lead to increased competition and a diversification of the services available.

(b) At national level, it is clear that the decentralisation legislation made regional development a responsibility shared between central government and the regions. The regions' role in this regard is laid down explicitly in the Act of 5 March 1982 and is one of the factors that differentiates the functions and responsibilities of the regions from those of the two other tiers of local government in France, namely the *départements* and *communes*. It should be noted that these lower two tiers are not inactive in matters of regional development. By virtue both of the decentralisation legislation and some of their own specific responsibilities, initiatives and policies, both the *départements* and the *communes* have also become important actors in regional development. Above and beyond these three tiers of government, many other actors and stakeholders can be identified, such as the so-called "inter" actors (intercommunal, interdepartmental, interregional, etc. see Dumont, 1994) that came into being initially as a result of the desire of autonomous actors to collaborate on joint programmes and to share their thinking on regional development policy. Finally, outside the local authorities and their various groupings, account should be taken of organisations with responsibilities and/or interests in regional development, such as chambers of commerce and industry at both local and regional level, whose important role in the development of local and regional economies has already been noted (see Gallouj, 2006).

The multiplicity of actors, briefly surveyed above, is certainly a good thing for regional and urban development, since it clearly demonstrates the extent of collective concern in such matters. Nevertheless, this same multiplicity also poses problems. After all, the actors in question, over and above differences in resource levels and the responsibilities conferred on them by law and regulation, also have very different strategies, objectives and human and cognitive resources. They also differ in their geographical area of activity. All things considered, therefore, this has led to a fragmentation of decision-making centres (due in part to the numerous transfers of competence) and to a general tendency for actors to wish to move out of their original spheres of operations. Thus local economic action is characterised by disjointed, poorly coordinated operations resulting in, for example, a proliferation of facilities poorly adapted to demand and often redundant—in other words to counterproductive competition between regions or cities (Demazière, 2002; Jayet & Paty, 2000; Madies, 2000).

The level of interest in services varies considerably, both between the various actors and within the same category of actors (for any given actor, this interest may also be very variable over time). Thus Demazière (2002) observes considerable differences in local practices between the conurbations of Tours and Lille. He notes that the public actors in the Tours conurbation suffer from an enormous shortage of information on and knowledge about the business services sector. For example, they have no data on the exact number of firms or jobs in business services in the conurbation. Demazière also notes a distinct lack of enthusiasm among the actors concerned (the general public, the chamber of commerce and industry and the development agency) for improving their knowledge of the issue. The Lille conurbation, in contrast, is said to have particularly detailed (sector by sector) information and data (especially from studies carried out by the chamber of commerce and industry) that include some very precise figures on real and potential demand. This may well be a result of the extensive professional and personal contacts and relationships built up over a long time between the Chamber of Commerce and Industry and the university, which have their roots in the establishment in the mid-1980s of a specialist research centre on the service economy.

3.3 What Type of Support: Direct or Indirect?

The question of the trade-off between direct and indirect assistance is an important one (see Madies, 2000). Whereas the former involves financial transfers from the local and regional authorities to firms, the latter is aimed more at improving the environment in which those same firms operate.

3.3.1 Direct Assistance

Direct assistance can be targeted at both supply and demand. When the concern is with regional policy on services, then the initial instinct is to consider direct assistance for the development of services. However, this type of policy is relatively rare in OECD countries. Focused solely on the supply side, such policies are problematic in various ways. Firstly, there is the problem of how existing service providers will react to a policy whose aim is to develop the regional supply, i.e. the competition (even though this is competition that can be deemed to be "fair"). Secondly, when this new supply is provided through public and "intermediate" bodies, there is the question of a possible reduction in the existing private supply as a result of "unfair competition" (or at least competition regarded as such). After all, the provision of certain services either at no cost or at a price lower than the market price will help to bring down the overall market price and thereby reduce supply as part of a process of "quantitative adjustment." In this case, the public supply would be replacing the private supply, a process known as "crowding out." Finally, there is the problem of the effective demand from SMEs, which still often regard business services as costs rather than investments (DRIRE, 2009; Euroquality, 2009). Thus unless it also tackles the demand side, the cost of a policy focused solely on the demand side seems high.

In fact, it is clear that the interaction between supply and demand in business services is too strong for a dichotomic analysis to be sufficient. That said, action to boost demand is still extremely important and its supposed inflationary influence on the price of services does not seem to be proven. After all, while it is true that this type of assistance (when it takes the form of subsidies for service users) is likely to increase the price of the service (at any given price, demand is greater than it would be without the subsidy), this price increase will attract new providers into the market which, through the workings of competition, will in the medium term help prices return to a more "normal" level.

Overall, it would seem that action to boost demand is the most appropriate strategy for regional policy on the development of services. It has the advantage of greater flexibility and is better able to meet the needs of firms, if only because, despite everything, they bear part of the cost of using the services.

However, such policies have also attracted considerable criticism. For example, subsidies paid to firms using consultants (such as those available under schemes such as FRAC and *Impulsion conseil*—see above) are frequently perceived as subsidies for consultants, since they are intended to expand supply but without guaranteeing competence or quality. Similarly, this type of support may play a part in the "devaluation" of consultancy services (since firms using them may be less involved than if they had to bear the full cost). Moreover, it may well have a negative impact on the possible future development of service departments within client firms. Finally, and this is undoubtedly the most significant criticism, *measures to boost demand, which are more frequently implemented in regions lagging behind, are likely indirectly to benefit the wealthy regions, which already have a dense, diversified and high-quality supply of business services.* As far as the Nord-Pas-de-Calais region is concerned, this takes place through what we have called the leakage (or short-circuiting) of demand. This phenomenon has been particularly well analysed with reference to the UK. As noted in an OECD report (1989) on that country and the BIS support programme that was put in place: "the BIS programme is known to have assisted the wealthy regions. The strategy underlying this project was to bring about structural change in economically disadvantaged areas rather than artificially to boost service activities there." Barbe (1991) reaches similar conclusions in the case of the Lower Normandy region of France when he shows that two thirds of the demand financed under the FRAC programmes was in fact captured by consultants from outside the region, particularly from Paris.

It is nonetheless the case that these support policies, and the *Impulsion conseil* (or FRAC) programmes in particular, can contribute to the development of multi-regional external supply. According to the DRIRE, the FRAC programme in Nord-Pas-de-Calais contributed to the founding, in various specialist areas (strategy, quality, design etc.), of a number of new consultancy companies and to the establishment by existing firms of regional offices. After all, "companies based in other regions realised that the type of relationships to be developed with SMEs required them to have a permanent local presence. They established or strengthened their regional presence."

3.3.2 Indirect Assistance

Indirect assistance partially covers what we have termed "linking supply and demand" (Sect. 2.3) and "infrastructure development and planning policies" (Sect. 2.4). The use of such measures increased remarkably in the second half of the 1980s and accelerated in 1990s (Corolleur & Pecqueur, 1996). Their success can largely be explained by the relative ease with which they can be implemented. From the point of view of the local and regional authorities, they are, after all, a means of evading the very strict regulatory framework imposed on direct assistance. However, this success might be explained more positively by noting, as Gerbaux and Muller (1992) do, that the predominance of indirect over direct assistance can be interpreted as "a transition from emergency measures targeted at firms themselves to longer-term strategies focused on the environment in which firms operate" (see also Demaziere, 2002).

It might be assumed, consequently, that indirect forms of assistance are likely to be less discriminatory towards business services than direct forms (which, as we have seen, tend to have a pronounced bias towards manufacturing industry).

3.4 Tangible or Intangible Support?

The distinction between intangible and tangible support largely reflects the two main conflicting arguments put forward in attempts to explain the location of business service providers and their clear preference for metropolitan areas (Philippe, 1991, 2008). According to the first of these arguments, business service providers seek out locations with certain geo-economic advantages (industrial fabric, access to an extensive labour market) and other advantages largely in the hands of the public authorities (communications network, good-quality housing supply, transport infrastructure etc.). According to the second argument, business service providers give top priority to the attractiveness of local economic environments, which should be characterised by opportunities for learning and innovation. Thus what providers want above all is places to meet other agents and opportunities to pool experience and produce a shared language (Gallouj, 2008; May, 1999).

The aid provided by traditional local or regional authorities, whether direct or indirect, is generally little concerned with the intangible aspects of regional development and planning. It focuses rather on the development of business parks and enterprise zones or on the commercial property business. This type of action generally perpetuates the tradition of local economic interventionism as it has been practised since the enactment of the decentralisation legislation. The adoption of such an approach can be explained mainly by the immediate visibility of this type of action, which is of direct value to the actors involved (particularly if they have an electoral mandate). As Madiès (2000, p. 179) notes, "in the eyes of elected representatives, aid allocated to land and to buildings often constitutes a credible and visible commitment to businesses. It is a form of aid that local authorities feel they control, as opposed to intangible forms of assistance that pose more problems of economic expertise." On the other hand, it might also be felt that certain entities (e.g. the *communes* or intercommunal bodies) lack the expertise and resources required to exercise economic promotion and leadership (Demaziere, 2002). And yet, it would seem that, in the case of business services, which are relational activities, what is important is precisely the intangible dimension of public support (whether national, regional or local). As we have shown (see Gallouj, 2008), the activation of networks based on personal relations and of groups of individuals who know each other or are getting to know each other and to work together constitutes a competitive advantage of the first order for business service providers (see Box 3).

Box 3. Business Service Providers' Networks and Relational Strategies
The networks in which business service providers are involved can be divided into four main categories:

- Networks of service providers in the same area or type of activity that carry out joint projects with a view to strengthening their position as a group.

- Networks of users set up to deal with the difficulties likely to be encountered in identifying, selecting and "managing" service providers.
- Networks and other links between service providers and users intended to facilitate information flows and mutual knowledge.
- Extended networks involving a multitude of other actors or stakeholders who have or may have influence at whatever level on the world of business services.

4 Conclusion

Those in charge of local development and planning seem to have numerous and various options open to them as they seek to encourage the emergence and stimulate the growth of services on a local or regional basis. They encompass various dimensions, not simply the supply and demand sides but also, and more cross-cuttingly, the general environment in which both supply and demand operate.

However, it is clear that most of these policy options cannot be understood unless the characteristics of the regional service markets are taken into account. Most of these markets are characterised by considerable informational asymmetries. The supply side is still generalist, selective and unchanging, while the demand side is atomised, fluctuating, relatively undiversified and, in many cases, focused for its service needs on providers outside the locality or the region.

Under these circumstances, the decision on which policies to put into practice poses problems. Should the supply side be supported in order to establish a supply-side dynamic? Conversely, should the demand side be supported in order to establish a demand-side dynamic? These are important questions, since the growth dynamics in services and the policies likely to flow from them may sometimes give rise to contradictory effects (see Gallouj et al., 2006).

The latest trends give us reason to suppose that the most promising prospects are offered by those measures that seek to link supply with demand and, more generally, by all measures that seek to reduce the informational asymmetries in regional markets for services (and hence increase the transparency and clarity of those markets).

Some economic actors, notably the chambers of commerce and industry, have on their own initiative taken steps in this direction and it can be shown, following Jouve (2002), that they have helped to accelerate the expansion of the service sector in metropolitan and regional economies. Nevertheless, as the large (multi-regional) network service firms spread and increase in influence and gradually begin to dominate certain regional markets (this is the case with auditing, software houses, as well as temporary employment agencies and even cleaning companies), it might be asked whether publicly financed incentives and the strategies adopted by local institutions and authorities are not likely to be of benefit primarily to regions that are already well endowed.

References

Aguilera, A. (2003). La localisation des services aux entreprises dans la métropole lyonnaise: Entre centralité et diffusion. *L'espace géographique, 4*, 128–140.

Bailly A., Maillat D., & Rey M (1984), le rôle des activités de type tertiaire dans le développement des petites et moyennes villes, Lausanne, CEAT.

Barbe, D. (1991). Fonds régionaux d'aide au conseil et stimulation des relations tertiaire-industrie. *Revue d'Economie Régionale et Urbaine, 1*, 71–89.

Corolleur, F., & Pecqueur, B. (1996). Les politiques économiques locales en France Durant la décennie 1980. La construction territoriale comme enjeu stratégique pour les collectivités locales. In C. Demaziere (Ed.), *Du local ou global* (pp. 99–117). Paris: L'Harmattan.

Cuadrado Roura J., Maroto-Sanchez A. (2010), regional productivity growth in European countries: the role of services, 50th ERSA congress, 19–23 August, Jönhöping, Sweden.

Darnige, A. (2010, April). *L'innovation dans les PME en 2009: Tertiaire grand public et professionnel, synthèse sectorielle* (6 pp.). Paris: OSEO.

De Bandt, J., & Gadrey, J. (1994). *Relations de services, marches de services* (p. 360). Paris: CNRS Editions.

Drire Ile de France. (2009). *Le conseil aux entreprises en Ile de France, Etude menée par Euroquality* (8 pp.).

Demazière, C. (2002). *Politiques locales d'aménagement et de développement, services aux entreprises et centralité* (Rapport pour le PUCA, 272 pp.). Ministère de l'Equipement.

Devisme, L. (2000). Centres directionnels, centres d'agglomérations. In L. Devisme (Ed.), *La dimension territorial de la centralité*. Tours: Pôle d'initiative régionale en région centre, MSH Ville et Territoire.

Djellal, F., & Gallouj, C. (2007). *Introduction à l'économie des services*. Grenoble: PUG. 147 pp.

Djellal, F., & Gallouj, F. (2010). Innovation in services and sustainable development. In Kieliszewski, Maglio et Spohrer (Eds.), *The handbook of service science* (pp. 533–557). Springer, New York.

Dumont, G. (1994). *L'aménagement du territoire* (94 pp.). Paris: Les Editions d'Organisation.

Euroquality. (2009, November). *Le conseil aux entreprises en Ile de France* (8 pp.). Ile de France: DRIRE.

Fourcroy, C., Gallouj, F., & Decellas, F. (2012). Energy consumption in service industries: Challenging the myth of non-materiality. *Ecological Economics. 81*, 155–164.

Gadrey, J. (2010). The environmental crisis and the economics of services: The need for revolution. In F. Gallouj & F. Djellal (Eds.), *The handbook of innovation and services* (pp. 93–125). Cheltenham: Edward Elgar.

Gallouj, C. (1996). Le commerce interrégional des services aux entreprises: Une revue de la littérature. *Revue d'Economie Régionale et Urbaine, 3*, 567–596.

Gallouj, C. (2006). Les politiques locales (et régionales) de développement des services: Le cas de la Chambre Consulaire de Lille Métropole. *Géographie, Economie et Société, 8*, 385–406.

Gallouj, C. (2008). *Economie des services à la personne*. Paris: Wolters Kluwer. 211 pp.

Gallouj, F., & Djellal, F. (2010). *The Handbook of Innovation and Services* (p. 800). Cheltenham: Edward Elgar.

Gallouj, C., Leloup, F., Merenne-Schoumaker, B., & Moyart, L. (2006). *Services et développement régional* (380 pp.). Bruxelles: De Boeck.

Gallouj C., Vidal M. (2010), Le conseil aux administrations : spécificité du produit et critéres d'èvaluation, Revue sciences de gestion, ISEOR, n° 74, pp 109–126.

Gerbaux, F., & Muller, P. (1992). *Interventions économiques des collectivités locales* (Pouvoirs Locaux, no 6, Octobre 1990, reproduit dans Problèmes économiques, no 2275, pp. 7–13).

Illeris, S. (1989). *Services and regions in Europe*. Aldershot: Avebury, Gower. 222 pp.

Jayet, H., & Paty, S. (2000). Les zones d'activité françaises sont-elles trop nombreuses ? In A. Menemenis (Ed.) *Comment améliorer la performance économique des territoires*?, Paris, CDC, 373 p.

Jouve, B. (2002). Chambre de Commerce et d'Industrie et développement local: Le cas de Lyon. *Sociologie du Travail, 44*, 521–540.

Knoepfel, P., Larrue, C., & Varone, F. (2006). *Analyse et pilotage des politiques publiques* (p. 398). Basel: Helbing & Lichtenhahn.

Lacour, C., Delamarre, E., & Thoin, M. (2005). *40 ans d'aménagement du territoire* (152 pp.). Paris: La Documentation Française.

Lasch, F. (2005, November–December). *Le secteur TIC en France: Délimitation, localisation et création d'entreprise* (Gestion 2000, no 6, pp. 221–245).

Levet J.-L. (2006), Pas d'avenir sans industrie, Paris, Economica, 184 p.

Madies, T. (2000). Les interventions des collectivités territoriales en faveur des entreprises. In A. Menemenis (Ed.) *Comment améliorer la performance économique des territoires, les 3èmes entretiens de la Caisse des Dépôts sur le développement local* (pp. 171–214, 373 pp.). Paris: Collection les Hexagonales.

Madiot, Y., & le Mestre, R. (2001). *L'aménagement du territoire* (4èmeth ed.). Paris: Masson. 221 pp.

Marshall, J. N., & Bachtler, J. (1987). Service and regional policy. *Regional Studies, 25*(1), 471–475.

Mas Verdu, F., Ribeiro, D., & Roig Dobon, S. (2010). Governement policy and services: An approach to the international context. *The Service Industries Journal, 30*(1), 1–10.

May, N. (1999). Districts et métropolisation, réflexions à partir de deux figures spatiales de réorganisation productive. In J. M. Fontan, J. L. Klein, & D. G. Tremblay (Eds.), *Entre métropolisation et le village global: Les scènes territoriales de reconversion.* Montréal: Presses de l'Université du Québec.

Merenne-Schoumaker, B. (2002). *Géographie des services et du commerce* (p. 239). Rennes: Presses Universitaires de Rennes.

Monod, J., & Castelbajac, P. (2010). *L'aménagement du territoire*. Paris: PUF. 125 pp. (First edition 1971).

Mouhoud, E. (Ed.). (2009). *Economie des services et développement des territoires* (Rapport pour la DATAR, no 11). La Documentation Française.

Moyart, L. (2003). La diffusion spatiale des services aux entreprises en Belgique. *Economie et Société, 5–11*, 2043–2066.

Muller, E., & Zenker, A. (2001). Business services as actors of knowledge transformation: the role of KIBS in regional and national innovation systems. *Research Policy, 30*(9), 1501–1516.

Noin, D. (2009). *L'espace français*. Paris: A. Colin. 187 pp. (First edition 1995).

OCDE. (1995). *Dynamiser les entreprises: les services de conseil* (222 pp.). Paris: Publication de l'OCDE.

Philippe, J. (1991). Mercapoles: Un enjeu métropolitain pour les services aux entreprises. *Economies et Sociétés F, 32*, 149–162.

Philippe, J. (Ed.). (2008, October). *Les villes moyennes face à la tertiarisation de l'économie* (108 pp.). Rapport pour le PUCA.

Portnoff, J. Y. (2010). Le retour de l'industrie. *Futuribles, 364*, 5–23.

Pouyanne, G. et al.(Ed.). (2008). *Un renouveau des villes moyennes par les services aux entreprises* (286 pp.). Rapport pour le PUCA.

Romano, J. (1995). *La modernisation des PME, l'expert, le patron et le politique*. Paris: PUF. 297 pp.

Rubalcaba, L. (2006). Which policy for innovation in services. *Science and Public Policy, 33*(10), 745–756.

Toivonen, M. (2007). Innovation policy in services: The development of KIBs in Finland. *Innovation Management Policy and Practice, 9*(3–4), 249–261.

Viljamaa, A., Kolehmainen, J., & Kuusisto, J. (2010). For or against? An exploration of inadvertent influences of policies on KIBS industries in the Finish policy setting. *The Service Industries Journal, 30*(1), 71–84.

MIX
Papier aus verantwortungsvollen Quellen
Paper from responsible sources
FSC® C105338

If you have any concerns about our products,
you can contact us on
ProductSafety@springernature.com

In case Publisher is established outside the EU,
the EU authorized representative is:
Springer Nature Customer Service Center GmbH
Europaplatz 3, 69115 Heidelberg, Germany

Printed by Libri Plureos GmbH
in Hamburg, Germany